ISBN 978-1-5285-1089-9
PIBN 10910574

Technical and Bibliographic Notes / Notes technique

The Institute has attempted to obtain the best original copy available for filming. Features of this copy which may be bibliographically unique, which may alter any of the images in the reproduction, or which may significantly change the usual method of filming are checked below.

L'Institut a microfil
été possible de se
plaire qui sont peι
ographique, qui pe
ou qui peuvent ex
de normale de film.

- [x] Coloured covers /
  Couverture de couleur

- [ ] Covers damaged /
  Couverture endommagée

- [ ] Covers restored and/or laminated /
  Couverture restaurée et/ou pelliculée

- [ ] Cover title missing / Le titre de couverture manque

- [ ] Coloured maps / Cartes géographiques en couleur

- [x] Coloured ink (i.e. other than blue or black) /
  Encre de couleur (i.e. autre que bleue ou noire)

- [ ] Coloured plates and/or illustrations /
  Planches et/ou illustrations en couleur

- [ ] Bound with other material /
  Relié avec d'autres documents

- [ ] Only edition available /
  Seule édition disponible

- [ ] Tight binding may cause shadows or distortion along interior margin / La reliure serrée peut causer de l'ombre ou de la distorsion le long de la marge intérieure.

- [ ] Blank leaves added during restorations may appear within the text. Whenever possible, these have been omitted from filming / Il se peut que certaines pages blanches ajoutées lors d'une restauration apparaissent dans le texte, mais, lorsque cela était possible, ces pages n'ont pas été filmées.

- [ ] Coloured pa

- [ ] Pages dama

- [ ] Pages restor
  Pages restaι

- [x] Pages discol
  Pages décol

- [x] Pages detac

- [x] Showthrougr

- [ ] Quality of pri
  Qualité inéga

- [ ] Includes sup
  Comprend d

- [ ] Pages wholl
  tissues, etc.,
  possible in
  partiellement
  pelure, etc.,
  obtenir la me

- [ ] Opposing p
  discolouratio
  possible ima
  colorations \
  filmées deux
  possible.

L'exemplaire filmé fut reproduit grâce à la
générosité de:

Bibliothèque nationale du Canada

Les images suivantes ont été reproduites avec le
plus grand soin, compte tenu de la condition et
de la netteté de l'exemplaire filmé, et en
conformité avec les conditions du contrat de
filmage.

Les exemplaires originaux dont la couverture en
papier est imprimée sont filmés en commençant
par le premier plat et en terminant soit par la
dernière page qui comporte une empreinte
d'impression ou d'illustration, soit par le second
plat, selon le cas. Tous les autres exemplaires
originaux sont filmés en commençant par la
première page qui comporte une empreinte
d'impression ou d'illustration et en terminent par
la dernière page qui comporte une telle
empreinte.

Un des symboles suivants apparaitra sur la
dernière image de chaque microfiche, selon le
cas: le symbole ⟶ signifie "A SUIVRE", le
symbole ∇ signifie "FIN".

Les cartes, planches, tableaux, etc., peuvent être
filmés à des taux de réduction différents.
Lorsque le document est trop grand pour être
reproduit en un seul cliché, il est filmé à partir
de l'angle supérieur gauche, de gauche à droite,
et de haut en bas, en prenant le nombre
d'images nécessaire. Les diagrammes suivants
illustrent la méthode.

2   3   1

MICROCOPY RESOLUTION TEST CHART

(ANSI and ISO TEST CHART No. 2)

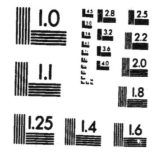

APPLIED IMAGE Inc

1653 East Main Street
Rochester, New York 14609 USA
(716) 482 - 0300 - Phone
(716) 288 - 5989 - Fax

Terminal Boundary Monument set by the first International Commission
at the Pacific Shore.

CANADA
DEPARTMENT OF MINES
GEOLOGICAL SURVEY
Hon. Robert Rogers. Minister; A. P. Low. Deputy Minister;
R. W. Brock. Director.

MEMOIR No. 38

GEOLOGY

OF THE

NORTH AMERICAN CORDILLERA

AT THE

FORTY-NINTH PARALLEL

BY

Reginald Aldworth Daly.

IN THREE PARTS

PART I.

OTTAWA
GOVERNMENT PRINTING BUREAU
1912

8164-

No. 1203

INTRODUCTORY.

Through the courtesy of W. F. King, C.M.G., LL.D., B.A., D.T.S., Chief Astronomer, Department of the Interior, the Geological Survey is enabled to publish this Memoir. The field work was done under, and at the expense of, the International Boundary Commission, and appears as an appendix to the report of Mr. King, the Canadian Commissioner. As the report constitutes a most important contribution to the geology of western Canada, and as in its 'Blue Book' form it would not be available for many libraries and individuals that would have use for it, Dr. King kindly consented to allow the Geological Survey to print it as a Geological Survey Memoir, and thus secure for it adequate distribution in geological quarters.

The Geological Survey is pleased to be able to add to its list of Memoirs this work that deals with the geology of such a long and important section through the Western Cordillera. It must be referred to constantly in future work dealing with the geology of British Columbia, and were it not available in the publications of the Survey great loss and inconvenience would result.

(Signed)   **R. W. Brock,**
Director.

GEOLOGICAL SURVEY,
OTTAWA, October 21, 1912.

APPENDIX 6.

REPORT OF THE CHIEF ASTRONOMER, 1910

GEOLOGY

OF THE

NORTH AMERICAN CORDILLERA

AT THE

FORTY-NINTH PARALLEL

BY

REGINALD ALDWORTH DALY.

IN THREE PARTS

PART I

LETTER OF TRANSMITTAL.

MASSACHUSETTS INSTITUTE OF TECHNOLOGY,
Boston, Mass., April 30, 1910.

W. F. KING, Esq., C.M.G., B.A., LL.D.,
   Commissioner for Canada, International Boundary Surveys,
         Ottawa.

   SIR,—I have the honour to submit the following report on the Geology of the mountains crossed by the international boundary at the Forty-ninth Parallel. The report is based on field-work carried on during the seasons of 1901 to 1906, inclusive. To yourself, under whose direction the whole work has been done and from whom I have received help in many ways, I beg to tender my sincere thanks.

                I have the honour to be, sir,
                        Your obedient servant,

                        REGINALD A. DALY.

TABLE OF CONTENTS.

PART I.

CHAPTER I.

2 GEORGE V, A. 1912

CHAPTER V.

SESSIONAL PAPER No. 25a

CHAPTER VI.

CHAPTER VII.

CHAPTER VIII.

CHAPTER IX.

CHAPTER X.

CHAPTER XI.

CHAPTER XII.

2 GEORGE V, A. 1912

CHAPTER XIII.

CHAPTER XIV.

2 GEORGE V, A. 1912

CHAPTER XV.

CHAPTER XVI.

2 GEORGE V, A. 1912

CHAPTER XVIII.

SESSIONAL PAPER No. 25a

PART II

CHAPTER XIX.

CHAPTER XX.

CHAPTER XXI.

CHAPTER XXII.

CHAPTER XXIII.

8364—C

2 GEORGE V, A. 1912

CHAPTER XXVI.

CHAPTER XXVII.

CHAPTER XXVIII.

SESSIONAL PAPER No. 25a

APPENDIX 'A.'

APPENDIX 'B.'

ILLUSTRATIONS.

PLATES.

2 GEORGE V, A. 1912

11. Casts of salt-crystals in Kintla argillite.
12. Looking east across Flathead Valley fault-trough to Clarke range.
13. Head of Lower Kintla Lake.
14. A.—Cliff in Siyeh limestone, showing molar-tooth structure; at cascade in Phillips Creek, eastern edge of Tobacco Plains.
    B.—Concretion in dolomite; lower part of Gateway formation, Galton range.
15. A.—Limonitized, simple and twinned crystals of pyrite, from Gateway formation at summit of McGillivray range.
    B.—Similar pyrite crystals in metargillitic matrix.
16. Exposure of the massive Irene conglomerate in head-wall of glacial cirque.
17. A.—Ripple-marks in Ripple quartzite; positives.
    B.—Ripple-marks in Ripple quartzite; negatives (casts).
18. Negatives of ripple-marks in quartzite. Summit of Mt. Ripple.
19. Mount Ripple and summit ridge of the Selkirk mountain system.
20. Columnar sections of the Summit, Purcell, Galton, and Lewis series.
21. Diagrammatic east-west section of the Rocky Mountain Geosynclinal at the Forty-ninth Parallel.
22. A.—Molar-tooth structure in Siyeh limestone (weathered), Clarke range.
    B.—Molar-tooth structure in Castle Mountain dolomite (unweathered) on main line of Canadian Pacific railway.
23. A.—Porphyritic phase of the Purcell Lava; from summit of McGillivray range.
    B.—Quartz amygdule in the Purcell Lava.
24. Secondary granite of a Moyie sill, fifty feet from upper contact.
25. Phases of the Moyie sill: specimens one-half natural size.
26. Looking eastward over the heavily wooded mountains composed of the Priest river terrane, Nelson range.
27. A.—Contrast of normal sericite schist of Monk formation (left) and contact-metamorphosed equivalent in aureole of summit granite stock, a coarse-grained, glittering muscovite schist (right).
    B.—Spangled, garnetiferous schist characteristic of Belt E of Priest River terrane.
28. Typical view of Bonnington-Pend d'Oreille mountains of the Selkirk system.
29. Percussion marks on quartzite boulder in bed of Pend d'Oreille river.
30. A.—Sheared phase of the Rykert granite, showing concentration of the femic elements of the rock (middle zone).
    B.—Massive phase of the Rykert granite, showing large phenocrysts of alkaline feldspar.
31. Tourmaline rosettes on joint-plane of quartzite; from contact aureole of summit granite stock, Nelson range.
32. Felsenmeer composed of Rossland volcanics. Record Mountain ridge, west of Rossland.
33. Two views of shatter-belt about the Trail batholith, Columbia river.
34. Sheared Cascade granodiorite, showing banded structure.
35. Park land on Anarchist plateau east of Osoyoos lake.
36. Fossil plants in the Kettle River sandstone.

SESSIONAL PAPER No. 25a

FIGURES.

37. Plunging contact surface between intrusive granodiorite of Castle Peak stock and Cretaceous argillites and sandstones of Pasayten series.
38. Plunging contact surface between intrusive granodiorite and Pasayten formation.
39. Plunging contact surface, Castle Peak stock south side.
40. Intrusive contact between granodiorite and nearly vertical Pasayten argillite.
41. Diagrammatic section showing of a 'winter-talus ridge.'
42. Illustrating two methods by which basic contact-shells in a stock or a dike might be formed.

TABLES

APPENDIX 'A.'

PART III

Containing seventeen geological maps, with structure sections (sheets 1 to 17), and two sheets of photographic panoramas (sheets 18 and 19).

CHAPTER I.

INTRODUCTION.

Area Covered.—In 1901 the writer was commissioned by the Canadian Minister of the Interior to undertake the geological examination of the mountains crossed by the Boundary Line between Canada and the United States, at the Forty-ninth Parallel. Field work was begun in July of that year and continued through the different summer seasons to and including that of 1906. During the summer of 1901 reconnai-sance surveys on the American side of the same line were led by Messrs. Bailey Willis, F. Leslie Ransome, and George Otis Smith, members of the United States Geological Survey. No further geological work in connection with the Boundary survey was carried on by the United States Government, and the map sheets prepared by the United States topographers were placed at the disposal of the writer as geologist (for Canada) to the International Boundary Commi-sion. The present report represents the principal results of the study made during the six field seasons.

The geological examination covered a belt along the Forty-ninth Parallel, from the Strait of Georgia to the Great Plains. The belt is 400 miles long and varies from 5 to 10 miles in width, with a total area of about 2,500 square miles. Its width was controlled, in part, by that of the map sheets prepared by the topographers of the Commission parties; in part, by the necessity of depending on the trails which those parties built into the Boundary belt. As a rule, this mountainous belt is heavily wooded and, without trails, is almost inaccessible to pack-animals. During the first three seasons accurate topographic maps on the required scale were not available, and in 1902 and 1903 the writer used, as topographic base, an enlarged copy of the West Kootenay sheet of the Canadian Geological Survey. In that relatively accessible part of the Boundary belt (from Grand Forks to Porthill, eighty miles to the eastward) it was found possible to cover a zone ten miles in width.

Conditions of Work in the Field.—No geologically trained assistant was employed in any part of the field. The work was, therefore, slow. Each traverse generally meant a more or less taxing mountain climb through brush or brulé. The geology could not be worked out in the detail which this mountain belt deserves. For long stretches the rock exposures were found to be poor. Such was the case for the heavily drift-covered mountains between Osoyoos lake and Christina lake, and, again, for nearly all of the 60-mile section between the two crossings of the Kootenay river, at Gateway and Porthill. Some confidence is felt in the maps and structure sections of the Rocky Mountains proper (from Waterton lake to Gateway), of part of the Selkirk

2 GEORGE V., A., 1912

mountain system (from the summit to the Columbia river), and of the Okanagan and Hozameen ranges of the Cascade mountain system. Elsewhere, the maps and sections, in their rigid lithography, suggest more certainty as to the run of contacts and as to underground structures than the writer actually feels. As a whole, the results lie half-way between those of a reconnaissance survey and those of a detailed survey.

One of the leading difficulties felt by all workers in this part of the Cordillera is the remarkable lack of fossils in the sedimentary rocks. The writer has been able to discover but few fossiliferous horizons additional to the small number already known to Canadian and United States geologists. Many of the correlations offered in the following report are to be regarded as strictly tentative and should not be quoted without reference to the many qualifications noted in the running text.

Some large proportion of the inaccuracy in maps and sections is due to the fact that for half of the field seasons the writer was provided either with no topographic map or merely with the four-miles-to-one-inch West Kootenay reconnaissance sheet of the Canadian Geological Survey. This sheet is excellent for its purpose, but was manifestly not intended for the use of the structural geologist, whose topographic-map scale should be at least one mile to one inch in the Selkirk and Columbia mountain systems. First in 1904, the writer was able to use copies of the manuscript Boundary Commission plane-table maps, on the scale of 1:63,360. Sheets 1, 2, 3, 4, 5, 12, 13, and 14 were constructed on that basis and are superior in accuracy of geological information to the other sheets. Between the Gulf of Georgia and the western limit of sheet 17 the Boundary line crosses a continuous thick deposit of Pleistocene gravels and sands. No other formation is there exposed in the five-mile belt and the broad plain is not represented in the maps.

Acknowledgments.—The writer was efficiently aided in the physical work of carrying on the survey, during five seasons, by Mr. Fred. Nelmes of Chilliwack, British Columbia. His faithfulness in many a tedious place was worthy of his sterling work as a mountaineer. During the season of 1903 the writer was similarly assisted in able manner by Mr. A. G. Lang, of Waneta, British Columbia. In the field many courtesies and much help were extended by Mr. J. J. McArthur, chief topographer for the Canadian branch of the Boundary Commission; by Mr. E. C. Barnard, chief topographer for the United States branch, and by his colleagues.

In the office work the writer was aided by many members of the Geological Survey of Canada, and owes much to the personal encouragement of Honourable Clifford Sifton, Minister of the Interior, during the progress of the survey. In numberless ways the work was forwarded by the able and most generous help of the Canadian Commissioner, Dr. W. F. King, to whom the writer owes the greatest debt of acknowledgment. Professor D. P. Penhallow of McGill University has made thorough study of the collections of fossil plants. The collections of fossil animal remains were, with much generosity, carefully

SESSIONAL PAPER No. 25a

studied and determined by Drs. T. W. Stanton, G. H. Girty, and C. D. Walcott of the United States Geological Survey, and by Dr. H. M. Ami, of the Geological Survey of Canada. Professor M. Dittrich, of Heidelberg, Germany, and Mr. M. F. Connor, of the Canadian Department of Mines, performed valued services in making the large number of chemical analyses noted in the report. The draughting has been performed with zeal and care by Mr. Louis Gauthier, of the Chief Astronomer's office at Ottawa, and by C. O. Senécal and his assistants of the Geological Survey. A number of professional geologists have discussed theoretical matters and thus markedly assisted in the composition of the report. To each of these gentlemen the writer tenders his thanks for all their kind and efficient help. Equally sincere thanks are due to the president and corporation of the Massachusetts Institute of Technology, who for more than two years have granted every available facility for the preparation of this report.

Special mention at this place may also be made of the fact that chapter XIV is largely a direct quotation from Mr. R. W. Brock's report on the Geology of the Boundary Creek Mining District.‡ The corresponding part of sheet No. 10 has been compiled from the map accompanying Mr. Brock's report. In view of the care spent on this part of the Boundary belt by this able investigator, it seemed inadvisable to spend much of the limited time allotted to the transmontane section on the Boundary Creek district. Accordingly, the present writer made no more than a couple of rapid east-west traverses across the district, corroborating, so far, the accuracy of Mr. Brock's mapping in a particularly difficult terrane.

Professor Penhallow's paper on the collection of fossil plants an appendix to the present report.

Collections.—During the survey 1.525 numbered specimens, with many duplicates, were collected. Each of the localities, whence the specimens chemically analysed were taken, is noted on the map sheets with a small cross and the collection number. Some 960 thin sections of the rocks were prepared and studied. Sixty rock analyses and one feldspar analysis were made for the report. Thirteen hundred photographs were taken by the writer, besides which many hundreds of others were taken by the photo-topographic parties operating for the Canadian branch of the Boundary Commission.

Previous Publications by the writer on the Forty-ninth Parallel Geology.—After each field season a brief account of the ground covered was published either in the summary report of the Director of the Geological Survey of Canada or in the annual report of the Chief Astronomer of Canada. As the work progressed it was thought advisable to publish separate papers on certain general and theoretical problems, which had arisen during the survey of the Boundary belt. The list of these papers, some of which, in more less amplified form, form parts of this report, is as follows:—

‡ R. W. Brock, Annual Report, Geological Survey of Canada, Vol. 15, 1902-3. Part A. pp. 98 to 105.

1. The Geology of the Region adjoining the Western Part of the International Boundary: Summary Report of Geological Survey Department of Canada for 1901, in Annual Report, vol. 14, 1902, Part A, pp. 39-51.

2. Geology of the Western Part of the International Boundary (49th Parallel): Summary Report Geological Survey Department of Canada, for the year 1902, in Annual Report, vol. 15, Ottawa, 1903, Part A, pp. 136-147.

3. The Mechanics of Igneous Intrusion: Amer. Jour. Science, vol. 15, 1903, pp. 269-298. .

4. The Mechanics of Igneous Intrusion (Second Paper): *ibid.*, vol. 16, 1903, pp 107-126.

5. Geology of the International Boundary: Summary Rep. Geol. Survey Department of Canada for 1903, in Annual Report, vol. 16, Ottawa, 1904, Part A, pp. 91-100.

6. Geology of the International Boundary: Summary Rep. Geol. Survey Department of Canada for 1904, Ottawa, 1905, pp. 91-100.

7. The Accordance of Summit Levels among Alpine Mountains; the Fact and its Significance: Jour. Geology, vol. 13, 1905, pp. 105-125.

8. The Secondary Origin of Certain Granites: Amer. Jour. Science, vol. 20, 1905, pp. 185-216.

9. The Classification of Igneous Intrusive Bodies: Jour. Geology, vol. 13, 1905, pp. 485-508.

10. Report on Field Operations in the Geology of the Mountains crossed by the International Boundary (49th Parallel): (1) in Report of Chief Astronomer for Canada for 1905, Ottawa, 1906, pp. 278-283; (2) in Rep. of Chief Astronomer for Canada for 1906, Ottawa, 1907, pp. 133-135.

11. The Nomenclature of the North American Cordillera between the 47th and 53rd Parallels of Latitude: Geographical Journal, vol. 27, June, 1906, pp. 586-606.

12. Abyssal Igneous Injection as a Causal Condition and as an Effect of Mountain Building: Amer. Jour. Science, vol. 22, Sept., 1906, pp. 195-216.

13. The Differentiation of a Secondary Magma through Gravitative Adjustment: Festschrift zum siebzigsten Geburtstage von Harry Rosenbusch, Stuttgart, Germany, 1906, pp. 203-233.

14. The Okanagan Composite Batholith of the Cascade Mountain System: Bull. Geol. Soc. America, vol. 17, 1906, pp. 329-376.

15. The Limeless Ocean of pre-Cambrian Time: Amer. Jour. Science, vol. 23, Feb., 1907, pp. 93-115.

16. The Mechanics of Igneous Intrusion (Third Paper): Amer. Jour. Science, vol. 26, July, 1908, pp. 17-50.

17. The Origin of Augite Andesite and of Related Ultra-basic Rocks: Jour. Geology, vol. 16, 1908, pp 401-420.

18. First Calcareous Fossils and the Evolution of the Limestones: Bull. Geol. Soc. America, vol. 20, 1909, pp. 153-170.

19. Average Chemical Composition of Igneous-rock Types: Proceedings Amer. Acad. Arts and Sciences, vol. 45, January, 1910, pp. 211-240.

SESSIONAL PAPER No. 25a

20. Origin of the Alkaline Rocks: Bull. Geol. Soc. America, vol. 21, 1910, pp. 87-118.

Earlier Work on the Geology of the Forty-ninth Parallel.—The British and United States governments attached geologists to the parties of the first International Boundary Commission appointed (1857-61) to mark the Forty-ninth Parallel across the Cordillera. The geologist, George Gibbs, traversed the Boundary belt for the United States government and published his results in the third and fourth volumes of the Journal of the American Geographical Society, New York, 1873-74. The late Hilary Bauerman was the geologist for the British government. His brief report was not published until 1884, when, at the suggestion of George M. Dawson, it appeared as a part of the Report of Progress of the Geological and Natural History Survey of Canada for 1882-3-4, Part B, Ottawa, 1884.

Dawson himself entered the same transmontane belt at its eastern end during his work as geologist t the British North American Boundary Commission. His report published in 1875, at Montreal, bears the title 'Report on the geology and resources of the region in the vicinity of the forty-ninth parallel from the Lake of the Woods to the Rocky Mountains.' Since then Dawson continued his memorable reconnaissance of British Columbia and, in the Boundary belt, was accompanied or followed by McConnell, McEvoy, Brock, Leach, Young, LeRoy, Camsell, and other members of the Geological Survey of Canada.

On the United States side of the line many other workers have similarly added to our knowledge of the formations cro sed by the Forty-ninth Parallel, though comparatively few of them, other than those already mentioned, have actually reached the Boundary line in their detailed work. Reference to the publications on British Columbia, Alberta, Montana, Idaho, and Washington geology can be readily found in the bibliographic bulletins of the United States Geological Survey and in the general index to the reports of the Geological Survey of Canada (published in 1908). Special note should be made of the papers published by the American geologists attached to the present Boundary Commission, namely:—

Stratigraphy and structure, Lewis and Livingston ranges, Montana: by Bailey Willls: Bull. Geol. Soc. America, Vol. 13, 1902, pp. 305-352.

A geological reconnaissance across the Cascade range near the Forty-ninth Parallel: by George Otis Smith and Frank C. Calkins, Bull. 235, U.S. Geol. Survey, 1904.

Continuation of the Forty-ninth Parallel Section.—Mr. Charles H. Clapp is now employed by the Canadian Geological Survey on a structural study of Vancouver island, and it is hoped that materials will soon be in hand for a continuation of the Forty-ninth Parallel section across to the open Pacific.

General Sketch of the Subject Matter.—The 400-mile section crosses the grain of the Cordillera and accordingly includes a high proportion of all the Cordilleran formations to be encountered in these latitudes. The structural

2 GEORGE V., A. 1912

complexity, like the stratigraphic complexity, is near its maximum for the given area n such a straight cross-section. A preliminary sketch of the different geological provinces traversed by the Boundary belt will aid the reader in understanding and grouping the mass of observations to be detailed.

In the first approach, the Cordillera at the Forty-ninth Parallel may be regarded as divisible into great zones. These are called the Eastern Geosynclinal Belt and the Western Geosynclinal Belt. The two overlap in the vicinity of the Columbia river. From the summit of the Selkirk range, just east of that river, to the Great Plains, sedimentary formations are dominant and are almost entirely included in one huge structure, hereafter named the Rocky Mountain Geosynclinal Prism (or simply Geosynclinal). The prism extends from Alaska, through the Great Basin, to Arizona. These rocks are so nearly unfossiliferous that their correlation with the standard systems is a matter of difficulty. Reasons will be shown for the belief that the whole conformable group ranges in age from the Mississippian to a great unconformity at the base of the Belt terrane, or Beltian system, as recently named by Walcott.* Near the western crossing of the Kootenay river, the prism rests on an elder group of metamorphic rocks, here called the Priest River terrane. The basement on which the Rocky Mountain Geosynclinal rests is nowhere else exposed on the Forty-ninth Parallel.

Younger and much more local geosynclinal prisms, of Cretaceous and Tertiary dates, have been laid down on the Rocky Mountain Geosynclinal along its eastern border, in Alberta, Montana, and farther south. None of these younger prisms of great thickness is represented in the section at the International Boundary, but it is convenient to refer to the whole compound belt of heavy sedimentation under the one name, the Eastern Geosynclinal Belt.

Similarly, the dominant sedimentaries west of the Columbia river, of Pennsylvanian, Triassic, and Cretaceous age, have been accumulated in great thicknesses. The Pennsylvanian strata have been recognized at many points, from Alaska to Southern California, and it appears probable that late Paleozoic sedimentation on a geosynclinal scale took place throughout that long stretch. More local Mesozoic and Tertiary geosynclinals were imposed upon the Pacific border of the prism developed in the Pennsylvanian period. Rocks apparently representing this older group of deposits crop out at intervals all the way from the Columbia river to the Gulf of Georgia. A part of one of these Mesozoic prisms was found in an enormously thick mass of Cretaceous strata largely composing the Pasayten mountain range between the Pasayten and Skagit rivers. A thick Triassic series is known on Vancouver island and forms part of the western slope of the Skagit mountain range, which lies between the Skagit river and the Strait of Georgia. The edge of the Tertiary geosynclinal composed of the Puget beds is, apparently, represented in the Fraser valley. To the entire composite mass of post-Mississippian sediments occurring in the western half of the Cordillera, the name Western Geosynclinal Belt may be given.

* C. D. Walcott, Smithsonian Miscellaneous Collections, vol. 53, No. 5, 1908, p. 169.

SESSIONAL PAPER No. 25a

The Eastern Geosynclinal Belt is characterized by open fold-, fault-blocks, and overthrusts, with but moderate regional metamorphism and quite subordinate igneous action. The Western Belt is characterized by close folding, mashing, strong regional metamorphism, and by both batholithic intrusion and volcanic action on a grand scale. From the western crossing of the Kootenay river to the Kettle river at Grand Forks the section crosses the West Kootenay Batholithic province, which is partly overlapped by the Rossland Volcanic province. West of Grand Forks is the Midway Volcanic province. The Okanagan (eastern) division of the Cascade mountain system is composed of the Okanagan Composite Batholith, and the heart of the Skagit range is made up of the Skagit Composite Batholith.

These various geological provinces are treated in the order of succession as they are encountered in passing from east to west. Numberless problems have arisen during the progress of the work. Special studies have been made on the relations and origin of the igneous rocks, which occur in the section on a scale not often surpassed in other mountain chains.

A chapter on the nomenclature of the Cordilleran ranges at the International Boundary illes ates the need of a systematic attack on the difficult problem of names. The long discussed but ever new question as to the origin of limestone and dolomite, coupled with that as to the cause of the rarity of fossils in pre-Silurian sediments, has prompted a theoretical chapter which, like the chapter on nomenclature, has already in largest part been published. Other subjects, including glaciation and physiography, were more inevitably to be considered and need no special introduction at this place.

CHAPTER II.

SYNOPSIS OF THE REPORT.

For the convenience of the reader a brief abstract of each of the following chapters is here offered. As a rule, the petrography, which forms a large part of each systematic section dealing with the rock formations, is not summarized.

CHAPTER III.—The necessity of subdividing the western mountain chain of North America into relatively small orographic units is felt by the naturalist who covers any large section of these mountains and then attempts to describe the results of his observations. Such subdivision for a belt lying between the Forty-seventh and Fifty-third parallels of latitude is suggested.

For scientific writing the well recognized name 'Cordillera of North America,' with the alternative, 'Pacific Mountain System of North America,' is preferred for the chain as a whole. The many other alternative names for the chain are listed.

The existing nomenclature for the ranges crossed by the Forty-ninth Parallel is inadequate and to some extent in confusion. An amplified nomenclature, based as far as seemed possible on prevailing usage, is offered. The main principle adopted is that of existing topographic relations, largely irrespective of the genetic history or rock composition of the different ranges: Specifically the lines of delineation are the axes of the greater valleys and 'trenches' in the mountain complex. The Rocky Mountain Trench, the Purcell Trench, the Selkirk Valley, and the Lower Okanagan Valley represent partial boundaries of the Rocky Mountain system, and the Purcell, Selkirk, Columbia, and Cascade Mountain systems at the Forty-ninth Parallel. The suggested subdivisions of these systems for the region adjacent to the International Boundary include the Lewis, *Clarke*, MacDonald, *Galton*, Flathead, *McGillivray*, Yahk, *Moyie*, Cabinet, *Nelson*, Slocan, *Bonnington*, Valhalla, *Pend D'Oreille*, *Priest*, *Kaniksu*, *Rossland*, *Christina*, *Midway*, Colville, *Sans Poil*, Okanagan, *Hozomeen*, and *Skagit* ranges or groups.

A small area of the *Belt of Interior Plateaus* is also represented in the Boundary line section. In the preceding list the names in italics are proposed by the present writer. The others date from the expeditions of Palliser, Dawson, Willis, Smith, Calkins, and MacDonald. The subdivision is illustrated with sketch maps.

CHAPTER IV.—The geological description begins with an account of the Rocky Mountain Geosynclinal Prism, of which nearly all the mountains between the Great Plains and the summit of the Selkirks are composed. Chap-

9

2 GEORGE V., A. 1912

ter IV, discusses the stratigraphy and structure of the Clarke range (Livingston range of Willis), the most easterly of the mountains covered by the survey. Willis' results on the succession of formations were confirmed by detailed study. The oldest formations in this part of the Rocky Mountain system are the Altyn and Waterton magnesian lime-tones and dolomites. The former is believed to be considerably thicker than the minimum estimate given by Willis, in whose traverse the lower part of the Altyn formation was not visible. The base of the Waterton formation is concealed. At Waterton lake a boring has located the plane of the Lewis overthrust at a depth of about 1,500 feet below the lake-level. At that level the bit of the machine entered soft shaly rocks assigned to the Cretaceous.

The fossil *Beltina danai* was found in the Altyn formation. No other determinable fossils were found in this range, the sediments of which were assigned by Willis to the Belt terrane of the Algonkian. They are here alluded to as the Lewis series. At the Flathead river, a local fresh-water, fossiliferous deposit of clays and sands –the Kishenehn formation– occurs; it is assigned to the Miocene.

The Clarke range forms a dissected broad syncline, which is accidented with a few faults and some lava warps. The valley of the North Fork of the Flathead river is an eroded graben or fault-trough. The range has been moved eastward at least eight miles along the great Lewis thrust. The writer favours the view that this thrust, as well as nearly all the other deformation represented in the range, dates from the close of the Laramie, but this has not been finally proved.

Chapter V.—Continuing westward, the older members of the geosynclinal, a'l unfossiliferous, were found to make up the greater part of the MacDonald and Galton ranges. The lithology has, however, changed and in some cases new names are given to the constituent formations. The whole conformable group, corresponding to the Lewis series, is called the Galton series.

On the east and west sides of the Galton-MacDonald mountain group downfaulted blocks of fossiliferous limestone, upper Devonian to Mississippian in age, make contact with some of the lowest members of the much older Galton series.

The dominant structural unit of the twin ranges is the fault-block.

Chapter VI.—West of the Rocky Mountain Trench the geosynclinal rapidly assumes a lithological character markedly different from that found in the four ranges just mentioned. The Purcell system is largely composed of massive quartzites and metargillites, forming the Purcell series, which is the more silicious equivalent of the dominantly argillaceous and calcareous or dolomitic sediments of the Lewis and Galton series. The Purcell series is of much more homogeneous composition than the other two series.

An interbedded volcanic formation, of the fissure-eruption type, has been followed from the Great Plains to the summit of the McGillivray range, where the lava is thickest. It is named the Purcell Lava. A special feature of the

SESSIONAL PAPER No. 25a

Purcell system is the presence of thick sills of a peculiar [...] These eruptive formations are described in later chapters.

The Purcell system is also characterized by numerous [...] faulting, though the McGillivray range shows a broad anticline [...]

CHAPTER VII.—At the Purcell Trench the continuity of the geosynclinal mass is effectively broken. From the alluviated floor of the trench [...] about sixteen miles farther west the rocks chiefly belong to the [...] River terrane, on which the geosynclinal was deposited. At the summit of the Nelson range the nearly entire thickness of the geosynclinal is exposed, the prism having here been upturned to a vertical position. Its sedimentary members are heterogeneous, including conglomerates, grits, coarse and fine sandstone (quartzites), and metargillites. A very thick volcanic formation, older than the Purcell Lava, is interbedded. A great unconformity at the base of the geosynclinal is exposed. The name Summit series is given to the whole conformable group of formations, from the basal unconformity to the horizon corresponding to the youngest member of the Purcell series. West of the great monocline the Summit series makes an apparently conformable contact with a younger metamorphosed mass of sediments named the Pend D'Oreille group. West of that contact the Rocky Mountain Geosynclinal rocks do not reappear in the Boundary section.

CHAPTER VIII.—In this chapter the detailed description of the Selkirk geology is interrupted, and the correlation of the Lewis, Galton, Purcell, and Summit series is discussed. The systematic variation in the lithology of the geosynclinal, as it is crossed from east to west, is noted in some detail, and the conclusion is drawn that the source of the clastic materials lay to the westward, probably not far from the present location of the Columbia river. Notes on the metamorphism of the prism and on its average specific gravity are entered.

The lithological correlation of the geosynclinal with the Cambrian formations described by McConnell and Walcott on the Canadian Pacific railway is then discussed. The result is to point to the probability that the geosynclinal at the International Boundary is largely Cambrian in age, though its basal members belong to pre-Olenellus horizons (Beltian of Walcott). Similar correlation with sections described in Montana and Idaho suggests a similar conclusion as to the age of the sediments in the four Boundary series, and it is held that a considerable thickness of the 'Belt terrane' is possibly, if not probably, of Middle and Lower Cambrian age.

The chapter closes with an outline of the argument that the eastern half of the Cordillera, from Alaska to Arizona and including the Great Basin of the United States, has been the scene of specially heavy sedimentation during the Beltian, Lower Cambrian, and Middle Cambrian periods. The lower part of the Rocky Mountain Geosynclinal, as defined, has an axial trend faithfully parallel to the main Cordilleran axis of the present day. This geosynclinal suffered a local deformation during an early Middle Cambrian period, and, at the Middle Cambrian Flathead stage, was generally depressed. The area of deposition [...]

2 GEORGE V., A. 1912

was thus enlarged and deposition was generally continuous throughout the Cordilleran belt until near the close of the Mississippian. Upon the Paleozoic beds thick Cretaceous and Tertiary prisms of sediment were locally laid down. These local geosynclinals and the master Rocky Mountain Geosynclinal compose the Eastern Geosynclinal Belt of the Cordillera.

CHAPTER IX.—Returning to the systematic description of the rocks, the important Purcell Lava formation is here considered. Its characters in the McGillivray, Galton, Clarke, and Lewis ranges are outlined. Certain associated dikes and sills are described and the relation of this fissure-eruption to the thick sills of the Yahk and Moyie ranges is discussed.

CHAPTER X.—The intrusive gabbro sills of the Purcell mountain system have already been described in preliminary papers. The matter of these publications, together with some new material, is presented in chapter X. It is largely petrographic. A group of the most important intrusive bodies discovered has been given the name, Moyie sills. It illustrates the ability of some very thick magmatic sheets to assimilate their country-rocks—quartzites in the case of the Moyie sills. The proofs of this are discussed in detail, and similar cases are briefly compared. The principle of gravitative differentiation of magmas is evident in all the cases.

CHAPTER XI.—The sedimentary rocks of the Nelson range, other than those of the Summit series and some others intercalated in the Beaver Mountain volcanic formation, include: the Kitchener quartzite, a small outlier of which seems to be represented along the western edge of the Purcell Trench; the P.iest River terrane; and the Pend D'Oreille group.

The pre-Cambrian Priest River terrane, the oldest rock-group identified in the Boundary section, is composed of micaceous schists, quartzites, quartz schists, dolomites, and metamorphosed greenstones, arranged in meridional bands, but so complex in structure as to defy all attempts at deciphering their true relation to one another. The petrography of the different bands is described, and a note is added on the correlation of the terrane with others found in the Cordillera north and south of the Boundary line.

The Pend D'Oreille group is divided into the Pend D'Oreille limestone and the Pend D'Oreille schist. As exposed in the Boundary belt, these rocks occur in the batholithic province of West Kootenay, a fact which helps to explain the heavy metamorphism of this group. The limestone is locally unfossiliferous and, with some doubt, is correlated with the similar marbles of definitely Carboniferous age at Rossland. The schistose division includes phyllite, sheared quartzite, amphibolite, and massive greenstone, which are intimately associated with the limestone and are therefore tentatively referred to the upper Paleozoic.

Then follows a brief analysis of the structure of the Nelson range (Selkirk system). The Purcell Trench is located at the Forty-ninth Parallel on a fault trough representing great vertical displacement. Horizontal shifts and a powerful overthrust, with rotation of the thrust-plane, are among the more important structural elements in this area of strong deformation.

CHAPTER XII.—The Priest River terrane, the formations of the Summit series, and the Pend D'Oreille group are cut by batholiths, stocks, and dikes of igneous rocks of varied nature. Chapter XII is largely devoted to the petrography of these eruptive bodies as exposed in the Selkirk range. In addition, certain minette dikes of the Rossland mountains, being closely related to others occurring in the Selkirks, are described in this chapter. A thoroughly abnormal 'granite,' probably a hybrid rock, cuts the Kitchener quartzite at the edge of the Kootenay river alluvium. A tentative correlation of all the formations composing the Selkirk mountains within the Boundary belt is given in tabular form.

CHAPTER XIII.—Though the Rossland mountain group is a small subdivision, the ten-mile belt crossing it shows an extensive variety of formations, chiefly igneous. Fossiliferous Carboniferous limestones, and Cretaceous (?) shales occur near Rossland; and conglomerate bearing fossil leaves (Cretaceous or Tertiary in age) was found on Sophie mountain. The areally important formations include the Rossland and Beaver Mountain groups, (latites, andesites, and basalts), the Trail batholith (granodiorite), the Coryell batholith (syenite), the Rossland monzonite, and the stocks of Sheppard granite. A peculiar 'olivine syenite,' occurring also in the Bonnington range, and a dike of the rare petrographic type, missourite, are described. The structural and time relations of the formations are discussed.

CHAPTER XIV.—Between Christina lake and Midway the bed-rocks form a complex, which is very similar to that in the Rossland mountains. The Christina range is chiefly composed of plutonic rocks, which include the gneissic granite (sheared granodiorite) of the Cascade batholith and the aplitic granite of the Smelter stock. The origin of the banding in the batholith is briefly discussed and a lateral-secretion hypothesis favoured.

Across the north fork of the Kettle river the formations have been studied in detail by Mr. R. W. Brock, from whose report liberal quotations are made. For purposes of convenience in later correlations the present writer gives special names to two of the formations described by Mr. Brock. These new names are: Attwood series, and Phœnix Volcanic group. The usual tentative correlation table is appended.

CHAPTER XV.—Just east of Midway the section enters the Midway volcanic province, representing thick Tertiary lavas and pyroclastic deposits. West of the volcanic mass is a broad band of metamorphosed Paleozoic sediments extending to the Osoyoos batholith. This chapter describes the two provinces, the Midway province demanding the greater detail of statement. The Paleozoic sediments, with included greenstone and basic schists of igneous origin, are named the Anarchist series. This series is unfossiliferous, but on lithological grounds, is correlated with the Cache Creek series and other upper Paleozoic groups described north and south of the Boundary. Unconformable upon it is the fossiliferous (Oligocene) Kettle River formation, composed of conglomerates, sandstones, and shales. These sediments are conformably overlain by thick

2 GEORGE V., A. 1912

masses of basalts and andesites. Younger than any of these formations is an alkaline suite of extrusive and intrusive masses. These intrusions include dikes and irregular injected bodies (chonoliths). Rhomb-porphyry and pulaskite porphyry are the intrusive types. A less crystalline rhomb-porphyry, alkaline trachyte, and 'shackanite,' an analcitic lava, are the extrusive types. Various pre-Tertiary intrusives are also described.

CHAPTER XVI.—The Okanagan Composite Batholith extends from the eastern slope of Osoyoos Lake valley to the Pasayten river, where it is covered by unconformable Cretaceous rocks forming the Pasayten series. The component batholiths and stocks, with their country-rocks, are described. The whole body is by far the largest continuous mass of plutonic rock in the Boundary belt. The petrographic types represented have wide range of composition, and the dates of eruption vary from late Paleozoic to the late Tertiary or the Pleistocene. A general idea of the order of eruption and nature of the different bodies may be quickly obtained by an inspection of the general table of contents under 'Chapter XVI'. The reader is also directed to the summary at the end of the chapter itself.

CHAPTER XVII.—The Hozomeen range forms a distinct geological province, being principally made up of an extremely thick geosynclinal mass, the Pasayten series. Its arkose, conglomerate, sandstone, and shale were deposited in a local, rapidly deepening downwarp of Cretaceous date. An important deposit of andesitic breccia forms the basal member of the series and lies on the eroded surface of the Remmel batholith at the Pasayten river. The Cretaceous rocks are fossiliferous at various horizons. They compose a faulted and otherwise deformed monocline with westward dips steepening toward the west. At Lightning creek canyon a great fault brings the youngest Cretaceous beds into contact with the much older Hozomeen series, which is tentatively correlated with the Anarchist, Attwood, Pend D'Oreille, and Cache Creek series.

Intrusive bodies with the relations of stocks, dikes, and chonoliths cut the Pasayten series. Special attention is paid to the Castle Peak granodiorite stock, since its structural relations are clearer than those of any other great intrusive mass in the Boundary belt. The evidences of its downward enlargement and of its having replaced or absorbed the Cretaceous sediments are believed to be clear.

CHAPTER XVIII.—West of the Skagit river, which is located on another master fault, the Hozomeen series is again represented in small patches. On the Pacific slope of the Skagit range a thick body of argillite, sandstone, and limestone, with a heavy mass of interbedded volcanics, is fossiliferous (upper Carboniferous) and under the name Chilliwack series is correlated with the Hozomeen series. So far as known, these are the oldest rocks locally developed in the Skagit range. Fossiliferous Triassic argillite, included in the Cultus formation, was found near Cultus lake. A thick mass of sandstone, etc., to the southward is called the Tamihy series. It is unfossiliferous as yet, but on lithological grounds is correlated with the Cretaceous Pasayten series farther east.

On Sinius mountain coal-bearing, obscurely fossiliferous sandstones and conglomerates are included in the Huntingdon formation, which is probably equivalent to the Eocene Puget beds of Washington. A very thick volcanic pile (Oligocene?) occurs on the eastern slope and is called the Skagit Volcanic formation.

Rocks assigned to the Hozomeen series are cut by the Custer gneissic batholith (sheared granodiorite), outcropping at the summit of the Skagit range. It is possibly of Jurassic age. It is cut by the Tertiary Chilliwack batholith of granodiorite, which is genetically connected with a batholithic mass named the Slesse diorite. Other intrusive masses are also described. The chapter closes with notes on correlation and on the structure of the Skagit range.

CHAPTER XIX.—Deals with the correlation of all the bed-rock formations encountered in the Boundary section between the Purcell Trench and the Strait of Georgia, the approximate limits of the Western Geosynclinal Belt at the Boundary line. The correlation of the Forty-ninth Parallel rocks with those described in sections ranging from Alaska to California is then briefly discussed and thrown into tabular form. A summary history of the Western Belt of the Cordillera closes the chapter.

CHAPTER XX.—Having described the many formations in the Boundary section, an attempt is here made to summarize the geological history of the Cordillera of the Forty-ninth Parallel. That necessarily brief statement is followed by a note on the theory of mountain-building.

CHAPTER XXI.—This chapter gives a sketch of the observations made on the glacial geology of the section. The limits of the great Cordilleran ice-cap at the Forty-ninth Parallel, as to ground-plan and depth, are noted. The two double rows of valley glaciers draining the Rocky Mountains and the Cascade system during the Pleistocene are described. The glaciation of each range is then considered, beginning with the Clarke range on the east. The résumé of the chapter is to be found at its closing page.

CHAPTER XXII.—Discusses certain of the physiographic problems connected with the section. A note on the origin of the master valleys is followed by a division of the Boundary zone into physiographic provinces, listed in a table. A running account of the morphology of the successive provinces, beginning with the Front range synclinal area, is accompanied by a theoretical discussion of the question as to Tertiary peneplanation of the Front ranges and of the Cascade mountains. The cause for the accordance of summit levels in alpine mountains (large extracts from a preliminary paper on that subject) is considered. The chapter closes with a statement of general conclusions on the physiographic development of the Cordillera of the Forty-ninth Parallel.

CHAPTER XXIII.—Is a theoretical chapter dealing primarily with the explanation of the fact that fossils are relatively rare in pre-Ordovician formations, and of the related fact that the great majority of those fossils are not calcareous like n·· of the post-Cambrian fossil remains. The favoured explanation was given in two preliminary papers and the argument as a whole is here presented for the first time. A summary on this highly complex subject is given in the chapter. The origin of the thousands of feet of limestone and dolomite found in the Rocky Mountain geosynclinal and in the Priest River terrane is attributed to direct chemical precipitation on the floor of the open ocean. Statistics show that the limestones of the earlier geological periods were originally more magnesian than those of the later periods. This evolution of the limestones is paralleled with a chemical evolution of the ocean.

CHAPTER XXIV.—Is an introduction to a general theory of the igneous rocks, the statement of which occupies the rest of the report. The Mode classification is preferred and a table showing the average chemical composition of each rock type is inserted. Magmatic heat in the earth is believed to be chiefly a primitive inheritance, though some of it is due to radioactivity. The argument for a general basaltic magma (perhaps highly rigid at the depth of the substratum) is presented, and is followed by the argument for a primary acid shell at the earth's surface. All igneous action is preceded by abyssal injection, whereby the basalt of the substratum mechanically displaces the lower part of the earth's crust and rises to an average level which is at moderate depth below the surface. A note on the essential mechanism of central-eruption volcanoes as distinct from fissure-eruption volcanoes closes the chapter.

CHAPTER XXV.—Discusses the classification of igneous intrusive bodies. The favoured primary division is into injected and subjacent bodies, the former group being largely satellitic to the subjacent masses, which are incomparably the more important as to volume.

CHAPTER XXVI.—The genetic problem of the eruptive rocks is, at its heart, also the problem of the batholith. This chapter discusses the processes by which batholiths are believed to have been formed. Their typical field and chemical relations are sketched. The older hypotheses as to the methods of intrusion are compared with the stoping-abyssal injection hypothesis. Abyssal assimilation of sunken roof-blocks is a prominent element in the favoured explanation of batholithic magmas. The chapter is largely a reprint of three preliminary papers, the matter of which is here systematically assembled.

CHAPTER XXVII.—Considers briefly certain points in the wide subject of magmatic differentiation. The dominating control of gravity is emphasized.

CHAPTER XXVIII.—The principles stated in the last four chapters are here applied to a genetic classification of magmas, and then to rocks actually found in the Forty-ninth Parallel section. The rock families specially discussed are the granites, granodiorites, diorites, andesites, gabbros, basalts, complementary dikes, pegmatites, and the alkaline types, including the syenites.

PLATE 2.

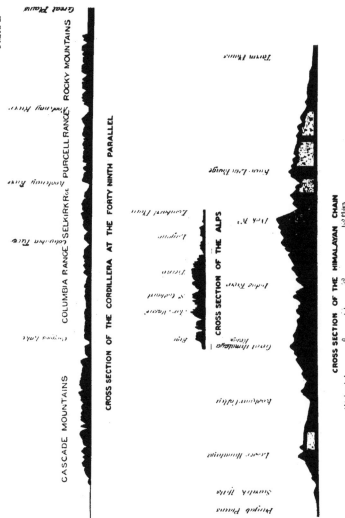

CROSS SECTION OF THE CORDILLERA AT THE FORTY NINTH PARALLEL

CROSS SECTION OF THE ALPS

CROSS SECTION OF THE HIMALAYAN CHAIN

Profile sections showing relative reliefs of the Alpine chain, the Himalayan chain, and the part of the Cordillera of North America between the Gulf of Georgia and the Great Plains. All three sections drawn to same scale.

CHAPTER III.

NOMENCLATURE OF THE MOUNTAIN RANGES CROSSED BY THE FORTY-NINTH PARALLEL.

INTRODUCTION AND OUTLINE.

Although the section covered by the Boundary Commission does not extend to Vancouver island, it is about as long as the longest line of cross-section of the entire Himalayan group of ranges from peninsular India to the Tibetan plateau. If the whole of Vancouver island were included in the Forty-ninth Parallel section, it would be nearly one hundred miles longer than any section-line crossing the Himalayan complex at right angles. Plate 2 shows the reliefs of Himalayas and Alps at their broadest as compared with the partial section of the North American chain covered by the Boundary Commission. The great size of the North American chain is further indicated by a comparison of areas. The chain of the Himalayas, using that term in its larger sense, covers about 300,000 square miles; the Alps of Europe from Nice to Vienna, not more than 70,000 square miles; the Andes, about 1,000,000 square miles; and the western chain of North America, over 2,300,000 square miles. (See also Plate 3.)

The vast mountain region crossed by the International Boundary between Canada and the United States has always been very sparsely inhabited. In the orographic features it is generally complicated, often to the uttermost. Its exploration is only well begun. There are thus excellent reasons why the mountain units of this region are so inadequately named and systematized in geographical works, whether issued as official Government reports, as educational textbooks or atlases, or as popular records of travel. Yet, whether he will or not, the explorer responsible for a report on any part of this region must confront the question of names. He returns from his rugged field, and, to tell of his findings, must use common nouns to indicate what kinds of land-relief he has found, and proper names to aid in individualizing and locating those reliefs in the huge backbone of the continent.

This duty has fallen to the writer in the task of reporting on the geology of the mountains crossed by the Forty-ninth Parallel. Though the same transmontane section has been described by the geologists attached to the 1857-61 Commission, though it occurs along the most thickly settled part of British Columbia, and though it is nowhere very far from the lines of two transcontinental railroads, a complete and systematic grouping of the mountains on the Boundary has never been made. The difficulty of supplying the lack was felt by the writer in the first of the six seasons devoted to the geology of the Boundary, but the difficulty was more fully realized as the confusion of the nomenclature already in vogue became apparent.

It is manifest that any attempt to develop a constructive view of the Boundary mountains should be founded as far as possible upon established units already understood and named. The literature has, therefore, been carefully searched to furnish this required foundation. The result has shown a truly surprising variety of usages in names and in concepts of the topography. The course of compilation inevitably led to the study of the nomenclature of western ranges even far away from the Forty-ninth Parallel of latitude. Examples of the differences of usage are recorded in the first part of this chapter. The record may serve in some degree to illustrate the need of a consistent scheme of nomenclature, possibly to suggest partial grounds on which uniformity may some day be established.

The second part of this chapter is concerned with a brief account of the nomenclature that seems most appropriate for the ranges crossed by the International Boundary.

DIFFERENT NOMENCLATURES IN USE.

The search for the variations of nomenclature was made both among authorities responsible on the ground of priority and among authorities influential as standard compilers from original sources. For the present purpose of indicating the lack of uniformity and the confusion into which the great mass of the people may be led by consulting existing works of reference, it is not sufficient to record names found only in Government map or careful scientific monograph. Perhaps more important still in this connection is the record to be made from standard atlases, from school geographies, and from standard influential guide-books. In reality, it has required the examination of but a very limited number of each kind of authoritative wor'
the moral. With few exceptions the only works consulted were printed in the English language.

DIVERSE NAMING OF THE WESTERN MOUNTAIN REGION AS A WHOLE.

The question of the best general title for the western mountains may be considered as trite by those who do not feel the immediate need of its solution in their professional work. The writer by no means believes it to be trite, as he now completely realizes the wide latitude in naming among the recent influential publications dealing with North American geography. It is scarcely to the credit of our geographical societies and alpine clubs that they will publish at length the statement of one traveller, that he found mosquitoes in Newfoundland, of another that his hotel accommodation in Manila was bad, and leave undiscussed the suggestive paper of Prof. Russell and his correspondents on the names of the larger geographical features of North America.* There would be no advantage

---

* Bull. Geog. Society of Philadelphia, 1899, Vol. 2, p. 55.

to the European geographers if the Alps masqueraded under a dozen different
general titles dependent on the personal tastes of individual writers on those
mountains.

It is well known that one of the first designations of the entire mountain
group lying between the Pacific and the Great Plains was due to Humboldt.
His 'Cordillerns of the Andes' extended from Cape Horn to the mouth of the
Mackenzie river. Humboldt occasionally used the singular form 'Cordillera of
the Andes' for the same concept. In view of the general restriction of the
term 'Andes' to the mountains of South America, Whitney, in 1868, proposed
that the name 'Cordilleras,' with variants, 'Cordilleran System' and 'Cordilleran Region,' be retained to designate the North American equivalent of the
Andes. This name was adopted in the United States census reports for 1870 and
1880, and by a great number of expert geologists and geographers since 1868.
In process of time, however, the singular form, 'Cordillera' and variants,
became used in the same sense. In one of these forms the Humboldt root word
with Whitney's definition has entered many atlases. It appears on numberless
pages of high-class Government reports, geographical, geological, and natural
history memoirs, and of such works as Baedeker's 'Guide-book to the United
States,' Stanford's 'Compendium of Geography,' etc.

The time-honoured, erroneous, similarly inclusive name 'Rocky Mountain,'
with variants, 'Rocky Mountain System,' 'Rocky Mountain Belt,' etc., has,
however, held the dominant place in the popular usage. Its inappropriateness
for the heavily wooded Canadian mountains west of the Front ranges is abundantly evident. For the United States, Clarence King wrote a generation
ago:—

'The greatest looseness prevails in regard to the nomenclature of all
the general divisions of the western mountains. For the very system itself
there is as yet only a partial acceptance of that general name Cordilleras,
which Humboldt applied to the whole series of chains that border the
Pacific front of the two Americas. In current literature, geology being no
exception, there is an unfortunate tendency to apply the name Rocky mountains to the system at large. So loose and meaningless a name is bad
enough when restricted to its legitimate region, the eastern bordering chain
of the system, but when spread westward over the Great Basin and the
Sierra Nevada, it is simply abominable."*

The following table summarizes the above-mentioned variants along with
others more recently introduced, and still other general names now only of
historical interest. The names of prominent authorities and the leading dates
when they have published the respective titles are also entered in the table.
The authority for some of the older n ...es is Whitney's work on the United
States, published in Boston, 1889.

| Mountains of the Bright Stones | General use, end of eighteenth century. |
| Shining Mountains | Morse, Universal Geography, 1802. |

*U.S. Geol. Exploration. 40th Parallel. Systematic Geology, 1878, p. 5.
25a—Vol. II—2½

2 GEORGE V., A. 1912

Stoney or Stony Mountains	Arrowsmith, 1795; President Jefferson.
Columbian (sic) Mountains	Tardieu, 1820.
Chippewayan Mountains	Hinton, 1834.
The Cordilleras of the Andes (in part)	Humboldt, 1808.
The Cordillera of the Andes (in part)	Humboldt, 1808.
The Cordilleras	Whitney, 1868; many authors since.
The Cordillera	G. M. Dawson, 1884; Gannett, 1898; Rand-McNally, 1905.
The Western Cordillera of North America	J. D. Dana, 1874, 1880.
The Cordilleras of North America	Hayden, 1883; Leconte, 1892.
The Cordilleran Region	Whitney, 1868; Hayden, 1883; Shaler, 1891.
The Cordilleran System	Whitney, 1868; King, 1878; Baedeker, 1893.
The Cordillera System	Hayden, 1883.
The Cordillera Belt	G. M. Dawson, 1879; Rand-McNally, 1902.
The Pacific Cordillera	Russell, 1899, 1904.
The Cordilleran Plateau	Hayden, 1883.
The Cordillera of the Rocky Mountains	J. D. Dana, 1895.
The Rocky Mountain System	Leconte, 1892; Heilprin, 1899; many others.
The Rocky Mountain Region	Powell, 1875; G. M. Dawson, 1890; Gannett, 1899.
The Rocky Mountain Belt	Rand-McNally, 1902.
The Rocky Mountains	Lewis and Clarke; popular.
The Pacific Mountains	Russell, 1899, 1904; Powell, 1899.
The Western Highland	Baedeker, 1893; Keith Johnston Atlas, 1896; Davis, 1899.
The Rocky Mountain Highland	Frye, 1895, 1904.
The Western Plateau	English Imperial Atlas, 1892.

In most technical writings, of both governmental and private origin, the suggestion of Whitney has been followed with varying fidelity during the last forty-four years. It is clear that the inherent connotation of 'Cordilleras' is different from that of 'Cordillera.' The one emphasizes the compound nature of the orographic unit; the other, the singular form of the word, emphasizes the organic union of members. Hayden used both forms of the word. In recent years there has been a rather widespread adoption of the term in the singular number. In 1874, J. D. Dana proposed that the great mountain systems of North America be referred to as the 'Western Cordillera' and the 'Eastern Cordillera,' the latter thus synonymous with what is now commonly called the Appalachian system. Russell, in 1899, proposed 'Pacific Cordillera' and 'Atlantic Cordillera' with respectively the same significance. Usage has.

SESSIONAL PAPER No. 25a

however, declared that there is but one Cordillera in North America. The expression 'Pacific Cordillera' is, according to such established usage, redundant. 'The Cordillera of North America,' 'The Cordilleran system,' 'The Cordilleran Region,' or, with the proper context, simply 'The Cordillera,' seem to be to-day the best variants on the Humboldt root-word.

The fine, dignified quality of the word, convenient in adjective form as in noun form, its unequivocal meaning and its really widespread use in atlas and monograph make 'Cordillera' incomparably the best term for technical and even for the more serious popular works. In fact, there seems to be no good reason why the name should not be entered in elementary school atlases. The objection that the word is likely to be mispronounced by teacher or scholar would equally exclude 'Himalaya' and 'Appalachian' from school-books. In teaching or learning what is meant by 'the Cordillera,' the teacher or scholar would incidentally learn so much Spanish. If, in the future, this should be deemed an intolerable nuisance, speakers in English could, in their licensed way, throw the accent back to the second syllable and avoid the unscholarly danger. The second objection that a cordillera is hereby made to include the extensive plateaus of Utah and Arizona or the great intermontane basins of the United States is more serious. It will, however, hardly displace the word from its present technical use as designating a single earth-feature ruggedly mountainous as a whole, but bearing subordinate local details of form and structure not truly mountainous. If this objection be regarded as invalid by advanced scientific workers, it will have still less weight for popular or educational use.

The ordinary connotation of the term 'highland' makes it unsuitable as part of the name indicating the world's vastest mountain group. Like Powell's name 'Stony Mountains,' suggested for the majestic Front ranges north of the Union Pacific Railroad, 'highland' is 'belittling.' To most readers it would inevitably suggest Scotland's relief. If the word be raised to the dignity proposed in 'Western Highland' or 'Rocky Mountain Highland,' the writer on the natural features of the Cordillera runs the risk of ambiguity in employing the indispensable common noun 'highland,' while dealing with local problems of geology, geography, or natural history.

For popular use, the best title alternative with 'Cordillera' is, in the writer's opinion, 'The Pacific Mountain System.' As suggested by Russell's 'The Pacific Mountains.' The addition of the world 'system' seems advisable as stating the unity of the whole group. The proposal of J. D. Dana to restrict the common noun 'system' to mean merely the group of ranges formed in a single geosyncline has to face overwhelming objections. The usage of generations is against it; the difficulty of actually applying it in nature is, perhaps, yet more surely fatal to the idea.

The restriction of the titles 'Pacific Ranges' (Hayden), 'Pacific Mountains' (Powell in his earlier use of that term; he later applied it to the whole Cordillera), and 'Pacific Mountain System' (A. C. Spencer and A. H. Brooks) to the relatively narrow mountain belt lying between the ocean and the so-called 'Interior Plateau' of the Cordillera, seems particularly unfortunate.

2 GEORGE V., A. 1912

If there is one grand generalization possible about the entire Cordillera, it is that the Cordillera is, both genetically and geographically, a Pacific feature of the globe. The Rocky Mountain ranges proper, the Selkirks, and the Bitterroots bear the marks of interaction of Pacific basin and continental plateau as plainly as do the Sierra Nevada, the Coast ranges, or the St. Elias range. The large view of the Cordillera assuredly claims the word, 'Pacific' for its own, and cannot allow in logic that 'Pacific Mountain System' shall mean anything less than the entire group of mountains. . The artificial nature of the narrower definition would be equally manifest if it were applied to a topographic or genetic unit forming a relatively small part of the Andes along the immediate shore-line of South America. The Andes mountains form the Pacific mountain system of South America as the whole North American Cordillera forms the true Pacific mountain system of North America.

Yet the term 'system' is itself so elastic that it is fitly applied to a subdivision of the Cordillera. For example, the Rocky Mountain System expresses an unusually convenient grouping of the northern ranges in Alaska, and of the eastern ranges of the Cordillera in Canada and the United States. Popular, as well as scientific, usage has once for all recognized the propriety of there being in name, as well as in fact, system within system in the grouping of mountains.

DIVERSE NAMING OF RANGES CROSSED BY THE FORTY-NINTH PARALLEL.

There is a double difficulty in describing the mountains along the International Boundary. The same range may bear different names with different authorities. or may be differently delimited by different authorities. Some examples chosen from recent atlases and texts will illustrate this point.

1. *Cascade range* (also called Cascade chain or Cascade mountain chain), according to different authorities:—

 (*a*) Extends from Mount Shasta into the Yukon territory;

 (*b*) Extends from Mount Shasta to the British Columbia boundary;

 (*c*) Extends from Mount Shasta to the Fraser river, and east of it to the Thompson river;

 (*d*) Forms the extreme northern part of the British Columbia Coast range north of Lynn canal, the real Cascades being mapped as the 'Coast Range' (Johnson's Cyclopædia).

2. *Coast range* of British Columbia, also called the 'Alpes de Colombie' (Atlas Vidal-Lablache) and 'See Alpen' (Stieler's Handatlas, which continues the 'Cascaden Kette' across the Fraser river). See also usages under 'Cascade Range.'

3. *Selkirk mountains*, according to different authorities:—

 (*a*) Lie west of Kootenay lake, entirely in Canada, or extending into the United States;

 (*b*) Lie west of Kootenay lake, and entirely in Canada, or extending into the United States;

    (c) Extend on both sides of Kootenay lake, but entirely in Canada;

    (d) Do not extend south of the northern extremity of Kootenay lake;

    (e) Contrary to all of the above-mentioned usages, extend across the Columbia river north-westward to Quesnel lake in 53° N. lat. (Brownlee's Map, 1893).

   4. *Purcell range,* according to different authorities -

    (a) A local rangelet in the West Kootenay district, British Columbia;

    (b) Includes all the mountains between Kootenay lake and the Roky Mountains proper, entirely in Canada;

    (c) Includes the same mountains as under (b), but extends into the United States as far as the great loop of the Kootenay river.

   5. *Bitterroot mountains* (also spelled 'Bitter Root') used—

    (a) In the larger sense of most maps; or

    (b) In a much narrower sense, a small range overlooking the Bitterroot river (Lindgren).

   6. *Rocky Mountains* or . . . . *'n Mountain system,* also called the Front range, and Laramide range; often alternative for 'Cordillera.'

   7. *Gold range* of British Columbia, a name applied to a local range crossed by the main line of the Canadian Pacific railway, and west of the Columbia river; also applied to a much greater group, including the Selkirk, Purcell, Columbia, Cariboo, and Omineca ranges (Gold ranges, an alternative form of the title in this latter meaning).

   . The confusion of the nomenclature is aggravated, in the case of certain atlases, which in different map-sheets give different titles to the same range: Thus, on one map of the new Rand-McNally 'Indexed Atlas of the World,' the western mainland member of the Cordillera in British Columbia is correctly named the Coast range and, on another sheet, incorrectly named the Cascade range. The same indefensible carelessness even appears in certain Canadian school atlases. In the Rand-McNally map of British Columbia, the Selkirks are represented as ending on the south at the head of Kootenay lake, and are continued to the eastward of that lake by the 'Dog Tooth Mountains,' the latter name being little familiar to the people of British Columbia. In the general map of the United States published in the same Atlas, the Selkirks are represented as quite defined to the westward of Kootenay lake. The area thus inconsistently mapped has a width equal to the average width of the Alps.

ADOPTED PRINCIPLE OF NOMENCLATURE FOR THE BOUNDARY MOUNTAINS.

On the line of the Forty-ninth Parallel, the Cordillera has already assumed what may be called its British Columbia habit as contrasted with its Fortieth Parallel habit. The division of the whole into orographic units is relatively simple in Colorado, Utah, Nevada, and California, where the building and erosion of the Cordillera have resulted in a comparatively clear-cut separation of the

2 GEORGE V., A. 1912

component ranges by broad intermontane plains of mountain waste, or of lava filling vast structural troughs or basins. Nothing quite comparable is to be seen anywhere in the Canadian portion of the Cordillera. Near the latitude of Spokane, the whole mighty group of ranges is marshalled into a solid phalanx of closely set mountains which sweep on in substantial unity north-westward through Yukon territory into Alaska. The area of British Columbia alone would enclose twenty-four Switzerlands. For purposes of exposition this mountain sea must be divided and subdivided. How shall it be done?

The remarkable insight and generalizing power of the pioneer in British Columbia geology, G. M. Dawson, early supplied what seem to be the only fruitful principles. His classification applies chiefly to southern British Columbia, but it is probable that its principles must be extended throughout the Canadian Cordillera. In 1879 Dawson announced the possibility of a natural division of the mountains between the Forty-ninth and Fifty-fifth parallels into three broad belts paralleling the coast.

The middle belt, the ' Interior Plateau,' afterwards described in some detail, has the special style of topography characteristic of closely folded mountains once reduced by denudation to mere rolling hills or an imperfect plain, since uplifted and cut to pieces by streams. In other words, the Interior Plateau is, by Dawson's definition, an uplifted, dissected peneplain, a region of plateaus and hills remnant from the old surface of denudation. Yet Dawson himself concluded that, while many of these tabular reliefs may be correlated into the ancient facet of denudation, other similar reliefs in the belt are structural, and due, namely, to the erosion of wide, flat-lying lava flows that flooded the country after the peneplanation. Another and simpler explanation of the topography makes the lava flooding anterior to peneplanation. Still, a third history may, on further investigation, turn out to be the true one. At the present time it is impossible to decide between the rival views.

A safer definition of the region is purely topographic; it may thus be called the Belt of Interior Plateaus, or, briefly, the Interior Plateaus. (Plate 3.) This slight change in Dawson's name lays stress upon the individual tabular reliefs so characteristic of the region. These reliefs are facts; the peneplain and the involved assumption that the myriad individual reliefs belong to a physiographic unit, a single uplifted peneplain, are matters of theory. The pluralizing of the word ' plateau ' in the title not only changes the emphasis, but, in so doing, restores the term to its more advisable definition of a tabular relief bounded by strong downward slopes. The Interior Plateau as defined by Dawson is bounded on all sides by the strong upward slopes of the enveloping mountain ranges.

The belt of Interior Plateaus having thus been differentiated on special grounds, we may pass to the subdivision of the remaining two parts of the British Columbia complex. Those two belts separated by the plateau belt are rugged, often alpine, and, as a rule, do not show tabular reliefs. Present knowledge of the vast field cannot provide a rational treatment of these mountains rigidly on the basis of either rock composition or structural axes or

PLATE 3.

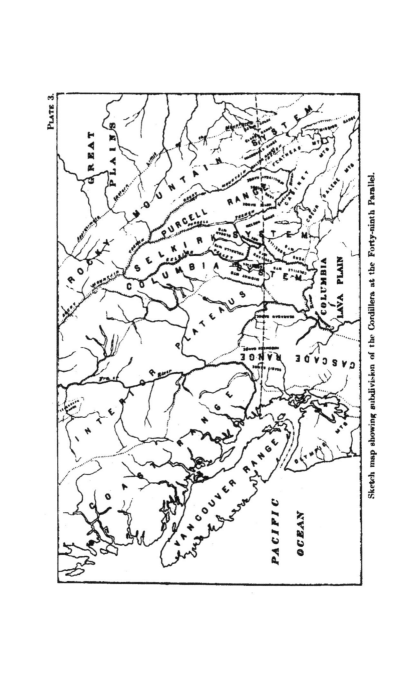

Sketch map showing subdivision of the Cordillera at the Forty-ninth Parallel.

geological history. It is possible, if not indeed probable, that the ranges immediately bounding the belt of Interior Plateaus have had a common history with it; they certainly include the same rock formations as occur in the interior plateaus. If the peneplain theory be finally accepted for the latter, it may ultimately prove best to treat the Coast range and other ranges in terms of the same theory. The only feasible scheme of subdivision at the present day must be based on topography only.

Mere hypsometry will not serve alone; the ranges of summit altitudes is too slight, their variation too unsystematic, for that. Dawson found that continuity of crest-lines and the position of the greater erosion valleys formed the most available basis of classification. As field work progresses in British Columbia, it becomes more and more certain that this double principle is the best that could be devised for present use. Many of the larger valleys are undoubtedly located on structural breaks, but it is evident that the strength of most of the valleys is the more direct result of fluviatile and glacial erosion. Owing to the peculiarly complicated rearrangements in the drainage of the Cordillera, whether due to glacial, volcanic, or crustal disturbances, or to spontaneous river adjustments, the valleys of British Columbia are in size very often quite out of relation to their respective streams. For example, the longest depression in the whole Cordillera is occupied by relatively small streams, the headwaters of the Kootenay, Columbia, Fraser, etc. Each of the rivers named, in its powerful lower course, flows through narrow canyons. Erosion-troughs rather than rivers have, therefore, been selected by Dawson and other explorers as the natural lines of demarcation between most of the constituent ranges of the Cordillera in these latitudes. The procedure is not new, but it is noteworthy as the most wholesale application of the principle on record. It stands in contrast to the more structural treatment, not only possible, but enforced by the orographic conditions in the United States.

. In the course of his own work, the writer has become convinced of the permanent value of Dawson's early and consistently held general view of the British Columbia mountains. But there has arisen the necessity of extending it to cover the Boundary mountains which, for the most part unvisited, were left unnamed by Dawson. The task of systematizing them is simple only in the stretch from the Great Plains to the Kootenay river at Tobacco Plains, a width of about seventy-five miles. The remainder, or five-sixths, of the Cordillera on the international line is not generally grouped into organic units at all; or, where so grouped, the names of the groups are not universally accepted. In attempting to supply this lack of system, the writer's aim has been to develop a system of grouping and nomenclature largely founded on names and concepts already in use, but not generally applied to the mountains so far south as the Boundary.

TRENCHES AND GREATER VALLEYS.

A point of departure is readily found. Within the Cordillera on the Forty-ninth Parallel, there are four principal longitudinal valleys which serve as convenient lateral boundaries for leading members of the system.

2 GEORGE V., A. 1912

(Plate 3.)   The whole valley occupied at the Boundary by the Kootenay river is the easternmost and much the longest.   It is a part of a single Cordilleran feature easily the most useful in delimiting the Canadian ranges.   From Flathead lake to the Liard river, a distance of about 800 miles, this feature has the form of a narrow, wonderfully straight depression lying between the Rocky Mountain system and all the rest of the Cordillera.   Unique among all the mountain-features of the globe for its remarkable persistence, this depression is in turn occupied by the headwaters of the Flathead, Kootenay, Columbia, Canoe, Fraser, Parsnip, Finlay, and Kechika rivers, and is therefore not fairly to be called a valley.   It may for present purposes be referred to as the 'Rocky Mountain Trench.'   The term 'trench' throughout this report means a long, narrow, intermontane depression occupied by two or more streams (whether expanded into lakes or not) alternately draining the depression in opposite directions.   An analogy is found in a military trench run through a hilly country.   (See Plate 4.)

The first-rank valley next in order on the west is also occupied at the Boundary by the Kootenay river, returning into Canada from its great bend at Jennings, Montana.   This valley begins on the south near Bonner's Ferry, Idaho, and is continued north of Kootenay lake by the valley of the Duncan river.   Recently, Wheeler has shown that the singular 40-mile trough occupied by Beaver river, which enters the Columbia river at the Canadian Pacific railway, is precisely *en axe* with the Duncan river valley.*   The whole string of valleys from Bonner's Ferry to the mouth of the Beaver, a distance of approximately 200 miles, forms a topographic unit that may be called the 'Purcell Trench.'   (Plate 5.)

The third of the first-rank valleys is drained southward by the Columbia river, expanded upstream to form the long Arrow lakes.   At its northern extremity near the Fifty-second Parallel of latitude, this valley is confluent with the Rocky Mountain Trench.   The southern termination of the valley regarded as a primary limit for these mountain ranges occurs about sixty miles south of the Forty-ninth Parallel, where the Columbia enters the vast lava plain of Washington.   To distinguish this orographic part of the whole Columbia valley between the points just defined, it may be called the 'Selkirk Valley.'

A glance at the map will show that the two primary trenches and the Selkirk Valley are in simple mnemonic relation to three principal mountain divisions of the Cordillera.   They lie respectively to the westward of the Rocky Mountain system, the Purcell range, and the Selkirk system.

The fourth of the first-rank valleys carries the south-flowing Okanagan river, with its various upstream expansions, including Osoyoos and Okanagan lakes.   The latter lies wholly within the belt of Interior Plateaus, a primary Cordilleran division.   Important as Okanagan lake is, no one has yet suggested that the plateau belt itself be subdivided into named portions separated by the lake.   It appears, on the other hand, wiser to recognize in the nomenclature

---

* A. O. Wheeler. The Selkirk Range, Gov't Printing Bureau, Ottawa, map in Vol. 2, 1905.

PLATE 4.

Boundary slash across the Rocky Mountain Trench at Gateway. Kootenay River visible in the slash ;
McGillivray Range in background.

PLATE 5.

Looking east across the Purcell Trench, from western edge of Kootenay River delta near Corn Creek. McKim Cliff is about nine miles distant and from three to four thousand feet high, above level of the river.

the essential unity of the belt. The southern portion of the Okanagan valley stretching from the mouth of the Similkamoon river to the confluence with the Columbia, has, however, a decided function in separating the Cascade range from the very different mountains east of the Okanagan river. This portion may be called the Lower Okanagan valley.

SUBDIVISION OF THE ROCKY MOUNTAIN SYSTEM.

The Rocky Mountain system, where it crosses the Forty-ninth Parallel, is very definitely bounded: on the east, by the great plains; on the west, by the Rocky Mountain Trench. (Plate 4.) This great element of the Cordillera is itself so vast that, for the purpose of presenting the facts of its stratigraphy and general geological history, the system must be subdivided into convenient units. By a kind of international co-operation this is being accomplished.

In Dawson's reconnaissance map of the Rockies, published in 1886, he designates as the 'Livingstone Range' the long Front range, stretching from the Highwood river at 50° 25′ N. Lat. southerly to the North Kootenay Pass at 49° 35′ N. Lat. On the west it is bounded, for many miles, by the straight valley of the Livingstone river and, in general, by the low mountainous area covered by the Crowsnest Cretaceous trough. The name had appeared in Arrowsmith's map of 1862, and in Palliser's of 1863, but Dawson gave the range its first definition.[*] Sixteen years later Willis made his admirable reconnaissance of a part of northwestern Montana and proposed that the 'Livingston Range' be considered as extending across the International Boundary as far as Lake McDonald.[†] There are, however, certain objections to making this change of definition. These may be briefly stated.

The crests of the Livingstone range, as delimited by Dawson, are composed almost entirely of Devono-Carboniferous limestones. Midway in the range-axis these rocks are interrupted, for a distance of about two miles, by a transverse band of Cretaceous beds, but this local variation in geologic structure involves no marked break in the line of crests. On the other hand, Dawson's map and accompanying text indicate clearly that the range unit ends a few miles north of the North Kootenay Pass. At that point a broad area of Cretaceous rocks squarely truncates the Devono-Carboniferous limestone and forms comparatively low mountains of the foothills type. The independent rangelet of which Turtle mountain is a part, is also composed of the Devono-Carboniferous limestone and is in a similar manner cut off by the zone of Cretaceous hills. The zone is fully twelve miles broad on the line of the axis of the Livingstone range as mapped by Dawson. On the south of the zone, lofty mountains of the Front range type are again to be found and these continue in strength to and beyond the International Boundary. The rocks composing these mountains south of the broad, transverse Cretaceous belt are, however, not of Devono-Carboniferous age but belong to a much older. Cambrian series underlain by conformable pre-Cambrian strata.

---

[*] Annual Report, Geol. Survey of Canada, 1885, Part B, p. 80.
[†] Bull. Geol. Soc. America, Vol. 13, 1902, p. 312.

2 GEORGE V., A. 1912

It is thus seen that the Cretaceous zone at the North Kootenay Pass makes a complete structural and topographic break in the Front range of the Rockies. To the north of the zone Dawson's Livingstone range forms a well-defined unit, its summits being composed of the later Paleozoic limestone. To the south of the zone the Front range, also rugged and in strong topographic contrast to the Cretaceous hills, is essentially composed of quartzites, argillites, and magnesian limestones of pre-Cambrian and earliest Paleozoic age. It seems, therefore, inappropriate to extend the Livingstone range any farther south than the North Kootenay Pass.

From that Pass south to McDonald lake in Montana the great range lying between the Flathead and the Great Plains in Canada and between the Flathead and the Lewis range on the American side of the international Boundary, needs a special name. Such a name has not hitherto been suggested. To supply the need the title 'Clarke range' may be proposed. The name is taken from that of the colleague of Meriwether Lewis who led the famous Lewis and Clarke expedition into the region in 1806. This splendid range is worthy of the able explorer and his memory is worthy of the range. The new designation for these mountains is in simple mnemonic relation to the name of the adjacent Lewis range, a name which is officially recognized by the United States Geological Survey.[*] (See Figure 1.)

After a review of the topographic and geologic relations Mr. Willis has expressed his own belief that the proposed change of nomenclature is advisable. In a letter to the writer he states:

'I took the name of Livingstone range from a Canadian map without particularly investigating the topography north of the boundary. It sufficed for my study at the time to know that there was a range in the United States which was in alignment with one called the Livingstone range in Canada. . . . . .

'Your proposition to give a distinct name to the range in the United States is, I think, fully justified, and the one you select is a most happy counterpart to the name Lewis. I should be glad to have you publish the nomenclature as you suggest, namely, giving to the range west of the Lewis range, from McDonald lake northward to the Kootenay Pass, the name of Clarke range.'

On the Canadian side of the Boundary for a distance of thirty miles the Clarke range is the Front range. Just north of the Boundary line it runs behind, to the westward of. the equally important member of the system, called the 'Lewis range' by Willis. At the Forty-ninth Parallel the wide valley occupied by Waterton lake and its affluent, Little Kootna creek, forms a definite boundary between the Clarke and Lewis ranges, which, further south, are separated by the head-waters of McDonald creek. According to Willis, the Lewis range extends southeastward almost to latitude 46° 45'. On the north it ends in Sheep mountain, a couple of miles beyond the International

---

[*] See Chief Mountain sheet of the Topographic Atlas, U.S. Geol. Survey.

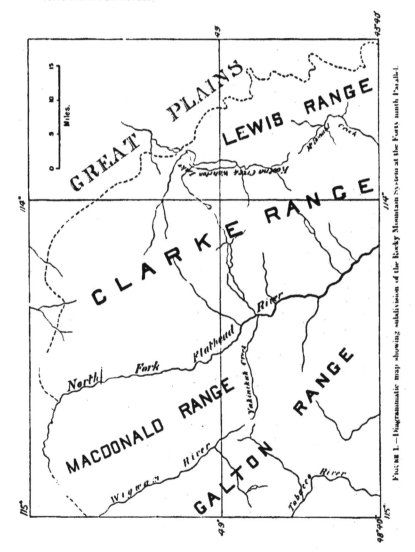

FIGURE 1.—Diagrammatic map showing subdivision of the Rocky Mountain System at the Forty-ninth Parallel.

2 GEORGE V., A. 1912

line. Dawson gave the name 'Wilson range' to a limited group of mountains in which Sheep mountain occurs. However, the title 'Lewis range' is to be a permanent feature of geographical nomenclature in Montana and must include the Wilson range, which is but a part of a whole first recognized in the scientific exploration of Montana. (Figure 1.)

West of the Flathead river and east of the Kootenay river, Dawson (following Palliser) recognized two distinct ranges as including the mountains along the Forty-ninth Parallel. On his 1886 map the more easterly range bears the name 'MacDonald range', the other bearing the name 'Galton range.' These ranges are separated, for a part of their length, by the straight valley of Wigwam river. Willis appreciated, the undoubted fact that the Galton range continues, with relatively unbroken crest-continuity far to the south of the Boundary line. In his 1902 map of northwest Montana, this range is represented as extending to the main Flathead river at Columbia Falls, the southwestern and western limit being fixed at the valleys of Stillwater creek, Tobacco river, and Kootenay river; and the northeastern limit in Montana being fixed at the valley of the North Fork of the Flathead river. Between the North Fork and the Wigwam the mountains are not named on Willis' map, but, apparently, were considered by him to belong to Dawson's 'MacDonald range.' In this view the MacDonald range is limited on the south by the strong transverse valley of Yakinikak creek. According to Dawson's map the northern limit of the Galton range seems to have been fixed at the Elk river and the northern limit of the MacDonald range at the Cretaceous area along the North Kootenay Pass.

Combining the views of Dawson and Willis we have a convenient subdivision of the western half of the Rocky Mountain system at the Forty-ninth Parallel into the two ranges, the Galton and the MacDonald, each of which, according to the law of crest-continuity, is a fairly distinct unit.

The sketch-map of Figure 1 illustrates the conclusions reached by the writer as to the most desirable topographic subdivision of this part of the Rocky Mountain system. It is very possible that further mapping of the region may show the necessity of modifying this orographic scheme. In its present form it will be found useful for the purposes of this report and seems to have the advantage of meeting the views of the few trained observers who have penetrated these mountains.

PURCELL MOUNTAIN SYSTEM AND ITS SUBDIVISION.

Westward from Tobacco Plains, on the Forty-ninth Parallel, we cross, in the air-line, sixty miles of ridges belonging to a range unit which is almost as systematic as the great group on the east. (Plate 4 and Figure 2.) The crests of this second group are in unbroken continuity from the wide southern loop of the Kootenay river at Jennings to the angle where the Purcell Trench is confluent with the Rocky Mountain Trench. Throughout this area the drainage is quite evenly divided by the easterly and westerly facing slopes of the unit-

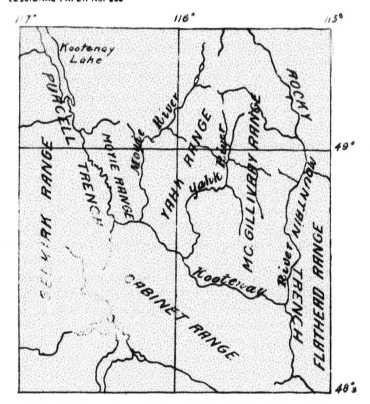

FIGURE 2. —Diagrammatic map showing subdivision of the Purcell Mountain System at the Forty-ninth Parallel.

relief. . This strong and extensive range has, in its northern part, been generally regarded as part of the Selkirk Mountain group. The middle and southern part, though broader and including most of the area, has, as a whole, never been authoritatively placed in the Selkirk system. Palliser gave the name 'Purcell Range' to a single component of the unit, namely, the group of summits lying between Findlay creek and St. Mary river. Dawson extended the name to cover all the mountains between Kootenay lake and the Rocky

2 GEORGE V., A. 1912

Mountain Trench, these mountains forming the 'Purcell division of the Selkirk system'; but he did not fix either a northern or a southern limit to the group so named.

The same usage appears in the maps and texts of most geographers publishing during the last twenty-five years. It was officially adopted by the Canadian Geological Survey, and by the British Columbia Government (1897). It appears in the general geological map of the Dominion, edited by Selwyn and Dawson, and issued by the survey in 1884. The name was accordingly entered in most of the American and European atlases of the world. For some unknown reason, the second edition of the official geological map of the Dominion (1901) represents the Purcells as constituting merely Palliser's original small group of summits, and this tradition has been followed in the new general map of the Dominion issued by the Canadian Department of the Interior (1902). Both official and general previous usages conflict with this quite recent official return to Palliser's mapping. In reality, the Palliser usage is not familiar to the people of British Columbia; it is subject to the criticism that the range let mapped by Palliser is not defined on the west by natural limits. The lack of definition in Palliser's exploratory sketch-mapping is such that it may even be doubted that Dawson really broke the law of priority in giving 'Purcell range' its broader meaning. The name is practically useless if it be not so extended. The long-established tradition of the influential atlases following the lead of Dawson makes it expedient to use the title in the broader meaning.

The question remains as to the northern and southern limits of the Purcell range. As a result of compiling all the available information, the writer has concluded that the range has no natural boundary to the northward, short of the confluence of the Purcell and Rocky Mountain trenches. The conclusion has been strikingly corroborated by the detailed studies of Wheeler along Beaver river. There is, similarly, no natural boundary on the south, short of the great bend of the Kootenay river in Montana. However vaguely supported by definite knowledge of the field, the latter conclusion has been anticipated by the editors of the Century Dictionary Atlas (map of Montana), of the Encyclopedia Americana (maps of British Columbia, Montana, and Canada), of Bartholomew's English Imperial Atlas, of Keith Johnson's Royal Atlas, and of Stieler's Handatlas. Maps occurring in all of these works represent the Purcell range as continuing southward into the United States as far as the Kootenay river. So far as known to the writer, there is no popular or official designation for the mountains lying between that river and the Canadian Boundary. The Cabinet mountains lie entirely south of the Kootenay river.

The first attempt on the part of the United States Geological Survey to name, in published form, the natural subdivisions of this extensive group of mountains was made in 1906. In Bulletin No. 285, published in that year and bearing the title 'Contributions to Economic Geology, 1905,' an outline map of northern Idaho and northwestern Montana was issued in connection with Mr. D. F. MacDonald's report on mineral resources of the district (page 42). On this map all the area enclosed between the International line and the Kootenay river as it swings through the great bend between Gateway and Porthill is shown as occupied by the 'Loop mountains.' That subdivision lying to the west of the Moyie river is mapped as the 'Moyie range'; a middle subdivision lying between the Moyie river and the Yahk river is mapped as the 'Yaak range'; an eastern division, lying to the eastward of the Yahk river is mapped as the 'Purcell range.'

No discussion of this scheme of nomenclature is given in MacDonald's paper, which was apparently written about the time when the preliminary paper of the present writer was in preparation. The name for the eastern subdivision of the Loop mountains was evidently given in the belief that the local Purcell range, as mapped by Palliser and Dawson, should be extended southward across the Boundary. A serious objection to this proposal is that the unit mapped by Palliser as the 'Purcell range' is, at the south, cut sharply off by the strong transverse valley of St. Mary river and by the wide plains about Cranbrook, nearly forty miles north of the Boundary line. If, then, it were thought expedient to limit the name 'Purcell' to an elementary range unit, as suggested though not enforced in Palliser's map, it is hardly possible to carry the Purcell range south of St. Mary river. On the other hand, we have seen that some official usage and the usage of several influential atlases have familiarized us with the idea of giving the old name 'Purcell range' to the entire mountain group occupying the area between the Rocky Mountain Trench and the Purcell Trench. This view implies that the rangelet limited on the east by the local mountain wall seen by Palliser as he looked across the Rocky Mountain Trench and mapped as belonging to the 'Purcell range,' should receive a special definition and a special name as soon as its extent as an orographic individual is known through actual mapping.

The general name 'Loop mountains' was presumably suggested by the loop of the Kootenay river, which bounds the whole group on the south. This great bend in the river is so remarkable a feature that the name is certainly appropriate on the United States side of the Boundary line. It is, however, true that four-fifths of the area and five-sixths of the length of the orographic unit involved, lie to the north of the Boundary and in no immediate relation to the bend of the Kootenay. For the greater part of the unit the name 'Loop mountains' is not appropriate. It is clear that the political boundary should, ideally, have no influence in fixing the general name. Systematic orography, supplemented by priority of usage, seem to declare for the older general name 'Purcell range' for the mountains considered, whether north or south of the line.*

In summary, then, the great range unit here called the Purcell range is bounded by the Rocky Mountain Trench, the Purcell Trench, and the portion of the Kootenay valley stretching from Jennings, Montana, to Bonner's Ferry, Idaho.

---

* Since the last paragraphs were written, Calkins has published Bulletin 384 of the United States Geological Survey, in which (Plate I) the "Loop mountains" are re-named the "Purcell mountains."

2 GEORGE V., A. 1912

In the present report MacDonald's name 'Moyie range' will be used to include all the mountains bounded by the Purcell Trench and by the strong valleys of the Moyie and Goat rivers. Similarly the name ' Yahk range' will be used with limits as follows: on the west and north by the Moyie river; on the south, by the Kootenay; on the east, by the Yahk river from source to mouth. The largest subdivision, the eastern one, will here be called the Mc-Gillivray range, a title taken from one of the earliest names of the Kootenay river.[*] This range is bounded on the east by the Rocky Mountain Trench; on the south, by the loop of the Kootenay river; on the west, by the Yahk river and the Moyie lakes; on the north, by the Cranbrook plains. (Plate 6.) This three-membered part of the Purcell system is there marked off by two huge trenches and by deep and wide transverse notches, faithfully followed by the two transmontane railroads, the Canadian Pacific and the Great Northern. (Figure 2.)

SELKIRK MOUNTAIN SYSTEM AND ITS SUBDIVISION.

The Selkirk Mountain system next on the west likewise forms a range unit considerably longer than the area generally ascribed to the Selkirk group. (Plate 3.) On principles similar to those adopted for the Purcell range, the Selkirk system may be defined as bounded on the east by the Purcell Trench; on the north and northeast by a portion of the Rocky Mountain Trench; on the west by the Selkirk Valley; on the south by the Columbia lava plain, Pend D'Oreille lake, and a short unnamed trench extending from that lake to the Purcell Trench at Bonner's Ferry. For a short stretch the Selkirk system is apparently confluent with the Cœur D'Alene mountains, though a short trench followed by the Great Northern railway may separate them. This extension of the Selkirks across the Boundary has already been indicated on maps of the Encyclopedia Americana, Stieler's Handatlas, and the Vidal-Lablache atlas.

The whole mountain complex embracing the Purcell range and Selkirk system, as just defined, may be viewed in another way. The Purcell range is thereby considered as part of the Selkirk system, and that division of the whole lying to the westward of the Purcell Trench, might be called the Selkirk range. The Selkirk system would thus include the Selkirk range and the Purcell range. As already noted, Dawson seems to have adopted this alternative view. An objection to it is the chance for confusion in using 'Selkirk' to mean now a component range, now the inclusive system. In favour of Dawson's view is the fact that in rock composition, structural axes, and geological history, the mountains lying between the Rocky Mountain Trench and the Selkirk Valley form part of a natural unit. On the other hand, the Selkirk range is, structurally and lithologically, as closely allied to the Columbia system as to the Purcell range; the Purcell range is, lithologically and historically, as closely allied to the Rocky Mountain system as to the Selkirk range. The practicable orographic classification, being based upon erosion troughs, recognizes the

---

[*] In J. Arrowsmith's map of British Columbia in British Government Sessional Papers relative to the affairs of British Columbia 1859.

dominant importance of the Purcell Trench. That superb t   ire of the Cordillera cleaves the mountains in so thoroughgoing a manner that a logical grouping must regard the Purcell range as a member co-ordinate with the Selkirk range.

In the map the latter division is called the Selkirk system, because it includes subordinate ranges. If, for purposes of exposition, this comprehensive character is not fixed for emphasis, the same Cordilleran member may be called the 'Selkirk range.' Similarly, when the Purcell range is, in the future, subdivided into its orographic units, it may bear the name 'Purcell system.' 'Cascade range' and 'Cascade system,' 'Coast range' and 'Coast system,' for example, may be profitably employed with the same distinctions. In all these cases it is a matter of emphasis.

The value of this distinction in common nouns, the great orographic significance of the Purcell Trench, and the weight of much authority in previous usages have caused the writer to suggest that the whole Purcell range be considered as co-ordinate with, and not part of, the Selkirk system.

No systematic subdivision of the system has ever been attempted. In discussing the geology of the system at the Boundary line there will be found to be much advantage in recognizing its subdivision into units of more convenient size. A tentative scheme will therefore be proposed.

Just north of the Forty-ninth Parallel a strong, though subordinate trench runs meridionally along the middle part of the system. This trench is occupied by the main Salmon river and by Cottonwood creek, which enters the West Arm of Kootenay lake at Nelson. It divides the system into two broad ranges, both of which are cut off on the north by the transverse valley enclosing the West Arm and the outletting Kootenay river. The eastern range, for which the name 'Nelson range' is proposed from the name of the chief town of the district), is bounded on the east by the Purcell Trench and on the south by a trench occupied by Boundary creek, Monk creek, and the South Fork of the Salmon river. The western range may be called the 'Bonnington range,' from the well-known falls which break the current of the Kootenay river. The southern limit of this range is the Pend D'Oreille valley; the western limit, the Selkirk Valley. (Figure 3.)

In the preliminary paper the 'Slocan mountain group was stated to be 'separated off definitely by the Slocan Trench, which is a longitudinal depression occupied by Slocan river, Slocan lake, and the creek valley mouthing at Nakusp, on Arrow lake.' The definition was framed partly on the ground 'that this mountain group includes the valley of Little Slocan river. On maturer consideration the writer wishes to recall this definition and to propose the name, 'Slocan mountains' for the group east of Slocan river and Slocan lake. The group west of the Slocan valley should probably have the name 'Valhalla mountains,' which was entered by Dawson, in 1890, on his 'Reconnaissance map of a portion of the West Kootanie District, British Columbia,' as the title for the complex of high peaks west of Slocan lake.

25a—3½

1 GEORGE V, A. 1911

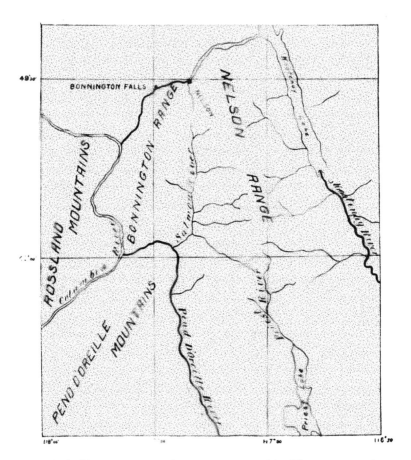

FIGURE 3.—Diagrammatic map showing extension of the Selkirk Mountain System at the Forty-ninth Parallel.

The mountain group lying southwest of the Pend D'Oreille river was called in the preliminary paper, the Pend-D'Oreille mountains. It may further be proposed that the two groups separated by Priest River valley be named the Kaniksu range (on the west) and the Priest range (on the east). 'Kaniksu' is the old Indian name for Priest lake. Though these names may not prove finally satisfactory, the writer believes that the naming of these groups in an authoritative and systematic manner would be a geographic gain. In passing, the question may be raised as to the advisability of regarding the mountains lying between Priest river, Pend D'Oreille lake, and the Kootenay river, as part of the Cabinet mountain range. The bulk of the Cabinet range, as now generally recognized, lies to the southeast of the strong trench running from Bonner's Ferry to Sandpoint. To the writer it seems both easy and expedient to consider this trench as bounding the Cabinets on the northwest and the distinct range, hitherto unnamed, on the southeast. The limits of the latter range are: Boundary creek on the north, Priest river valley on the west, the Purcell Trench on the east, and the Bonner's Ferry-Pend D'Oreille trench on the south and southeast.

COLUMBIA MOUNTAIN SYSTEM AND ITS SUBDIVISION.

The principal range unit adjoining the Selkirk system on the west is here called the Columbia system. (Plate 3.) It is definitely limited on the east by the Selkirk Valley and by a part of the Rocky Mountain Trench, the latter truncating the northern end of the Columbia system as it does the Selkirk and Purcell groups. On the south the Columbia system is limited by the Columbia lava plain. On the west the limit is determined by the lower Okanagan valley, and, to the northward, less well by the eastern edge of the belt of Interior Plateaus. That edge may be located for about thirty miles in the line of the main Kettle river valley. North of the main line of the Canadian Pacific railway, the belt of Interior Plateaus seems to reach, but not cross, Adams lake and Adams river. Still farther north, the western limit of the Columbia system is fixed by a trench occupied by the headwaters of the North Thompson river, and by an affluent of the Canoe river. Northwest of this trench begins the great system including the Cariboo mountains.

Apparently the first official (Governmental) name for the mountains explored on the Canadian Pacific railway line west of the Columbia river was 'Gold Range.'[*] The group so named extends from the latitude of Shuswap lake to the narrows between the Arrow lakes. This usage has been adhered to by the Government of British Columbia.[†] In 1874, the Dominion Department of Railways and Canals introduced the name 'Columbia range' for the much larger mountain group including the 'Gold range,' and extending from

---

[*] Map of British Columbia, compiled under the direction of the Hon. J. W. Trutch, Chief Commissioner of Lands and Works and Surveyor-General. Victoria, 1871.
[†] Map of the province of British Columbia, compiled by J. H. Brownlee by direction of the Chief Commissioner of Lands and Works. Victoria, 1893.

2 GEORGE V., A. 1912

the headwaters of the North Thompson river southward to Lower Arrow lake.‡
This usage was confirmed by Selwyn and Dawson, each in turn Director of the
Geological Survey of Canada.§ Nevertheless, the new general map of Canada
issued by the Department of the Interior at Ottawa (1902) gives the name
'Gold range' to this larger group. The extension of the limits of the Gold
range is a departure from the official tradition. of both the provincial and
Dominion governments. It appears best to hold the name 'Gold range' to its
original designation of a local mountain group. and retain the title 'Columbia
range' with a broader meaning.

For the immense Cordilleran unit stretching from end to end of the Selkirk
Valley, and bounded on the east by the Columbia river, there is no question
that the name 'Columbia range' is more significant and appropriate than the
name 'Gold range.' The latter name has a special disadvantage worthy of
note. Although Dawson, in his later writings, used the name 'Gold range'
in its original sense of a local mountain group, he as often used 'Gold range'
or 'Gold ranges' to include the Selkirk, Purcell, 'Columbia', Cariboo, and
Omineca ranges. This inconsistent usage robs the title 'Gold' range' of even
that modicum of value which it has as an alternative for the more significant
title. As already stated, the name 'Columbia range,' with its comprehensive
meaning, has the priority.

The extension of the apposite title, 'Columbia range.' (with variant
'Columbia system'), to cover the larger area described in the foregoing paragraphs is, it is true, not according to tradition, but, as in the case of the Selkirk
system and Purcell range, the widening of the meaning is justified by the lack
of definition as to the true areal extent of the 'Columbia range' in its original
use, and is enforced by the fact of crest continuity within a fairly well delimited
belt of the Cordillera.

The southern third of the Columbia system is characterized by comparatively low mountains, which in rock composition are allied both to the northern
part of the system and to the belt of the Interior Plateaus. These southern
mountains commonly show uniformity in summit levels; yet there are no
remnant plateaus or very few of them, and it is advisable to regard these
mountains as forming a group distinct from the Interior Plateaus. A convenient name for part of the group. 'Colville mountains,' was given as early
as 1859-60 by the members of the Palliser expedition. In the preliminary
paper it was proposed that the 'Colville group should include the mountains
lying between the two forks of the Kettle river as well as all the part of the Columbia system south of the river. Further study and the test of actual convenience
in description have since suggested the expediency of recognizing the mountains between the two forks of the Kettle river as forming an independent
subdivision, and to them the name 'Midway mountains' is given. Further-

‡ S. Fleming. Exploratory Survey, Canadian Pacific Railway report, Ottawa, 1874,
Map-sheet, No. 8.
§ Forest Map of British Columbia, published by G. M. Dawson in Report of
Progress. Geol. Surv. of Canada, 1879-80.

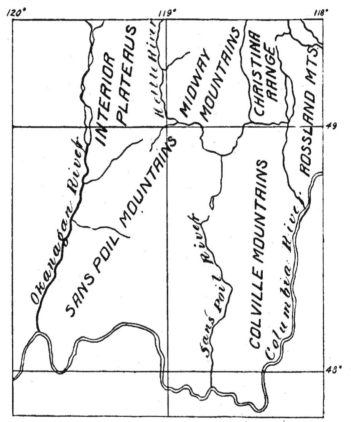

Figure 4. –Diagrammatic map showing subdivision of the Columbia Mountain System at the Forty-ninth Parallel.

more, the north-south trench occupied by the Sans Poil river. Curlew lake, and Curlew creek, divides the mountains south of the line into two distinct parts. To the eastern part, which is that nearer the site of old Fort Colville, the name 'Colville mountains' may be restricted; while the western division, bounded by the Kettle river, the Sans Poil-Curlew trench, the Columbia river, and the Okanagan river, may be called the 'Sans Poil mountains.' (Figure 4.)

2 GEORGE V., A. 1912

Another important, though small, natural subdivision of the system is limited on the north, east, and south by the Selkirk Valley; on the west, by the lower Kettle valley, and by a short trench running from Lower Arrow lake to Christina lake and the Kettle river at Cascade. This group may be called the Rossland mountains. (Plate 3.)

Again for convenience, the mountains occurring between Christina lake and the North Fork of Kettle river will be referred to as the Christina range. (Figure 4.)

The more northerly part of the Columbia system is yet too imperfectly known to permit of subdivision in a systematic way.

BELT OF INTERIOR PLATEAUS.

As we have seen, the belt of Interior Plateaus is of primary importance in the systematic orography of the Cordillera. (Plate 6.) It is difficult of delimitation. On nearly all of its boundaries the belt fades gradually into the loftier, more rugged ranges encircling it. Its limits have been compiled and drawn on the map (Plate 3.) after a study of Dawson's numerous reports of exploration. The limits are to be regarded as only approximate. The plateau character is obscure at the Forty-ninth Parallel, but the roughly tabular form and considerable area of Anarchist mountain, immediately east of Osoyoos lake, seem to warrant the slight extension of the belt across the International line. The southernmost limit of the belt is an irregular line following—(1) the main Kettle river valley; (2) a quite subordinate trench occupied by Myer's creek and Antoine creek, in the state of Washington; (3) a part of the lower Okanagan valley; and (4) the Similkameen-Tulameen valley.

CASCADE MOUNTAIN SYSTEM AND ITS SUBDIVISION.

Usage, both official and popular, has gone far toward finally establishing the nomenclature for the immense ranges lying west of the Columbia lava plain, Midway mountains, and belt of Interior Plateaus. The Cascade range is now defined on the principle of continuity of crests, not on the basis of rock-composition. At the cascading rapids of the Columbia river the range is a warped lava plateau; in northern Washington it is an alpine complex of schists, sediments, granites, etc. In British Columbia, Dawson adopted the name 'Coast range' to enforce the view that the granite-schist British Columbia mountains on the seaboard should be distinguished from the lava-built Cascades, as originally named, at the Columbia river. It has, however, become more and more evident, as the study of the Cordillera progresses, that rock-composition can never rival crest continuity as a primary principle in grouping the western mountains. Meanwhile, the name 'Coast range' has survived, and is, in fact, the only name officially approved by the Geographic board of Canada for any principal division of the Cordillera

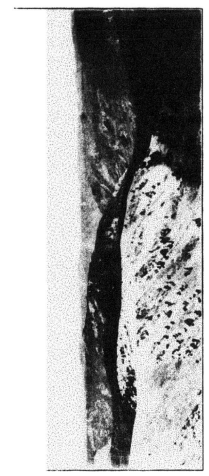

Edge of the Interior Plateaus: looking north from near Park mountain, Okanagan Range, over Ashnola River valley.

Dawson did not fix a southern limit for the Coast range. General usage has not fixed the northern limit of the Cascade range. The solution of the problem is obvious if the principle of limiting units by master valleys and trenches be applied. The Fraser river valley clearly supplies the required boundary between the two ranges. There seems to be no other simple adjustment of the two usages, which undoubtedly sprang up because of the existence of a political boundary at the Forty-ninth Parallel. It is important to note that the delimitation here advocated is not new, since it appears on two of the earliest official maps of British Columbia those accompanying the 1859 British Blue Books, entitled 'Papers relative to the affairs of British Columbia.'

The remaining boundaries of the Cascade and Coast ranges, as well as the boundaries of the Olympics and of the Vancouver range, are at once derived from the map, and need no verbal description. These natural boundaries seem in large part to be located along structural depressions, and belong, therefore, to a type unusual in the Canadian Cordillera.

The subdivision of the system where it crosses the Forty-ninth Parallel has already been recognized by Bauerman and, more in detail, by Smith and Calkins.* With these authors the present writer is in accord on the matter and a quotation from the report of Smith and Calkins will suffice to indicate such subdivision as seems necessary for the present report.

'In northern Washington, where the Cascade mountains are so prominently developed, the range is apparently a complex one and should be subdivided. This was recognized by Gibbs, who described the range as forking and the main portion or 'true Cascades' crossing the Skagit where that river turns west, while the 'eastern Cascades' lie to the east. Bauerman, geologist to the British commission, recognized three divisions, and as his subdivision is evidently based upon the general features of the relief it will be adopted here. To the eastern portion of the Cascades, extending from mount Chopaka to the valley of Pasayten river, the name of Okanagan mountains is given, following Bauerman. To the middle portion, including the main divide between the Pasayten, which belongs to the Columbia drainage, and the Skagit, which flows into Puget sound, Bauerman gave the name Hozomeen range, taken from the high peak near the boundary. For the western division the name Skagit mountains is proposed, from the river which drains a large portion of this mountain mass, and also cuts across its southern continuation. It will be noted that the north-south valleys of the Pasayten and the Skagit form the division lines between these three subranges, which farther south coalesce somewhat so as to make subdivision less necessary.

'The Okanagan mountains form the divide between the streams flowing north into the Similkameen and thence into the Okanagan and those flowing south into the Methow drainage. In detail this divide is exceedingly irregular, but the range has a general northeast-southwest trend, joining

* G. O. Smith and F. C. Calkins. Bull. 235, U. S. Geol. Survey, 1904, p. 14.

MICROCOPY RESOLUTION TEST CHART

(ANSI and ISO TEST CHART No. 2)

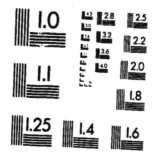

APPLIED IMAGE Inc

1653 East Main Street
Rochester, New York 14609 USA
(716) 482 - 0300 - Phone
(716) 288 - 5989 - Fax

2 GEORGE V., A. 1912

the main divide of the Cascades in the vicinity of Barron. The highest peaks, such as Chopaka, Cathedral, Remmel, and Bighorn, have a nearly uniform elevation of 8,000 to 8,500 feet and commonly are extremely rugged. Over the larger portion of this area the heights are above 7,000 feet, and below this are the deeply cut valleys.'

The respective east and west limits of the three ranges are clearly and definitely fixed by the longitudinal valleys of the Similkameen, Pasayten, and Skagit rivers, and by the partially filled depression of the Strait of Georgia. The northern and southern limits cannot at present be determined: that further step may be made when, in the future, the cartography of the rugged system is completed. (Plate 3.)

SUMMARY.

The writer keenly feels the responsibility of suggesting many of the changes and additions proposed in the cartography of this large section of the Cordillera. The attempt to describe the geology of the Boundary belt without some kind of systematic orography on which to hang the many facts of spatial relation, is truly the making of bricks without straw. The scheme outlined above has thus developed out of a clear necessity.

The orography of the International Boundary cannot profitably be treated without reference to longitudinal Cordilleran elements, often running many hundreds of miles to northward and southward of the Boundary. For this reason the accompanying map is made to cover all of the Cordillera lying between the forty-seventh and fifty-third parallels of latitude. (Plate 3.)

The terms 'range' and 'system' are used in their common elastic meanings, with 'system' more comprehensive than 'range.' The Cordilleran system, or Cordillera, includes the Rocky Mountain system, the Selkirk system, etc. The Cascade range includes the Okanagan range, Skagit range, etc. A system may include among its subdivisions a mountain group without a decidedly elongated ground-plan; thus the Columbia system includes the Rossland mountains. But both 'range' and 'system,' used with their respective broader or narrower meanings, involve the elongation of ground-plan and a corresponding alignment of mountain crests. The great weight of popular and official usage seems to render it inadvisable to attempt any more systematic organization of the common nouns in this case. It has been found almost, if not quite, as difficult to organize the proper names in an ideal manner.

The basis of mountain grouping is purely topographical, and is, in the main, founded on established usage. A primary grouping recognizes within the Cordilleran body two relatively low areas, characterized by tabular reliefs, accompanied by rounded reliefs, generally accordant in altitude with the plateaus. These two areas are the belt of Interior Plateaus in British Columbia and the Columbia lava plain of the United States. The remainder of the Cordillera—ridged, peaked, often alpine—is divided into systems, ranges, and more equiaxial groups, either by 'trenches,' by master valleys, or, exceptionally, by structural depressions.

The Cascade range, the Olympic mountains, the Vancouver range, and the Coast range of British Columbia, with their continuations north and south, compose what may be called the Coastal system. All the ranges east of the Rocky Mountain Trench, with their orographic continuations north and south, constitute the Rocky Mountain system. The Columbia lava plain and the belt of Interior Plateaus form the third and fourth subdivisions. A fifth more or less natural group, yet lacking a name, includes the Bitterroot, Clearwater, Cœur D'Alene, Cabinet, Flathead, Mission, and Purcell ranges, the composite Selkirk system, and the composite Columbia system, with the unnamed system including the Cariboo mountains.

LEADING REFERENCES.

Texts—

Baedeker's Guide-Book to the United States. 1893.

BAUERMAN, H.—Report on the Geology of the Country near the 49th parallel of latitude. Geol. Survey of Canada, Report of Progress. 1882-3-4, section and page 8B.

British Government Blue-Books.—Papers relative to the explorations by Captain Palliser in British North America. 1859.

Papers relative to the affairs of British Columbia, 1859.

Further papers relative to the Palliser exploration, 1860.

Further papers relative to the affairs of British Columbia, 1860.

Map to accompany above-mentioned papers: Stanford, London. 1863.

BROOKS, A. H.—Professional Paper No. 1, United States Geological Survey. 1902, pp. 14, 15.

National Geographic Magazine, vol. 15, 1904. pp. 217, 218.

Proc. Eighth International Geographical Congress. St. Louis, 1904, p. 204.

CANADIAN PACIFIC RAILWAY.—Pamphlets.

Century Dictionary, article 'Cordillera.'

DANA, J. D.—Manual of Geology. 2nd edit. 1874, and 3rd edit., 1880, pp. 15, 16, in each edition; 4th edit., 1895, pp. 389, 390.

DAVIS, W. M.—In Mill's International Geography. 664-668, 671, London, 1900.

DAWSON, G. M.—*Quart. Jour. Geol. Soc., London,* vol. 34, 1877, p. 89.

Report of Progress, Geol. Survey of Canada, 1877-78, part B, pp. 5-8.

Canadian Naturalist, new series. vol. 9, 1879, p. 33.

Report of Progress, Geol. Survey of Canada. 1879-80, part B, pp. 2, 3, and map.

Report, British Assoc. Advancement of Science, vol. 50, 1880, p. 588.

Geological Magazine, new series, vol. 8, 1881. pp. 157 and 225.

Ann. Rep. Geol. Survey of Canada. new series. vol. 1, 1885, part B, pp. 15, 17, 22; vol. 3, 1888, part B, p. 12 and map, and part R, pp. 7-11; vol. 5, 1889, part B, pp. 6, 7; vol. 8, 1894, part B, pp. 3, 4 and map.

2 GEORGE V., A. 1912

Transactions, Royal Society of Canada, vol. 8, sect. iv., 1890, pp. 3, 4.
The Physical Geography and Geology of Canada,' reprinted from the
Handbook of Canada, issued by the publication committee of the
local executive of the British Association, p. 7 and p. 40ff.
Toronto, 1897.

Bulletin Geol. Society of America, vol. 12, 1901, pp. 60, 61.

DAWSON, SIR J. W.—Handbook of Geology, pp. 202-204. Montreal, 1889.

DAWSON, S. E.—In Stanford's Compendium of Geography: North America,
vol. 1, 1898, map and text.

DE LAPPARENT, A.—Leçons de Géographie Physique, pp. 629-632. Paris,
1898.

FLEMING, S.—Exploratory Survey, Canadian Pacific Railway Report, map-
sheets 8 and 12. Ottawa, 1874.

Report on Surveys on preliminary operations of the Canadian Pacific
Railway up to January, 1877, map-sheets 1, 6, and 7. Ottawa,
1877.

FRYE, A.—Complete Geography. Boston, 1895, and later editions, various
maps and diagrams.

Grammar School Geography, map of North America and relief map,
Boston, 1904.

GANNET, II.—In Stanford's Compendium of Geography: North America,
1898, vol. 2, p. 26, and maps of United States and North America.

Geographic Board of Canada, Annual Report, 1904.

GOSNELL, R. E.—Year-Book of British Columbia, p. 5. Victoria, 1903.

HAYDEN, F. V.—In Stanford's Compendium of Geography: North America,
1883, pp. 40, 57, 58, 110, 111 and map.

HOPKINS, J. C.—Canada: An Encyclopædia of the Country, vol. 1, map,
Toronto, 1898.

JOHNSTON, KEITH.—'Geography,' p. 426. Stanford: London, 1896.

KING, CLARENCE.—United States Geol. Exploration, 40th parallel, Systematic Geology, 1878, pp. 1, 5, and 15.

LECONTE, J.—Elements of Geology, p. 250. New York, 1892.

LINDGREN, W.—Professional Paper, U.S. Geol. Survey, No. 27, 1904, p. 13.

RÉCLUS, E.—Nouvelle Géographie Universelle: Amérique Boréale, 1890,
pp. 260, 264, 281.

RUSSELL, I. C.—*Bull. Geog. Society of Philadelphia,* vol. 2, 1899, p. 55
(with letters from David " Davis, Dawson, Heilprin, and
Powell).

North America, pp. 60, 61, 120-125, 147, 165-168. Appleton: New
York, 1904.

SELWYN, A. R. C.—In Stanford's Compendium of Geography, North
America, 1883; map of British Columbia.

(SELWYN, A. R. C., and) G. M. DAWSON.—Descriptive Sketch of the
Physical Geography and Geology of the Dominion of Canada, pp.
33, 34. Montreal, 1884.

SESSIONAL PAPER No. 25a

SMITH, G. O., and CALKINS, F. C.—*Bull. U.S. Geol. Survey*, No. 235, 1904 pp. 12-14.

SHALER, N. S.—Nature and Man in America, pp. 250-256. New York. 1891.

WHEELER, A. O.—The Selkirk Range, 2 vols. Government Printing Bureau: Ottawa, 1905.

WHEELER, G. M.—U.S. Survey west of the 100th Meridian, Geographical Report, 1889, p. 11.

WHITNEY, J. D.—The United States, pp. 22-30, 68, 79, 122, and references. Boston, 1889.

General Atlases and Encyclopedias—

Atlas Universelle; Hachette: Paris, 1904. Atlas Vidal-Lablache; Paris, 1904. Century Atlas; New York. Encyclopaedia Americana, 1904. Encyclopaedia Britannica Atlas. English Imperial Atlas; Bartholomew: London, 1892. Home Knowledge Atlas; Toronto, 1890. Johnson's Universal Cyclopaedia; edited by Guyot, 1891. Keith Johnston's Royal Atlas; Edinburgh, 1885. Rand-McNally Indexed Atlas of the World, 1902. Rand-McNally Indexed Atlas of the Dominion of Canada, 1905. Stanford's London Atlas, folio edit., 1904. Stieler's Handatlas, 1897. Times Atlas; London, 1900. Universal Atlas; Cassell: London, 1893.

Official Maps—

Map of the Dominion of Canada, geologically coloured from surveys made by the Geological Corps, 1842-1882; two sheets, scale 45 miles to one inch. Geol. Survey Department, 1884. (Topography based on wall-maps issued by the Department of Railways and Canals.)

Geological Map of Dominion of Canada, western sheet; scale, 50 miles to one inch. Geol. Survey Department, 1901.

Map of the Dominion of Canada and Newfoundland, James White, F.R.G.S., geographer; scale, 35 miles to one inch. Department of the Interior, Ottawa, 1902.

Map of British Columbia to the 56th parallel N. lat., compiled and drawn under the direction of the Hon. J. W. Trutch, Chief Commissioner of Lands and Works and Surveyor-General. Victoria, 1871.

Map of the province of British Columbia, compiled by J. H. Brownlee, by direction of the Hon. F. G. Vernon, Chief Commissioner of Lands and Works. Victoria, 1893.

Map of the west division of Kootenay District and a portion of Lillooet, Yale and East Kootenay, B.C. Compiled by direction of the Hon. G. B. Martin, Chief Commissioner of Lands and Works. Victoria, B.C., 1897.

PLATE 7.

Looking down Kintla Lakes valley. Upper lake in middle ground; Mount Thompson, composed of Siyeh limestone, on left; scarp of Purcell Lava on right; Flathead valley and MacDonald Range in background.

CHAPTER IV.

STRATIGRAPHY AND STRUCTURE OF THE CLARKE RANGE.

ROCKY MOUNTAIN GEOSYNCLINAL PRISM.

One of the least expected results of the Boundary survey consists in the discovery that almost all of the mountains traversed by the Commission map between the Great Plains and the summit of the Selkirk range—an air-line distance of one hundred and fifty miles—are composed of a single group of conformable strata. These rocks are as yet largely unfossiliferous but all of them are believed to be of pre-Devonian age. For the most part they are water-laid, well-bedded sediments but contain one important sheet of extrusive lava which extends quite across the whole Rocky Mountain system and the eastern part of the Purcell system. Though the sedimentary group is a unit, it has been found that noteworthy lithological differences appear in the rocks as they are followed along the Boundary line from the Front ranges westward. These differences are due to gradual changes of composition and no two complete sections taken five miles apart on an east-west line would be identical. Nevertheless it has been found possible to relate all the essential features of these varying strata to four standard or type sections.

The most easterly type section was made in the Clarke range. It agrees very closely with the section already described by Willis from the Lewis range at localities lying on the tectonic strike from the localities specially studied in the Clarke range by the present writer. The rocks thus found to compose both the Lewis and Clarke ranges belong to what may be called the Lewis series. The type section constructed from traverses made in the Galton and Mac-Donald ranges include strata which are here grouped as the Galton series. The equivalents of the same series compose the entire Purcell mountain system at the Forty-ninth Parallel and belong to a sedimentary group which may be called the Purcell series. The fourth type section was constructed from magnificent exposures occurring in the eastern half of the Selkirk mountain system. This assemblage of beds will be referred to as the Summit series. The name is taken from Summit creek along which a great part of the series is exposed; the creek was itself named from the fact that it heads on the water-divide of the Selkirks. Analogy with the other three series names suggests 'Selkirk series' for this fourth group of strata, but that designation has already been used by Dawson for the related but lithologically distinct group described in his traverse on the main line of the Canadian Pacific railway.

47

2 GEORGE V., A. 1912

The retention of these four series names implies some slight tax on the memory but that drawback is much more than offset by the ease of grasping and systematizing the many petrographic and stratigraphic facts which must be reviewed before the constitution of the great geosynclinal prism is understood. In view of the general lack of fossils throughout the belt, the differentiation and correlation of the beds must be based on lithological properties. The following description of each series includes a statement of the facts on which is founded the writer's belief in the integrity of the whole sedimentary field, one huge sedimentary prism constituting the staple rocks in the eastern third of the Cordillera at the Forty-ninth Parallel. The summary of the individual facts, as they are clustered in describing the four series, will further well illustrate the systematic variation in the geosynclinal prism as it is crossed from east to west.

In each type section the formations will be considered in their natural order, beginning with the oldest. The description will, in each case, be made concisely and will be shorn of many items of fact which do not appear of importance in the larger stratigraphic problem. The Purcell Lava formation will be treated in chapter IX.

The description of the four type series will be found nearly to cover the stratigraphy of the different ranges from the Lewis on the east to the Yahk on the west. In the Galton and MacDonald ranges there are bodies of fossiliferous Devonian and Mississippian limestone which are properly parts of the prism, but, having generally been eroded away, now form only quite subordinate masses within the Boundary belt. These will be described in connection with the account of the Galton series. The only other bed-rock sedimentary formation occurring between the Great Plains at Waterton lake and the Purcell Trench at Porthill is a thick but local deposit of Tertiary fresh-water clays and sands flooring the Flathead valley. This occurrence will be noted in connection with the topographic description of the Clarke and Lewis ranges. The stratigraphy of the Selkirk system is much more highly composite than any of the eastern ranges; its description will, therefore, be detailed only so far as the Summit series and the underlying terrane are concerned, and will then be interrupted by a chapter giving the results of correlating so   this gigantic stratified unit, the Rocky Mountain Geosynclinal.*

As an aid to clearness it may be noted, in anticipation of a later chapter, that the Rocky Mountain Geosynclinal includes all the sedimentary formations from the base of the Belt (pre-Olenellus) terrane up to and including the Mississippian formation, as these beds are developed in the eastern half of the Cordillera. The Lewis, Galton, and Purcell series represent only a part of the whole prism, in each case the youngest exposed bed being

---

* Following Dana (Manual of Geology, 4th edition, p. 380) the writer distinguishes the geosyncline, the large-scale down-warp of the earth's surface, from the load of sediments which may accumulate on the down-warped area. In the present report the load of sediments will be referred to as a "geosynclinal prism" or, more briefly, as a "geosynclinal."

not far from an Upper Cambrian horizon, and the oldest exposed bed being located well above the base of the Belt terrane. The Summit series includes the entire Belt terrane and a vast thickness of conformably overlying strata which may represent the whole Paleozoic succession up to and including the Silurian. Overlying the Summit series, apparently conformably, is a very thick and massive limestone which is probably Carboniferous but may, in its lower part, belong to the Devonian. In other words, it seems possible that a complete geosynclinal prism is represented in the exposures of the Boundary belt where it passes through the southern Selkirks. The name 'Summit series' refers only to the unfossiliferous formation making up the lower and greater part of the prism in this mountain system.

LEWIS SERIES.

The writer has carefully studied the Lewis series only within the limits of the Clarke range. Since the Commission map extends but a mile or two to the eastward of the summit monument, a close mapping of the different formations between that monument and Waterton lake was not feasible. In this stretch of fifteen miles the field work was confined to the measurement of a few sections. These, however, occurred in areas of unusually complete rock-exposure and much light on the composition of the lower one-third of the series was derived from their examination. In the Lewis range the writer had no opportunity for close work and his experience there was limited to rapid traverses from Waterton lake to Chief mountain and thence, by way of Altyn and the Swift Current Pass, to Belton, Montana.

Limited as that opportunity was, it sufficed to corroborate the belief—already reached after reading Willis' paper on the 'Lewis and Livingston Ranges'—that the stratified sequence in the Lewis range is essentially identical with that in the Clarke range. It will, in fact, appear in the following account that the columnar section constructed by the writer from data obtained wholly within the Clarke range, matches well, member for member, with Willis' columnar section derived almost entirely from observations in the Lewis range. Partly in order to emphasize this identity the name 'Lewis series' has been selected to cover the whole group of strata in the Clarke range, the group now to be described. (See Plate 7.)

Beginning at the top the formations included in the _____ have been listed in the order of the following table:

Formation.	Thickness in feet.	Dominant
	Top, erosion surface.	
Kintla..	860+	Argillite.
Sheppard..	600	Silicious dolomite.
Purcell Lava..	60	Altered basalt.
Siyeh..	4,100	Magnesian limestone and argillite
Grinnell..	1,600	Metargillite.
Appekunny..	2,600	Metargillite.
Altyn..	3,500	Silicious dolomite.
Waterton..	200+	Silicious dolomite.
	13,720	
	Base concealed.	

2 GEORGE V. A. 1912

Excepting the Purcell Lava these rocks will be described in the present chapter. The volcanic formations of the range are described in chapter IX.

WATERTON FORMATION.

The lowest member of the Lewis series as exposed within the Boundary belt was seen at only one locality—at the cliff over which the waters of Oil creek (Cameron Falls brook of the older maps) tumble from the hanging valley of Oil creek into Waterton lake (Plate 8). This member may be called the

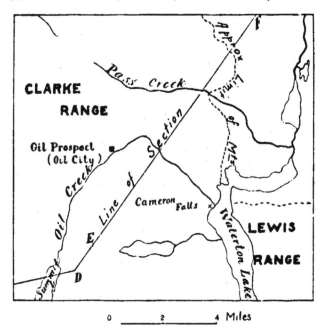

FIGURE 5.—Diagrammatic map showing position of the Structure-section (See Fig. 6) east of the Rocky Mountain Summit.

Waterton dolomite. At the cascade it is seen to be conformably overlain by the Altyn limestone. From the sharp bend below Oil City the creek faithfully follows the axis of a strong anticline which pitches gently northward. As one descends the creek he also descends in the stratified rock-series, and at Cameron Falls walks upon the Waterton dolomite, the visible core of the anticline. The

PLATE X.

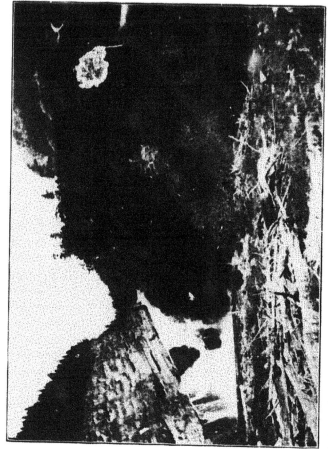

Cameron Falls on Oil Creek, at low water season. View illustrates character of Waterton dolomite and liquid beds in southwest limb of Oil Creek anticline

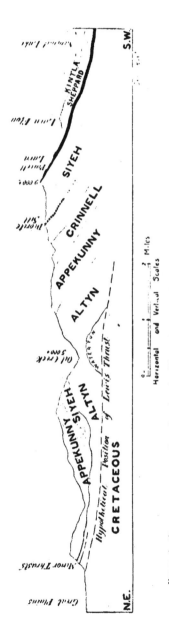

FIGURE 6.— Structure section across the strike, along the ridge southeast of Oil Creek, eastern slope of the Clarke Range. Looking southeast. For location see Fig. 5.

2 GEORGE V, A. 1912

section of the dolomite is incomplete; with all its typical characters it disappears, to the eastward, beneath the stream gravels and glacial deposits surrounding the lake. (Figures 5 and 6.)

Throughout its whole observed thickness of 200 feet the formation consists of an exceptionally strong and massive, dark gray carbonate rock, weathering dark gray to brownish gray and sometimes buff. In the field the rock has a most deceptive resemblance to a homogeneous, thick-bedded argillite. It effervesces but slightly in cold dilute acid, and the essential carbonate character

FIGURE 7.—Diagrammatic drawing from thin section of Waterton dolomite, showing middle part of a lense of orthoclase in interlocking granules. Rhombohedra of dolomite are embedded in the anhedral dolomite which forms most of the rock. The cleavages shown in the carbonate are diagrammatic only and in reality are seldom visible. The black spots represent carbonaceous matter. Highly magnified; diameter of circle 0·15 mm.

was not suspected until the more careful laboratory study was put on the rock. The thin section showed immediately that it is largely composed of carbonate grains. Their size is very small, the diameters steadily averaging about 0·02 mm., with a few grains reaching twice or thrice that diameter. These grains are sometimes knit together in a thorough, interlocking manner but more often show a tendency to assume the rhombohedral form, the habit characteristic of the grains in true dolomites.

In many of the laminae of the rock (0.2 mm. to 1 mm. in thickness) the minute rhombohedra are embedded in a compound, colourless to pale-brownish base. It is composed in part of very minute, anhedral grains of glass-clear substance. These range in diameter from 0.01 mm. to 0.05 mm. A few of them are undoubtedly quartz; the great majority have the single and double refraction of orthoclase. In addition to the rhombohedra of carbonate, the base is charged with abundant black, opaque dust. The particles of the dust average under 0.01 mm. in diameter. Since the rock decolourizes before the blow-pipe it seems clear that the dust is largely carbon, though hematite and probably magnetite are also represented in some amount.

Some laminae of the rock are seen to be specially charged with roundish clumps and lenses of minute orthoclase crystals. (Figure 7.) These are interlocked and in all of three thin sections made from two different hand-specimens, show no trace of a clastic origin. They give the writer the impression of having been introduced and crystallized from solution, or at least segregated in their present positions from the general mass of the rock. The few quartz grains interlock with the orthoclase and are just as clearly not of clastic origin.

Professor M. Dittrich analyzed a typical specimen of the rock, (No. 1338) with the result shown in Col. 1 of the following table. The extraordinary abundance of potash prompted a second determination of the alkalies in the same rock-fragment; this time the potash showed 6.12 per cent and the soda, 0.25 per cent. A different fragment of the same large hand-specimen gave Mr. M. F. Connor 5.54 (also 5.71) per cent of potash and 0.24 (also 0.18) per cent of soda. The average of all four determinations is entered in Col. 2, the other oxides being given in the amounts shown in Professor Dittrich's total analysis. Col. 3 shows the molecular proportions corresponding to Col. 2.

Analysis of the Waterton dolomite.

	1.	2.	3. Mol.
SiO_2	30.46	30.46	.508
Al_2O_3	6.86	6.86	.068
Fe_2O_3	4.53	4.53	.028
FeO	1.89	1.89	.026
MgO	10.07	10.07	.252
CaO	16.02	16.02	.286
Na_2O	.87	.38	.006
K_2O	5.71	5.77	.062
H_2O, at 110°C	.11	.11	....
H_2O, above 110°C	1.31	1.31	....
CO_2	22.55	22.55	.513
	100.38	99.95	
Sp. gr	2.749		

Insol. in hydrochloric acid	42.80%
Soluble in hydrochloric acid:	
Fe_2O_3	4.91
Al_2O_3	2.03
CaO	16.23
MgO	9.69

2 GEORGE V, A. 1912

The average analysis of Col. 2 has been calculated, on the assumption that the alkalies are referable to the orthoclase and albite molecules and the iron oxides to magnetite. All of the lime is referred to the carbonate. The result, noted below, is probably not far from representing the actual composition of the rock.

Orthoclase..	34·17
Albite..	3·14
Quartz..	6·00
Magnetite.	6·36
Magnesium carbonate..	21·17
Calcium carbonate..	28·60
	99·64

Half of the rock is composed of the two carbonates, in which the ratio of Ca to Mg is 1·87:1, indicating but a very slight excess of calcium over that found in normal dolomite. The remainder is chiefly silicious, especially feldspathic matter. The rock is apparently unique among analyzed dolomitic sediments in showing such a high percentage of potash. This alkali is without doubt contained in the orthoclase, which is probably somewhat sodiferous. The concentration of so much of this feldspar in a dolomitic sediment is hard to understand. If the microscopic relations permitted the view that the orthoclase, like the feldspars of the Altyn beds, were of clastic origin and derived from a granitic terrane, one would still be at a loss to understand the relative poverty in quartz. The suggestion due to optical study, that the feldspathic material has really been introduced in solution offers obvious difficulties but seems to be a more promising hypothesis to explain the presence of most of the feldspar. This more probable view itself suffers from the doubt arising from the fact that the rock shows no evidence of having been recrystallized or notably metamorphosed, as we might expect if it had been penetrated by solutions to the extent demanded. A third hypothesis, that, under special conditions, the potash was introduced into the original carbonate mud in the form of the soluble aluminate of potassium, which during burial and lithifaction, reacted with dissolved silica in the mud-water to form orthoclase, is perhaps worthy of mention; but it faces the obvious objection that no conditions in nature are known by which the aluminate is formed from the potassium salts in sea-water. Another suggestion may be drawn from the fact that isomorphous mixtures of calcium and potassium carbonates can be prepared in the laboratory. If such isomorphic mixture were thrown down from the sea-water of the Waterton time, the one constituent of the orthoclase would be added to the mud but the presence of alumina in its exact proportion to potash (and soda) would be hard to explain. Finally, as suggested to the writer by Professor C. H. Warren, the presence of so much alkali may possibly be due to the original precipitation of glauconite in the mud, in which the feldspar was formed by recrystallization under peculiar conditions. In view of its obvious difficulties the problem of this extraordinary rock must be left unsolved.

SESSIONAL PAPER No. 25a

The carbon dioxide shows some deficiency if, as seems necessary, practically all of the magnesia and lime are to be referred to the normal carbonate forms. From the fact that a similar deficiency is found in all of the analyzed carbonate rocks from the overlying Altyn, Siyeh, and Sheppard formations, it is reasonable to suppose that it is not due to the necessary error of analysis. In all these cases the deficiency may be hypothetically explained by the presence of small amounts of hydromagnesite, $(MgCO_3)_3 \ Mg \ (OH)_2 + 3 \ H_2O$. The large proportion of water expelled above 110° C. might also be referred in large part to the basic carbonate. It is thus possible to conceive that from five to seven per cent of the rock is made up of that substance.

The specific gravities of three type specimens of the impure dolomite were found to be respectively, 2·749, 2·777, and 2·782; the average is 2·769. These values show that magnesia must be high in all three specimens.

Though the dolomite occurs in massive plates from six to eight feet thick, and though it is highly homogeneous from top to bottom of the section at Cameron Falls, yet a close inspection of the ledges shows that the rock is made up of a vast number of thin, often paper-thin, beds. Scores or hundreds of such laminae can be counted in a single hand-specimen of the massive dolomite. Their surfaces are generally parallel, and cross-bedding, ripple-marks, or other evidences of shallow-water deposition are absent. The character of the rock, on the other hand, indicates that the carbonate was deposited quietly, persistently, on a sea-floor not agitated by waves or strong currents nor receiving coarse detritus from the lands. The minute bedding and the exceeding fineness of grain, point to an origin in chemical precipitation. The presence of the carbonaceous dust suggests that the precipitation took place in the presence of decaying animal matter and that the dolomite is thus analogous to the chemically precipitated, powdery limestone now forming in the deeper parts of the Black Sea. The theoretical questions regarding this and the other carbonate rocks of the geosynclinal prism will be discussed in chapter XXIII.

The Western Coal and Oil Company have made a boring a few hundred yards from Cameron Falls and in the middle of the Oil creek anticline. The log shows that the bore-hole penetrates 1,500 feet of hard limestones interstratified with subordinate beds of quartzite and silicious argillite (metargillite). All these rocks are fine-grained and, so far as one may judge from the drillings, many are similar to common phases of the Waterton formation. The beds all seem to underlie the Waterton dolomite conformably. We have, therefore, in addition to the exposed members of the Lewis series, at least 1,500 feet of still older beds which should be considered as belonging to the series. Until these strata are actually studied at surface outcrops they cannot be described adequately and for the present report, the Lewis series is considered as extending downward only to the bottom bed of the Waterton formation where it crops out at the cascade.

At the depth of about 1,600 feet the bit of the boring machine passed from the hard limestones into soft shales which persisted to the lower end of the bore-hole about 2,000 feet from the surface. These shales are referred to the

2 GEORGE V, A. 1912

Cretaceous. In other words, the oldest sedimentary beds visible in the Rocky mountains at the Forty-ninth Parallel here overlie one of the youngest formations of the region. The relation is plainly one of overthrusting, which will be discussed in the section devoted to the structure of the Clarke range.

ALTYN FORMATION.

General Description.—The Altyn formation, immediately overlying the Waterton dolomite, was · named by Willis, who described it from a typical section near the village of Altyn, Montana, fifteen miles south of the Boundary line.

The Altyn is not exposed within that part of the Clarke range which is covered by the Commission map. The writer studied the formation chiefly in a fine section on Oil creek and thus on the Atlantic side of the Great Divide. The exposures are there excellent for the greater thickness of the formation. In this section the eroded edges of 3,000 feet of Altyn strata can be seen on the long ridge running southwestward from the bend in Oil creek two miles below Oil City (Figure 6). At least 500 feet of additional, basal beds are exposed along the lower course of the creek and it is these which have been referred to as conformably overlying the Waterton dolomite.

Calcium and magnesium carbonate are the dominant constituents of the formation. With these are mixed grains of quartz and feldspar in highly variable proportion. The rock types thus include arenaceous magnesian limestones, dolomitic sandstones, dolomitic grits, and pure dolomites, named in the order of relative importance. The character of the bedding and the colours of the rocks were often found to vary in sympathy with the rock composition. On this threefold basis the thick formation as exposed along Oil creek, has been subdivided, though only approximately, as follows:

Columnar section of the Altyn formation, showing thicknesses.

Top, conformable base of the Appekunny formation.

a 300 feet.—Medium-bedded, light gray, sandy, magnesian limestone, weathering generally pale buff or, more rarely, strong brownish buff; a few interbeds of magnesian limestone.

b 950 " Thin-bedded, light gray and greenish gray magnesian limestone, weathering buff; subordinate interbeds of sandy limestone.

c 550 " Massive, homogeneous, light gray, sandy limestone, weathering yellowish white; in some horizons bearing cherty nodules and large, irregularly concentric silicious concretions.

d 50 " Thin-bedded, buff-weathering magnesian limestone.

e 750 " Massive, highly arenaceous or gritty, gray magnesian limestone, weathering white or very pale buff.

f 650 " Thin-bedded, relatively friable, gray or greenish gray magnesian limestone, weathering buff or yellowish white.

g 250 " Light gray, thick-bedded, sandy and gritty magnesian limestone, weathering pale buff; occasional thin intercalations of thin-bedded magnesian limestones bearing cherty nodules and silicious concentric concretions.

3,500+ feet.

Base, conformable top of Waterton formation.

SESSIONAL PAPER No. 25a

The total thickness of the Altyn as shown in this Oil creek section is much greater than that seen farther south by Willis (1,400 feet). The difference is not to be explained by overfolding or overthrusting. Individual beds and groups of beds are, it is true, considerably crumpled, especially in the lower part of the section; but the average southwesterly dip of about 30° is preserved throughout. The evidence of original conformity from top to bottom seems as clear.

Lithologically, the sediments here differ from those described by Willis in carrying a notable proportion of rounded grains of quartz and feldspar. The sandy and gritty strata occur chiefly in the middle of the section, there totalling nearly 1,000 feet in thickness. It is thus convenient to recognize a tripartite division of the Altyn as exposed along the International Boundary:— An upper member .(a and b) of thin-bedded, silicious dolomite 1,250 feet thick; a middle member of thick-bedded, massive arenaceous dolomite and calcareo-magnesian sandstones (c, d and e), 1,350 feet thick; and a lowest member of generally thin-bedded, silicious dolomite (f and g), at least 900 feet thick, containing sandy beds toward the base. Nowhere in the formation were there found sun-cracks, rill-marks, ripple-marks or any other indication that the sediments were laid down in very shallow water or on a bottom laid bare between tides.

A visit was paid to Chief mountain and to the original locality at Altyn, Montana, where the rocks were found to correspond to Willis' description except in being often distinctly arenaceous.' Willis' brief summary of the facts observed by him reads as follows:—

'Limestone of which two members are distinguished: an upper member of argillaceous, ferruginous limestone, yellow, terra-cotta, brown, and garnet red, very thin-bedded; thickness, about 600 feet; well exposed in summit of Chief mountain; and a lower member of massive limestone, grayish blue, heavy-bedded, somewhat silicious, with many flattened concretions, rarely but definitely fossiliferous; thickness, about 800 feet; type locality, the cliffs of Appekunny mountains, north of Altyn, Swift Current va '*

As the formation is followed southeastward the uppermost member shows a decided darkening of tint—to terra-cotta, red, and brown of various deep shades, which then dominate the lighter buff colour characteristic of that member at the Boundary. It seems clear that the whole of the lowest member and part of the middle member of the Altyn at the Boundary are not exposed in the sections studied by Willis. On the whole the field relations in the Oil creek section are more favourable to giving one an accurate idea of the whole Altyn formation than are the field relations at either Chief mountain or at Altyn itself.

As already noted this great formation is heterogeneous but every bed of it seems to carry a notable percentage of carbonates. The cement of even the

* Bull. Geol. Soc. America, Vol. 13, 1902, p. 317.

most sandy and gritty layers is dolomitic. Hand-specimens representing the principal phases were collected; each of them has a cement soluble in hot hydrochloric acid. The weathered surfaces of the arenaceous beds are always roughened by the clastic grains of quartz and feldspar standing out above the carbonate, the constituent more soluble in rain-water and soil-water. None of the many specimens collected shows other than the feeblest effervescence with cold, dilute acid. The specific gravity of thirteen specimens ranges from 2·688 in the most silicious phase, to 2·814 in the least silicious phase. The average for all thirteen is 2·763. These facts, together with the characteristic buff tint of the beds on weathered surfaces, of themselves indicated that the formation is throughout highly magnesian. That conclusion has been greatly strengthened by the chemical analysis of three specimens which respectively represent the staple rock-types in the lower, middle, and upper members of the formation. The analyses will be described in connection with the microscopic petrography of the three members.

Lower Division.—Thin sections from the dominant rock of the lowest members, a very homogeneous, compact, thin-bedded limestone, show that the carbonate occurs in the form of an exceedingly fine-textured aggregate of closely packed, anhedral, colourless grains averaging from 0·01 mm. or less to 0·02 mm. in diameter. The largest of the grains may run up to 0·03 mm. in diameter. A very few minute, angular grains of quartz and unstriated feldspar, and some dust-like, black particles (probably both magnetite and carbon) are embedded in the mass. The bedding is well marked in ledge or hand-specimen but is yet more conspicuous under the microscope. The laminæ are bounded by sensibly plane surfaces, affording in section parallel lines often only 0·2 mm. apart. This bedding lamination is brought out rather by small differences of grain among the layers than by admixture of material other than carbonate.

The specimen chemically analyzed has the microscopic characters just outlined. It was collected at the 5,050-foot contour on the spur running southwestward from the right-angled bend in Oil creek on the south side of the creek and about one mile below Oil City. The analysis made by Professor Dittrich (specimen No. 1322) showed weight percentages as follows:—

Analysis of type specimen, lower Altyn formation.

		Mol.
SiO_2	13·46	·224
Al_2O_3	1·56	·015
Fe_2O_3	1·05	·006
FeO	·18	·007
MgO	17·81	·445
CaO	25·08	·448
Na_2O	·28	·005
K_2O	1·08	·012
H_2O at 110°C	·04	....
H_2O above 110°C	1·23	·070
CO_2	38·08	·865
	100·15	
Sp. gr.	2·805	

A second analysis gave the proportions of the oxides entering into solution in hydrochloric acid and also the percentage of insoluble matter, as follows:—

Insoluble in hydrochloric acid	16·02
Soluble in hydrochloric acid:	
Fe₂O₃	1·70
Al₂O₃	·37
CaO	25·16
MgO	16·83

In the soluble portion CaO: MgO = 25·16: 16·83 = 1·495: 1, a ratio only very slightly higher than the ratio for true dolomite, namely, 1·4:1. The carbon dioxide required to satisfy those bases is 38·28 per cent, which is close to the percentage actually found.

Considering that the alkalies belong to the feldspars and the iron oxides to magnetite, the proportions of the various constituents have been calculated to be:—

Calcium carbonate	44·9
Magnesium carbonate	35·3
Quartz	8·0
Orthoclase molecule	5·6
Albite molecule	2·6
Magnetite	1·4
Remainder	2·2
	100·0

As in the case of the Waterton dolomite it is difficult to understand the high proportion of combined water. It may occur with the silica alone or it may occur in a hydrous silicate of magnesium. About 80 per cent of the rock is composed of carbonates in the form of true dolomite.

Middle Division.—A specimen characteristic of the middle member (zone c), though not of its most sandy part, was collected at the low cliffs four hundred yards east of the derrick at Oil City.

This rock on the fresh fracture has the typical pale gray colour of the formation and weathers whitish to pale buff. On the weathered surface the glassy wind-worn or water-worn, rounded to subangular quartz and feldspar grains stand out like white currants in a flour paste. The grains are of varying size up to 0·3 mm. in diameter, averaging about 0·2 mm. The quartz grains are the more abundant. The feldspar is chiefly a fresh and characteristic microperthite, with orthoclase in more subordinate amount. No soda-lime feldspar could be demonstrated. In this analyzed specimen as in the majority of the thin sections from all three members of the formation, round grains of chalcedonic or cherty silica, with diameters also averaging 0·2 mm., occur in considerable number. These small bodies are probably of clastic origin. Oolite grains with poorly developed concentric and radial structure are likewise rather abundant in both the analyzed specimens and others. Dr. H. M. Ami has noted that some of these grains have a certain resemblance to radiolaria, but regards their inorganic, concretionary origin as more probable.

2 GEORGE V., A. 1912

All of these various bodies are embedded in a carbonate base which in all essential respects is similar to that found in the dolomite of the lower member. The carbonate again forms a compact aggregate of anhedral grains, varying from 0·01 mm. or less to 0·025 mm. in diameter, averaging about 0·015 mm.

The total analysis of the specimen (No. 1320) afforded Professor Dittrich the following result:—

Analysis of type specimen, middle Altyn formation.

		Mol.
SiO₂...	18·89	·315
Al₂O₃...	0·49	·005
Fe₂O₃...	0·72	·004
FeO...	not det.	
MgO...	16·79	·420
CaO...	23·86	·426
Na₂O...	0·47	·007
K₂O...	0·57	·00
H₂O at 110 C...	0·18	....
H₂O above 110 C...	1·57	·87
CO₂...	36·89	·38
	100·43	
Sp. gr...	2·802	

A second, partial analysis of the same specimen gave the following data:—

Insoluble in hydrochloric acid...	21·13
Soluble in hydrochloric acid:	
Fe₂O₃ + Al₂O₃...	6·58
CaO...	24·62
MgO...	16·24

The table of molecular proportions shows that the alumina is too low to match the alkalies of the feldspars actually present. Another determination of alumina and iron oxides of a part of the same specimen gave Al₂O₃, 1·22 per cent; Fe₂O₃, 1·01 per cent; and FeO, 0·33 per cent.

The ratio of CaO to MgO in the soluble portion is 1·48 : 1, closely approximating the ratio in true dolomite. Calculation gives the following mineral proportions in the rock:—

Calcium carbonate...	42·6
Magnesium carbonate...	35·3
Quartz and chert...	14·2
Orthoclase molecule...	3·3
Albite molecule...	3·7
Magnetite...	·5
Remainder...	·4
	100·0

The rock is plainly an essentially normal dolomite rendered impure by the simple admixture of clastic grains of quartz and feldspar and by the presence of some silica and iron oxide, both of which may be of chemical origin.

Upper Division.—A specimen typically representing the chief phase of the upper member of the Altyn was collected at the 7,300-foot contour on the back of the ridge south of Oil City. The bed was situated about 100 feet vertically below the top of zone *b* of the columnar section.

Microscopically this phase is like the one just described but is very much strongly charged with water-worn, eminently elastic grains of quartz and feldspar. The latter stand out conspicuously on the weathered surface; they vary from 0.25 mm. to 1.0 mm. in diameter.

In this section the feldspars were determined as microperthite, microcline, orthoclase, and very rare plagioclase. One grain of the latter, showing both Carlsbad and albite twinning was found to be probably andesine, near Ab An. A few oolite-like grains of carbonate 0.5 mm. or less in diameter, a number of round grains of chert, and a very few small specks of magnetite complete the list of materials other than the general carbonate base. In grain and structure the base is practically identical with that of the specimens above described.

Professor Dittrich has analysed the rock (specimen No. 1326) with results as here noted:--

Analysis of type specimen, upper Allyn formation.

		Mol
SiO₂	25.50	.437
Al₂O₃	2.25	.023
Fe₂O₃	.62	.004
FeO	.34	.005
MgO	14.77	.369
CaO	21.65	.387
Na₂O	.86	.014
K₂O	1.27	.013
H₂O at 110°C	.12	...
H₂O above 110°C	.42	.023
CO₂	32.03	.724
	99.87	
Sp. gr	2.769	
Insoluble in hydrochloric acid		29.215
Soluble in hydrochloric acid:		
Fe₂O₃		.05
Al₂O₃		.19
CaO		21.92
MgO		14.29

The fact that sensibly all the lime is soluble shows that basic plagioclase can be present only in extremely small amount. The ratio of CaO to MgO in the soluble portion is 1.534 : 1, showing that here too the calcium carbonate is but slightly in excess of the proportion required for true dolomite.

The approximate mineral composition of the rock has been calculated with the following result:--

Calcium carbonate	39.1
Magnesium carbonate	30.9
Quartz and chert	14.7
Orthoclase molecule	7.3
Albite molecule	7.3
Magnetite	.9
Remainder	.4
	100.0

2 GEORGE V., A. 1912

Comparison and Conclusions.—Comparing the three analyzed specimens with all the others collected from the Altyn, as well as with the rock-ledges encountered during the different traverses, it appears probable that the average rock of the whole 3,500 feet of beds is composed of about 75 per cent of pure dolomite, about 4 per cent of free calcium carbonate, about 10 per cent of quartz and chert, and about 10 per cent of microperthite, orthoclase (with microcline), plagioclase (only a trace), magnetite, kaolin, and carbonaceous dust, with, possibly, a small proportion of hydrous magnesium silicate. No beds of pure calcium carbonate could be found anywhere in the section, nor any beds of ideally pure dolomite. Even in the most compact specimens examined a notable percentage of clastic quartz and feld-par never failed. Though these rocks are thus impure, like the Waterton formation, they may conveniently be referred to as dolomites.

One of the most noteworthy facts concerning all these beds, including both the Altyn and Waterton, is the constant size of grain in the dolomitic base. (Figure 8.) The minute anhedra of carbonate everywhere range from 0.005 mm. to 0.03 mm. in diameter, with an average diameter a little under 0.02 mm. This is true, no matter what may be the size of the clastic quartz or feldspar. The quartz grains vary from scarcely discernible specks to small pebbles 5.0 mm. or more in diameter. As regards the relative amounts and individual size of these silicious materials, the Altyn formation is quite variable in composition. But its essential base of dolomite is remarkably uniform in grain and in composition.

This contrast between carbonate base and enclosed clastic materials is worthy of close attention. The quartz grains in different phases of the formation vary from those as small as the average grain of dolomite to those several million times greater in volume, while through all the thousands of feet of strata, the grain of the dolomite itself is rigidly held below an extremely low limit. In most of the slides the quartz and feldspar fragments are thus gigantic compared to the granular elements of the base and, in most cases, there are very few silicious grains giving the full transition in size between the sand grains and the carbonate grains. Such transitions are to be seen but, as a rule, most of the silicious grains are enormously bigger than the carbonate grains. This steady contrast of size suggests very strongly that the mode of deposition of the quartz and feldspar grains was quite different from that of the enclosing carbonate. The former were unquestionably rolled and rounded by wind-action or under water and were then deposited from water-currents as mechanical precipitates.

The purity and homogeneity of the carbonate base, its remarkably fine grain, and its perfectly regular microscopic lamination of bedding all point to an origin in chemical precipitation. The sea-water must have been free from mud, the shores furnishing pure sand to the undertow and marine currents of the time. If the carbonate were the result of the mechanical breaking up of shells, coral reefs or older limestones, we should inevitably expect the detrital grains of carbonate to be much larger, or at least much more variable in

SESSIONAL PAPER No. 25a

size than the actual particles. There must have been changes of sea level or of depth of water, or changes in both during the accumulation of these 3,500 feet of sediment. It is virtually inconceivable that, throughout such changes, the size of carbonate particles broken off from either shells or bed-rock and brought hither by currents, should always average from 0.01 to 0.02 mm. in

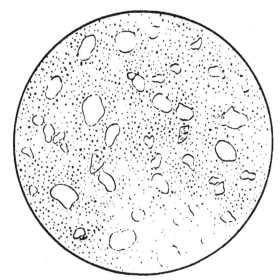

FIGURE 8.—Diagrammatic drawing from thin section of typical sandy dolomite of the Altyn formation. Round (wind-blown?) grains of quartz (clear white) and much less abundant microperthitic feldspar (transverse lines) drawn to scale. The dots represent, on the same scale, the size of the extremely minute carbonate granules composing the matrix of the rock. Diameter of circle 4·5 mm.

diameter and never reach diameters above 0·05 mm. or thereabouts. If the carbonate base were of detrital origin one should expect to find variations in its grain as he approached or receded from the source of detrital supply. Such variation is not to be found at any of the sections yet studied in the Altyn formation. So far as the pre-Altyn rocks are known there seems to have been in the adjacent Cordilleran region no magnesian limestones of anything like the volume required to furnish, from their mechanical disintegration, the material for the thousands of cubic miles of carbonate represented in the Altyn. The same may be said of the pre-Altyn formations underlying the Great

2 GEORGE V, A. 1912

Plains, for we doubtless have in the cores of the Belt mountains and Black Hills uplifts, average samples of such rocks as were washed by the sea waves during Altyn time.

In some thin sections the carbonate is often balled up in spherical or spheroidal bodies, averaging from 0.25 mm. to 0.5 mm. in diameter. These sometimes have an obscure concentric structure, recalling oolite grains. *More rarely an imperfect radial arrangement of the minute granules making up each spheroid, is discernible. In all of the observed cases these granules accord in size with those making up the general base of the rock. Certain of the spheroidal bodies recall the 'coccoliths' such as are precipitated by the action of decomposing albumen on the calcium sulphate of sea water.* There is no evidence that they are foraminiferal s.

Often associated with these carbonate concretions are fairly abundant spheroids of cherty matter, averaging about 0.25 mm. in diameter. These may be due to the silicification (replacement) of the carbonate spheroids, or they may also be due to direct chemical precipitation from sea-water. The former interpretation seems the more probable, therewith correlating these microscopic bodies to the manifestly secondary, large nodules of chert found at many horizons in the formation.

The field and laboratory studies of the Altyn rocks seem, thus, to show that the dolomites and the carbonate base of the subordinate sandy beds are alike the product of chemical precipitation from sea-water. The same may be said of the massive, underlying Waterton dolomite as above described.

The cause of the precipitation will be more fully discussed in chapter XXIII. on the theory of limestones. For the present it suffices to state, that the cause may possibly be found in the bacterial decay of animal remains on the sea-bottom. The ammonium carbonate generated during such decay reacts on the calcium and magnesium salts dissolved in sea-water, throwing down calcium carbonate and magnesium carbonate. The strong content of carbonaceous matter in the Waterton dolomite and its occasional occurrence in certain phases of the Altyn formation, may represent the residue of animal carcases.

Special note should be taken of the nature of the clastic feldspar. It is always remarkably fresh and is mostly a microperthite with typical characters. In view of the unusual nature of this dominant feldspar, it may be used as a sort of fossil in correlating the formation. On stratigraphic grounds it was concluded in the field that the Altyn formation is the equivalent of a part of the Creston formation in the Purcell range and of the Wolf grit and associated members of the Summit series in the Selkirks. The discovery that this special and far from common feldspar is an abundant constituent in all these formations to a degree corroborates this correlation.

The wonderful freshness of the feldspars suggests that these clastic fragments were derived from a terrane undergoing mechanical rather than chemical disintegration. One naturally thinks of an arid climate as supplying the

* G. Steinmann, Berichte der Naturforschenden Gesellschaft, Freiburg i.B., Band 4, 1889, p. 288.

SESSIONAL PAPER No. 25a

necessary condition and recalls the observation of McGee, who describes parts of the Gulf of California as being floored with quartz and fresh feldspar sand washed into the Gulf, during the cloud-burst seasons, from the adjacent arid land.*

In no ca e is there any evidence of pronounced metamorphism of the sediment. The tendency of metamorphism would be rapidly to increase the grain of the rock and to obliterate the delicate structure of bedding. The persistence of the extremely fine grain and of thin bedding seems to show that we have the sediment scarcely more changed from its original state than was necessitated in the act of consolidation.

Fossils.—Very abundant chitinous or calcareo-chitinous plates or films of highly irregular forms were found at a horizon about 975 feet below the top of the Altyn formation. These were seen at only one locality, namely, on the back of the ridge south of Oil City, at a level barometrically determined to be 6,875 feet above sea. In the well exposed ledges at that point thousands of the fragments can be readily laid bare by splitting the thin-bedded, silicious dolomite in which they occur. At least 200 feet of the series is, at intervals, characterized by the fragments. In spite of the formless nature of the fragments they were at once suspected to be of organic origin and to belong to the pre-Cambrian genus, *Beltina*, described by Walcott as occurring in the Greyson shales of the Belt mountains in Montana.† A collection of the fragments was sent to Dr. Walcott, who kindly determined them to have the essential features of *Beltina danai*. The resemblance of the material to that collected at Deep creek in the Belt mountains extended even to the character of the rock. No other species were discovered among the fragments nor did the formation prove fossiliferous elsewhere. At several horizons but particularly in the lowest member of the Altyn, large concentric concretions suggesting Cryptozoon were found but no evidence of their being of organic origin has been forthcoming.

The *Beltina* horizon at Oil City must be close to that which had been found by Weller at Appekunny mountain, near Altyn. Reporting on his collection Dr. Walcott wrote:—

' The mode of occurrence of the material is similar to that found in the Greyson shales of the Algonkian in the Belt mountains, Montana. Hundreds of broken fragments of the carapace of the crustaceans are distributed unevenly through the rock. Occasionally a segment or fragment of what appears to be one of the appendages is sufficiently well preserved to identify it.'‡

The repeated occurrence of the *Beltina* bed at three widely separated localities shows their very considerable importance as a horizon-marker in this little known part of the Cordillera. The fossils themselves have intrinsic interest in representing one of the oldest species yet described.

* W. J. McGee, Science, Vol. 4. 1896, p. 962.
† C. D. Walcott, Bull. Geol. Soc. America, Vol. 10, 1899, pp. 201 and 235.
‡ Bull. Geol. Soc. America, Vol. 13, 1902. p. 317

2 GEORGE V., A. 1912

APPEKUNNY FORMATION.

The formation immediately and conformably overlying the Altyn limestone
has been named the 'Appekunny argillite' by Willis. His original description
applies to these rocks as they crop out in the Boundary belt and it may be
quoted in full:—

'The Appekunny argillite is a mass of highly silicious argillaceous
sediment approximately 2,000 feet in thickness. Being in general of a
dark-gray colour, it is very distinct between the yellow limestones below
and the red argillites above. The mass is very thin-bedded, the layers
varying from a quarter of an inch to two feet in thickness. Variation is frequent from greenish-black argillaceous beds to those which are reddish and
whitish. There are several definite horizons of whitish quartzite from 15
to 20 feet thick. The strata are f: quently ripple-marked, and occasionally
coarse-grained, but nowhere conglomeratic. An excellent section of these
gray beds is exposed in the northeastern spur of Appekunny mountain,
from which the r᷎ is taken, but the strata are so generally bared in the
cliffs throughout ᷎ Lewis and Livingston [Clarke] ranges that they may
be examined with equal advantage almost anywhere in the mountains.

'The Appekunny argillite occurs everywhere above the Altyn limestone along the eastern front of the Lewis range from Saint Mary lakes
to Waterton lake and beyond both northward and southward. It also
appears at the western base of the Livingston range above Flathead valley
and is there the lowest member of the series seen from Kintla lakes southward to McDonald lake.'*

At the eastern end of the South Kootenay pass the lower part of the Appekunny includes a 75-foot band of thin-bedded magnesian limestone which is
identical with the staple rock of the Upper Altyn. Several other bands, each
a few feet in thickness, are dolomitic sandstones and grits, quite similar to the
beds of the Middle Altyn. The two great formations are thus transitional
into each other. On the other hand, the top of the Appekunny is rather sharply
marked off from the overlying Grinnell red beds.

The formation as a whole is not exposed in any one section within the belt
covered by the Commission map. The best exposures studied occur on the
southern slope of King Edward peak and on the mountain slopes north and
south of Lower Kintla lake. A complete section was found on the ridge south
of Oil City, and thus outside the area of the Commission map. The total
thickness in these sections was estimated to be 2,000 feet.

The dominant rock o᷎ ᷎he Appekunny is gray or greenish-gray and silicious,
weathering lighter gray or more rarely light greenish-gray or light rusty brown.
The content of silica is often so great that the rock might well be called an
impure quartzite. As noted in Willis' description the thin-bedded 'argillite' is
often interleaved with more massive strata of gray, whitish, and rusty-weathering

* B. Willis, Bull. Geol. Soc. America, Vol. 13. 1902. p. 322.

quartzitic sandstone. All these rocks are very hard, and were it not for the fissility incident to thin bedding, the formation would be exceptionally resistant to the forces of weathering.

The alternation of the quartzites and more argillaceous beds is so common and the graduation of the one rock-type into the other is throughout so persistent that it has proved impossible to make a useful minute subdivision of the formation. In the section at King Edward peak the uppermost 200 feet are very thick-bedded and are composed chiefly of typical quartzite. In the sections between Oil City and the north end of Waterton lake at least 100 feet of blackish, red, and reddish gray shaly beds are interbedded with the magnesian limestones and sandstones at the base of the formation. None of these types was noted in sections farther west, and, in its lower part at least, the formation seems to become more dolomitic or more ferruginous as it is followed eastward. Sun-cracks and ripple-marks, especially the former, were seen at many horizons from top to bottom of the Appekunny. No fossils have yet been found in it.

Collecting all the information derived from the Boundary belt, a composite columnar section of the formation as exposed in the Clarke range, has been constructed and may be described in the form of the following table:—

Columnar section of Appekunny formation.

Top, conformable base of the Grinnell formation.

200 feet.—Thick-bedded quartzite with subordinate interbeds of gray and rusty metargillite.

2,025 " Light gray to rather dark gray (dominant) silicious metargillite and quartzite, weathering gray and rusty-gray, thin-bedded; many relatively massive beds of whitish and rusty quartzite occur among the staple thin beds of the rapidly alternating metargillite and quartzite; sun-cracks and ripple-marks common.

75 "   Thin to medium-bedded, buff-weathering silicious dolomite.

300 "   Highly variegated, gray, green, reddish, and black metargillite and quartzite, weathering in tones of brown, red, and gray; a few interbeds of buff dolomitic sandstone and grit; sun-cracks and ripple-marks.

_____

2,600 feet.

Base, conformable top of the Altyn formation.

Thin sections of typical phases have been examined microscopically. They revealed an even higher percentage of free quartz than was in the field suspected to characterize the rock. This mineral occurs in very minute angular individuals from 0·005 mm. or less to 0·03 mm. in diameter, with an average diameter of about 0·01 mm. No certain trace of an originally clastic form was anywhere observed. The quartz is intimately intermixed and interlocked with a nearly colourless to pale-greenish mineral, which, on account of the extremely small dimensions of its individuals, is difficult to determine. The single refraction is notably higher than that of quartz; the birefringence is apparently low but, in reality, may be high, the common low polarization tints being due either to the section's passing across the optic axes or to superposition of differently

25a—5½

2 GEORGE V., A. 1912

orientated crystals in the exceedingly fine-grained rock. Many fine shreds and thin scales of similar material with needle-like cross-sections, have all the optical characteristics of sericite, and it seems highly probable that it is this mineral which forms the base of the dominant rock.

A whole thin section may, then, be made up of a homogeneous, intimate mixture of quartz and sericite with accessory grains of iron oxide; or, as is more commonly the case, the slide shows a well-defined banding representing original bedding. In the latter type of section the bedding is marked by alternation of more quartzose and more sericitic material or, yet more clearly, by long lines of limonitic and carbonaceous particles. A few grains of ilmenite or magnetite seem never to fail, but no other accessories, such as feldspars, have been observed. The specific gravity of four specimens, taken to represent the average types of these quartz-sericite rocks and thus the greater volume of the whole formation, varies from 2·708 to 2·760, with a mean value of 2·740. This comparatively high density shows that the sericite is fairly abundant.

The almost complete recrystallization of the original rock is evidenced in the intimate interlocking of the quartz and micaceous mineral and in the entire absence of amorphous argillaceous matter. The sericite is slightly more developed in the bedding plane than elsewhere, thus somewhat aiding the fissility of the rock in those planes. On the other hand, many scales of this mineral have their longer diameters developed at high angles to the bedding planes. Rarely is there a marked sheen on any surface of a hand specimen, nor has true schistosity been developed except in a few very local areas. The recrystallization of the typical rock is clearly the result of slow molecular rearrangement incident to age-long deep burial without true dynamic metamorphism.

Although evident only after microscopic study, this change is so pronounced that it is scarcely correct to speak of the normal phase of the Appekunny as an argillite at all. It is as much a crystalline rock as is a granitoid gneiss. We shall see that the same difficulty of nomenclature adheres to the description of many thousands of feet of beds, in each of the Lewis, Galton, and Summit series. It is convenient to have a term to represent these argillites, recrystallized, yet neither hornfelses (due to contact metamorphism), nor true mica slates or schists (due to dynamic metamorphism with development of notable cleavage or schistosity, generally cutting across bedding planes). For such once-argillaceous rocks, recrystallized merely by deep burial the name 'metargillite' will be used in the present report.

The meaning of the term may be made clear through a comparison with the names now in general use for pelitic rocks. An unconsolidated pelite is a clay or mud. If consolidated but not extensively recrystallized, it is an argillite. A thin-bedded, unaltered argillite which readily splits along the bedding planes because those are planes of original weakness, is a shale. If recrystallized during dynamic metamorphism only to the extent that easy cleavage following the planes of similarly orientated microscopic mica plates is developed, the rock is a slate. If an argillite becomes phanerocrystalline and foliated

SESSIONAL PAPER No. 25a

through dynamic metamorphism, it is a quartz-mica schist or a quartz-feldspar schist or gneiss. If an argillite has been more or less completely recrystallized by thermal action on igneous contacts it is a hornfels. If, finally, an argillite retaining the bedding structure has been essentially recrystallized by deep burial and without being affected by direct magmatic influence or by the notable development of cleavage or schistosity, it may be called metargillite.

At no point within the Boundary belt was true argillite (shale or slate) found in the Appekunny formation. Everywhere the once-pelitic phases belong to the metargillite type as just defined. On the mountain slopes running up eastward from the Flathead valley both metargillite and quartzite have been sheared and cleaved, the former giving local phases of slaty metargillite. The crystallinity never rises to the degree of true mica schist.

GRINNELL FORMATION.

The Grinnell formation was named by Willis and described by him thus:—

' A mass of red rocks of predominantly shaly argillaceous character is termed the Grinnell argillite from its characteristic occurrence with a thickness of about 1,800   .ount Grinnell. These beds are generally ripple-marked, exhibit mud-c, ks and the irregular surfaces of shallow water deposits. They appear to vary considerably in thickness, the maximum measurement having been obtained in the typical locality, while elsewhere to the north and northwest not more than 1,000 feet were found. It is possible that more detailed stratigraphic study may develop the fact that the Grinnell and Appekunny argillites are really phases of one great formation, and that the line of distinction between them is one diagonal to the stratification. The physical characters of the rocks closely resemble those of the Chemung and Catskill of New York, and it is desirable initially to recognize the possibility of their having similar interrelations.

' The Grinnell argillite outcrops continuously along the eastern side of Lewis range and its spurs, occurring above the Appekunny argillite and dipping under the crest of the range at the heads of the great amphitheaters tributary to Swift Current valley. About the sources of the Kennedy creeks it forms the ridge which divides them from Belly river. Mount Robertson is a characteristic pyramidal summit composed of these red argillites. The formation occurs in its proper stratigraphic position between the forks of Belly river and west of that stream in the Mount Wilson range of the Canadian geologists, the northernmost extremity of the Lewis range; and it dips westward under the valley of Little Kootna creek and Waterton lake. On the western side of Livingston [Clarke] range the Grinnell argillite was recognized as a more silicious, less conspicuously red or shaly division of the system, occurring about Upper Kintla lake.'*

* B. Willis, Bull. Geol. Soc. America, Vol. 13. 1902. p. 322.

2 GEORGE V. A. 1912

The formation is admirably exposed on the southwest side of King Edward Peak, which overlooks the Flathead valley about three miles north of the Boundary line.

In this section the total thickness is 1,600 feet, distributed as follows:

Columnar section of Grinnell formation.

Top, conformable base of the Siyeh formation.

355 feet.—Thin-bedded, red metargillite with intercalations of red, quartzitic sand-
         stone.
20 "   Flow of basic, amygdaloidal lava.
75 "   Thin-bedded, red metargillite.
100 "   Thick-bedded, gray metargillitic quartzite, weathering light rusty brown.
1,050 " Thin-bedded, red to reddish gray metargillite and quartzitic sandstone.
_____
1,600 feet.

Base, conformable top of the Appekunny formation.

Sun-cracks and ripple-marks are common in all the sedimentary members. In a section on Oil creek the ripples at one horizon measured from four to twelve inches from crest to crest, indicating currents of great power, such as the heavy tidal rips occurring in the Bay of Fundy and other estuaries of the present day. If these ripples were caused by wind-wave current, the waves must have been of very large dimensions.

Under the microscope the sandstone specimens are seen to be composed essentially of rounded quartz grains, averaging 0·25 mm. in diameter. Many of the grains are secondarily enlarged and to such an extent that the rock has the fracture and the strength of true quartzite. A small amount of amorphous, apparently argillaceous matter, tinted with the red oxide of iron, forms the rest of the cement. No feldspar was seen in this section.

The metargillites are made up of a very compact mass of sericite, quartz, chlorite, and abundant iron oxide. In one thin section minute crystals of a pale brownish carbonate, probably dolomite, are distributed through the mass. The carbonate seem also to be an original constituent and may have been chemically precipitated along with the mechanically deposited mud, the dominant original component of these beds.

The specific gravity of a type specimen of the red quartzite is 2·678. The specific gravities of two specimens of the metargillite are 2·740 and 2·757. The average for the whole formation is about 2·725.

The amygdaloid l is a dark green-gray, compact rock, which both macroscopically and microscopically, is similar to non-porphyritic phases of the overlying Purcell lava (described in chapter IX). It is much altered, but minute, thin tabular crystals of labradorite with the same abundance and mutual arrangement which this essential mineral has in the Purcell formation amygdaloid, still represent the original microphenocrysts. The base was once glassy but is now mainly composed of the usual secondary chlorite, quartz, and calcite. Numerous small crystals of original ilmenite are now represented only by pseudomorphs of yellowish leucoxene. The pores of the rock seem to be entirely filled with deep green chlorite. The rock is too greatly altered to afford a useful analysis but it is evidently a common type of basaltic lava.

SIYEH FORMATION.

In Willis' original description of the 'Algonkian' rocks of the Lewi- range, the following concise account of the Siyeh formation occurs:

'Next above the Grinnell argillite is a conspicuous formation, the Siyeh limestone, which rests upon the red shales with a sharp plane of distinction, but apparently conformably. The Siyeh is in general an exceedingly massive limestone, heavily bedded in courses 2 to 6 feet thick like masonry. Occasionally it assumes slabby forms and contains argillaceous layers. It is dark blue or grayish, weathering buff, and is so jointed as to develop large rectangular blocks and cliffs of extraordinary height and steepness. Its thickness, as determined in the nearly vertical cliff of mount Siyeh, is about 4,000 feet.

'This limestone offers certain phases of interna..structure which may be interpreted as results of conditions of sedimentation or as effects of much later deformation. Some layers exhibit calcareous parts separated by thin argillaceous bands, which wind up and down across the general bedding and along it in a manner suggestive of the architectural ornament known as a fret. It is conceived that the effect might be due to concretionary growths in the limestone, either during or after deposition, or to horizontal compression of the stratum in which the forms occur. Other strata consist of fragments of calcareous rock from minute bits up to a few inches in diameter, but always thin, constituting a breccia in a crystalline limy cement. Again, other strata consist of alternating flattish masses of calcareous and ferruginous composition, which rest one upon another like cards inclined at angles of 30 to 45 degrees to the major bedding. At times the lamination is so minute as to yield a kind of limestone schist. These internal structures suggest much compression, but the apparent effects are limited by undisturbed bedding planes, and it is possible that the peculiarities are due to development of concretions and to breaking up of a superficial hard layer on the limestone ooze during deposition of the beds. Walcott has described similar structure as intraformational conglomerates.

'The Siyeh limestone forms the mass of Mount Siyeh, at the head of Canyon creek, a tributary which enters Swift Current at Altyn from the south. It constitutes the upper part of all the principal summits of Lewis range north of Mount Siyeh, including Mounts Gould, Wilbur, Merritt, and Cleveland. It extends beyond Waterton lake westward into the Livingston [Clarke] range and forms the massive peaks between Waterton and North Fork drainage lines. Above Upper Kintla lake it is sculptured in the splendid heights of Kintla peak and the Boundary mountains.'*

---

*B. Willis, Bull. Geol. Soc. America, Vol. 13, 1902, p. 323.

2 GEORGE V, A. 1912

The Siyeh is the great cliff-maker of the Front ranges. Quite apart from the fact that it is capped by the resistant Purcell Lava, the limestone is itself strong enough to stand up in precipices thousands of feet in height. (Plate 9.) Among the many admirable exposures one of the best within the Boundary belt is that at the head of Starvation creek canyon; this section is typical of the formation as it occurs in both the Clarke and Lewis ranges.

The formation is notably homogeneous for hundreds of feet together; yet, as shown in the columnar section, it is divided into five zones of contrasted lithological character.

Columnar section of Siyeh formation.

Top, base of the Purcell Lava.

150 feet.—Medium to thin-bedded, reddish metargillite with subordinate, thin interbeds of buff-weathering, magnesian metargillites; sun-cracks, rain-prints and ripple-marks abundant.

950 " · Gray and greenish, thin-bedded, often calcareo-magnesian, metargillites; weathering buff (dominant), fawn, and gray; sun-cracks, rain-prints and ripple-marks common.

100 " Gray, concretionary, silicious and non-magnesian limestone, weathering light gray, with a five-foot band of buff-weathering dolomite at thirty feet from the top.

2,000 " Massive, thick-bedded, dark gray or dark bluish gray, impure magnesian limestone, weathering buff; thin interbeds of dolomitic metargillite, weathering buff, and a few thin beds of gray sandstone. Molar-tooth structure characteristic of limestone; metargillites bear sun-cracks and, rarely, obscure ripple-marks.

900 " · Thin to thick-bedded, light to dark gray and greenish gray calcareo-magnesian metargillites and quartzites, weathering fawn and buff, with a few interbeds of buff-weathering dolomite without molar-tooth structure.

4,100 feet.

Base, top of the Grinnell formation.

Notwithstanding their lithological variations, the strata form a natural unit; the peculiar buff tint of the weathered surface contrasting with the deep browns and blacks of the Purcell Lava and with the strong purplish red of the Grinnell formation below, is a common feature for most of the strata in the Siyeh. The uppermost beds weather reddish but they are so intimately interleaved with buff-weathering strata that a clean-cut separation of the red beds is impossible. It has thus appeared best to follow Willis in including all the strata between the Grinnell and the Purcell Lava under the one formation name. It is, on the whole, a single magnesian group. A second general characteristic is the massiveness of the beds. This is most prominent in the 2,000 feet of dolomite and limestone composing the middle part of the formation.

The table and Willis' statement afford a sufficient general description of the strata in this section. There are, however, certain structures in the buff magnesian limestone and the thick band of gray limestone which merit special notice. At the outcrop of the buff rock the observer's eye is struck with a repeated colour variation in the rock. The cause is speedily apparent. The calcareous constituent of the rock is seen to be segregated, sometimes irregul-

PLATE 9.

Mount Thompson, seen across Upper Kintla Lake ; summit 5,500 feet above lake.
Illustrates massive character of Siyeh formation.

arly, sometimes systematically. The buff-tinted (weathered) general surface of the ledge is thus variegated with many small masses of pure light gray to bluish gray limestone. These masses are in the form of roundish nodules and pencils, flat lenses, or irregular stringers of no definite shape. They are essentially composed of pure calcite; they effervesce violently with cold dilute acid. As in so many dolomites the calcareous segregation is often quite unsystematic.

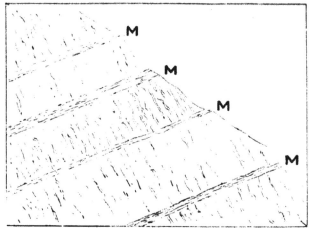

FIGURE 9.—Section showing common phase of the molar-tooth structure in the Siyeh formation. The calcitic segregations are lenticular and stand perpendicular to the plane of stratification, as shown by the metargillitic interbeds (M). The middle layer of limestone is two feet thick. Locality, north fork of the Yahk River.

but, on the average, it is definitely related to the two master planes of structure in the limestone. Where the rock is uncleaved, the bedding-plane has been selected as the favoured locus of growth of the segregation. The stratification may thus be marked by many small, independent lenses of lime carbonate completely surrounded by the magnesian matrix. There is transition between such isolated lenses and entire, uninterrupted beds of gray limestone conformably intercalated in the buff magnesian rock. Such beds may in many cases be due to original sedimentation.

The conclusion that the lime-carbonate lenses, pencils, and irregular bodies, and even some of the continuous bed-like masses, are due to secondary segregation within the dolomite, is clearly upheld by the relation of another kind of lime-carbonate partings. These were long ago observed by Bauerman and later by Willis. In localities where the dolomite has been specially nipped and

2 GEORGE V., A. 1912

squeezed or somewhat sheared, a cleavage was developed. Since the dip of the bedding is generally low, this cleavage runs at high angles to the plane of stratification. The cleavage planes have permitted easy passage to circulating waters. Owing to their activity, there has been a wholesale segregation of the more soluble lime-carbonate in the cleavages which have been thereby healed so as to restore much of the rock's original strength. In regions of formerly strong lateral pressure the rock is now converted into a laminated rock composed of thin, alternating, more or less continuous layers of pure lime-carbonate. These layers are highly inclined to the stratification planes and often run parallel to cleavage planes in argillites above and below the limestone and, like those planes, may be crumpled. (Plate 10 and Figure 9.) That the gray calcareous partings are due to secondary chemical deposition is shown also by the fact that where the original rock was argillaceous or sandy, these impurities remain entirely within the magnesian parts of the rock. On a weathered surface the latter may be quite gritty to the feel while the lime-carbonate partings are smooth and marble-like.

Bauerman described this rock as 'an impure limestone, in which the carbonate of lime is intermingled with argillaceous patches in folds resembling the markings in the molar tooth of an elephant.'* This appearance is most striking on the weathered ledges, the stringers of the more soluble, gray calcite locating numerous channels and pits which are separated by the brownish, projecting ribs of the more resistant magnesian and silicious parts.

Using Bauerman's simile, the structure may be called the 'molar-tooth' structure, whereby will be understood, in general, the internal modification of the original limestone by the secondary segregation of the calcium-carbonate. The term will also be used for the very common case where the weathered surfaces do not show the chance imitation of a worn molar-tooth; the last is best shown in the cleaved phases. The name is thus conveniently generalized as it may then be applied to the concretionary limestone even when cleavage has not been developed. The structure is of importance as an aid in the recognition of the Siyeh formation over great distances.

Under the microscope the contrast between the calcitic and magnesian parts of the molar-tooth rock is marked. The calcite, light gray in the hand-specimen, is colourless in thin section. It forms a compact aggregate of polygonal, sometimes interlocking grains varying in diameter from 0.005 mm. to 0.03 mm. and averaging about 0.01 mm. Very seldom, if ever, do these grains show the rhombohedral or other crystal form. A few minute cubes of pyrite are embedded in the mosaic, but, otherwise, the lenses and stringers are made up of practically pure carbonate.

The buff-weathering main part of the rock is sharply distinguishable under the microscope. It has a decided, pale yellowish-gray colour and a mixed composition. Anhedra and rhombohedra of carbonate, which is doubtless high enough in magnesia to be called dolomite, form more than half of the

* Report of Progress, Geol. Nat. Hist. Survey of Canada, for the years 1882-3-4, Pt. B, p. 26.

PLATE 10.

Sheared phase of Siyeh limestone, Clarke Range. Light parts highly magnesian; dark parts nearly pure calcium carbonate. Three-fourths natural size.

Sheared phase of dolomitic lense (weathered) in Kitchener formation, at Yahk River. Illustrates molar-tooth structure; sunken parts, calcium carbonate, and projecting ribs, siliceous magnesian limestone. Two-thirds natural size.

volume. The rhombohedral and subrhombohedral crystals average about 0.01 mm. in diameter. They are embedded in a base which is partly composed of numerous anhedral granules of the same carbonate but of much smaller size, with diameters varying from 0.001 mm. or less to 0.005 mm. A few minute angular grains of quartz and feldspar and a few dust-like particles of carbon and pyrite are associated with the carbonate. The remainder of the base is a colourless, amorphous to subcrystalline cement whose diagnosis is extremely difficult. Its single refraction is low and its double refraction either nil or extremely faint. These properties are like those noted for the silicious base of the Altyn dolomite. The chemical analysis shows a content of silica and alumina considerably in excess of the amounts required for the little clastic quartz and feldspar in the rock. It thus appears that the cement carries both silica and alumina, and it must carry the combined water. True argillaceous matter may be present, as well as amorphous and chalcedonic silica. In none of the thin sections could sericite be detected.

The specific gravity of six type specimens varies from 2.657 to 2.760, with an average of 2.702. These low values indicate the impurity of the carbonate. The lightening of the rock must, in largest amount, be attributed to the cement of the magnesian part.

Owing to the extensive and highly irregular rearrangement of the carbonates in the molar-tooth rock it is not easy to secure a specimen which shall faithfully represent its average composition. A specimen approximating to this ideal was taken from the cliffs of Sawtooth ridge, 1.5 miles east of Lower Kintla lake. Material from this specimen was so selected as to contain magnesian base and calcitic segregation in about their average proportions in nature, and the powdered mixture (specimen No. 1306) was analyzed by Professor Dittrich. The result is as follows:

Analysis of Siveh impure limestone.

		Mol.
SiO₂	35·58	·593
Al₂O₃	3·40	·033
Fe₂O₃	1·56	·010
FeO	·87	·012
MgO	10·09	·252
CaO	19·72	·352
Na₂O	·51	·008
K₂O	1·21	·013
H₂O at 110°C	·17	....
H₂O above 110°C	2·93	·163
C	·03	....
CO₂	23·80	·541
	99·87	
Sp. gr.	2·741	
Portion insoluble in hydrochloric acid	40·69	
Portion soluble in hydrochloric acid·		
Fe₂O₃	2·17	
Al₂O₃	1·87	
CaO	19·76	
MgO	8·95	

2 GEORGE V., A. 1912

This analysis cannot be calculated quite as readily as those of the Altyn dolomites; the alkalies are here, not assignable with certainty to definite feldspars. For the purpose of comparison, however, the same method of calculation has been applied here, giving a 'norm' wherein the soda is assigned to the albite molecule and the potash to the orthoclase molecule, just as the alkalies are assigned in calculating the 'norms' of igneous rocks. The carbonates have been calculated directly from the analysis of the soluble portion. The results are given in the following table, which shows the 'mode' for the rock as far as the carbonates are concerned, the other constituents being more arbitrarily treated:

Calcium carbonate..	35.3
Magnesium carbonate..	18.8
Silica..	28.0
Orthoclase molecule..	7.2
Albite molecule..	4.2
Magnetite.	2.6
Remainder..	3.9
	100.0

The proportions of the carbonates correspond to 41.2 per cent of normal dolomite and 12.9 per cent of free calcium carbonate.

This excess of calcium carbonate is probably not due to its having been introduced into the molar-tooth rock from other beds. The magnesian portions of the molar-tooth rock effervesce somewhat with cold, dilute acid; it seems simplest to believe that the calcium carbonate was there originally in excess and dates from the time of the deposition of the sediment. The two carbonates together are seen to make up about 54 per cent of the rock.

The high percentage of water (above 110°C) is of interest as showing that this metamorphic agent, even at the present time, is enclosed in sufficient amount to explain the solutional effects illustrated in the   tooth structure. The content of carbon, low as it is, is partly responsible   normally dark tint of the fresh rock.

Different as the Siyeh dolomite limestone and the Altyn dolomite are in field-habit, the two types are yet similar in several important respects. In each the carbonate base has a remarkably fine and homogeneous grain, with no suggestion in either case that the carbonate is of clastic origin. In each case the dolomitic grains tend to assume the rhombohedral form. Their average diameter is sensibly identical in size with that of the average calcite grains composing the lenses, pencils, and stringers of the molar-tooth rock. This average diameter is also practically equivalent to that characterizing the granules which compose each of the egg-like bodies forming occasional thin beds of oolite in the Siyeh and neighbouring formations. There is no doubt of the chemical origin of the latter, nor can there be doubt that the calcitic partings of the molar-tooth limestone were gradually crystallized out from water solutions. It would seem next to incredible that these three associated rocks, characterized by the same average size of constituent carbonate particles, could have in two cases an origin in chemical precipitation and, in the third, an origin in the deposition of land detritus on the sea-floor. A study of all the available facts has, thus, forced the writer to the belief that the huge Siyeh and Altyn formations are chiefly the product of long continued throwing down of calcium and magnesium carbonate from sea-water, from which there was a likewise slow deposition of silicious muddy matter brought from the lands. The molar-tooth structure of the Siyeh is secondary and was developed after burial.

SHEPPARD FORMATION.

General Description.—Conformably overlying the Purcell Lava in the Clarke and Lewis ranges is a group of strata which has been named the 'Sheppard quartzite' by Willis. He speaks of it as belonging to a

'distinctly sandy phase of deposition................a quartzite which is very roughly estimated to have a thickness of 700 feet. It forms the crest of Lewis range in the vicinity of Mount Cleveland and Sheppard Glacier between Belly river and Flattop mountain [type locality]. It has not been studied in detail but is recognized as a distinct division of the series.'*

The lithological character of the beds occurring in the Boundary sections and equivalent to the strata at Willis' type locality differs somewhat from the character stated in his brief description. This lack of accordance may possibly be explained through actual differences in the beds as they are encountered at different points along the axis of the Clarke range. It may be noted, however, that the present writer, during a rapid traverse across the Lewis range via the Swift Current Pass, found that there the beds of the Sheppard formation are extremely like those studied in the Boundary belt. The staple rock of the Sheppard is not easy to diagnose in the field. It was only after microscopic study that one could be sure of the true nature of the sediment. Its colour, compactness, and general habit are those of an impure, flaggy quartzite. The thin section shows that the rock is largely composed of carbonate (dolomite) and that quartz occurs as minute grains rather evenly distributed through the mass of carbonate. The staple rock of the Sheppard is, thus, in the Boundary belt and probably also farther south, a silicious dolomite or dolomitic quartzite. More typical quartzite occurs as a subordinate constituent of the formation, as shown in the following columnar section of the formation where exposed just north of the Boundary monument on the Great Divide:—

Columnar section of Sheppard formation.

Top, conformable base of Kintla formation.

580 feet.—Thin-bedded, light gray, highly silicious dolomites, weathering buff—a homogeneous member occasionally concretionary; some of the more silicious beds approximating magnesian quartzite.

20 "   Reddish, interbedded quartzite and silicious argillite.

600 feet.

Base, conformable top of Purcell Lava.

*B. Willis, Bull. Geol. Soc. America, Vol. 13, 1902, p. 324.

2 GEORGE V., A. 1912

At the head of Starvation canyon the section is slightly different:—

Top, conformable base of Kintla formation.

500 feet.—The staple, thin-bedded, buff-weathering silicious dolomite.
35 "   Basic, amygdaloidal lava.
50 "   Medium to thin-bedded reddish and gray, interbedded sandstone and
           argillites, with thin intercalations of buff-weathering dolomitic rock.

585 feet.

Base, conformable top of Purcell Lava.

Of the two columnar sections the former is to be regarded as the more typical for the Boundary belt. The intercalated bed of lava represents a quite local outflow, not found in sections a few miles to the eastward, nor anywhere in the Galton range.

The basal red beds of the formation are similar in character to the uppermost strata of the Siyeh formation and are like common phases of the Kintla. Ripple-marks and especially sun-cracks are common here as in all other strata below the base of the Kintla formation.

Under the microscope the most common rock of the Sheppard is seen to be a highly impure, very finely granular, homogeneous mass of carbonate. It occurs in the form of pale brownish grains, varying from 0·005 mm. or less to 0·03 mm. in diameter and averaging about 0·02 mm. The larger grains often show rhombohedral outlines. These are enclosed in a fine-grained base of anhedral carbonate grains, quartz, sericite, and probably feldspar fragments of minute size. Along with these fairly determinable constituents the base carries a considerable amount of colourless, nearly isotropic material, which seems to be identical with the cement found in the Siyeh limestone and certain phases of the Altyn dolomite. As in the latter rocks, this material must carry much of the combined water, which here forms nearly two per cent of the whole rock.

The specific gravity of the dominant phase, as represented in three fresh specimens, ranges from 2·695 to 2·785.

A type-specimen (No. 1301) collected at the head of Starvation creek, has been analyzed by Professor Dittrich, with result as follows:

Analysis of Sheppard impure dolomite.

		Mol.
SiO_2	24·61	·410
Al_2O_3	6·84	·067
Fe_2O_3	·58	·004
FeO	2·01	·028
MgO	13·34	·334
CaO	19·11	·342
Na_2O	·62	·010
K_2O	2·07	·022
H_2O at 110°C	·24	...
H_2O above 110°C	1·76	·098
CO_2	28·89	·656
	100·10	
Sp. gr.	2·779	

Portion insoluble in hydrochloric acid....................	32·23%
Portion soluble in hydrochloric acid.	
Fe₂O₃..	2·36
Al₂O₃..	2·29
CaO...	18·86
MgO...	12·96

The carbonates could be rather closely calculated if it were known how much of the ferrous iron is present in the sideritic molecule. Since the carbon dioxide is no more than sufficient to satisfy the lime and magnesia of the soluble portion, it is probable that iron carbonate is present in but very small amount. A definite assignment of the alkalies is here impossible. The soda is arbitrarily assigned to the albite molecule, although it is possible that paragonite is present. A partial calculation yields the following result:

Calcium carbonate...............................	33·7
Magnesium carbonate.............................	27·2
Albite molecule.................................	5·2
Free silica......................................	12·0
Sericite, potash feldspar, iron oxide, etc...........	21·9
	100·0

The ratio of CaO to MgO in the soluble portion is 1·455:1, a value very close to that in normal dolomite.

Dolomite forms about 61 per cent of this specimen. Probably nowhere in the formation does it form more than 75 per cent of any bed. The percentage in the more silicious beds may run far below 40 per cent, as shown by the specific gravity, 2·630, of one fresh specimen.

Inasmuch as the mineral dolomite dominates over the free quartz, the staple rock of the formation may be classified as a highly silicious dolomite. Chemically this rock notably resembles the Siyeh magnesian limestone and the Altyn dolomite. Also important are the similarities of structure and size of grain, implying like conditions of origin for the essential carbonates.

Interbedded Lava.—The amygdaloidal lava bed near the base of the formation is similar in composition to the Purcell Lava and doubtless represents a local, somewhat later flow from the same basaltic magma. This lava is vesicular throughout, much more so than the Purcell Lava. The vesicles are, as usual, particularly large and numerous near the upper surface of the flow. The larger ones approach 1 cm. in diameter. In the highly vesicular phase the vesicles average about 2 mm. in diameter and compose from one-quarter to one-third of the rock's whole volume. The vesicles are generally completely filled with well crystallized calcite, less often with granular or radially crystallized quartz, and a few are filled with both calcite and quartz. There is nothing specially noteworthy concerning the silicious amygdules, but the large majority of the pure-calcite amygdules present a remarkable phenomenon which, so far as known to the writer, has not been described in petrographic literature.

In the highly vesicular phase of the lava the calcite of many associated amygdules is all rigidly orientated so that there is simultaneous reflection of light from a cleavage-surface in each exposed amygdule. Careful examination

2 GEORGE V., A. 1912

has showed, in fact, that the calcite of many hundreds of amygdules together composes a single crystallographic individual. The appearance of such an interrupted individual is very similar to that of a coarsely poikilitic structure in a plutonic rock. There is, of course, nothing more than an analogy between the two cases, since the calcite crystallized from infiltrating water, but the parallel will serve, perhaps, to make the phenomenon better understood by the

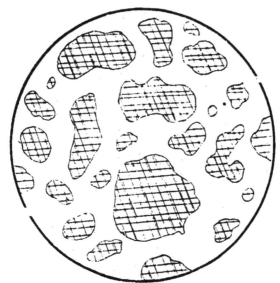

FIGURE 10. -Diagrammatic drawing to scale, from thin section of amygdaloidal basalt in the Sheppard formation, Clarke Range. The section shows twenty-four vesicles filled with calcite. The uniform orientation of the calcite is shown by the parallelism of cleavages (and by simultaneous extinction under crossed nicols). The basaltic matrix in which the amygdules lie (diabasic and very fine-grained) left blank. Diameter of circle 20 mm.

reader. The poikilitic calcite crystals are often of great size, diameters of 10 cm. (about 4 inches) being observed; in such cases the fillings of several thousand vesicles compose a single crystal of glassy calcite. The average crystal is about 5 cm. in diameter. These large crystals never appear to possess crystal outlines but lock together irregularly or are bounded by quartz amygdules and non-vesicular parts of the rock. (Figure 10.)

The nature of the process by which the carbonate was thus crystallized offers an interesting problem, as yet unsolved. Most of the amygdules show no intercommunication either to the unaided eye or under the microscope. The water from which the calcite was precipitated undoubtedly entered each vesicle through openings in the walls but such openings must have been generally of subcapillary size. It is difficult to imagine the play of forces or the history of the crystallization which grouped the calcite molecules in one amygdule so as to give its filling a common crystallographic axis with hundreds of neighbouring amygdules. It looks as if the force of crystallization had operated directly through the rock wall of each vesicle.

KINTLA FORMATION.

In the field the Kintla formation is a conspicuous element of the Lewis series. Stratigraphically the highest known member of the series, the Kintla commonly occurs on the higher summits and thus above tree-line. The fine exposures and a striking deep red colour, contrasting with the bright buff beds of the Sheppard, render the argillite visible for many miles. Beautiful colour effects in the rugged Lewis and Clarke ranges are controlled by the rich tints of the argillite as it lies in place on the mountain-crests or, by its streaming talus, lends broad slashes of colour to the lower slopes.

The best studied sections in the Boundary belt are both north of the Boundary line; one at the head of Kintla creek canyon, the other at the head of Starvation creek. The rocks in the former section have been described by Willis, to whom we owe the name of the formation:—

'The highest beds of the ancient sequence of strata found in this part of the range are deep red argillaceous quartzites and silicious shales, with marked white quartzites and occasional calcareous beds. They are named the Kintla formation from their occurrence in mountains on the 49th parallel, northeast of Upper Kintla lake. They also form conspicuous peaks west of Little Kootna creek. The Kintla formation closely resembles the Grinnell, and represents a recurrence of conditions favourable to deposition of extremely muddy, ferruginous sediment. The presence of casts of salt crystals is apparently significant of aridity, as the red character is of subaërial oxidation. The formation has an observed thickness of 800 feet, but no overlying rocks were found. Its total thickness is not known, and the series remains incomplete.'[*]

To Willis' account the following details may be added for this section. The basal member is sixty feet thick, consisting of red, sandy argillite interstratified with thin beds of bright gray silicious and magnesian limestone and magnesian quartzite, each type weathering buff. These interbeds are identical in character with the principal phase of the Sheppard, showing that the two formations are dovetailed together. Overlying these red beds is a forty-foot flow of basic vesicular lava, lithologically similar to both the Purcell Lava and

[*]B. Willis, Bull. Geol. Soc. America, Vol. 13, 1902, p. 324.

2 GEORGE V., A. 1912

to the sheet occurring in the Sheppard. Though this lava bed of the Kintla has been traced six miles to the westward, it is known to have formed but a single, local outflow of magma, lacking the singular persistence of the Purcell Lava.

Above the lava, in the Kintla formation, is a thickness of 300 feet of mixed beds:—dominant thin-bedded and purplish sandstone and argillite, in which are intercalated thin beds of light gray silicified (cherty) oolite, weathering buff; and thin beds of grayish white compact magnesian sandstone also weathering buff. About 100 feet above the lava there are two conspicuous beds of gray concretionary limestone which weathers gray. Like the oolitic dolomite these gray bands have their complete homologues in the upper part of the Siyeh formation. Above the 300-foot band of variegated sediments the section disclosed 460 feet of more homogeneous bright red to brownish and purplish thin-bedded argillite and sandstone; the dominant rock is argillite.

As noted by Willis, erosion has removed an unknown portion of these sediments and 820 feet is, therefore, a minimum estimate of their thickness. The section at the head of Starvation creek canyon, six miles to the west-north-west, afforded only 610 feet of strata in addition to the forty-foot sheet of amygdaloid. In that section the basal sixty-foot variegated argillite is reduced to ten feet of reddish-brown quartzite. The succession is similar to that of the type section but the gray limestone bands were not found and the oolitic structure was not observed in the magnesian interbeds.

The columnar section for the type locality may be noted:—

Columnar section of Kintla formation.

Top. erosion-surface.

460 feet.	Relatively homogeneous, thin-bedded, bright red, purplish and brownish red argillite and subordinate quartzitic sandstone.
300 "	Heterogeneous, thin-bedded, red argillite (dominant) and sandstone, gray and brownish sandstone, magnesian, oolitic limestone and gray concretionary limestone.
40 "	Amygdaloid.
60 "	Thin-bedded red argillite, with thin intercalations of magnesian quartzite.
860 feet.	

Base, conformable top of Sheppard formation.

A special feature of the argillites is the great abundance of casts of salt-crystals described by Dawson and Willis. The casts represent both complete cubes and the hopper shape of skeleton crystals. (Plate 11.) The cubes are of all sizes up to those 4 cm. or more in diameter. Ripple-marks and sun-cracks, especially the latter, are likewise very abundant. Thin-bedding and minute jointing have rendered the argillites highly fissile; the mountain peaks composed of this formation are usually covered with a fine-textured, creeping fels-senmeer which often, over large areas, completely covers the ledges of rock in place.

PLATE II.

Casts of salt-crystals in Kintla argillite; about two-thirds natural size.

The specific gravity of three specimens of the argillite ranges from 2.616 to 2.697, averaging 2.652. The average for the whole formation (including the dolomitic interbeds (2.743) and the amygdaloid (ca. 2.900) is about 2.675.

Thin sections from different specimens of the argillite show, under the microscope, differences of grain but, in other respects, are similar. Angular, sherdy grains of clear quartz and of notably fresh microcline and microperthite, cloudy orthoclase, and a little indeterminable plagioclase lie embedded in an abundant cement of apparently true argillaceous matter, reddened with much hematite. Grains of magnetite and apatite, probably of clastic origin, are present. Ragged foils and shreds of sericite developed in the bedding-planes, and secondary kaolin represent some recrystallization, but the rock must be regarded as a true argillite. In none of the thin sections does it show the amount of recrystallization seen in the older, more deeply buried metargillites. In one section, minute, pale brown grains of carbonate, probably dolomite or ferrodolomite, are distributed as in the Grinnell metargillite.

The recurrence of the special feldspar, microperthite, in the Kintla sediments—a constituent which is found in most of the clastic beds through the whole Lewis series—shows that probably one great crystalline terrane furnished the detritus during the deposition of the series. The great freshness of the feldspars in most of the beds suggests that the erosion of that terrane and the process of sedimentation were rapid. One may well suspect also that the climate in which disintegration overtook chemical weathering was an arid climate. This suspicion is strengthened by the discovery of abundant casts of salt-crystals in the Kintla rocks. Barrell has shown reasons for believing that the Kintla rocks were laid down under continental conditions, as subaerial deposits.* For this one formation the writer can quite agree with the view, but he believes that the Sheppard and all the underlying formations, excepting, possibly, a limited thickness of Grinnell and Appekunny beds, were laid down on the sea-floor.

ABSENCE OF TRIASSIC AND JURASSIC FORMATIONS.

No strata referable to the pre-Cretaceous Mesozoic occur within the Boundary belt either in the Clarke range or in any of the other ranges between the Great Plains and the Columbia river. In his reconnaissance of 1875 Dawson assigned the red beds of the Kintla formation to the Triassic, basing his correlation on the lithological similarity of the argillite-sandstone to the Triassic of the states farther south, and on the belief that the underlying Siyeh limestone is of Carboniferous age.† Willis has concluded that the Kintla formation is much older than the Triassic and, as indicated, the present writer agrees in assigning a pre-Devonian age to the formation.‡

*J. Barrell, Journal of Geology, Vol. 14, 1906. p. 553.
† G. M. Dawson, Bull. Geol. Soc. America, Vol. 12, 1901, p. 74, where further references.
‡ B. Willis, Bull. Geol. Soc. America. Vol. 13. 1902. p. 325.

2 GEORGE V., A. 1912

CRETACEOUS FORMATIONS OF THE GREAT PLAINS AT THE FORTY-NINTH PARALLEL.

The writer has made no special study of the Cretaceous formations over which the eastern part, if not all, of the Lewis range and the eastern part of the Clarke range have been thrust. These rocks crop out nowhere within the area covered by the detailed survey and they were only cursorily examined during a rapid traverse to Chief mountain. The reconnaissance of Willis and Weller has yielded useful results, which may be described by full quotation from Willis' paper:[*]

'Cretaceous strata are but poorly exposed along the eastern base of Lewis range, although they form the subterrane beneath hundreds of square miles of the plains. The mantle of drift is widespread and often thick, and outcrops of rock in place are limited to occasional freshly scoured gullies or ledges of sandstone along hilltops. Such outcrops were noted, however, in traversing the plains from Cutbank river to Saint Mary lake, and others were found about the mountain slopes west of Saint Mary lakes, up Swift Current valley, on Kennedy creek, about Chief mountain, and on Belly river. Weller collected fossils sufficient to determine three horizons, namely, Dakota, Benton, and Laramie, and through the light thrown by fossils on their relations these occasional Cretaceous outcrops become interesting as elements of a structure which they do not suffice to make clear. Their distribution is such that the Dakota and Benton, while occupying normal relations one to another, are apparently above the Laramie. The significance of this from the point of view of structure is discussed under that head.

'No occurrences of rocks of Cretaceous age were observed west of the Front range of the Rockies, and it is probable that there are none south of the Crowsnest coalfields.

'*Dakota.*—Arenaceous and argillaceous shales and sandstones of Dakota age occur on North fork of Kennedy creek near its junction with South fork, 5½ miles east by south from Chief mountain, at an elevation of 4,800 feet. The exposures constitute a bluff 30 feet high, near the top of which are layers bearing fossil plants and freshwater shells. A collection of leaves, though badly broken up in transit, was examined by Mr. Knowlton, who reports *Ficus protcoides* (?) *Lesq.*, *Magnolia boulayana Lesq.*, *Liquidamba integrifolius Lesq.*, *Liquidamba obtusilobatum Lesq.*, *Diospyro rotundifolia Lesq.*, *Phyllites rhomboideus Lesq.* "The above species," says Knowlton, "are all characteristic Dakota group forms, and the beds at this locality are referred without hesitation to this age." The strike of these Dakota beds is nearly north and south and they dip at a low angle, 0—10 degrees westward.

'*Benton.*—Dark bluish-black to leaden gray shales constitute the mass of Cretaceous rocks west of Saint Mary lakes. With them are associated

---

* B. Willis, Stratigraphy and Structure, Lewis and Livingston Ranges, Montana; Bull. Geol. Soc. America, Vol. 13, 1902, pp. 315 and 326.

thin beds of limestone and ferruginous sandstone. Weller's collections from outcrops north of lower Sherburne lake in Swift Current valley, and from southern slopes of Chief mountain, were submitted to Mr Stanton, who identifies *Inoceramus labiatus* Schlotheim, *Prionotropis* sp, *Ostrea congesta* Conrad (t), *Camptonectes* sp., *Scaphites ventricosus* Meek and Hayden, *A mia* sp., *Tellina* sp. Among these the *Inoceramus*, *Prionotropis*, and *Scaphites*, are classed as characteristic Benton forms.

'The topographic relations of the Dakota outcrop on Kennedy creek and the highest Benton outcrops under Chief mountain are such that if the beds were strictly horizontal the thickness of Cretaceous rocks would be 2,700 feet. As there is a slight dip from the former beneath the latter, this may be increased to 3,500 feet or more. It is, however, possible that the overthrusts which traverse the Algonkian are paralleled by others in the apparently undisturbed Cretaceous beds, and, if so, no estimate of thickness can be based on the meagre data now available.

'Just northeast of the northern end of Lower Saint Mary lake Weller collected from a gray sandstone and according to Stanton's determination obtained *Inoceramus* sp., possibly young of *I. labiatus, Mactra emmonsi* Meek (t), *Tellina modesta* Meek, *Donax cuneata* Stanton, *Corbula* sp., *Turritella* sp., and *Lunatia* sp. Of these Stanton says: "Although the evidence of these fossils is not absolutely conclusive as to the horizon, it is probable that they are from the Benton or at least from some horizon within the Colorado group."

'*Laramie.*—Ten miles east of Lower Saint Mary lake, on the middle fork of Milk river, occur outcrops of thin-bedded and cross-bedded gray sandstone and arenaceous shale. Some of the layers contain scattered and fragmentary plant remains. Others are barren of fossils. Certain ones are composed of oyster shells. In a section measuring 70 feet Weller found five oyster beds, from which he collected *Ostrea glabra* Meek and Hayden, *Corbicula occidentalis* Meek and Hayden, and small specimens of an undetermined *Melania*, which may be the young of *M. wyomingensis* Meek. The *Ostrea* of the highest stratum is said by Stanton to approach more nearly to *O. subtrigonalis* Evans and Shumard. These are all classed as belonging to the Laramie fauna.'

In his summary of the geology of the Rocky Mountain region in Canada, Dawson writes:—

'The aggregate thickness of the Upper Cretaceous in the southern part of the Laramide range— Front range—(including the lower portion of the Laramie, which may be regarded as Cretaceous) is found to be about 10,000 feet. It is unnecessary, however, to do more than allude to this section here, as it is more properly to be regarded as the western margin of the Cretaceous of the plains than as characteristic of the Cordilleran region.'

* G. M. Dawson, Bull. Geol. Soc. America, Vol. 12, 1901, page 78.

2 GEORGE V., A. 1912

Fifty miles north-northwest of the Boundary section through the Clarke range, the Livingstone range is adjoined on the west by the well known Crowsnest geosynclinal, containing from 12,000 to 13,000 feet of Cretaceous rocks, with possibly some conformable Jurassic beds at the base of the series.[†] This intermont development of the Cretaceous has no equivalent at the Forty-ninth Parallel unless it be represented in deeply buried strata beneath the Miocene of the Flathead fault-trough now to be described. Elsewhere in the Boundary belt the pre-Mesozoic terrane of the Front ranges is nowhere seen to be overlain by the Cretaceous, though it underlies the planes of the Lewis thrust and other contemporaneous major thrusts of the region. (See page 90.)

In the Little Belt mountains of Montana the Cretaceous beds lie conformably upon the thin Ellis formation which is assigned to the Jurassic. The Ellis formation in turn rests with apparent conformity upon Mississippian group of strata.[‡] The relations are similar to those which obtain at the Crowsnest Pass section, to which reference has been made. A hundred miles farther north, at Moose mountain, and again at the Lake Minnewanka section, Dowling has recently found the conformable Cretaceous-Jurassic series resting directly upon the Mississippian limestone.[§] At the latter point and at many others McConnell and Dawson have described the Cretaceous as lying, always with apparent conformity, upon the Carboniferous formation. These occurrences on both sides of the Boundary line, comparatively near to it, and all located on the line of strike of the Front ranges at the Forty-ninth Parallel, make it highly probable that the same general relation holds within the Boundary belt or its immediate, eastward extension. It will be noted that the Jurassic beds are generally of no great thickness and there is no certainty that they are represented in the Boundary line section at any point. If they are absent the Cretaceous may be regarded as resting on an erosion-surface terminating the Carboniferous limestone; it is further probable that the Carboniferous beds were little, if any, folded before the Cretaceous beds were deposited. The pre-Cretaceous erosion of these Paleozoic rocks at the Forty-ninth Parallel seems, then, to have followed a gentle, broad upwarp of the eastern part of the Cordillera.

KISHENEHN FORMATION.

The floor of the wide Flathead trough at the Boundary line is generally covered with glacial drift of extraordinary thickness and continuity. In the northern half of the five-mile belt the river has cut through the drift sheet. The bed-rock thus exposed may be seen at intervals in the low bluffs extending about three miles northward from the Boundary slash. Throughout that stretch the bed-rock belongs to a Tertiary fresh-water deposit which is not known to have an exact stratigraphic equivalent anywhere else in the area covered by

---

[†] J. McEvoy, Annual Report, Canadian Geological Survey for 1900, Vol. 13, Ottawa, 1903, page 90.
[‡] W. H. Weed, Little Belt Mountains folio, U. S. Geol. Survey.
[§] D. B. Dowling, Bull. Geol. Soc. America, Vol. 17, 1906, p. 295.

PLATE 12.

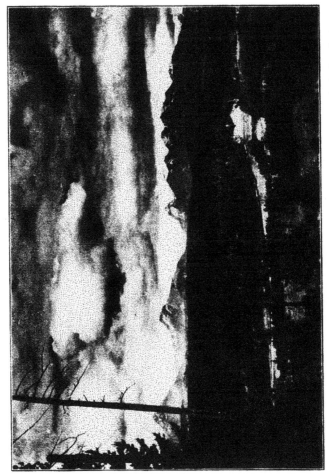

Looking east across Flathead Valley fault-trough to Clarke Range. Low cliff at river near locality of fossils in Tertiary Kishenehn formation. Winged-out local moraines farther in the background.

the Commission. The exposures are here fairly good but are not extensive enough to show the full thickness or relations of the formation. (Plate 12.)

The rocks consist chiefly of light to rather dark bluish-gray, often sandy, clays. In these there are numerous interbeds of hardened, light-gray sandstone, varying from two inches to a foot in thickness. The sandstone is very often characteristically nodular, with many concretions. A few seamlets of lignite up to 2 mm. in thickness and a few small, woody stems were observed in the clays. The latter usually very homogeneous and have the look of lake deposits.

At the river not more than 250 feet of different beds were actually seen, but it is probable that the total thickness represented in this section exceeds 500 feet. Ten miles down the Flathead valley, near the mouth of Kintla creek, the Kintla Lake Oil Company has drilled through 700 feet of soft 'shales' and sandstones bearing at intervals thin seams of coal. It is likely that these rocks form the southern continuation of the sediments at the Boundary line. Otherwise there is at present no hint as to the full extension of the lake beds.

Both clays and sandstones are at several horizons moderately fossiliferous. The fossils consist of small and extremely fragile shells. These have been examined by Dr. T. W. Stanton, who reported the collection to

'consist entirely of fresh-water shells belonging to the genera Sphaerium, Valvata (?), Physa, Planorbis, and Limnæa. Similar forms occur as early as the Fort Union, now regarded as earliest Eocene, but there is nothing in the fossils themselves to prevent their reference to a much later horizon in the Tertiary, because they all belong to modern types that have persisted to the present day, though it should be stated that their nearest known relatives among the western fossil species are in the Eocene.'

Dr. Stanton lists the species as follows:—

Sphaerium sp. Related to *Sphaerium subellipticum*, M. and H.
Valvata (?) sp. Resembles *Valvata subumbilicata* M. and H.
Physa sp.
Planorbis sp. Related to *Planorbis convolutus* M. and H.
Limnæa sp.

For convenience this group of Tertiary beds may be called the Kishenehn formation, the name being taken from that of the neighbouring creek. The same formation had been discovered near the mouth of the Kishenehn by Dawson who, in 1885, wrote:—

'Tertiary rocks resembling those assigned to the Miocene in the central plateau region of British Columbia, were met with in one or two small exposures in the bed and banks of the river, but poorly displayed and much disturbed by slides. They consist, so far as seen, of hard pale clays and sandy clays. It is probable that they underlie a considerable part of the width of this great flat-bottomed valley, though their extension to the north and south is quite indeterminate.'[*]

[*] G. M. Dawson, Ann. Report, Geol. Surv., Canada, 1885, Part B, p. 52.

2 GEORGE V., A. 1912

In his reconnaissance of 1901 Willis encountered the formation which he described in the following words:—

'On the North fork of the Flathead there are, as already stated, bluffs of clay with interbedded sandstones and lignites, in which no fossils were found. Details of constitution are summarized in the tabular statement of formations. The materials, degree of induration, and the lignitic condition of the carbonaceous deposits serve to indicate that they may be of Miocene or Pliocene age, as are beds near Missoula, which they resemble. These deposits are called lake beds because they are very distinctly and evenly stratified. They consist of fine sediment such as would settle from quiet water only, and they occur in a valley of such moderate width between mountains of such height that no simple condition of alluvial accumulation seems appropriate. It is possible that the lake was at times shallow like a flooded river. It is probable that it was some time reduced to the proportions of a river. It is certain that during considerable intervals some areas were marshes; but, admitting that a lake may pass through various phases of depth and extent, the term lake beds best describes these deposits.'[*]

At the Boundary line the dip is 18° to the eastward. Farther north the attitude is fairly constant in all the exposures, with strike north and south and dip, 40-45° east. The formation has evidently been disturbed by a strong orogenic force. The date of this particular phase of mountain-building cannot yet be fixed with certainty. It is pre-Glacial and post-Laramie. With some probability it may be referred to a mid-Tertiary stage, during which, according to Willis and Peale, crustal deformation took place in Montana.

It would be a matter of considerable interest to know the nature of the terrane underlying the lake beds. The fact that the drill at the Kintla creek oil-prospect struck continuous limestone at the depth of 1,290 feet suggests either that the lake beds lie directly on the Carboniferous or pre-Cambrian, or else, that only a very small thickness of Mesozoic strata (presumably Cretaceous) intervene between the lake beds and the pre-Mesozoic formations beneath the floor of the valley. This point will be considered again in connection with the dynamic history of the Rocky Mountains at the Forty-ninth Parallel.

POST-MIOCENE FORMATIONS OF THE GREAT PLAINS.

For the sake of completeness Willis' brief statement of the occurrence and nature of the youngest geological formations found on the plains in the immediate vicinity of the Boundary belt, may be given in summary. On pages 328 to 330 of his paper he gives some details concerning a Pliocene or early Pleistocene gravel fan to which the name 'Kennedy gravels' has been given.

[*] B. Willis, Bull. Geol. Soc. America, Vol. 13, 1902, p. 327.

For the purpose of this report, however, it will be expedient to reduce the summary of his findings to the table of formations given on page 315.*

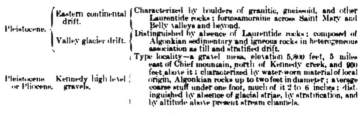

Pleistocene.	Eastern continental drift.	Characterized by boulders of granitic, gneissoid, and other Laurentide rocks; formsmoraine across Saint Mary and Belly valleys and beyond.
	Valley glacier drift.	Distinguished by absence of Laurentide rocks; composed of Algonkian sedimentary and igneous rocks in heterogeneous association as till and stratified drift.
Pleistocene or Pliocene.	Kennedy high level gravels.	Type locality—a gravel mesa, elevation 5,800 feet, 5 miles east of Chief mountain, north of Kennedy creek, and 900 feet above it; characterized by water-worn material of local origin, Algonkian rocks up to two feet in diameter; average coarse stuff under one foot, much of it 2 to 6 inches; distinguished by absence of glacial striae, by stratification, and by altitude above present stream channels.

The reader is referred to the fuller account of the Kennedy gravels, which the present writer has not specially studied except on the line of a single traverse. The view of Salisbury, quoted by Willis, that 'the high-level quartzite gravels on the plains east of the mountains are believed to be deposits made by streams at the close of the first epoch of baseleveling recorded in the present topography,' seems to be scarcely supported by the evidence of the gravels themselves, unless it is meant that uplift occurred at the close of the epoch. It is highly improbable that wide-spread clastic material of such coarseness could be formed during the closing stages of an erosion-cycle.

STRUCTURE.

FOLDS AND FAULTS.

On referring to map sheet No. 1, and especially to the general section, it will be seen that the Clarke range forms a great syncline which is accidented with a few faults and secondary warps. The eastern limb of the master fold shows the entire succession of rocks from the Waterton dolomite to the Kintla argillite. Every member of the lewis series is thus exposed, in its regular order, in the huge monocline stretching from the elbow of Oil creek (Cameron Falls brook) to the summit lake at the head of the creek. The dip slowly flattens from an angle of 30° in the Altyn formation to approximate horizontality at the water-divide of the range. The western limb of the syncline displays the complete series from the Kintla formation down nearly to the base of the Appekunny formation. The dips in the lower members there average about 20° to the northeast; those of the higher members, between upper Kintla lake and the Great Divide range from 3°· to 5° with variable strike.

At both sides of the master syncline the strata are rather sharply flexed down. There results, on the east, the narrow Cameron Falls (Oil creek) anticline, which plunges toward the northwest at a low angle. On the western side the down-warped strata are very poorly exposed but it is probable that the

* B. Willis, Bull. Geol. Soc., America, Vol. 13, 1902.

2 GEORGE V., A; 1912

observed down-flexure of the Appekunny at Starvation creek is but an incidental result of a great normal fault or system of normal faults which limits
the Flathead valley on the east. In this view the writer's observations agree
with those of Willis, who states that the structural relations at the western
side of the main syncline 'are those of a normal fault of great displacement,
and downthrow on the west. From the topographic relations the position of
this normal fault is inferred to be along the base of the Livingston [Clarke]
range, the downthrown block underlying Flathead valley.'*

The axis of the Akamina creek valley is located on a flat syncline, with
an axial trend parallel to those of the main syncline and the Cameron Falls
anticline, namely, northwest-southeast.

Normal faults are rare and none, except the North Fork fault, shows strong
displacement. The attitudes of the beds to north and south of the Kintla
lakes suggest a fault following the axis of the valley, with downthrow on the
north. The movement was slight but apparently sufficed to locate the erosion
channel of which the present valley is the greatly enlarged descendant. A
second, neighbouring fault has brought the red Grinnell beds down into contact with the lower beds of the Appekunny, indicating a moderate throw on the
east of the fault.

GREAT LEWIS OVERTHRUST.

The most important structural feature of the range is a great
thrust by which the eastern part of the main syncline, together with
the Mt. Wilson (Lewis range) block, has been driven eastward or northeastward over the Cretaceous formations of the plains. The nature and relations
of this thrust are quite similar to those described by McConnell at the Bow
River pass and Devils lake, 150 statute miles to the north-northwest, and to
those worked out by Willis at Chief mountain and elsewhere in Montana.† The
proof that the thrust-plane extends to Waterton lake suggests the possibility
that the one great dislocation extends from 48° 30′ N. Lat. to at least 51° 30′
N. Lat. (See Figure 6.)

The demonstration that the thrust-plane passes under the Clarke range
is not as full as that adduced by Willis for the Lewis range. In the more
westerly range the natural rock exposures in the Boundary belt are not of
themselves sufficient to show the fact. The reason for believing that at least
the eastern part of the Clarke range block has actually overridden the plains
strata is found in the log of the deep boring made by the Western Oil and
Coal Company at Cameron Falls, on the west side of Waterton lake. At that
point the drill penetrated 1,500 feet of silicious dolomites which, as above noted
(page 55), form the downward extension of the Lewis series. At that depth
the drill suddenly entered soft shale which continued for another 400 feet,

---

* B. Willis, Bull. Geol. Soc. America, vol. 13, 1902, p. 344.
† R. G. McConnell, Ann. Rep. Geol. Survey of Canada for 1886, Part D, p. 31; B.
Willis, Bull. Geol. Soc. America, Vol. 13, 1902, p. 331; D. B. Dowling, Bull. Geol. Soc.
America, Vol. 17, 1906, p. 296.

PLATE 13.

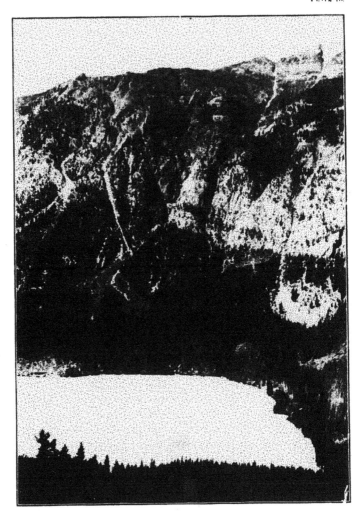

Head of Lower Kintla Lake. Cliff composed of Appekunny and overlying Grinnell formations. The small cleared patch across the lake, on the extreme left, is the site of
petroleum prospect with showings of oil and with steady blow of natural gas.
25a—vol. ii—p. 90.

SESSIONAL PAPER No. 25a.

when the work was stopped, and the bore-hole, for the time at least, abandoned. These shales have been examined by Mr. T. Denis, of the Canadian Department of Mines, and by the writer; the material proved to have the habit of typical Cretaceous, probably Benton, sediments. Their colour, softness, and carbonaceous character are quite different from those characterizing any phase of the Lewis series; on the other hand the shales are sensibly identical with fossiliferous Cretaceous beds occurring below the thrust-plane at Chief mountain.

How much farther west the thrust has caused the superposition of the Belt terrane on the Cretaceous can only be conjectured. It is not impossible that the entire Clarke range in this region represents a gigantic block loosed from its ancient foundations, like the Mt. Wilson or Chief mountain massifs, and bodily forced over the Cretaceous or Carboniferous formations. In that case the thrust would have driven the block at least forty miles across country. Such a speculation is of some interest in giving one explanation of the emanation of gas and petroleum in the Flathead valley and in the heart of the Belt-Cambrian rocks at lower Kintla lake. (Plate 13.) These hydrocarbons would thus be considered as originating in the Carboniferous limestone or in the Cretaceous sediments underlying the thrust-plane. Since the Carboniferous limestone is highly bituminous, that formation would naturally offer an original source for the oil and gas.

On the other hand, a second hypothesis may be framed, whereby the seepages in the Flathead valley are thought to originate in the Carboniferous limestone which was faulted down during the formation of the Tertiary fault-trough, while the seepages at Kintla lake are interpreted as emanations from Carboniferous limestone locally underthrust on the west side of the main syncline of the Clarke range. On this view the Waterton lake thrust need not extend much farther west than the lake itself.

Or, thirdly, one might conceive that the hydrocarbons originated directly in the Beltian rocks themselves (see page 53), so that the existence of the seepages would have no direct bearing on, or afford no proof of, any large-scale thrust-plane beneath the western slope of the range. There is as yet no decisive evidence forcing a choice among these three hypotheses. The known extent of the bodily movement represented in the Waterton lake thrust is, at a minimum, about eight miles, as measured on the perpendicular to the line tangent to Chief mountain and the outpost mountains of the Clarke range. The movement has probably been ten miles or more and may be as much as forty miles.

The thrust proved at Waterton lake is doubtless a northern continuation of the 'Lewis thrust' described by Willis as explaining the peculiar relations of pre-Cambrian and Cretaceous at Chief mountain and southward. Willis has, in fact, stated that he has traced the outcrop of the thrust surface around Mt. Wilson to Waterton lake. He concludes that ' according to these observations, the relation of the Lewis and Livingston [Clarke] ranges, en echelon at the 49th parallel, is an effect of step-like though very gentle flexure in the

fault-surface of the Lewis thrust.'* For his discussion of this and related points
the reader is referred to his paper.

Willis also gives a detailed account of the interesting structural effects
wrought in the overriding block, particularly as illustrated in Chief mountain.
Similar evidences of the mighty force involved were observed by the present
writer on the Canadian side of the Boundary. As in the case of the Altyn
strata on Chief mountain the Appekunny beds north of Pass creek were seen
to be separated by flat-lying, heavily slickened surfaces, showing the existence
of minor thrusts with movement from west to east. At Chief mountain the
writer had opportunity of seeing the truly spectacular effects on the overriden
Benton shales.

Willis has formulated a hypothesis according to which the existing structure
of the Clarke and Lewis ranges is attributed to two distinct periods of orogenic
movement. A summary of the hypothesis as it relates to the dating of the Lewis
thrust may be given in his own words:—

'Along the eastern base of the Rocky mountains in general the facts
of structure express the action of a compressive stress, the Cretaceous and
older strata being folded. The post-Cretaceous effects are commonly attributed to a single episode of compression; in what follows they are assigned
to two episodes, at least for the particular district under discussion.

'The first episode of compression began at some date not closely
determinable, but which may be placed not earlier than Laramie time,
nor later than early Tertiary. It is possible that flexure went on during
Laramie deposition. It is also possible that it did not begin till after that
deposition was completed. The distinction is not important to the present
thesis. Flexure in its early stages was an effect involving relatively great
stress, as the nearly flat Algonkian strata were exceeding'y inflexible. It
is probable that folds developed slowly. As the Laramie sea was shallow
and was succeeded by emergence of the area, the anticlines were subject
to erosion, whether they developed earlier or later, and the synclines received
their ute either as sediments beneath marine waters or in estuaries or in
lakes   as valley deposits.

e effect which for a time satisfied the compressive stress was one
of erate folding. The succeeding condition was one of quiescence and
it endured long enough for the planation of Cretaceous rocks to the
Blackfoot peneplain. The name Blackfoot may be extended to the topographic cycle ending in the development of the plain. The Blackfoot
cycle cannot be accurately dated by any evidence now available. It was
post-Laramie and probably earlier than the orogenic movements which,
in Montana, gave rise to ranges and lake basins. The latter having yielded
Miocene vertebrates, the movement may be placed in mid-Tertiary. That
it was preceded by the Blackfoot cycle is an inference based on general
observations of an extensive peneplain over the summits of the Rockies

* B. Willis, Bull. Geol. Soc. America, Vol. 13, 1902, p. 333.

of western Montana and Idaho, observations which leave no doubt in the writer's mind of the existence of such a peneplain, but which do not suffice positively to identify it as the Blackfoot plain. On the probability of that identification the Blackfoot cycle may be placed in early Tertiary time.

'At the close of the Blackfoot cycle the topographic features of the region under discussion were the peneplain on Cretaceous rocks and low hilly, past-mature relief on Algonkian rocks, such as is now presented by the summit hills of eastern Flattop..........

'Among the effects of folding and erosion, at the close of the Blackfoot cycle was the exposure of the edges of some Algonkian strata as outcrops; being gently inclined westward, these beds had probably wide extent underground. They were relatively stiff and lay with one edge free. Under these conditions, supposing that a compressive stress again became effective, a part at least of the Algonkian beds were so placed that they met but slight resistance in their tendency to yield by moving forward. So far as they were unopposed, or not sufficiently opposed to check and fold them, they did ride forward. That part which was thus overthrust separated from that which was not in general along bedding planes near the base of a particularly rigid stratum, such as the Altyn limestone. The Siyeh limestone, the Carboniferous limestone, or other stiff formation may elsewhere be found to have determined the thrust surface within the old rocks.'*

Willis then concludes this postulated history by noting the geological date deduced for the Lewis thrust and directly associated movements. He also briefly states the alternative view that the folding and thrusting were products of a single period of deformation. He writes:

'On the hypothesis of a single episode of compression, from which resulted all the phenomena of folding and thrusting in Cretaceous and Algonkian rocks in the district, the Lewis thrust and the associated structures must be assigned to a date closely following the Laramie deposition. The growth of the Front ranges and the development of the Blackfoot plain must be placed later, and the expression of the Lewis thrust must be considered subordinate at the surface to these later effects of orogeny and erosion.

'On the other hand, on the hypothesis of two episodes of compression, separated by the Blackfoot cycle, the Lewis thrust must result from the second episode, and falls probably in mid-Tertiary. Its orogenic effects are then dominant in the Front ranges, and the physiographic history is to be read in terms of structure as well as of erosion.

'It is concluded that the date of the Lewis thrust may be placed in either late Cretaceous or mid-Tertiary time, and the principal criteria for determining which date is correct are to be found in the relations of structure to physiography.'†

* B. Willis, Bull. Geol. Soc. America, Vol. 13, 1902, p. 339.
† Ibid, p. 343.

2 GEORGE V., A. 1912

Though inclined to favour the two-episode hypothesis Willis was careful to leave the question open. The problem is complicated by the apparent necessity of believing that, at a late Paleozoic (post-Mississippian) stage the whole Purcell-Rocky Mountain province was broadly upwarped; forming a geanticlinal area, from which the sea was excluded during the Triassic and Jurassic periods. There thus seems to be a possibility that the structural condition for the Lewis thrust—an antecedent flexure near the site of the present frontal escarpment of the Rockies—was established long before the Laramie or the Cretaceous period. To the present writer the view that the post-Cretaceous movements all belonged to a single orogenic episode has the merit of simplicity and does not seem to be contravened by any known fact. Moreover, one observation in the field seems to indicate directly that great thrusts were developed in the Clarke range either during the main post-Laramie folding, or at any rate, before that folding was completed.

This partial evidence is illustrated in the structure section of Figure 6 and map sheet No. 1. It will be observed that a heavy wedge of Siyeh limestone has been pushed eastward along a nearly horizontal thrust-plane, which has truncated a thick mass of the Altyn beds. The wedge has penetrated the overlying Appekunny beds, crumpling, mashing, and forcing them aside like an enormous plowshare. An inspection of the whole section seems to warrant the conclusion that this thrust which forced the younger Siyeh limestone into contact with the older Altyn and lower Appekunny beds, must have antedated the development of the Cameron Falls anticline in its present structural form. It is hard to believe that the limestone wedge was driven downwards to an actually lower level. A more reasonable view is that the Altyn-Appekunny beds on the eastern side of the present fold were formerly faulted or bent upwards so as to feel the thrust of the still flat-lying Siyeh limestone on the west. After the wedge was intruded and by the continuance of the same compressive stress, the whole series was flexed so as to show the existing anticlinal structure. On the probable supposition that this thrust and the Lewis thrust were contemporaneous, it follows that the latter was also contemporaneous with an early stage in the main folding of the Clarke range.

At other points in the Clarke range as well as in the Galton-Macdonald mountain group the geosynclinal rocks underwent a powerful compression before they were folded at all. At these localities the dip of the bedding is 20° or more, while marked slaty cleavage has been developed with its plane characteristically perpendicular to the bedding-plane. It seems most probable that this cleavage was the product of a compressive force which was directed along the bedding-planes and had developed the cleavage before significant upturning occurred. This phase of tangential compression may be that in which the great Siyeh wedge was thrust into the Appekunny metargillites and in which the much greater Lewis thrust was formed. Most of the folding may have been produced as a slightly later expression of the same but somewhat intensified, tangential force. On this view both thrust and fold belong to one orogenic episode.

Neither of the two hypotheses can as yet be proved. It remains for future students of the Front ranges to find the true solution to the dynamic problem. The question is important since it deeply affects our understanding of the later geological history of the Clarke, Lewis, and several adjacent ranges of the Rocky Mountain system.

CHAPTER V.

STRATIGRAPHY AND STRUCTURE OF THE MACDONALD AND GALTON RANGES.

GALTON SERIES.

From the Flathead river to the Kootenay river at Gateway the mountains are principally composed of the Galton series, the westward extension of the same stratified series that form the peaks and massifs of the Clarke and Lewis ranges. Two of the formations described as constituting the Lewis series—the Altyn and Siyeh—are very clearly represented on the west side of the Flathead and, in the following account of the Galton series, will bear the original names given by Willis. The other stratified members of the Galton series are related to the corresponding members of the Lewis series but are stamped with distinctly individual characters, and merit special names which will be employed in order to emphasize the contrasts between the two series. The columnar section for the strata actually visible in the Galton range has been supplemented at its base by the addition of formational units which crop out only in the MacDonald range.

The two groups of strata are so similar lithologically that the description of the Galton series scarcely needs great detail. Many features are simply repetitions of those already described for the Lewis series.

The Galton series includes the formations noted in the following table.

Formation.	Thickness in feet.	Dominant rocks.
	Top, erosion surface.	
Roosville..	600+	Metargillite.
Phillips..	550	Metargillite.
Gateway..	2,025	Metargillite and quartzite.
Purcell Lava..	310	Altered basalt.
Siyeh..	4,000	Magnesian limestone and metargillite.
Wigwam..	1,200	Sandstone and metargillite.
MacDonald..	2,350	Metargillite.
Hefty..	775	Sandstone and quartzite.
Altyn..	650+	Silicious dolomite.
	12,460	
	Base concealed.	

At the summit of the McGillivray range heavy blocks of Mississippian limestone have been faulted into contact with the older members of the Galton series. Neither top nor bottom of the limestone is exposed in the Boundary belt. It is certainly younger than the uppermost bed of the Galton series, but it is not known whether the relation was that of original conformity. The maximum thickness of the exposed limestone seems to be about 2,800 feet.

25a—vol. ii—7

2 GEORGE V., A. 1912

On the eastern edge of Tobacco Plains a smaller block of Devonian limestone and dolomitic quartzite, estimated to show a thickness of 1,600 feet, has
been faulted down into contact with the Gateway formation. Its relation to
the Galton series could not be directly observed. As elsewhere in this part of
the Cordillera the limestone doubtless passed gradually upwards into the
Mississippian lime-tone but no rock of that age was determined on the west
side of the range.

ALTYN FORMATION.

The oldest formation exposed in the MacDonald range where crossed by
the Boundary belt consists of strata essentially similar in stratigraphic relations
and in composition to the uppermost member of the Altyn formation of the
Lewis series. The identity is so complete that the same name may well be
used for these rocks of the MacDonald range. Though not exposed in the
Galton range at the Boundary, the equivalents of the same strata unquestionably underlie the surface rocks of that range as well.

West of the Flathead the Altyn crops out at only two localities within
the Boundary belt. On the ridge overlooking the Flathead from the west and
culminating in Mt. Hefty, the triangulation peak, this formation forms part
of a block faulted into contact with Carboniferous limestone. Just south of
the Boundary monument on the ridge, a thickness of 650 feet was measured
for the Altyn but the base was not seen, the lower beds having been faulted
away, out of sight. The second locality is that at a box-canyon six miles due
west of the Hefty ridge; there, only 120 feet of the uppermost beds are exposed.

The Altyn formation in the MacDonald range consists of a succession of
fairly homogeneous but very thin-bedded, silicious dolomites. The rock is
always compact and relatively hard, yet very fissile on account of the thin
bedding. The layers vary from 1 cm. to 10 cm. in thickness. When fresh it
is slightly gray or greenish gray; it weathers buff and bright brownish yellow.
Quite subordinate are more. massive beds (three to five feet thick) of gray,
calcareous quartzite. Towards the top, thin intercalations of red calcareous
sandstone and argillite indicate transition to the Hefty formation above. Certain of the magnesian limestone beds are somewhat argillaceous and then
commonly bear sun-cracks.

Microscopic and chemical analysis of the dominant rock, the silicious
dolomite, shows that it is similar to the principal phase of the upper Altyn in
the Lewis and Clarke ranges. The main mass of the rock is a very compact
carbonate, occurring in grains from 0.005 mm. to 0.03 mm. Angular particles
of quartz, averaging 0.02 mm. in diameter, are accessory constituents which,
in certain layers, may become quite abundant. This quartz was probably in
part of clastic origin but some of it may have been due to the recrystallization
of colloidal silica. When the carbonate is in contact with quartz, the former
often shows clean-cut, rhombohedral outlines.

Minute angular particles of orthoclase or microcline and of microperthite are other, probably original, clastic accessories. A few small shreddy foils of sericitic mica are rare metamorphic constituents.

The high specific gravity of the dominant phase (2.716—2.816) immediately suggests that it is highly magnesian. An analysis of a type specimen (No. 1270) from the box-canyon above mentioned has been made by Professor Dittrich:—

Analysis of upper Altyn impure dolomite.

		Mol.
SiO_2	26·07	·435
Al_2O_3	3·92	·038
Fe_2O_3	2·06	·013
FeO	2·68	·037
MgO	12·99	·325
CaO	19·58	·349
Na_2O	1·04	·017
K_2O	1·40	·015
H_2O at 110°C	·04	....
H_2O above 110°C	1·52	·083
CO_2	28·14	·662
	100·46	
Sp. gr.	2·816	
Insoluble in hydrochloric acid		30·80%
Soluble in hydrochloric acid:		
Fe_2O_3		4·25
Al_2O_3		2·17
CaO		19·38
MgO		12·69

In this case, the relative quantities of the different components may be roughly calculated. The proportions are approximately as follows:—

Calcium carbonate	34·6
Magnesium carbonate	26·7
Free silica	15·0
Albite molecule	9·9
Sericite and potash feldspar, about	9·7
Remainder (pyrite, limonite, etc.) about	5·0
	100·0

It is, however, possible that the mica bears some of the soda and that paragonite itself may be present. If so, the foregoing calculation would be manifestly incorrect. In any case the carbonates compose rather more than 60 per cent of the rock. The ratio $CaO : MgO$ is 1·527:1 and thus not far from the theoretical ratio in pure dolomite. In all essential respects this sediment is like that of the type of Upper Altyn in the Clarke range except in the higher proportion of quartz and silicate material in the Galton series type. In each case the rock is a silicious dolomite.

HEFTY FORMATION.

In the MacDonald range the Altyn formation is conformably overlain by a group of strata which are exposed at the same two localities where the Altyn

2 GEORGE V., A. 1912

was seen in the Boundary belt. From its occurrence at Mount Hefty, this assemblage of beds has been named the Hefty formation. The full thickness, estimated at 775 feet, was measured at both localities. In both cases the exposures are good. On the whole the formation is homogeneous and, as with the associated Altyn, useful subdivision did not seem possible in the field.

The staple rock of the formation is a heavily bedded, red or reddish gray, fine-grained sandstone. As a rule, it is not metamorposed to the condition of true quartzite. Its mass is interrupted by thin beds of red shale and by rarer, light greenish-gray, brown-weathering quartzites. Occasionally the sandstone is somewhat calcareous. While the group of beds is generally red, this colour, being mixed with gray, is not so striking as in the case of the overlying Wigwam formation. Sun-cracks and ripple-marks are common at various horizons. A further general characteristic of ledges and hand-specimen is the relatively abundant development of metamorphic sericitic mica in the bedding planes.

The formation passes upward with some abruptness, into the MacDonald sandstone. As already noted, there is some dovetailing with the underlying Altyn.

Under the microscope the dominant sandstone is seen to be composed of rounded grains of quartz and less abundant feld-par, with a compound cement of angular quartz and feld-par fragments, carbonate (probably dolomite or magnesian ferrocalcite), and a small amount of iron ore. As a rule the quartz is glass-clear and uncrushed; occasionally the grains show enlargement by the familiar addition of silica, crystallographically orientated. The feld-spars are again microperthite, microcline, orthoclase, and plagioclase (probably andesine) named approximately in the order of their relative abundance. As a rule the sections show the occurrence of well-rounded grains of cherty silica. The clastic grains vary from 0.1 mm. to 0.6 mm. in diameter, averaging about 0.2 mm. The rock is thus a typical fine-grained feld-pathic sandstone, sometimes calcareous and always more or less ferruginous.

The subordinate, argillaceous interbeds of this formation are usually in a more visibly metamorphosed condition, and recall the type phases of the Appekunny formation. The alteration (again by deep burial, not by mountain-building thrust) approaches the degree of true metargillite but it cannot be said that amorphous matter is entirely replaced. Grains of carbonate (probably dolomitic) are practically constant accessories, as they are in the more sandy phases.

From its position and its petrographic nature we may conclude that the Hefty formation is the coarser-grained equivalent of the lower Appekunny in the Lewis series, especially of its variegated basal beds as exposed at the eastern end of South Kootenay Pass.

The average specific gravity of seven hand-specimens of the sandstone is 2.646, and of three hand-specimens of the more argillaceous phase is 2.743. The average specific gravity for the whole formation is about 2.695.

SESSIONAL PAPER No. 25a

MacDonald Formation.

Above the Hefty formation in the conformable Galton series is a thick division of beds which betoken the long continued deposition of rather uniform sediment. Since these rocks underlie an extensive surface in the MacDonald range, they have been grouped under the name of the MacDonald formation. Good exposures are found among the ridges overlooking the Wigwam river. The whole thickness was, however, not observed in any section traversed within the Boundary belt. The estimated total, 2,350 feet, is doubtless not quite accurate but the error is believed not to be great.

The formation is notably homogeneous, so far as the main lithological features are concerned. It was found in the field that a subdivision into three members could be recognized with advantage. This subdivision is based largely on differences in the colours of weathering, rather than on any fundamental differences of composition or origin. The lowest member, 550 feet thick, weathers characteristically light brown or brownish gray; the middle member, 1,100 feet thick, weathers light gray, and the top member, 700 feet thick, weathers light brown or buff, though a few beds weather gray.

The principal rock type throughout the formation is a highly silicious argillite or metargillite, quite similar to the standard phase of the Appekunny. The colour of the fresh rock is a light gray, or more commonly, light greenish-gray. The bedding is usually thin but becomes more massive in the lowest member of the formation. Sun-cracks and ripple-marks, especially the former, are abundant in many horizons ranging from summit to base. The top member carries some thin intercalations of red, sandy, or argillaceous strata, indicating a transition to the overlying Wigwam formation. Along with these there occur a few dolomitic lenses, rarely over six inches thick, which bear a number of flattened concretions. The concretions range up to a foot in greatest diameter. They are composed of alternating concentric layers of different carbonates. The bulk of each concretion appears to be a ferruginous dolomite forming layers from one-eighth to one-quarter inch in thickness. These are separated by much thinner laminae of nearly pure calcite which, on the weathering of the rock, is dissolved away at a more rapid rate than the dolomite. At first sight the weathered section of one of the concretions suggests Cryptozoon or other possibly organic form, but the writer believes that the organic appearance is deceptive and that the structure is due to a physical and chemical rearrangement in magnesian limestone, analogous to that causing the molar-tooth structure of the Siyeh limestone.

About 500 feet above the base of the middle member a thin (two-inch) but remarkably persistent layer of gray (brown-weathering) somewhat magnesian oolite was seen in several traverses separated by distances of from two to eight miles. The spherical grains have the usual concentric and radial structure and average about 1 mm. in diameter. They are cemented by calcite and infiltrated quartz. Some of the quartz present is probably of clastic origin as it is associated with grains of microcline. A variable proportion of a grain

2 GEORGE V., A. 1912

is commonly seen, under the microscope, to be replaced by quartz in fine mosaic. A less common result of metamorphism is the generation of small but beautifully crystallized, idiomorphic plagioclase feldspars. Usually not more than one of these new crystals is developed in one of the 'eggs.' Optical tests seem to show that the feldspar is not the expected pure lime feldspar but the sodiferous labradorite, with possibly the more basic bytownite sometimes developed.

About 100 feet above the oolitic bed a similarly persistent zone of red shale, about 60 feet thick was observed on the ridge of Mt. Hefty and at the Boundary line on the high ridge immediately east of Wigwam river.

Much the greater volume of the formation, practically all of it except the relatively insignificant intercalations just noted, is made up of the silicion-metargillite, which merits a few words of detailed description.

The microscope shows that the rock was originally a typical argillaceous sediment. More than half of it was clayey matter, in which small, angular grains of clastic quartz, microcline, microperthite, and plagioclase were embedded. In its present condition the rock carries a highly variable amount of sericite, chlorite, and cryptocrystalline silica. There are, thus, all transitions from partially recrystallized argillite to true metargillite. The specific gravity correspondingly varies, from values as low as 2·687 to those as high as 2·754, the latter representing practically complete crystallinity with abundant sericite developed. The average specific gravity of seven selected specimens is 2·722.

In several of the thin sections grains of carbonate, causing liberal effervescence with cold dilute acid, are to be seen distributed through the silicious matrix. From the fact that a half dozen or more of these grains (calcite, sidérite, or ankerite?) extinguish together on rotation between crossed nicols, it may be concluded that they are of secondary origin, and, like the sericite, crystallized after burial of the sediment. At the same time there is no reason to doubt that the material of the carbonate was a component of the original mud.

Professor Dittrich has chemically analyzed one of the least recrystallized phases. The specimen (No. 1250) was collected at the top of the 6,700-foot summit about 2,300 yards southwest of the Boundary monument at Wigwam river. The analysis yielded the following proportions:—

Analysis of type specimen, MacDonald formation.

SiO₂	68·37
Al₂O₃	7·02
Fe₂O₃	4·41
FeO	3·99
MgO	4·41
CaO	3·89
Na₂O	·87
K₂O	1·34
H₂O at 110°C	·25
H₂O above 110°C	3·60
CO₂	1·91
	100·06
Sp. gr.	2·687

SESSIONAL PAPER No. 25a

Insoluble in hydrochloric acid..	74·08
Soluble in hydrochloric acid:·	
Fe₂O₃..	6·62
Al₂O₃..	5·96
CaO..	2·70
MgO..	3·64

The strong ' deficit ' in carbonate of CO_2 and the general absence of combined water leads one to suspect that the magnesium may be present in the form of the basic carbonate or in the form of a hydrous magnesium silicate, or possibly in both forms. The analysis evidently does not lend itself to useful calculation. The rock has the chemical composition of a somewhat dolomitic argillite, which is high in silica and iron oxides, and low in alumina.

WIGWAM FORMATION.

The MacDonald formation is succeeded above by the Wigwam formation, named from the river which receives part of the drainage of these mountains. On reference to the map sheet, it will be seen that a band of rocks referred to the subdivision follows a long, high ridge running south from the Boundary line at a point half-way between the Kootenay and Flathead rivers. A second extensive, though less perfect exposure of the Wigwam occurs about six miles to the eastward. Elsewhere in the belt these rocks have either been eroded away or lie buried beneath the Siyeh formation. The total thickness was measured in the more westerly hand and was found to be about 1,200 feet.

The Wigwam formation consists of a mass of fairly homogeneous red or brownish-red sandstones, interrupted by partings of red, silicious metargillite. Though the bedding is generally thin, the sandstones are often united into platy aggregates one to three or more feet in thickness. A few gray or brown-rusty metargillite beds and some red gritty layers form subordinate intercalations. Throughout the formation sun-cracked and ripple-marked, sometimes cross-bedded, horizons occur. A few markings, interpreted as annelide burrows, were seen, but no more useful fossils were discovered. These red beds are rather sharply defined against the gray or light brownish strata of the overlying Siyeh, which is, however, perfectly conformable. As already noted, the Wigwam and MacDonald rocks are dovetailed together through interbedding.

The principal pitase of the formation is a fine-grained sandstone, charged with a variable amount of ferruginous and once-argillaceous cement. Its essential constituents are quartz, in rounded and angular grains; abundant subangular grains of orthoclase, microperthite, microcline, and plagioclase (near andesine Ab_2An_1); and generally-rounded grains of ferruginous chert. The iron ore, probably hematite, is relatively abundant; it is finely divided and generally opaque. Kaolin and sericite are extensively developed. A few clastic zircons were observed under the microscope. The sericite is especially developed in the non-argillaceous beds and in the bedding planes. Rarely the grains of quartz show enlargement with new silica. On the whole the rock preserves the eminently clastic structure of true sand-stone and it cannot fairly be called a quartzite. On the other hand, the never-failing generation of

2 GEORGE V., A. 1912

sericitic mica on the bedding planes, through static metamorphism, shows that the rock has suffered some change. It may, for convenience, be called meta-sandstone. From that type all transitions to true metargillite are represented in the formation.

The specific gravity of the metasandstone varies from 2.649 to 2.653, averaging about 2.634; the specific gravity of a typical specimen of metar-gillite is 2.711. The average for the formation as a whole is about 2.65.

The Wigwam formation is evidently the western equivalent of the Grinnell formation of the Clarke and Lewis ranges. It differs from the Grinnell in the possession of coarser grain, in the greater predominance of sandstone, and in its smaller thickness.

SIYEH FORMATION.

The general equivalence of the Galton and Lewis series is, lithologically, most evident in the thick formations respectively overlying the Wigwam and Grinnell beds. The similarities of age, composition, structure, and origin are so manifest in the field that the same name is here adopted for the formation as in the Clarke-Lewis sections. The principal petrographic difference is found in the greater prominence of argillaceous matter in the Siyeh of the Galton range. The total thickness is estimated, with low limits of error, at 4,000 feet. This corresponds well with the thickness of 4,100 feet yet more closely deter-mined among the better exposures of the eastern ranges.

More or less complete sections of the Siyeh occur within the Boundary belt, on the eastern slope of the Rocky Mountain Trench. Others were studied along the head-waters of Phillips creek and of Wigwam river. A composite columnar section was constructed, showing the information derived from seven traverses, spaced several miles apart. It was found that, notwithstanding the great thickness of the formation, there are few lithological horizon-markers. The columnar section resolved itself into the following relatively simple scheme:—

Columnar Section of Siyeh Formation.

Top, conformable base of Purcell Lava.

1,200 feet.—Chiefly gray and greenish gray, medium to thin-bedded, silicious, often dolomitic metargillite, weathering light brown and buff. At the top some 250 feet of the beds have a general reddish cast, owing to abun-dant intercalations of red-gray, ripple-marked sandstone. Between 400 feet and 700 feet from the top, several beds of light gray lime-stone, weathering gray to whitish, occur: the thickest of these, 25 feet thick and about 600 feet from the top of the formation, was followed for several miles in the Galton range. Sun-cracks are abundant at many horizons in the metargillite.

2,000 "   Dark gray, argillaceous magnesian limestone or dolomite, in massive with typical molar-tooth structure. Occasional intercalations of argillite. The lower part of the member is more silicious than the average rock. Most of the beds weather brown or buff, a few weather-ing reddish. The individual beds vary in thickness from a fraction of an inch to two feet or more, but generally they are grouped or cemented together in massive plates three to ten feet in thickness.

PLATE 14.

Cliff in Siyeh limestone, showing molar-tooth structure; at cascade in Phillips Creek, eastern edge of Tobacco Plains. Hammer fourteen inches long.

SESSIONAL PAPER No. 25a

500 feet. Rocks like those of the upper division but without red beds or gray limestone; chiefly medium-bedded to thin-bedded green and greenish-gray, highly silicious, sometimes dolomitic metargillite, weathering light brown and, less often, gray. Many sun-cracks and some ripple marks occur at various horizons.

————
4,000 feet.

Base, conformable top of the Wigwam formation.

The strong chemical contrast between the middle member and either of the other two members might suggest the inadvisability of grouping all these rocks in one formation. The grouping has been made partly in the interests of correlation, partly on the ground that throughout the whole 4,000 feet of thickness the strata show nearly uniform compactness and character of bedding and fairly constant colour in both fresh and weathered phases, so that, in the field, it is not easy to distinguish the metargillite from the often highly argillaceous limestone.

There is nothing specially novel in the detailed characters of the upper and lower members but certain noteworthy conclusions follow from the facts derived from the microscopic and chemical examination of the limestone in the middle member. A close study has been made of type specimens collected on the spur just west of the cascade on Phillips creek, near Roosville post office. In colour, character of bedding, and other microscopic character, these rocks are essentially similar to the typical Siyeh limestone of the Clarke and Lewis ranges. They show the molar-tooth structure in notable perfection. (Plate 11, A). Not only the calcitic lenses and pencils but also the buff-weathering magnesian parts effervesce with cold dilute acid.

The calcitic partings have microscopic characters identical with those of the partings in the molar-tooth rock of the Clarke range; each is made up of aggregated granules of calcium carbonate averaging 0.02 mm. in diameter. Scarcely a grain of another substance is to be seen in these parts of the thin sections.

The magnesian parts are, on the other hand, quite highly composite. The pale brownish, often rhombohedral crystals of dolomite or magnesian calcite are distributed through an abundant matrix of quartz, feldspar, sericite, chlorite, and a thin cloud of black dust-particles, partly magnetite and partly carbon. The rhombohedral grains average about 0.02 mm. in diameter; the anhedral carbonate grains may be considerably smaller. The grain of the matrix is very fine; the quartz particles not surpassing the carbonate grains in average size. The feldspars are too small for specific determination and are recognized as such by their polarization tints, checked by the chemical analysis of the rock. Sericite (also paragonite?) is relatively abundant. Original argillaceous material cannot be demonstrated in the thin section; its recrystallization in the form of sericite, chlorite, quartz, iron ore, and possibly feldspar seems to be nearly perfect.

A large, characteristic specimen (No. 1221) of the molar-tooth rock, which was collected just west of the Phillips creek cascade near Roosville post office, was selected for chemical analysis. The difficulty of securing material carry-

2 GEORGE V., A. 1912

ing an average proportion of the calcitic lenses was again felt here, as it was in selecting material for the analysis of the Siyeh rock of the Lewis series. It was found that in spite of all care different powders intended to represent the average gave different analytical results. Two total analyses of such powders (A and B) were made by Professor Dittrich, who also determined the portions of A soluble and insoluble in acids. The mean of A and B, shown in the following table, is doubtless the average chemical composition of the rock more nearly than either A or B.

Analyses of type specimen, Siyeh formation.

	A.	B.	Mean of A and B.	Molec. prop. in mean.
SiO_2	36·64	36·97	36·80	·613
Al_2O_3	4·24	7·59	5·92	·058
Fe_2O_3	·99	1·82	1·40	·009
FeO	·57	1·12	·85	·012
MgO	4·38	8·38	6·38	·150
CaO	25·79	16·28	21·03	·376
Na_2O	·49	1·04	·76	·012
K_2O	·88	2·48	1·68	·018
H_2O at 110° C.	·22	·24	·23	....
H_2O above 100° C.	1·87	3·11	2·49	·139
C	....	·08	·08	....
CO_2	21·31	21·11	22·71	·516
	100·38	100·22	100·23	
Sp. gr.	2·748			

Portion of 'A' insoluble in hydrochloric acid	42·46%
Portion of 'A' soluble in hydrochloric acid:	
Fe_2O_3	1·33
Al_2O_3	·44
CaO	25·52
MgO	3·75

Practically all the lime is soluble in acid and may be directly assigned to the carbonate. It is probable that the soluble magnesia of the average rock occurs in nearly the same proportion as in A, so that of the 6·38 per cent of the mean credited to magnesia about 5·5 per cent should be regarded as assignable to magnesium carbonate. Arbitrarily assigning the soda to the albite molecule, the potash to the orthoclase molecule, and the iron oxides to magnetite, the 'norm' of the rock may be calculated in the same way as it was for the Siyeh rock of the Lewis series. The result gives col. 1 of the following table:—

	1.	2.
Calcium carbonate	37·6	35·3
Magnesium carbonate	11·6	18·6
Silica	26·0	28·0
Orthoclase molecule	10·0	7·2
Albite molecule	6·3	4·2
Magnetite	2·3	2·6
Remainder	6·2	3·9
	100·0	100·0

SESSIONAL PAPER No. 25a

These figures can not, of course, represent the actual composition of these or of the rock except in as regards the carbonates. The calculation has some value, however, in facilitating the chemical comparison of the dolomitic rock in the Galton series with that in the Lewis series. A reference to the table col. 2, showing the 'norm' of the latter rock, indicates how nearly equivalent the two rocks are in chemical composition. In both cases we are dealing with a silicious, strongly dolomitic limestone of peculiar history and structure.

In analysis B an appreciable amount of carbon was determined; here again the carbonaceous matter largely controls the dark tint of the fresh mud or tooth rock, which decolourizes before the blowpipe.

The specific gravity of three specimens of the nodar-tooth rock varies from 2·670 to 2·718; that of four specimens of the metargillites in the formation from 2·630 to 2·739. The average of all seven specimens is 2·700.

GATEWAY FORMATION.

A striking difference in the lithological character of the Lewis and Galton series is to be found in the nature of the beds conformably overlying the Purcell Lava in the respective ranges. We have seen that, in the Clarke and Lewis ranges, the Sheppard formation, occupying this position, is a homogeneous silicious dolomite and that it is overlain by the red beds of the Kintla. In the Galton range the beds intervening between the Purcell Lava and the red beds, equivalent to the Kintla, have a much greater total thickness that the Sheppard and a quite different composition. These strata are well exposed on the heights east of Gateway and overlooking Tobacco Plains; they may be grouped under the name Gateway formation. Its total thickness was found to be about 2,025 feet. It is . . . . members of unequal strength.

The lower member . . . . mediately upon the Purcell Lava contains beds which at . . . . dentity of origin with the Sheppard. This correlation is . . . . specially detailed columnar section of the member is here . . . . on the basis of field sections along good exposures north . . . .

Columnar section of . . . . ern formation (lower part).

i . . . . of 1,850-foot member.

5 feet.—	Massive, light gray dolomite, weathering buff and brown.
4 "	Massive, light gray quartzite.
6 "	Light gray magnesian and ferruginous limestone, weathering rusty brown.
10 "	Thin-bedded, light gray quartzite.
6 "	Highly silicious, gray metargillite.
4 "	Thin-bedded, gray dolomite weathering buff.
20 "	Thick-bedded, hard, light gray, often cross-bedded and ripple-marked quartzitic sandstone.
20 "	Thin-bedded, concretionary, light gray dolomite, weathering strong buff and brown.
50 "	Massive, dark gray, coarse, feldspathic sandstone, bearing locality lenses of grit and fine conglomerate one to two feet thick.

125 feet.

Base, conformable top of Purcell Lava

2 GEORGE V., A. 1912

Apart from the concretions found in certain layers, the limestones are in many respects similar to the staple phase of the Sheppard formation. The high specific gravity of some specimens, 2·826 to 2·871, shows that they are very high in magnesia or iron and probably approximate ideal, though somewhat ferruginous dolomite. In any case all the carbonate bands are rich in magnesia. .

The concretionary structure noted in the thickest dolomitic stratum is a constant feature but is not always typically developed. Though the concretionary masses strongly resemble type specimens of Cryptozoon, there seems to be no reason to regard them as of other than inorganic, metamorphic origin. They are spheroids or ellipsoids composed of dolomite in concentric layers separated by thin laminae of cherty silica. The diameter of the bodies varies from a few inches to a foot or more. (Plate 11, B.) Similar, though smaller concretions were found in the basal beds of the Sheppard formation in the Clarke range.

The upper member was estimated to be 1850 feet thick. It is a fairly homogeneous mass of thin-bedded, highly silicious metargillite, inter-stratified with subordinate, more or less sericitic metasandstone. On a fresh fracture both rock types are generally light gray or greenish gray, the metargillite naturally being of somewhat darker tint. The weathered surface may be gray, brownish gray, rarely red or reddish brown. The member is more ferruginous toward the top. Ripple-marks, rill-marks, sun-cracks, and casts of salt-crystals up to 2 cm. or more in diameter, are all exceedingly common throughout this member. The salt-crystal casts were not found in the lower member.

Under the microscope these rocks show great similarity to the chief phases of the MacDonald formation. In all the slides, though especially in the more quartzitic types, feldspar is seen to be present. Orthoclase, microperthite, and plagioclase (probably andesine) form a considerable percentage of the clastic grains. A few broken zircons and tourmaline crystals were observed. Sericite, chlorite, and secondary quartz have replaced the original, argillaceous matter. The specific gravity of seven type specimens varies from 2·643 to 2·701, with an average of 2·676. The average for the formation as a whole is about 2·680.

The stratigraphic position, chemical composition, and occasional concretionary structure of the lower, dolomitic member are features directly correlating that member with the Sheppard formation of the eastern ranges; the Sheppard thus thins rather rapidly to the westward. The thick upper member of the Gateway carrying abundant salt-crystal casts, is almost certainly of contemporaneous origin with the lower part of the Kintla and, like the Kintla, was doubtless deposited as a continental deposit in an arid climate.

PHILLIPS FORMATION.

The Gateway beds are specially ferruginous toward the top, where they gradually pass into a still more ferruginous mass of sediments. From its occurrence on two summits about two miles north of Phillips creek, this assembinge of strata may be called the Phillips formation. The exposures are not extensive and the formation crops out nowhere else in the Boundary belt. Two different traverses covered the formation; on both occasions, because of bad weather, the writer was not able to make a thorough examination of these beds. The essential facts of the lithology were obtained but it is not known whether these strata or those of the overlying Roosville formation are fossiliferous.

The Phillips consists, for the most part, of about 550 feet of dark, purplish or brownish red, fine-grained to compact metargillite and metasandstone in alternating thin beds. At the base three massive beds of gray quartzitic sandstone, respectively four, ten, and twenty feet thick, are intercalated. Sun-cracks and ripple-marks are again plentiful. No salt crystal casts were found, though they might, on more prolonged search, be found. Under the microscope, specimens of the red rocks proved to be always highly silicious. Small subangular to angular grains of quartz, orthoclase, microperthite, plagioclase, and cherty silica lie embedded in a variable base of sericitic mica and fine grains of magnetite and hematite. The mica is, as usual in the series, abundantly developed in the planes of bedding. According to the abundance of the once argillaceous material, the rock may be classed as a metargillite or metasandstone. The total thickness of the formation is about equally divided between these two rock-types.

The specific gravities of three specimens were found to be 2.652, 2.674, and 2.721. Their average, 2.683, is about the average for the formation as a whole.

The general composition, colour, and field relations of the Phillips are so similar to those of the upper part of the Kintla formation that one can hardly doubt that the two are in the main, stratigraphic equivalents. The chief lithological difference is that the Phillips appears to be slightly the more silicious and coarser grained of the two. It may be noted that neither in the Gateway nor in Phillips was any contemporaneous lava discovered.

ROOSVILLE FORMATION.

The Phillips formation is conformably overlain by the Roosville, the highest recognized member of the Galton series. The name is derived from the post office recently opened on Phillips creek. The Roosville outcrops at only one point within the area covered by the Commission map. It there forms the summit of a peak lying three miles east-northeast of Phillips creek cascade at the junction of the creek canyon with the great Kootenay trough. Erosion has removed the upper part of the formation, of which only about 800 feet of beds now remain. How much greater the total thickness may be is not known.

The formation as exposed at this one locality is essentially made up of thin-bedded, light green, light gray, and greenish gray silicious metargillite bearing thin, more quartzitic, interbeds. The colours of weathering are light gray or brownish gray. Sun-cracks and ripple-marks are common. In field

2 GEORGE V., A. 1912

habit and most lithological details the dominant phase of the Roosville is very similar to that dominant phase of the Gateway formation. It seems, however, that casts of salt-crystals are wanting in the younger formation. The metargillite is composed of angular quartz and feldspar grains (averaging only 0.02 mm. in diameter) in an abundant matrix of sericite, chlorite, iron ore, and possibly, in some beds, a little of the original argillaceous matter. The feldspars again include orthoclase, microperthite, and plagioclase. Bedding planes are well marked by glinting sericite in the form of innumerable minute foils and shreds.

Thus, at the top of the Galton series as at the bottom, static metamorphism has effectually changed the original clayey sediments into nearly or quite holocrystalline rocks, The mica foils developed in the Hefty or MacDonald strata are, at many horizons, larger than the micas characteristic of the Roosville, Phillips, or Gateway, and the top members of the series may have retained a greater quantity of original argillaceous matter. In these two respects the older formations have, through deeper burial, suffered a slightly more advanced metamorphism than the beds lying seven to ten thousand feet higher in the series. Nevertheless, the evidence is clear that the Roosville formation, like the Kintla of the Lewis series, has been buried beneath many thousands of feet of still younger strata, doubtless including the heavy Devonian and Carboniferous limestones; to that ancient burial the development of the metargillitic facies of the Roosville beds is due.

The specific gravity of a type specimen from the metargillite is 2.730. A somewhat weathered hand-specimen gave 2.675. The average for the formation is probably about 2.710.

The Roosville has yielded no fossils. The formation appears to be younger than any beds belonging to the Lewis series as above described. It may prove to be equivalent to an upper division of the Kintla which is not exposed in the Boundary belt, or may represent the westward extension of a distinct formation.

DEVONIAN FORMATION IN THE GALTON RANGE.

DESCRIPTION.

At the eastern edge of the drift-covered Tobacco Plains (115° 3′ W. Long.), a block of fossiliferous Devonian limestone has been faulted down into contact with the Gateway formation. On the west and south the limestone is covered by drift and alluvium. The main fault which limits the block on the east can be rather sharply located, the strikes of the limestone and Gateway metargillite being nearly at right angles to each other. This fault is marked on the map sheet, where it will be seen to run roughly parallel to other faults that are responsible for the local graben character of the Rocky Mountain Trench. The limestone is itself affected by numerous minor slips, so that it is impossible to be certain of the thickness. In general, the block is monoclinal, with an average north-easterly dip of about 45 degrees. The apparent thickness of all the strata is approximately 1,800 feet. Of this total 300 feet represents dolomitic quartzite, occurring at the base of the section.

The quartzite is white to cream-coloured on the fresh fracture, weathering yellowish or buff. Its beds are generally thick and massive. It bears no other fossils other than a few markings like annelide borings.

Conformably overlying the quartzite is the very massive limestone, which rarely shows bedding planes. This rock is usually fetid or bituminous on the fresh fracture. It weathers from the normal dark grey tint to a much lighter one. Cherty nodules up to three or four inches in diameter, are common in certain horizons.

FOSSILS.

Just above the contact with the underlying quartzite a collection of fossils, bearing the station number 1217 on its label, was made. These were determined by Dr. H. M. Ami, whose notes are here entered in full:—[*]

'*Station No. 1217.*—Boundary monument at eastern edge of the Tobacco Plains.

In a dark gray, impure crinoidal and at times semi-crystalline limestone.

Age: Upper Devonian.

Formation: Jefferson limestone.

Genera and species:

1. Crinoidal columns.
2. *Producte"a subaculeata.*
3. *Schizophoria striatula.*
4. *Athyris vittata.*
5. *Athyris vittata,* a narrower and more tumid form.
6. *Athyris vittata,* timbriate form.
7. *Athyris parvula,* Whiteaves, or allied species.
8. *Athyris aff. coloradoensis,* probably a new species.
9. *Athyris.*
10. *Trematospira* (?) sp. No species of this genus has as yet been obtained from these limestones in Montana.
11. *Puqnax pugnus,* a small diminutive form.
12. *Spirifer whitneyi,* compare Hall's *Spirifer whitneyi (Spirifer animascensis).*
13. *Spirifer disjunctus,* var. *animascensis.* A specimen with high area and fine plication on the costae, high and twisted beak. Resembles a form from S. W. Colorado.

[*] Both the writer and Dr. Ami are under special obligation to Dr. G. H. Girty and to Dr. E. M. Kindle of the United States Geological Survey for valuable aid in determining the material of these collections; also for excellent opportunities for Dr. Ami to compare the Canadian forms with specimens from various localities south of the International Boundary.

2 GEORGE V., A. 1912

14. *Spirifer utahensis.* Shows a plication in the ... Ventral valve with twisted beak. This is the only specimen found of *Spirifer utahensis* in Dr. Daly collections. This is eminently character-istic and abundant in the Jefferson limestone of the United States.

15. *Camarotoechia,* p.

16. *Pleurotomaria,* (probably a. t... or allied form), flat-valved.

17. *Gasteropod.*

18. Aviculoid (?) shell, too imperfect for identification.

19. *Orthoceras,* sp. Part of shell representing some twenty septh of a test rapidly increasing toward the aperture.

About 800 feet higher up in the apparent section are other fossils, taken at station No. 1218, are named by Dr. Ami as follows:—

'*Station No. 1218.*—150 yards north of Boundary line, eastern edge of the Tobacco Plains.

A dark gray coralline limestone; very similar to the characteristic rock of the Jefferson limestone of Montana. The identical association of forms and the general physical properties of the limestones of British Columbia and Montana are remarkable and leave no doubt as to the identity of the horizons.

Age: Upper Devonian.

Formation: Jefferson limestone.

Genera and species:

1. *Stromatoporoes.* Exhibit concentric lamina ... A. variolare, large and small masses.

2. *Favosites,* sp. A form very close to, if not identical with, *F. limitaris,* Rominger.

3. *Favosites,* sp. A form consisting of much larger fronds and smaller corallites than those of last species. (New species?)

4. *Favosites,* sp. Cf. *F. limitaris,* Rominger.

5. *Brachiopod,* ribbed ... too imperfect for identification.

6. *Atharis,* sp. Cf. *A. ...* a very obscure example of what appears to be this species.

7. *Atharis.* Small species resembling *A. parvula* W.

8. *Spirifer cha mans.* The same form occurs also in Montana

9. *Spirifer.* Cf. *S. argentana* ... The radiating lines which are pro-minent on ... fold constitute a rather distinct feature in this species.'

The quartzite is tentatively assigned to an Devonian ... Mississippian horizons are possibly represented in the fault-block, for the search for fossils has been by no means exhaustive. The greater part of the limestone is to be correlated with the Jefferson limestone of Montana.

SESSIONAL **PAPER** No. 25a

PALEOZOIC LIMESTONES OF THE MACDONALD RANGE.

DESCRIPTION.

With reference to the fault troughs respectively occupied by the Kootenay river (at Gateway) and the Flathead river, the Galton-MacDonald mountain system is a compound horst. We have seen that the Devonian lime-tone at Tobacco Plains now stands at a common level with strata as old as the base of the Siyeh. Much greater displacements at the western side of the Flathead trough have dropped Devonian and Mississippian limestones down into contact with the oldest members of the Galton series, including the Altyn formation. The result of this faulting is peculiar, since a long, slab-like block of Altyn, Hefty, and MacDonald beds is bounded on both sides by Mississippian limestone. The younger, fossiliferous limestones form two masses separated by the slab and may be referred to as the Western and Eastern blocks.

The western block is well exposed only at comparatively few points; elsewhere it is covered by heavy forest. The bounding faults are, therefore, mapped only approximately. This limestone is dark bluish-gray, weathering light gray to whitish. It is massive, rarely showing stratification planes; fetid under the hammer; semi-crystalline, with the larger calcite crystals blackened by films of bituminous matter. In one shear-zone the normal colour is changed to yellowish gray or brown. At other outcrops the fresh limestone is crystalline and white; the bituminous matter has there been distilled out.

The rarity of visible bedding-planes makes it impossible to make certain as to the attitudes assumed by the limestone throughout the block. The best exposures along the Commission trail, where it threads the canyon at the Boundary line, show a horizontal position, but farther to the northwest probable dips of about 30° to the southwest were observed. It is likely that the western block is compound and bears numerous local faults and shear-zones.

A few fossils were found at a cascade just south of the Commission trail at 114° 38' W. Long., and 400 yards south of the Boundary slash (Station No. 1278). Dr. Ami identified these as including a species of Menophyllum and an Athyroid form. The rock elsewhere bears crinoid stems. The horizon could not be determined but it is 'presumably upper Mississippian.'

The eastern block is composed of both Devonian and Mississippian limestones which are greatly broken by step-faults. In the field no lithological distinction could be made between the two limestones. Wherever fossils occurred the rock was massive, crinoidal, bituminous, and gray, corresponding in all respects to staple phases of the western block and of the limestone at Tobacco Plains. The proved Devonian beds, however, are specially rich in cherty nodules and are often mottled with irregular magnesian and dolomitic parts.

At the edge of the Flathead valley drift cover, 5,000-foot contour, and 1,000 yards north of the Boundary line, some 500 feet of unfossiliferous, pinkish-gray, sandy beds were noted. These are generally magnesian and include thin lenses of grit containing small, black pebbles of argillite. Cross-bedding was

25a—vol. ii—8

2 GEORGE V., A. 1912

seen in the more quartzose layers. As a rule these beds, like the main lime-stone, were fetid under the hammer. A strong cleavage affects them, with strike N. 12° W. and dip, 75—80° E., it suggests local faulting parallel to the trend of the Flathead valley.

On the slope immediately above the reddish zone the normal gray, massive limestone begins and continues westward to the great fault where the limestone and the MacDonald metargillite make contact. In that traverse the dip gradually steepens to a maximum of about 40° or more, to the south-westward. At the 6,500-foot contour and one-half mile north of the Boundary line, the dip abruptly changes to 55° S.W., with strike N. 55° W. The change of dip takes place at a meridional belt of intense shearing, where, for fifty feet across the belt, the limestone is a white, brecciated marble. East of this shear-belt the fossils collected are Mississippian in age; west of it the fossils are Devonian. The shear-belt seems, thus, to mark the outcrop of a strong fault along which the Mississippian limestone has been dropped down, relatively to the Devonian on the west.

That traverse is probably typical of a number which might be made across the eastern block. The relations are those of step-faulted blocks with down-throw to the east. Further remarks made on page 117 as to the local structures should be added to this brief account. Since the distribution and throws of the various faults are unknown, it is not possible to state the true thicknesses of the fossiliferous limestones. Either the Devonian or the Mississippian lime-stone is certainly many hundreds of feet in thickness; their combined thickness must be well over 1,000 feet.

Neither top nor bottom of the series has been discovered in the Boundary belt. In the Yakinikak valley, about five miles south of the Boundary line, in this same mountain range, Willis found a small mass of limestone carrying numerous fossils of the Saint Louis horizon of the Mississippian. He writes that the limestone—

'Is without upper stratigraphic limit, but rests conformably on a quartzite, which is unconformable on Algonkian strata. The quartzite is about 25 feet thick; and it and the limestone lie in a nearly horizontal position. The name Yakinikak is here applied to the limestone, exclusive of the quartzite, which may elsewhere develop independent importance. . . Its [the limestone's] occurrence on Yakinikak creek is apparently due to down-faulting, as it lies at a comparatively low level among mountains composed of the Algonkian argillites. Its presence in this locality, taken in connection with other occurrences north and south, may be considered evidence of the former extension of the upper Mississippian limestone over the entire region. The absence of earlier Mississippian strata is significant of an unusual overlap.'*

If this Yakinikak limestone were deposited unconformably upon the Galton ('Algonkian') series, there must have been strong deformation and extensive

* B. Willis, Bull. Geol. Soc. America, Vol. 13, 1902, p. 325.

SESSIONAL PAPER No. 25a

ceal erosion beforehand, for the Devonian limestone is represented in great thickness only three or four miles to the north. Many observations indicate with some certainty that such orogenic movements in this part of the Cordilleran region have not intervened between the deposition of the Jefferson limestone and the upper Mississippian limestone. Some other than Willis' interpretation of the Yakinikak contact seems legitimate, if not necessary. The writer knows of no facts which involve any notable erosion unconformity between the Devonian or Carboniferous sediments and the Cambrian-Beltian series of southeastern British Columbia or of Montana northwest of the Belt mountains and the Helena district.

FOSSILS.

The Devonian fossils were all found along the western edge of the eastern block. The exact localities and the faunal lists prepared by Dr. H. M. Ami, are here given.

Station No. 1276.—At 114° 38′ W. Long.; seven miles west of the Flathead river, and two miles and a half north of Boundary (6,400-foot contour); close to great fault mapped.

In a dark gray, impure, dislocated limestone, weathering peppery-gray, yellowish-gray, or buff; fractured and recemented, more or less altered by pressure. Surface marked by pitted structure in uniformly shallow rounded depressions, or cavities, in which a layer of silicareous (?) matter appears, as a thin lining on the inner wall.

Age: Devonian.

Formation: Jefferson limestone in the upper part of the Devonian system.

Genera and species:

1. *Chonetes* (?) sp. An imperfectly preserved specimen not recognizable. Crushed valve showing punctate structure.
2. *Atrypa aspera.* One specimen and two small fragments of this species characterize these limestones.
3. *Spirifer englemani.* Exfoliated specimen in a block of limestone. Another individual, partially exfoliated, represents one of the numerous types of *Spirifer englemani* with a high hinge area, and resembles very closely the *Spirifer englemani* from the Jefferson limestone of Utah, Montana, and Nevada as represented in the collections of the U. S. Geological Survey obtained by Dr. E. M. Kindle.

Station No. 1277.—Five hundred yards southeast of station No. 1276, and close to great fault.

The fossils occur in a dark, fractured and recemented limestone, weathering peppery-gray; calcite veins prevalent.

Age: Upper Devonian.

Formation: Jefferson limestone.

25a—vol. ii—8½

2 GEORGE V., A. 1912

Genera and species:

    1. Sponge-like organism. Long, slender cylindrical stem-like rods or spicules (?) *Hyalostelia* (?).

    2. *Favosites* sp. Compare *Favosites limitaris*. Rominger.

    3. *Cladopora*, one specimen.

    4. *Atrypa aspera*, Schlothurm.

Station No. 1290.—Near meridian of 114° 33' W. Long.; four miles west of the Flathead river, and one-half mile north of the Boundary line; 6,500-foot contour.

A limestone weathering rusty or brownish-yellow; specimens of fossils silicified.

Age: Upper Devonian.

Formation: Jefferson limestone.

Genera and species:

    1. *Zaphrentis* (?) sp.

    2. *Atrypa reticularis*, Linnaeus.

    3. *Atrypa aspera*, Schlothurm.

The Mississippian fossils of the eastern block are listed, with locality indications, by Dr. Ami, as follows:—

Station No. 1285.—At 114° : " 30" W. Long.; two and a half miles west of Flathead river, and one-half mile north of the Boundary line; 5,500-foot contour.

Age: Upper Mississippian.

Formation: Madison limestone.

Genera and species:

    1. *Lithostrotion*, sp.

    2. *Syringopora*, sp.

    3. *Zaphrentis*, sp.

    4. *Menophyllum*, sp.

    5. *Stenopora*.

    6. Monticuloporoid.

    7. *Composita*, sp. aff. *Composita trinuclea*.

    8. (?) *Reticularia*, sp.

    9. *Producta cora*.

    10. *Productus*, sp.; compare *P. giganteus*. Exhibit sculpture similar to that in English specimens.

    11. *Camarotoechia*, sp.

    12. *Spirifer*, sp.; aff. *S. Keokuk*.

    13. *Spirifer leidyi*, or a very closely allied species, agreeing with Norwood and Pratten's description in nearly every detail. Middle rib and sinus does not always extend to beak. Sometimes three ribs of sinus equal; at other times one larger and two smaller. etc

SESSIONAL PAPER No. 25.

14. *Clathurdina hirsuta.* Specimens large and lamellose. Ragged edge of lamellae like spines in previously described specimens. See also *Athyris hirsuta*, Hall; figs. 18-21 (Spergen Hill), and Pl. 6, Vol. I, Bull. No. 3, Amer. Museum of Natural History.

Station No. 128: Three hundred yards west of station No. 128A; 5,800-foot contour.

Age: Probably upper Mississippian.

Genera and species:

1. *Lithostrotion,* sp.
2. *Syringopora,* sp.
3. *Menophyllum,* sp.

STRUCTURE OF THE GALTON-MACDONALD MOUNTAIN SYSTEM.

The geosynclinal rocks between the Flathead and the Rocky Mountain Trench at the Forty-ninth Parallel are very much more deformed than are those of the Clarke range or the Lewis range. The exceedingly inflexible nature of the rocks has prevented the development of systematic folds; the structure all across the Galton-MacDonald system is almost entirely determined by faulting. At least twelve major fault-blocks are represented in the map sheets as occurring in the five-mile belt where it crosses the two ranges. Within the belt the dips range from 0° to 90°, averaging about 30°.

The most easterly and the most westerly blocks contain, respectively, Carboniferous and Devonian limestones which, excepting the Kishenehn lake beds, are the youngest bed-rock formations in the Rocky Mountain system at this latitude. These particular blocks are of special interest since they clearly show the magnitude of the displacements to which the Flathead trough and the Rocky Mountain Trench owe their origin. The Carboniferous limestone on the west side of the Flathead is on the same level with strata on the east side, belonging to the lower Appekunny. One may fairly estimate that a net displacement of at least 15,000 feet or possibly 20,000 feet is here indicated. The western part of the Clarke range has been lifted nearly or quite three miles higher than the most easterly block of the MacDonald range. The latter block is downthrown by an even greater amount with respect to the block next on the west. The Carboniferous limestone at the Flathead valley is, in fact, the visible upper portion of a broad block or series of parallel blocks which have been dropped a minimum of about three miles below the adjacent blocks of the Clarke and MacDonald ranges. The Flathead trough is thus structurally a typical fault-trough or 'graben.' It is also highly probable that the depression has always been a graben in a topographic sense. It has been partially filled with lake beds and has been deformed by the folding of these beds but there is no evidence that the initial trough form was ever quite destroyed.

It will be observed from the map sheet that the Carboniferous limestone of the MacDonald range occurs in two different fault-blocks separated by a

2 GEORGE V., A. 1912

narrow, slab-like block of strata ranging in age from the Altyn to the Mac-
Donald inclusive. It is not easy to understand the conditions under which
this narrow slab, composed of the oldest sediments in the range, can
make contact on each side with the youngest formation of the range. Two
hypotheses are conceivable. According to the first the relations were established
by simple normal faulting whereby the Carboniferous blocks were dropped
down. According to the second hypothesis one may postulate a local overthrust
of the older rocks upon the Carboniferous, followed by normal faulting which
dropped the Altyn-MacDonald block down into the Carboniferous limestones.
This second view would naturally correlate with the speculation that the whole
Clarke range has been thrust over Carboniferous or Cretaceous formations.
Extreme as this idea may be, the known facts do not exclude it and the two
hypothetical alternatives are still open. It may be noted that the outcrops of
the fault planes on the east side of the Altyn-MacDonald slab have been drawn
with considerable confidence. The fault line on the western side was not so
readily plotted in the field, but it is believed to be mapped with approximate
accuracy.

In four or more leading cases the fault has have been shown as following
stream courses among these mountains. The local valley of Wigwam river has
been determined in position by a break which has strongly affected the dips of
the MacDonald formation on either side of the river.

The five most westerly blocks of the Galton range show a progressive down-
dropping of blocks from east to west, with the result that the MacDonald,
Siyeh, Purcell, Galloway, and Devonian-limestone formations are successively
in lateral contact. The equivalence of level between the Devonian limestone
at Tobacco Plains and the lower MacDonald beds along the Wigwam shows
that the net relative displacement of the two blocks has been at least 10,000
feet and may have been several thousand feet greater. We are therefore prepared to find that the Rocky Mountain Trench at the Forty-ninth Parallel has
been located on a zone of strong faulting. This conclusion will be noted again
in the next chapter, on the Purcell mountain system.

The relation of the Devonian and carboniferous formations to the older
geosynclinal prism has been discussed in connection with the stratigraphy of
the younger limestones. The Boundary belt has furnished very little information on this subject.

CHAPTER VI.

STRATIGRAPHY AND STRUCTURE OF THE PURCELL MOUNTAIN SYSTEM.

PURCELL SERIES.

As one leaves the Rocky Mountain system and crosses the wide master trench to study the composition of the Purcell system along the Forty-ninth Parallel, he enters a much more difficult field. Between Gateway and Porthill the mountains seldom rise above tree-line and the forest cap is, throughout this stretch, of unusual density and continuity. Notwithstanding the steepness of the mountain slopes the timber generally stands thick upon them. Beneath the trees a heavy growth of brush and generally, a discouragingly thick layer of moss and humus, form an impenetrable cover over most of the bed-rock on the Boundary belt. During many traverses made during the season of 1904 outcrops absolutely failed for a mile, or even for several miles, at a time. Field work was further rendered unsatisfactory during that extraordinarily dry season on account of the thick smoke which hung over the mountains. For one period of seven weeks the smoke was dense enough to interfere seriously with the work of discovering outcrops.

In the Purcells the stratigraphic conclusions were rendered all the more delicate because of the remarkable uniformity of the sedimentary formations. It was found that much the greater part of the belt is underlain by the stratigraphic equivalent of the Galton and Lewis series. This equivalent has been named the Purcell series. Very seldom is there represented among its members anything like the lively contrasts existing, for example, between the Kintla and Sheppard formations, between the Siyeh and Grinnell, between the Appekunny and Altyn, or between the respective pairs of formations in the Galton series. For thousands of feet together the strata of the Purcell series exhibit a homogeneity that is bound to excite wonder in the mind of the geologist. In the Moyie and Yahk ranges not a single stratum of marked individuality has been discovered which is proved to persist throughout the ranges. In none of the three ranges has any formation yielded fossils. This failure of well defined horizon-markers in a region of considerable structural complexity is, perhaps, the greatest of the difficulties that confront the geologist in the Purcells.

For these reasons the writer has not felt justified in attempting to describe the Purcell series in the detail which is warranted in the case of the formations composing the Galton or Lewis series. It has seemed safer to express the stratigraphy of the Purcell mountains in terms of three very thick,

MICROCOPY RESOLUTION TEST CHART

(ANSI and ISO TEST CHART No. 2)

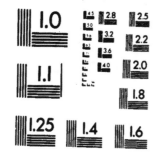

APPLIED IMAGE Inc

1653 East Main Street
Rochester, New York 14609 USA
(716) 482 - 0300 — Phone
(716) 288 - 5989 - Fax

2 GEORGE V., A. 1912

conformable sedimentary formations, each of which, on account of its homogeneity, as yet defies profitable systematic analysis into subdivisions of more usual thickness. Even these grand divisions of the thick series, called the Creston, Kitchener, and Moyie formations, are not always with ease separable from one another in the field. All are highly silicious in character; all are fine-grained to compact in texture; all show phases which are indistinguishable in the hand specimen or in the ledge. The three formations are, in fact, separated on the ground of comparatively subordinate lithological differences, such as colours of fresh fracture and weathered surface.

Microscopically and chemically the immensely thick Creston and Kitchener formations are proved to be almost identical in constitution. A prevailing and clearly minor difference between them, consisting in the fact that the Kitchener is the more ferruginous of the two formations, has been used as a principal means of distinguishing these two parts of the series in the field. In addition, the Kitchener is thinner-bedded than the Creston. With such criteria merely, it is clear that the mapping of these formations in the fault-riven mountain masses is a delicate matter. The geological boundaries as shown on the map sheets are thus to be considered as drawn, in many instances, with more doubt than is the case with the sheets located east of Gateway.

Two of the sedimentary formations have been named after stations on the Canadian Pacific railway; the third, the Moyie formation is só called after the river of that name. Their estimated thickness and general composition are noted in the following table:

Formation.	Thickness in feet.	Dominant rocks.
	Top, erosion surface.	
Moyie..	3,400+	Metargillite.
Purcell Lava..	465	Altered basalt.
Kitchener..	7,400	Quartzite.
Creston..	9,500+	Quartzite.
	20,765+	
	Base concealed.	

CRESTON FORMATION.

General description.—The lowest member of the Purcell series and the oldest formation seen in the Boundary belt within the entire Purcell mountain system has been named the Creston formation. Its best exposures include the one in and east of the lofty McKim cliff four miles from Porthill; a less complete one on the slope immediately east of the Moyie river; and, finally, the most favourable one of all, on the two sides of the wide valley occupied by the east fork of the Yahk river. In each of these exposures the formation preserves nearly constant characters to the lowest bed visible; it is thus highly probable that this gigantic sedimentary formation is, as a whole, yet thicker than the total mass actually measured in the field.

In different sections among the fault-blocks characteristic of the Purcell mountain system, estimates of from 6,000 to 9,900 feet were obtained for the

SESSIONAL PAPER No. 25a

whole thickness locally observed. The highest figure refers to the remarkably extensive outcrop of the Creston rocks at the Yahk river. High as the estimate appears, a minimum thickness of 9,500 feet is assigned to the formation. The estimate is the result of two complete traverses run across the great monocline at this locality.

It cannot be denied that there may be some duplication in this particular section, but, on the other hand, the writer, after careful study in the field, found not the least hint of duplication. Similarly, in each of a half-dozen other sections in as many different fault-blocks, as much as 5,000 to 7,000 feet of the upturned Creston quartzite were measured without any clue to repetition of the beds. In several fault-blocks the strata stand nearly vertical and errors of mensuration were reduced to a minimum. At McKim cliff, about 3,000 feet, of nearly horizontal, typical Creston are exposed to one sweep of the eye, with neither the summit or base of the formation to be found at that locality.

A further indication that the Purcell series, of which the Creston makes up nearly one-half, is enormously thick, is derivable from McEvoy's reconnaissance map of the East Kootenay District.* The map shows that at least 3,000 square miles of the Purcell mountain system north of the Boundary is almost continuously underlain by a silicious series evidently equivalent to that cropping out at the Forty-ninth Parallel. The continuity of the colour representing the series on the map is broken only by patches of gabbroid intrusions doubtless similar to the intrusions so plentifully found in the Boundary belt. When it is remembered that the rocks of the large area in East Kootenay are much faulted and otherwise disturbed so as to present all angles of dip even to verticality, we see certain proof that these conformable strata must have very great total thickness. This conclusion may be corroborated by information won from even the fleeting glance one can give to the rocks that are visible from the railway train on the stretch from Cranbrook to Kootenay Landing. In minor degree the estimated thickness of the formation may vary according to the somewhat arbitrary position assigned at each exposure to the upper limit of the Creston. In every case the formation gradually becomes more ferruginous and thus passes slowly into the overlying Kitchener. The doubtful intermediate band of strata often totals several hundred feet in thickness. The top of the Creston has been generally fixed within the band where the thinner bedding as well as the rusty character of the Kitchener becomes pronounced in the quartzitic strata.

In conclusion, then, the writer believes it to be best to trust the minimum estimate of 9,500 feet for the Creston as embodying the net balance of probabilities derived from the field study. It may be added that, in the opinion of the writer, this vast thickness for a single formation is not to be explained as only the apparent thickness of beds deposited in fore-set bedding as a submarine delta. The recent emphasis of geologists on this source of error in measuring the actual thickness of a clastic formation is certainly justified.

* Accompanying Part A. Ann. Rep. Geol. Surv., Canada, Vol. 12, 1899.

2 GEORGE V., A. 1912

In the present case, however, the criteria of inclined fore-set bedding, in contrast to practically horizontal bedding on a subsiding, flat sea-floor, do not seem to be matched by the facts. The prevalence of sun-cracks, ripple-marks and other shallow-water markings in the perfectly conformable Kitchener and Moyie formations, as well as in the Creston formation in less degree, appears to show that the sea bottom and the bedding planes of the sands and muds were nearly level throughout the deposition of the Purcell series.

The Creston formation is no more extraordinary for immense thickness than it is for its wonderful homogeneity in any one section. There is a signal absence of well-marked lithological horizon-markers. The nearest parallel to this homogeneity among the Boundary formations is that afforded by the basal arkose member of the Cretaceous section at the Pasayten river. The lack of strong horizon-markers is not to be explained by the lack of sufficient outcrops; the frequent recurrence of the Creston rocks among the fault-blocks, coupled with the excellence of exposure, for portions of the formation in each large outcrop, render it improbable that important bands of rock other than the staple quartzite have been overlooked in the Boundary belt. The forest cap interferes much more with determinations of total thickness and of the larger structural features such as faults and folds, than with the study of the details of composition. Neither in slide rock nor in gravels of the canyon-streams of the areas mapped, as underlain by the Creston, was any other rock discovered in large amount than those which are the dominant components of the Creston quartzite as hereafter described.

While relative homogeneity characterizes the formation from top to bottom **at any one exposure, noteworthy changes in its constitution were observed as the** Boundary belt was traversed from west to east. The Creston as outcropping in the Moyie range and western half of the Yahk range thus stands in a certain lithological contrast to the same formation where it crops out farther east. For the understanding of this important fact it is convenient to recognize two different phases of the formation in the Purcell mountain system—a western and an eastern phase.

Western Phase.—At McKim cliff and in the outcrops immediately east of the crest the material was largely gathered for the following description of the Creston formation in a typical section representing the western phase.

In the cliff itself the staple rock is a very hard and tough quartzite, breaking with a sonorous, almost metallic ring. The individual beds vary from a few inches to twenty-five feet or more in thickness, averaging perhaps three feet. Very often the more massive plates are seamed with thin dark-gray laminae of once-argillaceous quartzite or metargillite, but true shale or slate was never seen in this part of the section. For 2,500 feet measured vertically up the cliff the quartzite, which dips 3°-10° eastward, is specially massive, giving the effect of superb cyclopean masonry, broken horizontally by widely spaced bedding-planes and broken vertically only by master joints. Toward the top of the cliff the rock is somewhat thinner-bedded, but is still a strong, typical quartzite.

The dominant colour on fresh fractures is throughout gray or greenish-gray, weathering to a somewhat lighter tint of nearly pure gray. A few white or grayish-white beds occur irregularly through the formation and, also rarely greenish-gray beds weathered light rusty-brown or reddish-brown, so as to resemble typical Kitchener quartzite.

Heavier beds characterize the formation where its upper part crops out just east of the cliff. That seems to be the rule for the quartzite generally as it is exposed in the Purcell range; the bedding is thick and massive in the top and bottom divisions and thinner-bedded in the middle division of the strata. The exposures, however, are nowhere continuous enough to allow of a trustworthy estimate of the relative strength of these three divisions.

The sediment sometimes, though quite rarely, shows cross-bedding. Sun-cracks, rill-marks, ripple-marks, and annelide burrows were not identified in a single case among the strata exposed on McKim cliff. Elsewhere within the Boundary belt these markings were found; rarely in the dominant quartzite, but more particularly in the metargillitic horizons.

Already in the hand-specimens numerous glints of light from non-micaceous particles suggest that the rock is highly feldspathic. At the same time it is seen that the general greenish tint of the quartzite is due to disseminated minute plates and shreddy foils of mica. These observations are confirmed by microscopic examination. Interlocking quartz, feldspar, and mica are seen to be the essential constituents. Each of these minerals is glass-clear in the fresh specimens. Orthoclase, microcline, microperthite, oligoclase, and probably albite make up the list of feldspars. Of these orthoclase and microperthite are the most abundant, though it is not certain that, in any specimen, the other feldspars of the list are absent. The mica includes both highly pleochroic biotite and muscovite, the latter being either well developed in plates or in the typical shreds of sericite. In some specimens the biotite is the more abundant of the two micas but in others it becomes subordinate to muscovite and may disappear altogether.

Other constituents are very subordinate; they include rare anhedra of titanite, titaniferous magnetite, pyrite, epidote and zoisite.

The quartz and feldspar grains vary from 0·02 mm. to 0·2 mm. in diameter, averaging perhaps 0·06 or 0·08 mm. The lengths of the mica scales are usually much greater. Though few direct traces of clastic form are left among the minerals, it is probable that these dimensions represent approximately the size of the original grains. The texture of the quartzite is thus quite fine in the type specimens as, indeed, throughout all the exposures; in all the thousands of feet of thickness no conglomeratic, gritty, or even very coarse sandy bed was seen.

It is an open question, perhaps, whether this rock should be called a quartzite if by that term we mean an indurated sandstone. The average quartz grain is much too small to have formed originally a true sand. In fact the average grain of the rock is not more than one one-thousandth as large as the average grain of typical beach sand. The name 'quartzite', adopted

2 GEORGE V., A. 1912

by McEvoy, Dawson, and others for these rocks, has been retained because of the chemical composition and tough, massive field habit of the beds selected as the types of the formation. To the writer a distinct genetic problem remains. One cannot easily understand the conditions under which such an immense accumulation of fine quartz and feldspar particles has been made. The purely argillaceous material must have been quite subordinate through thousands of feet of the Creston formation. The question arises as to the mechanism by which residual clay has been thus separated from the more silicious matter. Such separation is very rare, if not unknown, in the muds now accumulating on the ocean-floor. The writer has failed to find in the 'Challenger' report on the deep-sea deposits an account of any mud which chemically or mineralogically matches the Creston type of deposit. The 'Blue Muds' of the report furnish the nearest parallels and yet show vital contrasts. This problem of genesis applies also to the rock forming the type of the Kitchener quartzite*

The micas and the accessories are chiefly the result of the crystallization of a small original admixture of micaceous, argillaceous, and ferruginous material in the sandy sediment. It is probable that most of the quartz and of the feldspars represent clastic material cemented together by secondary growths of the original crystal fragments. One of the plagioclases, referred with some doubt to albite, may be of metamorphic origin. The metamorphism which led to the crystallization or recrystallization was, almost without doubt, not dynamic but static in nature. As in the case of the metargillites of the Lewis and Galton series, these effects have resulted from deep burial with consequent increase of temperature and pressure.

Professor Dittrich's analysis of a typical specimen (No. 1125) of the homogeneous quartzite from McKim cliff gave the following result—

---

* While this chapter was going through the press the writer had opportunity to study the Shuswap terrane, from which the clastic materials of the Creston, Kitchener, and other formations composing the Rocky Mountain Geosynclinal were derived. Great thicknesses of phyllites, chlorite schists, green schists, greenstones, and fine-grained mica schists were found in this pre-Beltian terrane as exposed at the Shuswap lakes. In general these rocks are abundantly charged with secondary quartz developed in minute individual crystals (anhedra). During the secular weathering of such rocks the more soluble micas, chlorite, talc, uralite, etc., would be leached out and the more resistant quartz and alkaline feldspar would be washed out to sea. The writer is inclined to credit this explanation of the silicious muds which have been consolidated to form the thick, very dense quartzites of the Cambrian and Beltian formations. The rich content of microcline and microperthite repeatedly emphasized in the descriptions of the latter can be explained as due to the weathering and washing of the millions of aplite and pegmatite dikes and sills cutting the Shuswap sediments and green schists. These injections are associated with large batholiths of likewise pre-Beltian granite; its débris is also represented in the Rocky Mountain Geosynclinal.

SESSIONAL PAPER No. 25a

Analysis of Creston quartzite, Western Phase.

		Mol.
SiO₂	82·10	1·368
TiO₂	·40	·005
Al₂O₃	8·86	·087
Fe₂O₃	·49	·003
FeO	1·38	·019
MnO	·03	....
MgO	·56	·014
CaO	·82	·014
Na₂O	2·51	·040
K₂O	2·41	·026
H₂O at 110°C	·05	....
H₂O above 110°C	·37	·020
P₂O₅	·04	....
	100·02	
Sp. gr	2·681	

Assigning all of the soda to the albite molecule, one half of the lime to the anorthite, the other half of it to titanite, apatite, epidote, and zoisite, the weight percentages of the constituent minerals have been roughly calculated as follows:—

Quartz	54
Albite molecule	21
Orthoclase molecule	9
Anorthite molecule	4
Micas, ca	7
Accessories	5
	100

It is quite possible that a considerable fraction of the soda should be assigned to the sericite and even that paragonite itself is present. However, from microscopic evidence it is probable that the soda-feldspar molecule is the principal source of this alkali. The mineral percentages are, therefore, believed to be nearly enough accurate to give a fair idea of the composition of the average quartzite. It is clearly a quite highly feldspathic sediment.

The only notable variation from this average composition of the typical quartzite is found in the thin, darker, more micaceous and ferruginous laminæ which often interrupt the dominant light gray quartzite. These laminæ, varying from a centimetre or less to several centimetres in thickness, often have the habit of metargillite, but usually they are so acid as to rank among the impure quartzites.

The specific gravity of the analyzed specimen, 2·681, is near the average for the staple, light gray rock. The average for ten specimens typical of the whole-western phase is 2·698.

The monotony of the western phase is seldom broken by the appearance of any lithological novelties.

At a few horizons the micaceous material of the rock is segregated into flattened spheroidal or more irregularly shaped, concretionary masses of all sizes up to a foot or more in width. The greatest diameter of the concretion.

2 GEORGE V., A. 1912

almost invariably lie in the plane of bedding. The heart of each segregation is especially rich in biotite and sericitic muscovite. Their main mass is composed of a grayish white, granular base of interlocking grains of quartz and subordinate feld-spar, in which are embedded abundant, conspicuous foils of black biotite 5 mm. or more in diameter, and many highly poikilitic red garnets, with large anhedra of titanite. The material of the dark-coloured minerals has plainly migrated inward from the surrounding rock-mass, for each segregation is enclosed in a white, decolourized shell of quartzite, consisting of nearly pure quartz and feld-spar. In each of the larger segregations the mica and garnet are not regularly distributed with reference to the periphery but occur in numerous small clumpy aggregates within the main body of the segregation. In some of the smaller segregations the micas are more evenly distributed, in a manner similar to that observed in concretions in the Kitchener quartzite.

At a few other horizons the feldspathic quartzite is spangled with large biotite foils up to 1 cm. in diameter. These cut across the bedding plane at all angles. The cause of their growth and of their restriction to a very limited number of strata in the great, apparently homogeneous series is not understood. Neither special dynamic metamorphism nor the thermal metamorphism of igneous intrusives were feasible explanations for the spangled quartzite at the localities where it was actually discovered. As a rule the Creston quartzite is not cleaved, but in the fault-block at the Moyie river there is a distinct cleavage crossing the bedding planes at relatively low angles. In this case sericite is developed in the secondary planes as well as along the bedding.

Eastern Phase.—The western phase just described characterizes the formation as it crops out in the Boundary belt between Porthill and the Moyie river. Eastward of the river the Creston gradually assumes the features which are normal to the eastern phase. The latter is typically developed at the Yahk river, where the formation finally disappears beneath younger rocks.

The most important lithological contrasts with the western phase consist in:—first, a decided decrease in the average thickness of the beds, often leading to a fine lamination at many horizons; secondly, a pronounced increase in the amount of argillaceous matter which here forms many distinct beds and also occurs as a notable impurity in the still dominant quartzite; and thirdly, the appearance of calcium and magnesium carbonates as subordinate elements in both the quartzite and the more argillaceous strata. The increase of the carbonate manifests itself in the rock-ledges, which, on account of the special solubility of the carbonates, present, to sight and touch, a characteristic roughness on weathered surfaces. In general, the calcium carbonate seems to be in some excess over the magnesian carbonate, as shown by a certain amount of effervescence with cold dilute acid.

In order to obtain a definite idea as to the composition of the eastern phase, type specimens were collected at the Yahk river section and have been studied microscopically. One of these specimens, taken from a large outcrop

SESSIONAL-PAPER No. 25a

on the Commission trail, about one thousand yards west of the main fork of the river, has been chemically analyzed. Its description will serve to show the general nature of the typical calcareous part of the formation.

On the fresh fracture the rock is light gray, compact, and thin-bedded, though platy because of the cementation of many laminae of varying composition. The weathered surface is generally of a still paler gray colour, but for a depth of one or two millimetres below the surface there is usually a shell of altered rock of a brown or buff colour. The decolourization at the surface is doubtless an effect of leaching by vegetable acids.

Under the microscope the rock is seen to be composed of carbonates, quartz, feld-spar, sericitic mica, a little green biotite, and small grains of limonitized iron ore. These constituents are named in the order of decreasing abundance. The carbonate grains vary from 0·01 mm. or less to 0·03 mm. in diameter and average about 0·02 mm. They never appear to have rhombohedral development. The quartz and feldspar grains which are, doubtless, in largest part of clastic origin, vary from 0·02 mm. to 0·1 mm. or more in diameter, averaging about 0·06 mm. The dominant mica, sericite, is not distributed uniformly but is most abundant in rather sharply defined laminae of specially fine grain. Such laminae were evidently more purely argillaceous than the remainder of the rock. No true argillaceous material can be discerned in thin section; the sediment has been very largely recrystallized and its insoluble base is a metargillite.

The percentages in Professor Dittrich's chemical analysis (specimen No. 1179) are not very different from those roughly deduced from microscopic study:—

Analysis of type specimen, Creston formation, Eastern Phase.

		Mol.
SiO₂	51·65	·861
	7·85	·077
	1·74	·011
	·98	·014
O	3·67	·092
	15·02	·268
...O	2·69	·044
K₂O	1·38	·015
H₂O at 110°C	·09	
H₂O above 110°C	1·81	·100
CO₂	13·05	·297
	99·93	
Sp. gr.	2·654	

Insoluble in hydrochloric acid	66·21%
Soluble in hydrochloric acid:	
Fe₂O₃	1·92
Al₂O₃	2·02
CaO	12·88
MgO	2·41

Assigning the soluble lime and magnesia to the carbonates, the remainder of the lime to the anorthite molecule, the soda to the albite molecule, the

2 GEORGE V., A. 1912

potash to the orthoclase molecule, the iron oxides to magnetite, and the residual silica to quartz, the following 'norm' has been calculated for the rock:

Quartz	25·5
Albite molecule	23·0
Orthoclase molecule	8·3
Anorthite molecule	10·5
Magnetite	2·5
Calcium carbonate	23·0
Magnesium carbonate	5·0
Remainder	2·2
	100·0

In general this 'norm' is not far from representing the actual mineralogical composition of the rock. The unexpected abundance of the soda again raises the suspicion, here as in the study of the analyzed western phase, that paragonite is really present; how far the 'norm' deviates from the 'mode' in this respect cannot be declared.

Chemically and mineralogically the rock has certain features of each of the three different types:—a feldspathic quartzite like the type of the western phase of the Creston formation; a metargillite like that dominant in the Mac-Donald or Appekunny formations; and a magnesian limestone. The size of grain of the carbonate is close to that, characterizing the Altyn dolomite and other carbonate-bearing members of the Galton and Lewis series. This eastern phase may thus be a rock-type transitional between the western phase and the rocks composing the Waterton, Altyn, and Appekunny formations.

KITCHENER FORMATION.

At all the localities where the two formations have been seen in contact, the Creston passes quite gradually into the conformably overlying Kitchener formation. The change from one to the other is so gradual, and the lithological differences between the two are of so low an order that, as already noted, the mapping of these formations offered considerable difficulty at many points in the Boundary belt. Much additional field work and the discovery of more favourable sections will be necessary before the Kitchener formation can be described in detail.

It is convenient and instructive to group the facts known about the Kitchener into a statement regarding both a western and an eastern phase. Where outcropping in the Moyie and Yahk ranges, the dominant rock belongs to the western phase; the western slope of the McGillivray range bears thick masses of strata belonging to the eastern phase. Finally, on the eastern slope of the McGillivray range, the eastern phase of the Kitchener was found to be so far changed as to be, for hundreds of feet together, indistinguishable from the Siyeh formation. So, in fact, the rocks of the McGillivray range, have been mapped with the express recognition of the stratigraphic equivalence between the Siyeh and the main mass of the Kitchener. (See map sheets Nos. 3 and 4.)

SESSIONAL PAPER No. 25a

Western Phase. —The thickness of the formation as exposed in the Moyie and Yahk ranges was roughly measured at two sections nearly along the Boundary slash on the two sides of the Moyie river, the measurement gave approximately 8,000 feet; the other, 7,400 feet. In neither case was the base or top of the formation actually visible. In a section still farther west both base and top can be found in a nearly complete section of the Purcell series, but the poor exposures in the dense forest cap there conspire with the difficulties of mensuration, in an area of variable dips, to prevent a trustworthy measurement of total thickness. This third section has, however, offered sufficient data to render it probable that in the two former sections we have nearly the whole thickness represented. The smaller of, the two estimates, 7,400 feet, was won from the structurally very favourable section in the fault-block bearing the great Moyie sills at the Boundary line and immediately west of those sills. It is possible, however, that even this lower estimate is too high and that the true thickness might be more accurately placed at 7,000 feet. It appears certain only that the Kitchener in this area cannot be less than 6,500 feet thick or much more than 8,000 feet thick. For the present the original estimate of 7,400 feet may be accepted, with the understanding that it may be several hundreds of feet too great.

The dominant rock of the western phase is to be classed as a notably uniform quartzite. The bedding is, on the average, considerably thinner than in the typical Creston quartzite. Individual strata range from a minute fraction of an inch to six feet or more in thickness. A few whitish beds, up to twenty feet thick, were observed at various horizons, but they are rare. The average thickness of the individual bed seems to be about three inches. As in the Creston formation, many of the thinner strata may be grouped into strong, non-fissile plates several feet thick. The rock is regularly gray or greenish-gray on the fresh fracture, this tint being normally darker than that of fresh Creston quartzite. The weathered surface is strong rusty-brown, in characteristic contrast to the older quartzite. Cross-bedding, ripple-marks, and sun-cracks were seen at various horizons in both the quartzite and interbedded metargillite, but these features are not so common as in the overlying Moyie formation.

Microscopic study shows that from 50 to 75 or 80 per cent of the dominant quartzite is composed of grains of glassy quartz. The other essential constituents are the feldspars, including sodiferous orthoclase, microperthite, and probably untwinned albite; a variable but generally abundant quantity of sericite, biotite, and possibly paragonite. Secondary epidote and kaolin, along with magnetite, pyrite, zircon, and apatite grains are minor constituents. The rusty colour of the rock is due, 1. so much to the alteration of magnetite or pyrite as to the freeing of iron oxide from the weathering micas. The essential constituents are interlocked after the same thorough fashion observed in thin sections of the Creston quartzite. The grain of the rock is always fine, the average diameters of the quartz individuals varying, in different specimens

2 GEORGE V., A. 1912

from 0.03 mm. to 0.3 mm., with an approximate average of not more than 0.1 mm.

Mr. M. F. Connor made the following analysis of a typical specimen (No. 1135) of the quartzite, taken near the Boundary monument on the isolated mountain immediately west of the Moyie river:--

Analysis of Kitchener quartzite.

	1.	Mol. 1a.	2.
SiO₂	76.90	1.282	76.81
TiO₂	.35	.004	....
Al₂O₃	11.25	.111	11.78
Fe₂O₃	.69	.004	.55
FeO	3.04	.042	2.88
MnO	.02	....	tr.
MgO	1.04	.025	1.39
CaO	.88	.016	.70
Na₂O	3.28	.053	2.57
K₂O	1.36	.015	1.62
H₂O at 110° C.	.20	....	
H₂O above 110° C.	1.20	.007	1.87
CO₂	tr.	....	
P₂O₅	.45	.001	....
	100.33		100.18
Sp. gr.	2.680		

The oxide proportions correspond to about 52 per cent of free quartz. It is difficult to calculate for the other constituents, largely because the exact distribution of the alkalies is not known. If all the soda be assigned to the albite molecule, this feldspar would make up nearly 28 per cent of the rock. From the great excess of alumina, over that required to form the normal feldspar molecules from all of the potash, soda, and lime present, it appears almost certainly necessary to believe that the paragonite molecule is represented in relatively large amount. The excess of alumina, a notable fraction of the very high soda and a fraction of the high combined water can all be satisfactorily assigned on that supposition. Whether the paragonite exists in the free state or is in isomorphic mixture with the analogous potash molecule of sericite, cannot be readily determined. The excess of soda over potash is also characteristic of the two type analyses of the Creston formation; all three analyses are thus in contrast to the average analysis of sandstones or argillitic rocks generally, in which potash is certainly the more abundant oxide of the two. The microscopic evidence is against the view that the dominance of the soda is due to a corresponding abundance of albite or other plagioclase. It may be added that the analysis was most carefully made, a second complete determination of the alkalies agreeing very closely with the first.

The hypothesis that much of the mica is paragonite or a highly paragonitic muscovite, renders the analysis more clearly understood, but it complicates the calculation. Assigning ten molecules of potash to orthoclase, five molecules to sericite, ten molecules of lime to anorthite and six to epidote—propor-

SESSIONAL PAPER No. 25a

tions which are probably not grossly astray; the leading constituents have been roughly calculated as follows:

Quartz...............................	42.0
Albite...............................	24.0
Paragonite,.........................	7.0
Orthoclase..........................	5.5
Sericite,............................	4.0
Anorthite,..........................	3.0
Epidote,............................	1.0
Magnetite, titanite, etc............	4.5
	100.0

Obviously these proportions are only approximate, but they will serve to give a better conception of the rock than if no estimate of relative quantities were made. In the discussion of the Moyie sills (chapter X) this analysis is of primary importance, and it will be seen that even this estimate of weight percentages is of value in the discussion. The analysis illustrates a general characteristic of the Kitchener quartzite, -that it is highly feldspathic and is rich in the alkalies, especially soda. The rock is chemically very similar to pre-Cambrian graywacke from Wisconsin (analysis of col. 2 in the preceding table).*

It should be noted that this description of the typical quartzite differs, in several essentials, from its preliminary description, given on pages 189-90 of the writer's paper on 'The Secondary Origin of Certain Granites,' published in the American Journal of Science, Vol. 20, 1905. At the time of writing that paper no chemical analysis of the rock was available nor was an adequate number of thin sections at hand for the proper diagnosis of the rock. The general lack of twinning among the feldspar grains led the writer to the belief that orthoclase was the prevailing feldspar. The careful study of a larger number of thin sections, following the receipt of Mr. Connor's results, permitted the correction of the error in a second brief account of the quartzite, published in the Rosenbusch Festschrift, 1906, p. 224.

A prevailing characteristic of the formation is an abundant interbedding of silicious metargillite in thin intercalations. Compared to the quartzite, these are typically of a darker gray colour on the fresh fracture and weather to a yet stronger brown than belongs to the weathered quartzite. Minute, often interlocking grains of glassy quartz and clear or dusty feldspar are essential constituents but they are subordinate to the now very abundant micas, both biotite and sericite (or paragonite?). These dark interbeds simply represent the more argillaceous phase of the same sediment which now forms the micaceous, feldspathic quartzite. There is no trace of original clayey matter and the interbeds may be classed with the typical metargillites.

Both slaty cleavage and true schistosity are notably lacking throughout nearly the whole extent of the exposed Kitchener. Here again there can be little doubt that the complete crystallization of these ancient shales has been

*W. S. Bayley, Bull. 150, U.S. Geol. Survey, 1898, p. 87.

2 GEORGE V., A. 1912

brought about merely as a consequence of deep burial and without the help of tangential pressure or ordinary dynamic metamorphism. All transitions exist in the ledge or even in a single hand-specimen between the metargillite and the feldspathic quartzite. For this reason alone it would be difficult to form an accurate conception of the average composition of this western phase of the Kitchener. It is certainly less silicious and more micaceous than the Creston. Possibly one-third of the thickness is made up of silicious metargillite; the remainder, of the quartzite with its own subordinate admixture of once-argillaceous and feldspathic material with the essential quartz.

The average specific gravity of three type specimens of the quartzite is 2·705; that of three type specimens of the silicious metargillite is 2·738. The average for the whole western phase, estimated on the above-mentioned quantitative ratio of the two types in the formation, is about 2·716.

Variations on this relatively simple scheme of composition are extremely rare throughout the thousands of feet of beds composing the western phase. In a very few thin strata the quartzite is spangled with large plates of biotite, each about 1 cm. in diameter. As in similar beds of the Creston these foils run at all angles through the rock; they appear to be simply greatly enlarged equivalents of the staple biotite individuals of the normal rock.

At a few other horizons, especially toward the top of the formation, the quartzite contains conspicuous round, blackish concretions somewhat flattened in the planes of bedding. They measure from two to three inches or more in greatest diameter. With the microscope the concretions are seen to be composed of quartz and feldspar grains cemented by abundant biotite and limonite with a small amount of sericitic mica. The feldspars are usually glassy and belong to the usual species, orthoclase, microperthite, and a well-twinned plagioclase, probably andesine. These minerals and the quartz occur in grains from 0·04 mm. to 0·3 mm. in diameter and thus of the average size characteristic of the interlocked essential minerals of the enclosing quartzite. The most noteworthy feature is the plainly clastic form of all these grains of quartz and feldspar. Most of them are angular but the largest quartzes are often distinctly rounded. There is no sign of secondary enlargement. We appear to have, then, in the heart of these concretions the only surviving relics of the original clastic form. The destruction of the clastic outlines through static metamorphism has been arrested through the secretion of the mica and iron ore in which the quartz and feldspars now lie. These clastic grains are separated from the metamorphosed substance of the enclosing sandstone, and, as a rule, are separated from each other.

Eastern Phase.—In the heart of the Yahk range, for a distance of twelve miles, measured along the Boundary line, the Kitchener formation appears to have been completely eroded away, the mountains there being composed of the underlying Creston and of heavy masses of intrusive gabbro. To the westward of this area the Kitchener steadily preserves the lithological habit which has just been described. The first outcrops of the Kitchener on the eastern side of the area already shows signs of systematic variation in the staple rock-types. As the section is carried farther eastward the changes become more and more marked, until, in the angle between the two forks of the Yahk river, the formation has attained what may be called its eastern phase.

In the meantime the total thickness seems to diminish so that at the Yahk river, where the top and base are both represented, the Kitchener measures not more than 6,200 feet in thickness. As with the western phase the passage to the underlying and overlying formations is not abrupt and it is impossible to be certain of an exact figure. In any case the thickness is believed to be close to 6,000 feet at the Yahk river.

Along the west fork of the river great thicknesses of the strata assigned to this phase of the Kitchener are still so similar to the rocks of the western phase that there can be no reasonable doubt that it is the one great formation reappearing on the eastern side of the twelve-mile interval. This view was corroborated by finding these beds developed in their proper relations to the typical Moyie and Creston formations.

The chief differences between the eastern and western phases are two in number. Though the feldspathic quartzite still persists, its vertical continuity is yet more signally broken by intercalations of metargillite which gradually increase in importance as the sections lead eastward. At the same time a wholly new ingredient or pair of ingredients appears in the formation. Many beds of the metargillite and even some of the more quartzitic facies betray an accessory amount of calcium carbonate which causes effervescence on the application of cold dilute acid to the specimens. Magnesium carbonate is present but, as in the Creston formation, is not so abundant as the calcium carbonate.

Finally, on the ridge running south from the Boundary line along the right bank of the west fork of the Yahk river, strong interbeds of somewhat silicious, magnesian limestone with typical molar-tooth structure occur among the still dominant quartzites and metargillites. The exposures are nowhere all that could be desired but there seems to be no doubt that at this locality, the carbonate rock forms several beds. As the section is carried eastward, these beds increase rapidly both in number and thickness, with a simultaneous decrease in the amount of more purely silicious strata.

About halfway between the two main forks of the Yahk and three thousand yards north of the Boundary line, a thick bed of molar-tooth limestone, which crops out again at the line, has afforded the specimen illustrated in Plate 10. This bed is, microscopically, at any rate, a good type of the molar-tooth rock so characteristic of the Siyeh formation. Along the Commission trail on the west slope of the McGillivray range, several hundred feet of this more or less impure limestone are well exposed at several points.

Finally, the character of the Kitchener strata still further eastward has become so far modified that the molar-tooth limestone and highly calcareous metargillites must total 1,000 feet or more at the Kootenay river opposite Gateway.

2 GEORGE V., A. 1912

In fact it became strongly suspected toward the close of the season of 1904, that a large part of the Kitchener quartzite is the stratigraphic equivalent of the Siyeh formation of the Rocky Mountains proper. This suspicion was raised to a practical certainty when, in the following year, the succession and characters of the Galton series were marked out. So clear were the field evidences of the equivalence that that part of the Kitchener formation which covers the eastern twelve miles of the Boundary belt in the Purcell range, could be coloured in the map sheet as belonging to the more closely defined Siyeh formation, rather than to the more extensive division, the Kitchener quartzite. It was further deduced from a review of all the sections in the Purcell range, that the top of the Kitchener formation coincides, in stratigraphic position, almost precisely with the base of the Purcell lava formation.

Since the eastern phase stands about midway, lithologically, between the already described western phase and the Siyeh formation of the Galton range, there is no special need of describing the eastern phase in detail. A thin section of the molar-tooth limestone at the West Fork locality was specially studied for purposes of comparison with the Siyeh limestone. The limestone is here a light to medium gray or brownish gray, compact rock, weathering buff and interrupted by the usual irregular partings, lenses, stringers, or round, eye-like masses of much less magnesian, light-gray, compact limestone, weathering gray or, rarely, pale buff-gray.

The buff-weathering main part of the rock effervesces to some extent but is clearly magnesian and silicious; the gray partings effervesce violently and seem to be nearly pure calcite. The diameters of the carbonate grains in the magnesian part vary from 0.02 to 0.1 mm. with an average of perhaps 0.06 or 0.07 mm. They enclose a notable amount of clastic quartz, orthoclase, microperthite, and an indeterminable plagioclase, in grains averaging less than 0.1 mm. in diameter. The carbonate grains have the characteristic rhombohedral development seen in the Siyeh and Altyn limestones and often show clean-cut crystal outlines.

The gray calcite partings are extremely uniform in character and lack any significant admixture of silicious particles. The diameter of the calcite grains steadily averages 0.03 mm.; they are usually allotriomorphic. Here again the evidence of the thin section corroborates the field evidence that the calcitic partings are segregations or secretions, formed after the limestone was well buried. The systematic lamination of the specimen illustrated in Plate 10, B, is clearly due to segregation of the calcium carbonate along shearing planes. In the average case, the secretion seems to have accompanied a contraction of the magnesian portion of the rock, possibly occasioned by the dehydration of the original sediment.

It should be noted, further, that the grain of this rock is very similar to that found for the magnesian limestones of the Galton and Lewis series, an important point to which attention will be directed in the section on the origin of these limestones.

The average specific gravity of thirteen specimens of the limestone (range,

SESSIONAL PAPER No. 25a

2·658 — 2·773) is 2·710. This value indicates the admixture of the less dense quartz and feldspars in the limestone. The average specific gravity of the whole eastern phase is about 2·700.

MOYIE FORMATION.

The youngest member of the Purcell series is exposed on the western slope of the Moyie river valley, where it crosses the Boundary line, and again on a strong meridional ridge immediately east of the Yahk river at the same line. . In both cases the exposed top of the formation is an erosion surface. At the Yahk river section the base is cut off by a major fault. The thickness of the formation as a whole cannot, therefore, be stated. At the Moyie river a maximum thickness of 2,200 feet was observed; at the Yahk river the estimates varied from 3.100 feet to 3,700 feet. The safest of the larger estimates may be placed at about 3,400 feet, which is a minimum thickness.

The formation is here considered as including, at the summit, the Yahk quartzite, which was proposed as a formational name in the summary report for 1904. On later study of the sections it has appeared advisable to withdraw the name ' Yahk quartzite' from the list of Boundary formations. The rocks to which it refers crop out only at one place in the belt; in composition they are rather closely allied to the overlying beds; thirdly, they are not specially well exposed, are warped and broken, and are limited above by an erosion surface, so that, clearly, the whole thickness cannot be found in the Boundary belt.

The upper 400 feet of the Moyie formation as redefined are chiefly composed of whitish and gray quartzites, with metargillitic intercalations. The lower 3,000 feet form a somewhat heterogeneous assemblage of argillites, metargillites, and impure quartzitic or cherty rock in rapidly alternating beds. The strata are, on the average, much thinner than those of the underlying formations, running from a small fraction of an inch to a couple of feet in thickness. Though many of the thinner laminæ are often aggregated in plates six inches thick or more, these rocks are of a decidedly fissile habit.

The argillites are often true shales, but probably most of the beds must be referred to true metargillite. Their colour varies from light gray to very dark gray or black; the colours of weathering are brown and gray. At the Moyie river locality several hundred feet of the shales occurring at the base of the formation are sandy and have a dark purplish-red colour, owing to a special content of oxide of iron. A few, very thin (1 to 2 mm. thick) layers of red hematite were observed in these purplish strata. The latter merge gradually into the underlying conformable Kitchener formation. The interbedded quartzites are always very fine-grained or compact, of a light gray colour on fresh fractures and gray, brown, and light buff on weathered surfaces. Many of the quartzites are argillaceous. Some of the beds are charged with a variable amount of calcium and magnesium carbonates, which were also found, by tests in the laboratory, to characterize specimens of the gray shales. The

2 GEORGE V., A. 1912

distribution of the carbonate-bearing strata is rather general but they are probably most numerous in the lower part of the formation. Pure limestone or dolomite was nowhere found.

Sun-cracks are extremely abundant throughout the formation and ripple-marks are not uncommon. No casts of salt-crystals were observed in any section. The very different looking cuboidal casts of weathered-out pyrites occur at several horizons.

The specific gravity of eight typical specimens, ranged from 2·567 for the shales to 2·735 for the dolomitic quartzites. The average of all is 2·676, which is not far from the average for the whole formation.

The stratigraphic relation of this formation to the Kitchener suggests at once that it may be the equivalent of the Gateway, Phillips, and Roosville formations of the Galton range. The writer believes such to be the fact. There is a close lithological similarity, especially between the Gateway and Moyie formations, not only in composition and general habit, including thin-bedding and colours, but as well in the persistence of shallow-water features through the beds. The apparent absence of salt-crystal casts, so characteristic of the Gateway and Kintla formations, does not appear to be of vital signifi-cation in the correlation, for obviously the conditions for the development of a supersaturated brine would not extend over an unlimited area of contempor-aneous sedimentation. The deposition of the Moyie sediments may well have taken place in open-sea water. Some of the calcareo-magnesian quartzites and argillites have close resemblance to the impure Sheppard dolomite, the equivalent, in the Clarke and Lewis ranges, of the lower Gateway formation.

With longer study of the known outcrops of the Moyie formation and, above all, with the discovery of more favourable exposures, it may be possible in the future to subdivide this group of beds; at present, it seems best to recognize only the one inclusive formation name for the sediments overlying the Kitchener in the Moyie and Yahk ranges.

GATEWAY FORMATION IN THE McGILLIVRAY RANGE.

In the McGillivray range the conditions for immediate correlation with the Galton series are more favourable. The peaks of the highest ridges in the Boundary belt are almost all composed of the Purcell Lava formation, which, as we shall see, is the most perfect horizon-marker in the Rocky Mountain sections. At the summit of the McGillivray range this lava formation has been warped into a broken, north-pitching syncline. Considerably more than a thousand feet of thin-bedded, sun-cracked and much ripple-marked strata conformably overlie the lava.

The base of this group is formed of beds unquestionably equivalent to those in the lowest Gateway, while the main mass is lithologically transitional between the upper 1,850 feet of the Gateway strata and the more heterogeneous Moyie strata. The closer affinities of these strata at the summit of the range are distinctly with the Gateway formation and its colour has accordingly been

PLATE 15.

Limonitized simple and twinned crystals of pyrite, from Gateway formation
at summit of McGillivray range. Two-thirds natural size.

Similar pyrite crystals in metargillitic matrix. Same locality. Two-thirds natural size.

used in mapping the rocks overlying the Purcell lava in the McGillivray range. Erosion has there, within the Boundary belt, removed the equivalents of the Phillips and Roosville formations.

These Gateway beds are so similar in composition and habit to those across the Kootenay and already described that a special account of the former is not necessary. They are marked by an unusual wealth of ripple-marks and annelide trails and borings. Several beds of ferruginous and metargillitic quartzite, occurring some 300 feet above the Purcell Lava, carry remarkably large and perfect cubes of more or less limonitized pyrite. These range from 1 cm. to 4 cm. or more in diameter and form most conspicuous elements of the rock. (Plate 15.) They often form simple interpenetration twins. The crystals seem to have grown in the original mud either before or during the period of its consolidation. On any other supposition it would be difficult to understand how space was made for their growth; the lamination of the rock immediately surrounding each crystal is usually quite undisturbed and not crinkled or bowed around the crystal.

The specific gravity of eight hand-specimens, representing types for the whole Gateway formation in the McGillivray range, varies from 2·646 to 2·747, averaging 2·687.

STRUCTURE OF THE PURCELL MOUNTAIN SYSTEM.

As already remarked there are special physical difficulties in the way of discovering the structure of the Purcell system at the Forty-ninth Parallel; hence the details of structure are not as well understood as are the structures in the eastern ranges. Enough facts are in hand, however, to show that the Purcell system is, like the Galton-MacDonald mountain group, chiefly composed of great monoclinal fault-blocks. Of these twelve have been determined without much residual doubt. Most of them are found in the Yahk and Moyie ranges. The McGillivray range shows a tendency towards the structure of terranes characterized by open folds.

Between Gateway and the summit the Kitchener (Siyeh) and Purcell Lava beds are warped into a broad, unsymmetrical anticline. The dips average 35° N.E. on the eastern limb, a steepness of dip which would rapidly carry the top of the entire Purcell series of sediments far below the level of the Devonian limestone at Tobacco Plains. The distance between the limestone and the most easterly of the outcrops (Purcell Lava) across the drift-covered Purcell Trench is eight miles. We can only conjecture the structures beneath the drift cover. Those actually visible indicate that the Rocky Mountain Trench is, at the Boundary line, located on a zone of combined faulting and down-flexure. In all probability the faulting has had the dominant control in locating the trench.

The western limb of the broad anticline shows northwesterly dips of 15° to 20°. The convergence of strike lines on the two limbs shows that the fold pitches gently to the north.

2 GEORGE V., A. 1912

The anticline is succeeded on the west by the summit syncline which also pitches north at a low angle. Like the anticline this fold shows numerous local warps and, on the south, it is truncated by a strong east-west fault shown on the map sheet. The western limb of the syncline shows a section through nearly the entire Purcell series. Two miles east of the main fork of the Yahk river the Creston beds have a sharp reversal of dips, indicating an anticline broken by a longitudinal, north-south fault. The Yahk river is located in the heart of this anticline. It may have been originally placed on the line of fault, from which position the river has since slipped down, the dip an average distance of two miles. To the west of the main fork of the river the dips gradually change from an average of 45° W. to horizontality, and in the interval, a great part of the Creston formation, the whole of the Kitchener and some 3,000 feet of the Moyie formation are exposed in succession.

On the ridge overlooking the west fork of the river on the east, the dips in the Moyie beds again become easterly, showing a narrow syncline which is here only visible in this formation. Exactly on the line of the west fork the Moyie strata are dropped down into contact with a gabbro sill which is intrusive into the Kitchener formation. This west fork fault is remarkably straight in the six miles through which, with unusual certainty, the outcrop of the fault could be followed. The downthrow is, of course, on the east and may measure more than 2,000 feet.

From the west fork of the Yahk to Porthill nearly all suggestion of folding is wanting and the relations are those of many fault-blocks. The dips are highly variable, values from 5° to 80° or more being recorded. The dips are generally much the higher in the narrower blocks. Here as in the Galton-MacDonald system the fault-planes usually trend towards the north-northwest and their dips seem invariably to approach verticality.

The faults mapped between the west fork of the Yahk and the Moyie river are among the most obscurely exposed of all. Others not shown on the map sheet may be responsible for the duplication of the great gabbro sills in this part of the Boundary belt. Much additional time and labour must be expended before the full structure of this part of the belt will be declared. The two blocks immediately east of the Moyie river are shown as separated by a reversed fault along which the Creston quartzite has been driven up on the back of the likewise steeply dipping and apparently underlying Kitchener quartzite. A second interpretation is open, whereby the two formations are regarded as in normal contact but both overturned to the west.

The plane of the main fault at the Moyie river is nowhere exposed but the relations of dip and strike are such as to leave no doubt as to the nature of the displacement. The downthrow is to the west and is very great, probably approaching 8,000, if not 10,000 feet.

The fault running along the western base of the isolated mountain bearing the Moyie sills is also believed to be mapped correctly. The downthrow is again to the west but the displacement is probably no more than a couple of thousand feet.

From that point to the Kootenay river the faults shown on the map and section are not so certainly placed. The master-fault following the base of McKim cliff has not been directly observed but is postulated because of the fact that the great sill of gabbro on the west is underlain by rusty quartzite which is believed to belong to the Kitchener formation. If this be the correct interpretation the Purcell Trench is located along a displacement by which the Kitchener formation has been dropped down into lateral contact with strata near the base of the Creston quartzite as defined in this report. The total displacement of the fault or faults east of Porthill and west of the summit of McKim cliff would thus approach 10,000 feet. The geology of the Selkirk range shows, however, that the zone in which the trench lies has been the scene of still more profound faulting; the evidence is summarized in the next chapter.

The rocks of the Purcell mountain system have transmitted thrusts of enormous power and have been vigourously upturned at many points. Yet those rocks bear few traces of shearing, cleaving, or dynamic metamorphism. Only in one narrow zone at the Moyie river is cleavage notably developed and that structure is only conspicuous on the weathered ledges. This general failure of metamorphic structures in rocks which have undergone at least once the severe pressures of extensive mountain-building, is amply accounted for by the exceeding strength of the sediments. That strength is in part explained by the homogeneity of the formations and in part by their thorough welding by deep burial and static metamorphism during the immense interval between their deposition and deformation. To the inherent strength of the sedimentary prism has been added the reinforcement by the thick sills which formed so many new, relatively inflexible ribs in the whole mass. Where massive homogeneous quartzite and gabbro predominated (Yahk and Moyie ranges), folding is almost entirely absent and the orogenic pressures produced monoclinal blocks. Farther east, where relatively thin-bedded argillites entered the formation in greater number and where the gabbro sills were not intruded (McGillivray range), the mountain-building produced broad folds rather than upturned fault-blocks. Nevertheless, the rocks of the Purcell series seem everywhere to have much greater average strength than have geosynclinal sediments generally.

Note added during reading of proof.—Mr. S. J. Schofield has recently shown that a thick, ferruginous quartzite-metargillite series, named the Aldridge formation, underlies the Creston quartzite. It appears probable that the Aldridge is represented in some of the fault-blocks mapped west of the Yahk river. The writer now (1912) suspects that the succession in the sediments immediately east of the Moyie river is normal and that the reversed fault there mapped does not exist. If so, the "over-thrust" block of rusty quartzite really belongs to the Aldridge formation and not to the closely similar Kitchener formation.

Plate 16

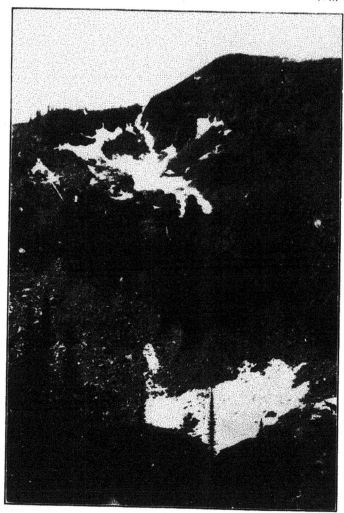

Exposure of the massive Irene conglomerate in head-wall of Boundary line and two miles west of Priest feet in height.
cirque; one mile north cliff is eleven hundred

CHAPTER VII.

STRATIGRAPHY OF THE SELKIRK MOUNTAIN SYSTEM (IN PART).

SUMMIT SERIES.

Excluding the igneous rocks, the principal formations encountered in the Nelson range within the ten-mile belt may be grouped in three divisions. The rocks belonging to the oldest division, called the Priest River terrane, are found only on the eastern slope of the range. The rocks of the youngest division are confined to the western side of the range and to the valley of the Pend D'Oreille river, to and slightly beyond its confluence with the Columbia. This younger principal division may be called the Pend D'Oreille group. Lying between these two divisions both geographically and stratigraphically, is the Summit series, a large part of which is the equivalent of the whole Purcell series. The present chapter is devoted to a summary description of the Summit series. It will be followed by a chapter of the correlation of all four of the great series so far discussed, and then the systematic account of the formations occurring in the Selkirks at the Boundary will be resumed.

In order to facilitate a rapid understanding of the Summit series a tabular view of the formations is here presented:--

Formation.	Thickness in feet.	Dominant rock
	Top, erosion surface?	
Lone Star..	2,000 +	Phyllite and quartzite.
Beehive..	7,000	Quartzite.
Ripple..	1,650	Quartzite.
Dewdney..	2,000	Quartzite, with conglomerate.
Wolf..	2,900	Silicious grit, sandstone, and conglomerate.
Monk..	5,500	Quartzite, phyllite, and conglomerate.
Irene Volcanics..	6,000	Effusive greenstones.
Irene Conglomerate	5,000 +	Conglomerate.

32,050 +

Base, unconformity with Priest River terrane.

IRENE CONGLOMERATE FORMATION.

The basal member of the Summit series is a conglomerate, outcropping on the summit and slopes of Irene mountain. It has, accordingly, been named the Irene Conglomerate formation. Excellent exposures are numerous along the outcrop from the International line to the Bayonne batholith, eight miles distant. (Plate 16.) The most instructive section was found on the long

2 GEORGE V., A. 1912

ridge running in an easterly direction from the triangulation station south of Monk creek on the Canadian side to the steep slope immediately overlooking Priest river canyon.

Through interbedding the conglomerate is transitional into the overlying Irene Volcanic formation; the base of the conglomerate marks a profound unconformity with the much older Priest River terrane. The width of the conglomerate belt, measured on the map, is about 1·5 miles. Everywhere the rock shows evidence of exceedingly intense crushing and shearing. The true bedding is thus masked by schistosity, especially in the coarser and more homogeneous phases of the ancient gravel. The two structures were sometimes found in the same ledge and then usually had the same strike but differed in dip from ten to thirty degrees. The average strike of the bedding, to the southward of the Dewdney trail is about N. 5° E.; its average dip is at least 60°. In spite of the obvious difficulties of mensuration the minimum thickness of the formation must be placed at a very high figure. The apparent thickness based on the average dip is nearly 8,000 feet. Since the beds of conglomerate were probably not laid down horizontally but were built out in imbricate fashion on a sloping sea-bottom, this estimate must be corrected by some, as yet unknown, amount. To what extent the bedding was originally inclined is a problem which, on account of the heavy subsequent metamorphism of the formation, it is doubtless impossible to solve in the area so far studied. Allowing for a strongly inclined deposition a conservative minimum estimate of the total thickness, an estimate based on three complete sections, is 5,000 feet; it should, perhaps, be many hundreds of feet greater.

Coarse conglomerate is the highly dominant constituent of the formation. It occurs in well-knit, very massive beds of squeezed pebbles, which, as a rule, were well water-worn when they finally came to rest in their respective beds. The pebbles range in size from coarse sand-grains to bouldery masses a foot or more in diameter. More than one-half of them are composed of gray vitreous or micaceous quartzite or of white sugary quartz. Next to them in abundance are pale gray or white compact pebbles of dolomite-marble (specific gravity 2·833 - 2·875), often silicious to some extent. A few pebbles of phyllitic slate and, yet more rarely, pebbles of a biotite granite may also be seen. The top-most beds bear small angular fragments of altered porphyrite and diabase which seem to have been directly derived from the contemporaneous, locally interbedded lavas and tuffs of the Irene Volcanic formation. Some of the larger, bouldery masses of the quartzites and especially of the dolomites, are subangular and apparently were not long rolled on a beach.

The majority of the pebbles have been deformed in the crush of mountain-building. They are commonly flattened into lenses much longer than the original pebbles. The mashing is wonderfully illustrated in the case of small pebbles examined microscopically in thin section. A notable biproduct of this metamorphism of the dolomitic pebbles is the common generation of many glass-clear, twinned crystals of basic plagioclase (probably acid bytownite) among the grains of carbonate.

The cement of the conglomerate is usually in large amount and rather uniform throughout the formation. Originally it must have been of the nature of a graywacke or very muddy sand. In its present condition it is a schistose, crystalline mass of various shades in gray and greenish-gray. Clastic grains of quartz of all sizes up to one or two millimetres in diameter, and very much rarer grains of orthoclase lie embedded in an extremely abundant fine-grained matrix of sericitic muscovite, biotite, and chlorite. The foils of mica are specially developed in the planes of schistosity. Grains of magnetite, leucoxene and pyrite are constant subordinate accessories, while anhedra and minute idiomorphic crystals of titanite are often very abundant in thin sections. Irregular or roughly rhombohedral, secondary crystals of calcium carbonate (probably somewhat magnesian), a millimetre or less in diameter, seldom fail to appear in the sections. They sometimes, though not always, enclose quartz and the micas poikilitically. Quite often the clastic quartz grains show the fam..e proofs of secondary enlargement.

The mass of the conglomerate may be interrupted by lenses of metamorphosed sandstones and pelites a few inches to several feet in thickness. These rocks have been metamorphosed to phyllitic schists of composition practically identical with that of the conglomerate cement.

The specific gravity of five type specimens of the conglomerate ranges from 2·680 to 3·753. Their average, 2·732, is believed to be nearly the average for the whole, fairly homogeneous formation.

After field and laboratory study of these rocks there can be little doubt as to the origin of some of the clastic materials. The colour, composition, and general field habit of the quartzite, phyllite, and dolomite pebbles clearly show their derivation from the underlying Priest River terrane. Nevertheless, the writer has not found a single pebble of the spangled quartz-mica schists so abundant in that terrane and, in general, the larger quartzite pebbles show a massiveness or lack of schistosity, which is more marked than that expected if they were derived from the Priest River terrane in its present lithological condition. It seems necessary to conclude that a large proportion of the metamorphism suffered by the older terrane, including the growth of the biotite spangles and some of the intense shearing and sericitization of the quartzites, has affected the terrane since the Irene conglomerate was rolled on the ancient beaches. One may naturally hold that the metamorphism of the Priest River terrane occurred simultaneously with the mashing and partial recrystallization of the Irene conglomerate as younger and older formations were upturned together. Even in the conglomerate there is striking proof of immense tangential pressure such as is nowhere given in the Purcell, Galton, or Lewis series of formations.

Since most of the material for the conglomerate was won from the older terrane, which in this region is not known to contain pre-Irene acid plutonic masses on any large scale, it is not surprising that neither the cement of the conglomerate nor the phyllitic interbeds are highly feldspathic. It is clear, on the other hand, that the feldspathic grits and sandstones of the over

2 GEORGE V., A. 1912

lying members of the Summit series, must have been formed from the ruins of coarse-grained, granitic rocks which, in post-Irene time, became exposed to erosion within this region. This contrast between the lower and upper formations of the series is particularly noteworthy in the case of the absence of clastic microperthite in the conglomerate, while that mineral is a prominent clastic component of the Wolf grit and still younger members of the Summit series. From the fact that this peculiar feldspar occurs in the oldest exposed beds of the Lewis, Galton, and Purcell series, there is already good presumptive evidence that the Irene conglomerate has no stratigraphic equivalent in the eastern series. There is abundant corroboration of this view in the general stratigraphy, as will be noted in the section on correlation.

IRENE VOLCANIC FORMATION.

The Irene conglomerate is conformably overlain by a great mass of lava flows which, for a thickness of a hundred feet or more, are interbedded with the conglomerate. These lavas crop out along the western slopes of Irene mountain, and they may be grouped under the name of the Irene Volcanic formation. As with all the other members of the Summit series, the band of lavas may be followed from the Boundary line northward across nearly the whole width of the ten-mile belt. The northern extremity of the band occurs at the cross-cutting contact of the Bayonne granite batholith. Complete sections were measured on the Dewdney trail, on Irene mountain, and on the ridge south of Monk creek. The best exposures of the formation as a whole were found in the last mentioned section.

The formation chiefly consists of a large number of thick basic lava flows, in which a few subordinate layers of basic tuff, a thick band of conglomerate-breccia, and a strong bed of dolomite are intercalated.

Like the conglomerate and the overlying Monk schists the whole mass has been greatly altered by dynamic metamorphism, with a general development of marked schistosity. The massiveness of the flows and the prevalence of this secondary structure render it often impossible to determine true dip at even extensive outcrops. Nevertheless, the attitude of the original layering has been discovered at so many horizons that an important generalization can be made,—the dip of bedding is always steep, varying from 70° E. to 70° W., with strikes varying from N. 7° E. to N. 30° E. Bedding and schistosity planes are in most cases nearly or quite coincident. The outcrop of the formation averages nearly 1·5 miles in width. Assuming an average dip of only 70° and considering the structure of the band as monoclinal throughout, the thickness of the formation is at least 6,000 feet. High as this figure is, it must be regarded as the smallest allowable estimate. Extensive duplication of the beds by folding or faulting within the area is highly improbable. The bed of conglomerate-breccia was followed for at least eight miles, through which distance it preserved its thickness, high dip, and proper horizon below the base of the Monk formation. The breccia and the associated dolomite are conspicuous members and could scarcely escape detection if they were repeated in the various sections, especially in the one traversed on the nearly treeless ridge south of Monk creek. The total thickness is, then, taken to be at least 6,000 feet; it may be 7,000 feet or more.

The rocks composing the Irene Volcanic formation, as it crops out in the Boundary belt have been grouped in divisions as here shown:-

Columnar section of Irene Volcanic formation.

Top, conformable base of Monk formation.

50 feet.	Greenstone schist, a crushed basic amygdaloid.
200 "	Angular conglomerate or breccia with phyllitic cement.
1,710 "	Greenstone schist with a few thin bands of phyllite toward the top.
40 "	Gray to white, fine-grained dolomite.
4,000	Sheared and greatly altered basaltic and andesitic lavas = largely greenstone schist.

6,000±feet.

Base, conformable top of Irene Volcanic formation.

The great bulk of the formation is composed of a notably uniform type of highly altered andesitic lava, now typical greenstone. It is a dark green or greenish gray, compact, schistose rock, in which, as a rule, there is scarcely a trace of the minerals originally crystallized out of the magma. A large proportion of the greenstone is amygdaloidal, the amygdules (composed of calcite or, much more rarely, of quartz) being mashed out into thin lenses parallel to the pronounced schistosity. While the greenstone has been essentially derived from surface lava flows, it is usually impossible to distinguish the limits of any one flow. The difficulty of doing this is evidently due in part to the intense mashing and metamorphism of the lavas. It appears probable that, while the great mass was accumulated by many successive flows, each flow was of considerable thickness.

From the study of over twenty-five thin sections cut from as many typical and relatively unweathered specimens, it has been found that throughout the entire thickness, the rock has a very homogeneous character. It is a confused, felted mass of uralite, chlorite, epidote, quartz, calcite, limonite, sericite, saussurite, and often biotite, with which pyrite, magnetite, and ilmenite (generally altered to leucoxene) regularly form accessories in variable amount.

For several thin sections this list exhausts the list of constituents; in their corresponding rocks metamorphism has evidently been thorough.

Excepting possibly the iron ores, the only original magmatic constituent is plagioclase, which with surprising regularity is represented in most of the sections only by a few, highly altered, broken crystals. The form and relations of these crystals show that they generally formed phenocrysts in the original lava, which had an abundant glassy or microcrystalline base. An exceptional holocrystalline, ophitic, fine-grained phase was found near the base of the formation on the ridge just north of the Boundary line. In two thin sections of this phase the plagioclase is better preserved and gave in the zone of

25a—vol. ii—10

2 GEORGE V., A. 1912

symmetry a maximum extinction of 20°; it appears thus, to be an acid labradorite. The phenocrysts of the porphyritic phases, though singularly hard to diagnose, seem to be of nearly the same species of feldspar. In not a single slide was there found the slightest trace of other phenocrysts. Even pseudomorphs of such possible original phenocrysts as pyroxene or amphibole entirely fail. Judging from the nature of the secondary minerals, the original lava was in all probability a rather basic andesite or andesitic basalt.

Some of the fine-grained, non-amygdaloidal greenstone may, at certain points in the field section, belong to dikes or sheets of the lava cutting slightly older flows. Largely on account of the profound metamorphism it has proved as yet impracticable to distinguish such possible intrusives in the field. They can, however, in any case, form but a small part of the whole mass.

The microscopic character of the long list of secondary minerals shows thorough banality and needs no special description.

The specific gravity of eleven type specimens ranged from 2·791 to 3·096, with an average of 2·919, which cannot be far from the average for all the greenstone.

About 2,000 feet below the top of the formation the greenstone is interrupted by a forty-foot interbed of compact, somewhat sheared, gray limestone weathering light yellowish or buff. Under the microscope the rock is seen to be a remarkably homogeneous granular aggregate of carbonate grains without other visible impurity than a little granular quartz occupying narrow, microscopic veinlets, cutting the rock proper. The carbonate grains are anhedral, roundish, and of nearly uniform size, averaging 0·015 mm. in diameter. The rock effervesces very slightly with cold dilute acid. The specific gravity is 2·853, indicating a nearly pure dolomite. The purity of this carbonate mass, coupled with its fineness and uniformity of grain, strongly suggests a chemical origin for the rock. It should be noted that the average size of the carbonate grains is very similar to the average size of the grain in the Altyn, Siyeh, Sheppard, and other magnesian formations of the eastern series.

The 200-foot breccia-conglomerate occurring near the top of this formation is of special value as a horizon-marker. Because of its high angle of dip and because of its power of resistance to the processes of general erosion, the conglomerate often projects in strong peaks or ridges above the surrounding greenstone. Fine exposures were found on the summits north of Monk creek and on the long northern slope of Summit creek valley. (Plate 72, B and C.) From the Boundary line to Summit creek this conspicuous rock-bed is always practically vertical and runs in a remarkably straight line, bearing a few degrees east of north. Throughout that stretch there seems to be no possibility of any important amount of dip-faulting in the Summit series as a whole. The persistence of this clastic bed, both in strike and dip, and its steady parallelism to the boundaries of the other nearly vertical members of the Summit series outcropping in this area, testify to the conformity of the whole Irene volcanic formation with the Irene conglomerate, and with the Monk and younger formations. Had it not been for the discovery of this band of conglomerate, the writer would not have the actual, strong belief that the volcanics form a part of one enormous, conformably bedded group upturned in a gigantic monocline.

Not only the structural relations but, as well, the composition of the breccia illustrates the propriety of regarding both it and the underlying and overlying greenstone as members of this conformable group. The rock is a very massive grouping of angular to subangular, very rarely rounded, fragments of dolomite-marble and of quartzites, embedded in an abundant phyllitic matrix.

The dolomite is compact and white, weathering the usual buff colour. It is silicious, carrying considerable clastic quartz which is strained and crushed. The specific gravity of a typical fragment is 2.804. Many fragments are highly pisolitic or coarsely oolitic, with grains of excellent concentric structure and of diameters from 1 mm. to 4 mm. The largest dolomite fragment seen was quite angular and measured seven feet by four feet by three feet.

The greatly sheared matrix is composed essentially of sericite and quartz, the latter often showing typical water-worn outlines. Small rounded grains of dolomite also appear in the thin section. The matrix is a carbonate-bearing phyllite, derived from a clay or mud. No trace of volcanic ash was seen in hand-specimen or in thin section. Notwithstanding the intimate field association with true lavas, the whole 200-foot band must be regarded as a water-laid, though not well sorted, angular conglomerate. Its detrital materials doubtless originated from the Priest River terrane. The specific gravity of a large type specimen of the breccia is 2.824.

Except for the relatively great abundance of dolomitic material both in the matrix and bouldery fragments of the breccia, the whole rock is extremely similar to coarser phases of the Irene conglomerate. The chief essential difference is that the latter has suffered yet more intense mashing than the 200-foot band, which, before the upturning, lay 6,000 feet or more nearer the earth's surface than the basal conglomerate. The amount of shearing and metamorphism in the 200-foot band is intermediate between that shown in the basal conglomerate and that in the similar conglomerate beds of the Monk formation overlying the volcanics. This appears to mean that shearing and recrystallization in similar rocks of the series have, as might be expected, progressed in direct proportion to the depth of their burial.

MONK FORMATION.

The formation immediately overlying the Irene volcanics is, of all the members of the Summit series, by far the most poorly exposed. Only two complete sections, furnishing even tolerable exposures, appear in the Boundary belt. One of these was crossed on the summits just north of Monk creek but it could not be used as a basis for a description of the typical formation, because most of the beds are there signally metamorphosed by adjacent batholithic granite. The following notes on the formation express the facts which were gathered chiefly on a traverse between Monk creek and the Boundary line along the top of the ridge running east-southeast from Mt. Ripple. Unfortunately, that ridge

2 GEORGE V., A. 1912

is heavily timbered through most of its extent. Blanks of three hundred feet or more occur at several points within the section. The composition and other salient features of the formation are, therefore, not known with anything like the certainty that attaches to the other members of the Summit series.

This group of sediments underlying the Wolf grit and resting on the Irene volcanics, may be called the Monk formation, after the name of the creek which cuts across its outcrop. The total thickness is very great; a minimum of 5,500 feet is estimated. There is also considerable heterogeneity in the mass. Nevertheless, it is considered advisable to group all these beds under the one formation name. The definite naming of the lithological subdivisions is not warranted until better exposures are found than those so far studied.

The subdivision shown in the following columnar section is to be considered as decidedly crude. The thickness of some of the members could only be conjectured, since the outcrops in such cases were discontinuous and quite insufficient to give assured conclusions as to the composition of the covered beds. The estimates then given were partly based on the character of the 'wash' and even that was often thoroughly buried under the dense forest cap. When, in the future, this mass of strata is stratigraphically well worked out, it will doubtless be profitable to recognize by distinctive names certain of the subdivisions; the name 'Monk formation' may then be restricted to the most important member recognized in the re-examination. The columnar section for the formation may be tentatively described as follows:

Columnar section of Monk formation.

Zones.	Thickness.	
	Top, conformable base of Wolf formation.	
a	120 feet. —	Sericite-quartz schist.
b	50 "	Quartz grit, little sheared.
c	650 "	Sericite-quartz schist.
d	20 "	Coarse grit, little sheared.
e	1000 ± "	Sericite-quartz schist, sometimes cyanitic.
f	600 ± "	Dark gray slate and phyllite.
g	1300 ± "	Chiefly sericite-quartz schist with interbeds of sheared grit and conglomerate : poor exposure.
h	550 "	Sheared quartz conglomerate.
i	700 ± "	Chiefly sericite-quartz schist and sheared quartzite : poor exposure.
j	60 ± "	Schistose conglomerate.
k	250 "	Phyllite.
l	200 ± "	Phyllitic slate.
	5500 ± "	
	Base, conformable top of Irene Volcanic formation.	

As a rule it is very difficult to determine the attitude of the bedding, so effectually is that structure masked by the never-failing schistosity. The most of the readings of true dip were obtained at the contacts of the grits and conglomerates with the schists. At such points the average strike was about N. 10° E. and the dip from 75° W. to 90°. The corresponding readings for schistosity gave, on the average, nearly the same strike, with dip ranging from 79° W. to 55° E., averaging nearly vertical. However, at one locality the bedding and schistosity of a slate-phyllite phase, though holding the regional strike, N. 12° E., gave, respectively, 30 E. and 55 E. for the dips. There is evidently some crumpling, especially in the fine-grained phases, but on the whole, schistosity and bedding seem to be very nearly coincident throughout the formation; their planes are seldom far from the vertical.

The natural suspicion that so great a thickness of fairly homogeneous, schistose rocks might be, in part, explained by duplication was not strengthened by the data secured during four different traverses over the section. The poorness of the exposure makes it unsafe to exclude the possibility that there is duplication, but the fact that the band of rocks belonging to this formation conserves its width as it is followed from the Boundary line northward for six or more miles, affords some evidence against the idea of repetition of beds. If the faulting or folding had repeated these particular beds to any great extent, we should expect the beds of the conformable Monk grit and Irene Volcanic formation to show strong local deviation from the regional strike. On the contrary, the contact-lines of these formations run remarkably straight for the whole six miles across a very mountainous area. The simplest, as well as the most probable, conclusion is that these three great formations all belong to one conformable series locally upturned in a single monocline and that in no one of them has there been duplication by either folding or faulting.

The greater part of the formation is composed of quartz and sericite in variable proportion. The original composition of the dominant fine-grained rocks ranged from compact quartz sandstone to argillite. For hundreds of feet together in each of zones c, d, g, and i, the beds are made up of sheared sericitic, light greenish-gray quartzite. This phase alternates with darker greenish-gray, highly fissile schist in which metamorphic mica (sericite and, much less abundantly, biotite) equals or dominates the quartz in amount. Within these limits there is great uniformity in the formation except for the occurrence of the gritty or conglomerate zones. The usual accessories, magnetite, pyrite, chlorite, etc., are present but are always quantitatively unimportant. Feldspar has not been observed and if, as is probable, it was originally accessory in the quartzitic phases, it has itself been sericitized. The monotony in the mineralogical composition of these schistose rocks is known to be broken only in zone e, where well crystallized cyanite in simple twins, has developed in some abundance.

Zones b, d, h, and j, totalling about 700 feet in thickness, are made up of detrital materials which are fairly uniform in composition though not in grain. Zone j is a greatly mashed, gray conglomerate with pebbles of quartzite and black slate, pressed or drawn out into lenses up to four or five inches in length. Pebbles of dolomite were not seen but this rock is very similar to common phases of the Irene conglomerate. The matrix of the pebbles is again phyllitic. Zone h is a conglomerate of the same type, though bearing sandy and gritty phases which are strongly feldspathic. A thin section from a coarse arenaceous specimen showed that glassy quartz, much typical microperthite, orthoclase, basic andesine, some microcline in a cement of shreddy muscovite, and a little chlorite formed the principal constituents. Euhedra of magnetite and

2 GEORGE V., A. 1912

much limonite disseminated through the cement, are the usual subordinate minerals. The clastic grains, large or small, are characteristically angular and the rock as a whole, may be classed as a metarkose. Many of the quartz grains, though several millimetres in diameter, are fragments of single crystals, showing that their source was doubtless a very coarse granite.

Zones *b* and *d* are in composition simply finer-grained, gritty equivalents of zones *h* and *j*. The former zones seem to be more massive than the latter and less sheared or mashed. Nevertheless, the thin sections are replete with evidences of the great stresses which have operated on all these rocks. The quartz grains and pebbles always show undulatory extinction or granulation. Owing to this minute fissuring and the resulting partial decomposition of light reflected from the interiors of the glassy grains, the quartz is commonly opalescent in bluish tones which are sometimes quite deep and pure.

The average specific gravity of two specimens of the conglomerate-sandstone zones is 2·640. The average of four specimens of the schists is 2·717. Allowing for the relative thickness of these rock-types, the average specific gravity of the whole formation may be placed at about 2·705.

WOLF FORMATION.

Zone *a* of the Monk formation is conformably overlain by a mass of very heavily bedded sandstones, grits, and fine-grained conglomerates, which in all essential respects are identical in character with the coarser-grained phases of the Monk formation. On account of its thickness and conspicuous nature this mass has been distinguished by a special name, the Wolf formation.

Its exposures are unusually perfect in the broad band crossing the ten-mile belt from Mt. Ripple northward to the headwaters of Wolf creek. The outcrops are especially extensive along the Dewdney trail at the summit of the range and, again, on the south-eastern flank of Mt. Ripple. At the last named locality the beds stand vertical or nearly vertical and there the formation can be best studied. Some uncertainty must attach to measurements of thickness, for this formation passes very gradually into the overlying Dewdney quartzite and in none of the sections is the actual base exposed. At the Mt. Ripple section the total thickness was measured at 2,900 feet and this seems to be steadily held throughout the Boundary belt.

The formation is more massive than any other sedimentary member of the Summit series; where most massive it consists chiefly of a feldspathic quartz grit or conglomerate which, for fifty or more feet of thickness at a time, shows no conspicuous plane of bedding. In the lower two-thirds of the formation and much oftener in the upper one-third, the grit or conglomerate is interrupted by thin beds of metamorphosed, more or less argillaceous sandstone. Practically without exception the beds are of a medium gray or, less commonly, greenish-gray colour on fresh fractures and weather a pure gray or brownish gray.

The larger pebbles of the conglomerate are composed of vitreous quartz; sugary, gray or white quartzite; much more rarely, dark gray to blackish slate. They may be as much as four or five centimetres in diameter but the average diameter is under one centimetre. Many are well-rounded but most were subangular at the time of deposition. Occasional phases show some flattening of the pebbles by orogenic pressure, though the degree of shearing and

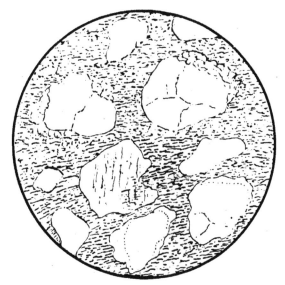

FIGURE 11.—Drawing from thin section of metamorphosed argillaceous sandstone, Wolf formation. Large grains are quartz except the partly shaded one in southwest quadrant (microperthite). Quartz shows cataclastic structure and some secondary enlargement. Ground-mass of quartz and sericitic mica. See text. Diameter of circle, 5 mm.

mashing never, even distantly, approaches that represented in the Irene conglomerate or in the lower zones of the Monk formation. Here, again, many of the pebbles (some as large as 5 mm. or more in diameter) are made up of fragments of single quartz crystals, apparently indicating the great coarseness of the granitic rock which furnished this immense body of silicious detritus. The single-crystal pebbles, as well as others of compound and granular texture, are greatly strained, with the result that they are often of the peculiarly rich blue or gray-blue opalescent colour noted in the Monk conglomerates.

2 GEORGE V., A. 1912

Though much fewer in number than the quartz pebbles, angular fragments of feldspar are seldom wanting from the conglomerates and coarser grit beds. The microscope shows them to be orthoclase (or microcline), microperthite, and basic andesine, named in their apparent order of importance. Where the feldspars are specially abundant, the grit has the look of a metarkose. The feldspar is usually more or less kaolinized or sericitized. Slate fragments are always relatively rare and probably never make up more than five per cent of the whole number in any one bed.

The cement of the conglomerate and grit is a variable mass of sericite, and fine-granular quartz, with which minute foils of biotite may be associated; magnetite forms a never failing though not abundant accessory; chlorite, zoisite, tourmaline, and sillimanite are other constituents, always in small amounts.

From the conglomeratic phases there are all transitions to the only less important interbedded sandstones and metamorphosed sandy argillites. The sandstone may, in fact, be regarded as but finer-grained equivalents of the conglomerate, while the altered argillites are more highly micaceous, compact analogues to the cement of the conglomerate. Feldspar grains appear to be very rare in these finer-grained phases. The well water-worn grains often afford beautiful examples of secondary enlargement whereby these rocks have become very strong and resistant both to the hammer and the weather.

Where the rock is fractured, the surface of fracture, as in a true quartzite, passes indifferently through quartz grain and cement. The minute mica plates and shreds strongly tend to be developed in planes of schistosity. These planes pass clear through the clastic grains of quartz in such a way that a large grain is flanked by two swarms of similarly orientated mica-foils, as shown in the accompanying Figure 11. Thus, the micas as a rule do not wrap around the clastic grains but are grouped in straight lines or zones which are cut off sharply by the grains. It is clear that in this case the schistosity produced by the common orientation of the micas is not due either to shearing of the rock or to the rotation of pre-existing sericite and biotite but is due to the crystallization of these minerals with their cleavages lying perpendicular to the direction of a compressive force.

The schistosity is almost always parallel to the bedding. Part of the metamorphism may have taken place after the old sediments were turned up on edge. However, the fact that the flat-lying sandstones and argillites of the Lewis, Galton, and Purcell series show similar fissility and recrystallization, seems to indicate that most of the recrystallization of the Wolf and overlying formations was completed before the upturning. In the present case tangential force simply completed a process which had been nearly finished under conditions of static metamorphism.

The microscope shows that the feldspar of the coarser sandstones is characteristically microperthite or microcline. Orthoclase and plagioclase are very rare and generally seem to fail altogether. The microperthite, like the micas (sericite and biotite) and much of the quartz, shows evidence of having developed during the recrystallization of the rock. As in so many other phases of the geosynclinal sediments, the abundance of this feldspar, which is so rare in normal sandstone, is an interesting problem.

That much of the microperthite is of metamorphic origin is suggested, not only by the microscopic relations, but also by the fact that this feldspar has been formed in special abundance and in clearly non-clastic forms within the metamorphic collars developed in the Wolf and Monk sediments where they are cut by intrusive granite. Nevertheless some of the microperthite has the outlines and relations of clastic grains similar to those found so abundantly in the sandy dolomites of the Lewis series, where there is little chance that the feldspar is of metamorphic origin.

The specific gravity of the conglomerate-grit phases varies from 2.630 to 2.683; that of the more micaceous, sandy, and argillaceous phases, from 2.729 to 2.895. The average of twenty specimens selected to represent the whole formation, is 2.720.

DEWDNEY FORMATION.

By insensible gradations the Wolf formation passes into the conformably overlying Dewdney formation. The plane separating them is thus an arbitrary one. In its typical development, however, the younger formation, while chemically very similar to the older, is finer-grained and thinner-bedded a banded quartzite. Excellent exposures through its whole thickness appear on both sides of the Dewdney trail, from which the formation has been named. Other complete sections were measured on traverses southeast and south of Mt. Ripple. The thickness seems to be tolerably constant throughout the Boundary belt. At the trail the following section was roughly measured:—

Columnar section of Dewdney formation.

Top, conformable base of Ripple formation.

375 feet.—	Medium to thick-bedded banded quartzite.
30 "	Coarse conglomerate.
120 "	Banded quartzite.
225 "	Coarse conglomerate.
1,250 "	Thick-bedded, banded quartzite.
2,000 "	Base, conformable top of Wolf formation.

The formation consists, in the main, of light gray and greenish-gray quartzite, well and rather uniformly banded. Interbedded with the quartzite are subordinate dark greenish-gray strata which were originally argillaceous, but are now felted aggregates of quartz, feldspar, biotite, sericite, and iron oxide. These rocks generally weather gray and only rarely brown. Thick bedding is the rule, each of the massive plates averaging three feet more or less in thickness. They are composed either of single strata of quartzite, or of well-knit composite masses of highly indurated sandstone and silicious metargillite in alternating layers.

2 GEORGE V., A. 1912

The steady occurrence of the dark-coloured, often-argillaceous beds in the sandstone suggested the name 'Lower Banded Quartzite' as an early designation for the formation in the field notes. The analogous name 'Upper Banded Quartzite' was similarly used for the Beehive quartzite which likewise shows marked alternation of dark and light silicious beds.

The quartzite is similar in composition to the fine-grained phases of the Wolf formation and needs no detailed description. The light-tinted, often ripple-marked beds are almost entirely made up of thoroughly interlocked quartz grains, between which a few sericite foils may be seen. These beds are, as a rule, apparently very poor in feldspathic material, though it must be said that the specimens collected are too few to afford complete microscopic evidence on this point. The darker bands, which vary from a fraction of an inch to several inches in thickness, are charged with some biotite as well as with the dominant sericite, while the accessory magnetite grains are abundant.

The conglomerate interbeds persist, with nearly constant thickness across the entire ten-mile belt. Throughout that long distance they stand almost exactly vertical and parallel to the banding of the quartzite. The exposures are often very fine (Plate 19). From the higher peaks the dark bands of the conglomerate can be followed with the eye for miles. The vertical dip explains the extraordinary straightness of the mapped outcrop of the formation as it traverses mountain and deep canyon alike. At several localities the pebbles of the 225-foot band are arranged in layers making angles of from 5° to 12° with the contact planes of the band, clearly showing the imbricated, fore-set bedding of the old beach.

The pebbles are water-worn; the diameters are of all lengths up to one foot, averaging three inches. They consist of glassy quartz, gray or greenish quartz schist and, rarely, black slate. The schistose dark green-gray cement is highly variable in constitution. Quartz grains and a few grains of altered feldspar are subordinate clastic ingredients; most of the cement is composed of sericite, biotite, chlorite, and accessory magnetite. One deep-green, compact specimen, without visible pebbles of any kind, proved on microscopic examination to be made up almost entirely of felted chlorite in which minute, angular, accessory grains of quartz could be seen.

On the southeast slope of Mt. Ripple the alumino-magnesian cement has been rather thoroughly recrystallized so as now to be a mass of intimately interlocking anhedra of cordierite, 0.2 to 0.4 mm. in diameter. This mineral encloses swarms of minute sericite foils and magnetite grains. Small lenticular areas of granular quartz here and there occur in the thin section. The development of cordierite at this point, three miles from the nearest intrusive granite, would hardly have been anticipated. It is probably the result of thermal metamorphism by the underlying batholithic magma, of which the granite stock at the Dewdney trail was a constituent part.

The composition of these conglomerates is, on the whole, like that of most of the conglomerates in the Wolf, Monk, and Irene formations; the younger beds are, however, much less sheared and schistose than the older ones.

PLATE 17

Ripple-marks in Ripple quartzite; positives. Summit of Mount Ripple. Dark patches are lichens. Hammer is two feet long.

Ripple-marks in Ripple quartzite; negatives (casts). Same locality and scale.

PLATE 18.

Negatives of ripple-marks in quartzite.   Summit of Mount Ripple.   Hammer two feet long.

The specific gravity of the conglomerate averages about 2.700; that of the quartzite and metargillite 2.670, and that of the whole formation about 2.675.

RIPPLE FORMATION.

Wherever the Dewdney formation crops out within the Boundary belt, it is conformably overlain by a heavily bedded mass of white quartzite which forms the summit of Mt. Ripple and has therefore been named the Ripple formation. Three complete sections were measured on as many ridge-summits lying between Wolf creek and the Boundary line. The whole massive formation is unusually resistant to the weather and its vertical strata compose some of the highest summits in the region; such outcrops are very favourable to study. The thickness seems to remain fairly constant at all the localities examined, the average of the measurements giving 1,650 feet as the most probable value for this region.

The Ripple formation consists of a remarkably uniform, hard, very heavily plated quartzite, breaking with a sonorous metallic ring under the hammer. There are practically no interbeds of other material. The dominant colour of the rock is white, but flesh-pink and light yellowish tones are common. The general colour of the weathered surfaces, including joints, is a bright buff-yellow which is characteristically decolourized to snow-white through the agency of lichens and other plants. The effects of these colours among the extensive felsenmeers above the forest-cap are as beautiful as they are striking. (Plates 17, 18 and, 71 B.)

A principal feat·   of the quartzite is the occurrence of extremely well-preserved ripple-m·   ·arious horizons. On Mt. Ripple itself these markings a·e exposed in   spectacular fashion. In bed after bed for a thickness of several hundr·   .·c together the surfaces of the old sand were moulded into typical ripples of highly varied orientation (Plate 18). As exposed on bedding-planes these marks are to-day apparently as sharply marked as they were when each bed was just covered by the next wash of sand. Whole cliffs are ornamented with the strong ridges and troughs of the ripples themselves or with their negative impressions. Occasionally a slab of the frost-riven rock shows the compound ripple pattern of pits and mounds where the same sand layer was subject to two succeeding currents setting from different directions. Sometimes the quartzite is fissile along the planes of such rippled beds, only a centimetre or so thick, but as a rule, the rock breaks out in large, massively constructed slabs a half metre to a metre or more in thickness. In the task of reducing the peaks formed of this stubborn rock, the frost uses joint-planes rather than bedding-planes. (Plates 70, B and 71, B.)

The quartzite is extremely simple in composition. Under the microscope it is seen to be essentially made up of subangular, or much more rarely, rounded grains of glassy quartz from 0·1 to 0·4 mm. in diameter. These are cemented by yet more granular quartz and some accessory shreds of sericite. The quartz grains are usually strained, if not actually fractured. Probably more than 90

2 GEORGE V., A. 1912

per cent, by weight, of the average rock is quartz. Not a grain of feldspar was seen in thin section and there is a singular lack of the accessories found in the surrounding formations. This quartzite is clearly the most highly silicious member of the Summit series.

The specific gravities of two type specimens were found to be respectively, 2·655 and 2·661; their average, 2·658, is very close to the average for the whole formation.

BEEHIVE FORMATION.

The Ripple quartzite passes with some abruptness into the conformably overlying Beehive formation, so named after its typical occurrence on Beehive mountain north of Lost creek. Of this formation two complete sections and four other partial sections were traversed. The best exposures within the belt were found on Beehive mountain and on the ridge overlooking, from the north, the south fork of the Salmon river.

The formation is heterogeneous, yet the recurrence of a rusty-weathering, quartzitic rock-type is so constant throughout the whole mass that it has seemed expedient to include many thousands of feet of these beds under one formational name. The total thickness is only roughly estimated but it is believed to be 7,000 feet at a minimum. At Beehive mountain itself there are over 9,000 feet of these strata well exposed, but at that section, there is possibly some repetition of beds by overthrust, g. As with the majority of the members of the Summit series, suitable horizon-markers for a definite and workable subdivision of the huge sedimentary mass, are very rare. On the western slope of Beehive mountain a 50-foot bed of limestone is included in the field section and will be noted in the columnar section of the formation, but it was not seen outcropping at other localities so as to be a really serviceable horizon-marker.

A further difficulty in giving a precise lithological description of the formation consists in the relatively high dynamic metamorphism which has affected the mass, especially in the upper part. The only tolerably good exposure of that part, within the belt, occurs on the western slope of Beehive mountain. This section was studied in bad weather and but a very limited time could be devoted to it, although it is the locality most favourable to the discovery of the principal facts concerning the upper one-third of the formation. At this locality there is apparent conformity with the Lone Star schists, but there is a chance that the appearance is due to the intense mashing which characterizes this local area, a dynamic effect whereby the conformity of the schistose structures in the two formations simulates conformity in the dips and strikes of the true bedding-planes. This question of conformity or non-conformity between the Lone Star and Beehive formations cannot be solved with information now at hand.

A compilation of the facts derived from the six field-sections led to the following columnar section. It will be understood that it cannot pretend to a high degree of accuracy.

SESSIONAL PAPER No. 25a

Columnar section of Beehive formation.

Top, base of Lone Star schist formation.

2,850 feet.—Thin-bedded, variegated (green, gray, brown, red and whitish) phyllite; silicious metargillite and quartzite; weathering rusty-brown; ripple-marks.
   50 "   Thin-bedded, light gray limestone, weathering gray.
  270 "   Light green-gray sericite-quartz schist.
1,500 "   Thin-bedded, greenish, silicious metargillite and interbedded quartzite; weathering brown; ripple-marks.
   30 "   Bed of massive white quartzite.
  180 "   Thin-bedded, light greenish-gray, silicious metargillite, weathering light rusty brown.
  120 "   Massive, hard, bluish gray quartzite, weathering brown.
2,000 "   Thin to medium-bedded, light greenish gray quartzite, weathering rusty brown, with thin, though numerous interbeds of dark greenish silicious metargillite, weathering dark brown or brown-gray. Ripple-marks, sun-cracks and annelide trails are plentiful. One hundred and seventy-five feet from the top, a bed of magnetite mixed with lenses of magnetitiferous quartzite; this bed from two inches to eight feet thick.

7,000 feet.

Base, conformable top of Ripple formation.

Ripple-and rill-marks, sun-cracks, and, less often, annelide trails and borings are common at many horizons.

The bed of magnetite, noted in the lowest member was found in the course of three different traverses, two of which were seven miles apart; the bed is notably persistent, but as yet does not promise a commercial quantity of iron ore. The maximum thickness of the magnetite was found on the summit of the ridge, 2,000 yards northeast of the Boundary monument at the south fork of the Salmon river.

Apparently at the same horizon a similar though much thinner (two-inch) zone of crystallized, granular magnetite was found on the ridge north of Lost creek and on the line of strike from the former locality.

Under the microscope the quartzites are seen to be composed of the usual clastic grains of quartz, often secondarily enlarged and regularly cemented by infiltrated silica and by subordinate sericite. Unfortunately, no specimen was collected from the feldspathic phases, so that the species of feldspars have not been determined. The clastic quartz grains average about 0·3 mm. in diameter. The metargillites of the lower members are yet more compact masses of quartz, chlorite and sericite, with accessory biotite and magnetite; the micaceous minerals, though all of metamorphic origin, lie with their basal planes parallel to the bedding, so that the metargillitic character is typically represented.

Higher up in the sections, where true dynamic metamorphism has locally affected the beds, the metargillites are largely replaced by phyllites. On the ridge running eastward from Lost mountain, the phyllites (probably because of the influence of the Lost Creek granite magma) are charged with numerous crystals of andalusite and with much metamorphic biotite. Southeast of Beehive mountain similar, undoubtedly thermal, metamorphism has developed much cyanite in small crystals disseminated through the phyllitic beds.

2 ·GEORGE V.. A. 1912

In general, the once-argillaceous material, now crystallized as sericite, chlorite, biotite, magnetite, etc., grows more abundant toward the top of the Beehive formation, which is there also somewhat thinner-bedded than in the lower, more quartzitic members.

The average specific gravity of ten selected specimens, 2·717, is believed to be near the average for the whole formation.

LONE STAR FORMATION.

The Beehive formation at its upper limit merges gradually into a division of sedimentary rocks which, everywhere in the ten-mile Boundary belt, have been so much disordered and metamorphosed that it has proved quite impossible to declare their exact thickness or their relation to the younger Paleozoic formations in contact with them. There is apparent conformity not only with the Beehive formation below but also with the Pend D'Oreille schistose sediments and limestones, strata which are believed to be mainly of Upper Paleozoic age. All of these formations have, however, suffered complete metamorphism, crumpling, faulting, and mashing, in consequence of which the apparent conformity may not exist. For the present, the schistose sediments immediately contacting with the Beehive quartzites are regarded as the youngest rocks in the Selkirk range which can, with any safety, be considered to be part of the conformable Summit series. This uppermost member is very roughly estimated as 2,000 feet in thickness and is given the name, Lone Star formation, so called from its exposure on the eastern slope of Lone Star mountain.

The formation consists principally of dark-gray or greenish-gray, often carbonaceous phyllite, along with some lighter tinted, greenish sericite-quartz schist and thin interbeds of light-gray quartzite. The dominant phyllite sometimes, though rarely, passes into true slate in which the well developed cleavage cuts across the bedding-planes. As a rule, the bedding is very obscure and the schistose structure is the dominant one. In nearly all the sections these schists, like the conformable Beehive quartzites, dip to the eastward at high angles, showing that the great monocline in which the Summit series has been studied, is overturned to the westward.

The extreme metamorphism of the Lone Star formation is due partly·to the original nature of the sediments, which were specially liable to alteration in orogenic crush, and largely, also, to the vicinity of intrusive granites and other igneous masses. It is unfortunate for the study of this upper part of the Summit series at the Forty-ninth Parallel that it is thus exposed only at the eastern edge of one of the greatest fields of intrusive rocks in the Cordillera. One must look to the other sections, particularly to those farther south, for a more satisfactory diagnosis of the Summit series in its relation to younger formations. Four miles east of the Salmon river the Lone Star schists dip under the Pend D'Oreille schists and therewith the entire series disappears from sight, so that no rocks referable to the great Rocky Mountain Geosynclinal prism are to be found in any part of the Boundary belt to the westward.

Mount Ripple and summit ridge of the Selkirk Mountain System.

SESSIONAL PAPER No. 25a

The Lone Star schists were traversed during bad weather at the close of the season of 1902, when, as yet, the existence of the Summit series monocline just described was unsuspected. No later opportunity was afforded for revisiting the few sections in which the schists are exposed in the belt. For these reasons the collection of specimens and of field data is especially meagre for the formation. The foregoing brief and very general account is all that is warranted from the writer's limited knowledge of these rocks.

CHAPTER VIII.

CORRELATION OF FORMATIONS IN THE ROCKY MOUNTAIN GEOSYNCLINAL.

CORRELATION ALONG THE FORTY-NINTH PARALLEL.

The stratigraphic equivalence of respective formations in the Lewis, Galton, Purcell, and Summit series is indicated in Table I, and in the plate bearing their columnar sections. This correlation is based on lithological similarities of individual members. Confidence in the general correlation is greatly strengthened by the fact that the lithological succession in one series is matched more or less closely by a similar lithological succession in all the other series

TABLE I.—*Correlation of the Rocky Mountain Geosynclinal rocks.*

SUMMIT SERIES.	PURCELL SERIES, WESTERN PHASE.	GALTON SERIES.	LEWIS SERIES
Conformity with Upper Palæozoic?	Erosion surface.	Erosion surface.	Erosion surface.
Lone Star, 2000' Phyllite and quartzite.	Movie, 3400'. Metargillite, with quartzite and shale.	Roosville, 600'. Metargillite with quartzite.	Kintla, 800 Argillite, sandstone and dolomitic limestone.
		Phillips, 550' Metargillite with quartzite.	Sheppard, 600' Silicious dolomite, with quartzite and argillite.
		Gateway, 2025'. Metargillite with quartzite and dolomite.	
	Purcell Lava, 465'	Purcell Lava, 310'	Purcell Lava, 250'
Beehive, 7000' Quartzite with metargillite and phyllite.	Kitchener, 7400' Quartzite, with metargillite.	Siyeh, 4000' Dolomitic limestone, with much metargillite and some quartzite.	Siyeh, 4100' Dolomitic limestone, with much metargillite and a little quartzite.
Ripple, 1650' Quartzite.		Wigwam, 1200' Sandstone, with metargillite. MacDonald, upper part, 700' Metargillite.	Grinnell, 1600' Metargillite, with quartzite.

2 GEORGE V., A. 1912

TABLE I.—*Correlation of the Rocky Mountain Geosynclinal rocks*—Con.

SUMMIT SERIES.	PURCELL SERIES. WESTERN PHASE.	GALTON SERIES.	LEWIS SERIES.
Dewdney, 3000' Quartzite, with conglomerate. Wolf, 2000' Grit, with conglomerate and sandstone. Moule, upper part, 2500' Quartzitic sandstone, with metargillite and conglomerate.	Creston, 8500' Quartzite, with metargillite.	MacDonald, lower part, 1050' Metargillite, with a little dolomite. Hefty, 775' Sandstone, with quartzite and a little metargillite.. Altyn, upper part, 650' Siliceous dolomite.	Appekunny, 2500' Metargillite, with quartzite and a little dolomite. Altyn, 3500' Siliceous dolomite, with dolomite grits and sandstone. Waterton, 2000' Siliceous dolomite.
Monk, lower part, 3000' Quartz schist and phyllite, with conglomerate. Irene Volcanics, 8000' Greenstone and greenstone schist, with a little phyllite and one bed of angular conglomerate. Irene conglomerate, 5000' Coarse conglomerate, with sandstone and grit and a little interbedded greenstone.			
Basal unconformity.	Base concealed.	Base concealed.	Base concealed.
Total, 32,050 feet.	Total, 20,765 feet.	Total, 12,000 feet.	Total, 11,720 feet.

The most useful horizon is that of the Purcell Lava formation. Probably no other geological horizon betokens contemporaneous events in distant localities more surely than such a lava flood. A sandstone bed or other product of sedimentation on the floor of a transgressing sea may belong to more than one geological period. A lava flood not more than a few hundred feet in thickness at any point is, on the other hand, developed with comparative rapidity. Even a great compound flood generally covers its whole field in but a small fraction of a geological period. Essential contemporaneity thus characterizes the surface of the sediments overrun by the Purcell Lava. Since the overlying strata are apparently in absolute conformity to the lava and to the strata underlying the lava, it is probable that several hundreds of feet of beds above and below the Purcell Lava are likewise practically contemporaneous. Since the lava formation has been traced from southeast of Altyn, Montana, and from the heights overlooking Waterton lake at the Great Plains, all the way to the eastern summits of the Purcell range on the Boundary, the value of this particular horizon-marker is evident. The mountains covering at least three thousand square miles show, at frequent intervals, the outcrops of the lava No other horizon is more competent to demonstrate the stratigraphic equivalence of the Lewis, Galton, and Purcell series in their respective upper portions.

The correlation of the three eastern series is further facilitated by the occurrence, in all three, of magnesian strata characterized by the peculiar molar-tooth structure, which becomes prominent at a horizon about a thousand feet or more below the Purcell Lava. This structure is dominant in the Siyeh dolomite, a formation unmistakably recognized in the Galton series (with about the same thickness as in the Lewis series) where the Siyeh formation was first described. The molar-tooth rock with all its typical features also occurs in the eastern half of the Boundary belt crossing the Purcell range. In that region the rock occurs in the form of relatively thin strata that interrupt the staple silicious sedimentaries of the Kitchener quartzite. The recurrence of such a highly special structure and the fact that the Kitchener quartzite and the Siyeh formation in the Galton and Lewis ranges are capped by conformable and contemporaneous flows of the Purcell Lava, are principal indications that the Siyeh formation must be correlated with the upper part of the thick Kitchener formation.

The equivalence of most of the members of the Lewis and Galton series is otherwise very manifest in the field. The thin-bedded, silicious dolomite of the upper Altyn on Oil creek is well matched in its leading lithological characters as well as in stratigraphic position by the thin-bedded, dolomitic quartzite and dolomite of the upper Altyn on Mt. Hefty and at other points in the Galton range. The reddish-brown beds of the Hefty formation match the lowermost, rusty beds of the Appekunny. The greater part of the Appekunny formation is almost identical in composition with the middle member of the MacDonald formation. The 1,580 feet of Grinnell red argillites and sandstones correspond to the 1,200 feet of red argillites and sandstones in the Wigwam formation and the rusty-brown and reddish beds of the upper member of the MacDonald. The homogeneous, dolomitic quartzite of the Sheppard formation is, in part, paralleled by similar strata in the lowermost 125 feet of the Gateway formation. The red argillites and sandstones of the Kintla match the red sandstones and argillites of the Phillips. Abundant casts of salt-crystals, sun-cracks and ripple-marks, showing special conditions of origin, are characteristic of the Gateway, Phillips, and Kintla beds at many horizons in each formation, and are also to be found in the red argillitic beds of the Sheppard formation. In the Lewis and Clarke ranges at the Boundary, erosion has destroyed the equivalent of the Roosville formation, if beds of that age were ever laid down in the region east of the Flathead river. Half of the upper Altyn beds, the whole of the middle and lower Altyn, and the Waterton argillite are not represented in the Galton section, because neither upturning or erosion has exposed these older rocks in the Galton range.

25a—vol. ii—11½

2 GEORGE V., A. 1912

The great homogeneity of the three huge formations in the Purcell range has rendered it as yet impossible to correlate in detail the 20,000 feet of strata there exposed with the well-marked members of either the Galton or Lewis series. As already noted, the fortunate exposure of the Purcell Lava conformably overlying the Kitchener quartzite in its typical, eastern phase, affords an invaluable datum-plane.

The Moyie argillite-sandstone formation is regarded as the equivalent of the whole Gateway-Phillips-Roosville group as well as of the Sheppard-Kintla group, though it is probable that the Moyie formation is stratigraphically a larger unit than the Sheppard and Kintla combined.

Apart from the occurrence of lenses or tongues of molar-tooth limestone in the upper Kitchener, there is no indisputable field evidence as to the exact relation of the Kitchener to the variegated rocks of the two eastern series. A probable but tentative correlation may be based on the fact that the Kitchener quartzite is typically ferruginous. The strata of the Galton series are dominantly ferruginous down to the base of the upper MacDonald; the strata of the Lewis series are dominantly ferruginous down to the base of the Grinnell. As illustrated by Table 1, and Plate 20, the base of the Kitchener is accordingly correlated with these two horizons, while the top is definitely fixed at the Purcell Lava.

The Creston quartzite was, in the field, differentiated from the Kitchener quartzite by the non-ferruginous character and lower stratigraphic position of the older formation. The gray quartzites and argillites of the Appekunny and MacDonald correspond, even in details of colour, composition, ripple-markings, etc., to the top beds of the Creston quartzite. The reddish beds of the Hefty and lowermost Appekunny are not paralleled, so far as known, by reddish beds in the Creston, but may be equated with the somewhat rusty-brown strata which occasionally occur in the Creston at horizons 1,500 feet or more below its summit. Similarly, there is no evident lithological equivalent of the Altyn anywhere within the Boundary belt.

The perfect conformity within each of the three great series is, however, a strong argument for considering even the strongly contrasting Altyn dolomite and Creston quartzite as stratigraphic equivalents. The massive Waterton dolomite is similar in field-habit to certain parts of the eastern phase of the Creston in the Purcell range. The vigorous upturning of the fault-blocks in that range has occasioned the exposure of a specially great thickness of beds filling the ancient geosynclinal; the lower one-half of the Creston formation as exposed in the Boundary belt seems to be older than the oldest beds exposed in the Galton, Clarke, or Lewis ranges in the same belt.

The lithological contrasts between the different members of the Summit series when compared with the members of the Purcell series, is almost as great as the contrasts existing between the Lewis and Purcell formations. Moreover, in the correlation of the western series we have no datum-plane of absolute contemporaneity such as the Purcell Lava affords in the eastern part of the wide geosynclinal.

PLATE 20.

Columnar sections of the Summit, Purcell, Galton, and Lewis Series; with correlations.

FESSIONAL PAPER No. 25a

Nevertheless, in the Summit series there is a very thick group of ferruginous, silicious sediments conformably overlying a yet thicker group of gray silicious sediments not specially ferruginous. The former group, the Beehive quartzite (7,000 feet thick), relatively thin-bedded, ripple-marked, and charged with thin interbeds of dark gray to brown argillite, is essentially similar to the Kitchener quartzite (7,400 feet thick), and are so shown in the correlation table and Plate 20. Plate 21 probably errs in correlating the Dewdney with the Kitchener; thus representing the writer's early view, abandoned since this diagram was drawn.

The Lone Star schists are to be correlated with the Moyie formation. Immediately beneath the Beehive quartzite is the remarkably rippled, white Ripple quartzite (1,650 feet thick), succeeded below by the gray Dewdney quartzite (2,000 feet thick), the gray Wolf grit and the huge mass of gray and greeni-h-gray Monk argillites, sandstones and interbedded conglomerates. Many of the quartzite beds of this huge group of gray-tinted sediments cannot, in ledge or hand-specimen, be distinguished from the dominant Creston quartzite.

In favour of correlating the Creston quartzite with these formations, excluding the Ripple quartzite, is the fact that the Dewdney, Wolf and Monk formations are, like the Creston, composed of dominant quartz, with which much essential feldspar is usually mixed. In both the Purcell (Kitchener and Creston) and Summit series (Beehive to Wolf inclusive), this feldspar is very commonly microperthite, finely lamellated in normal fashion. The recurrence of this feldspar, here essential yet so uncommon in such thick sedimentary masses, gives excellent corroboration of the conclusion arrived at in the field that within the Summit series, the equivalent of the Creston quartzite includes the relatively non-ferruginous formations below the Ripple quartzite. The feldspar is a kind of fossil. The base of this group, equivalent to the Creston as exposed in the Boundary belt, probably occurs some 2,500 feet below the summit of the Monk formation.

The equivalents of the lower and greater part of the Monk formation of the great Irene Volcanic formation and of the thick basal Irene conglomerate have nowhere, within the Purcell or Rocky Mountain systems, been thrust up to view in the Boundary belt.

If the foregoing correlation is correct, the monocline of the Selkirk range furnishes a key to the stratigraphy of the whole geosynclinal. The base of the geosynclinal prism is seen at the unconformable contact of the Priest River schists on Monk creek. The uppermost beds of the prism have, in all the eastern mountain ranges, been eroded away as a result of the repeated orogenic uplifts which have occurred since Cambrian time. In the Selkirks the whole prism may be represented, but the extreme mashing and metamorphism of the upper beds have made their stratigraphic relations at the Forty-ninth Parallel very obscure.

2 GEORGE V., A. 1912

SYSTEMATIC VARIATION IN ROCK-CHARACTER OF THE GEOSYNCLINAL AT THE FORTY-
NINTH PARALLEL.

A study of the correlation plate (Plate 20) and the foregoing descriptions show that all four of the sedimentary series fit into a single scheme
of rock-genesis. Distance from the ancient shore-line, off which the many beds
were deposited, is the main key to the scheme. It is about 120 miles from the
thick monoclinal section of the Summit series in the Selkirks to the spectacular
monoclinal section of the Lewis series on Oil creek in the Clarke range. The
east and west line joining the two monoclines is not only transverse to the
existing mountain ranges but is also the line of cross-section through what
seems to be the thickest part of the great stratified prism. In the western monocline the sediments are largely littoral deposits,—coarse and fine conglomerates;
coarse grits and coarse and fine sandstones. In the eastern monocline the sediments are those characteristic not so much of very deep water as of mere
distance from the immediate shore-line, the home of turbulent waves, strong
wave-erosion, and powerful transportation of coarse detritus. The members
of the Galton and Purcell series represent the expected transitional formations
between the two extremes.

In general the prism is lithologically homogeneous in its middle part and
highly heterogeneous in the zone of shore-deposits, and also highly heterogeneous in the eastern end of the section, far from the old shore-line.

An estimate has been made of the relative proportions of conglomerate,
grit, sandstone, argillite (metargillite), and limestone occurring in each
member of the four series. The results of the estimate have been tabulated as
follows:—

TABLE II.—*Showing general lithological character of the four standard sections in the Rocky Mountain Geosynclinal.*

	Carbonate rocks.	Argillite.	Sandstone.*	Grit.	Conglomerate.
Lewis Series :—	Feet.	Feet.	Feet.	Feet.	Feet.
Kintla....		720	100		
Sheppard............	580	10	10		
Siyeh............	2,200	1,700	200		
Grinnell................		1,200	380		
Appekunny......	75	1,800	725		
Altyn............	3,000		500		
Waterton....	200				
	6,055	5,430	1,915		
Total, 13,400 feet.					
Galton Series.—					
Roosville		500	100		
Phillips		350	200		
Gateway....	35	1,625	365		
Siyeh............	1,800	2,000	200		
Wigwam...........		300	900		
MacDonald............	50	1,450	850		
Hefty	25	100	650		
Altyn............	650				
	2,590	6,325	3,265		
Total 12,150 feet.					
Purcell Series :—					
Moyie................		2,400	1,000		
Kitchener............	100	1,300	6,000		
Creston............		500	9,000		
	100	4,200	16,000		
Total 20,300 feet.					
Summit Series :—					
Lone Star............		1,400	600		
Beehive............	50	1,450	5,500		
Ripple...........			1,650		
Dewdney............		250	1,500		
Wolf				2,000	900
Monk...........		2,000	2,600	200	700
Irene conglomerate..............		200	800		4,000
	50	5,300	12,650	2,200	5,850
Total 26,050 feet.					

* The "sandstone" here includes the quartzites, which, as already described in the case of the Creston and Kitchener types, are not strictly "sand" stones.

2 GEORGE V., A. 1912

Expressed in percentages the proportions of the different kinds of sediments, rated according to thickness, are:—

Rocks.	Summit Series.	Purcell Series.	Galton Series.	Lewis Series
Conglomerate............	24·5	0·0	0·0	0·0
Grit	9·2	0·0	0·0	0·0
Sandstone..........	50·1	78·8	26·8	14·3
Argillite	16·18	20·7	52·1	40·5
Carbonate rocks.	0·02	0·5	21·1	45·2
	100·0	100·0	100·0	100·0

The corresponding percentages for the respective parts of each series which are stratigraphic equivalents of the whole Galton series (the least complete section of the four) are approximately as follows:—

—	Carbonate rocks.	Argillite.	Sandstone.	Grit.	Conglomerate.
Lewis Series	30	55	15	....	....
Galton Series	20	52	28	....	....
Purcell Series (western phase)...	25	75		....	....
Summit Series..	under 1	18	61	12	9

As the formations are followed eastward from the summit of the Selkirks, the conglomerates and grits of the immediate shore-zone are replaced by sandstones. The sandstones are largely replaced by argillites. Finally, still farther east, argillites are largely replaced by more or less impure dolomite. These relations are illustrated in the synthetic diagram of Plate 21.

The systematic character of the chemical variations encountered along the east-west section of the prism is well shown in the analyses of Messrs. Dittrich and Connor. Those analyses which correspond to types of contemporaneous strata in the Purcell and Rocky Mountain systems have been entered in the two following tables (III. and IV.). None of these selected analyses exactly represents the average composition of a formation but each differs from the corresponding average in comparatively minor degree.

Table III. illustrates the chemical contrasts between the Kitchener quartzite and its eastern equivalent, the Siyeh magnesian limestone of the Galton and Clarke ranges. Cols. 1 and 3 refer to specimens collected at points about eighty-five miles apart.

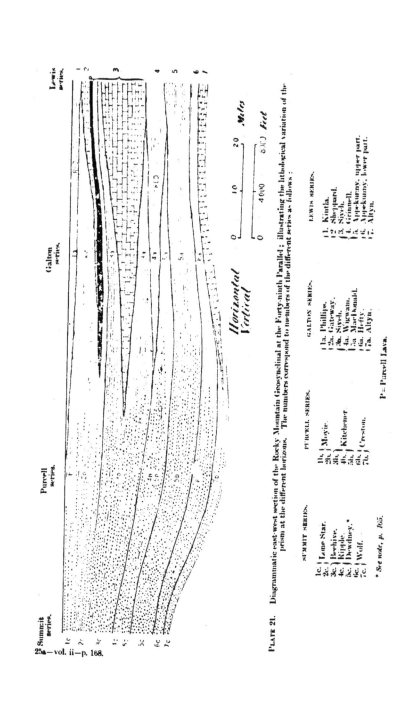

Summit series.

Purcell series.

Galton series.

Lewis series.

25a—vol. ii—p. 168.

Horizontal
Vertical

0   10   20 *Miles*

0   4000   8000 *Feet*

PLATE 21.   Diagrammatic east-west section of the Rocky Mountain Geosynclinal at the Forty-ninth Parallel; illustrating the lithological variation of the prism at the different horizons.   The numbers correspond to the numbers of the different series as follows :

SUMMIT SERIES.

1c. } Lone Star.
2c. } Beehive.
3c. } Ripple.
5c. } Dewdney.*
6c. } Wolf.
7c. }

PURCELL SERIES.

1b. } Moyie.
2b. } Gateway.
4b. } Kitchener.
5b. } Creston.
6b. }
7b. }

GALTON SERIES.

1a. Phillips.
2a. Gateway.
3a. Siyeh.
4a. Wigwam.
5a. MacDonald.
6a. Hefty.
7a. Altyn.

LEWIS SERIES.

1. Kintla.
2. Sheppard.
3. Siyeh.
4. Grinnell.
5. Appekunny, upper part.
6. Appekunny, lower part.
7. Altyn.

P = Purcell Lava.

* See note, p. 165.

SESSIONAL PAPER No. 25a

TABLE III. *Showing composition of equivalent formations.*

	1.	2.	3.
	Kitchener; western phase.	Sixth of Galton series; mean of two analyses.	Sixth of Lewis series; mean of two analyses
SiO_2	76.90	36.80	32.37
TiO_2	.55	....	.
Al_2O_3	11.95	5.92	3.51
Fe_2O_3	.69	1.40	1.47
FeO	3.04	.85	1.15
MnO	.02	....	....
MgO	1.01	6.38	5.98
CaO	.88	21.63	22.55
Na_2O	3.28	.76	.51
K_2O	1.36	1.68	1.15
H_2O-	.20	.23	.12
H_2O	1.20	2.49	2.48
P_2O_5	.15	....	....
C	....	.08	.63
CO_2	tr.	22.71	25.16
	100.33	100.33	100.03
Sp. gr.	2.680	2.748	2.741

Table IV. shows the similar contrasts between the western and eastern phases of the Creston formation when compared with each other and with the synchronous Altyn formation of the Galton and Clarke ranges. Cols. 1 and 4 refer to specimens collected at points about 100 miles apart.

TABLE IV.—*Showing composition of equivalent formations.*

	1.	2.	3.	4.
	Creston, western phase.	Creston, eastern phase.	Upper Siyeh of Galton	Altyn of Lewis series; mean of three analyses.
SiO_2	82.10	51.65	36.07	19.28
TiO_2	.40	....	....	1.43
Al_2O_3	8.86	7.85	3.92	1.43
Fe_2O_3	.49	1.74	2.08	.80
FeO	1.38	.98	2.68	4.3
MnO	.03	....	....	....
MgO	.56	3.67	12.99	16.46
CaO	.82	15.02	19.58	23.53
Na_2O	2.54	2.69	1.04	.54
K_2O	2.41	1.38	1.40	.97
H_2O-	.05	.09	.04	.11
H_2O	.37	1.81	1.72	1.07
P_2O_5	.04	....	....	....
CO_2	....	13.05	29.14	35.67
	100.02	99.93	100.46	100.29
Sp. gr.	2.681	2.644	2.816	2.792

2 GEORGE V., A. 1912

In both series of analyses the absolute amount of each oxide, except CaO, MgO, and CO_2, varies almost in simple inverse proportion to the quantity of (calcium and magnesium) carbonate which enters into the different rocks. It is to be noted, however, that in the non-carbonate portion of each rock, the iron oxides increase as the content of carbonate increases. The non-carbonate portions have been calculated to 100. The new percentages of the iron oxides in the respective rocks are noted in the following table:—

	Fe_2O_3	FeO
Kitchener, western phase..	·69	3·04
Siyeh, Galton series..	2·74	1·67
Siyeh, Lewis series..	3·30	2·60
Creston, western phase..	·49	1·38
Creston, eastern phase..	2·43	1·97
Upper Altyn, Galton series..	5·44	7·00
Altyn, Lewis series..	3·20	1·70

The fact expressed in the table goes far to explain the much stronger rusty or buff tint of the weathered rock throughout the Galton and Lewis series, as compared with that of the equivalent strata in the Purcell or Summit series. Under weathering conditions the carbonate of the eastern rocks is dissolved out, leaving the more insoluble, ferruginous material in the weathered crust. The pigmentation of the buff-weathering dolomites is, in part, also probably due to the presence of a small amount of the siderite molecule in the carbonate.

At many horizons the non-dolomitic sediments likewise tend to become more ferruginous in direct proportion to their respective distances from the old shore-line in the west. Thus, the deep-red metargillites and quartzites of the Kintla and Grinnell formations are connected, through the transitional Phillips and Wigwam formations, with the rusty-brown Moyie metargillite and Kitchener quartzite, both of which are much less charged with iron compounds. In their respective series, 550 feet of Phillips red beds (Galton series) correspond to the 800 feet of Kintla red beds (Lewis series); similarly, 1,200 feet of Wigwam red beds (Galton series) correspond to over 1,500 feet of Grinnell red beds (Lewis series). The Kintla and Phillips together form a sub-prism of red rocks which feathers out to nothing somewhere about the medial line of the Purcell range. The Grinnell and Wigwam form a second sub-prism of red rocks which also runs to a feather-edge in the Purcells. The uppermost beds of the Siyeh formation redden strongly as the sections are followed eastward from the Yahk river. Finally, the equivalents of the gray to rusty-gray upper strata of the Creston are dark reddish-brown in the Hefty sandstone of the Galton range and are either deep red shales and sandstones or buff-weathering, impure limestones in the lowermost Appekunny.

It is clear that the great geosynclinal prism is a very heterogeneous body. It is composed of a large number of formerly horizontal sub-prisms of stratified rock. These are intimately dove-tailed together and some of the sub-prisms have complicated multiple edges. As a rule these edges are not sharp, since the rock of one prism merges gradually into the contemporaneous material of a sub-prism of a different rock type.

For example, the Moyie-Gateway sandstone sub-prism thins rapidly eastward and does not appear at all in the Clarke and Lewis ranges, its place being taken by the contemporaneous Sheppard magnesian quartzite and the red rocks of the Kintla. The thick magnesian limestone of the middle Siyeh thins toward the westward, being dovetailed first into argillite and then, farther westward, into sandstone, both of which rocks are contemporaneous with the limestone. The limestone in its most westerly outcrops occurs in the form of several thin tongues running out westward from the main limestone sub-prism into the Kitchener quartzite. The sub-prisms of red beds have already been described. The thick sub-prism of silicious and magnesian limestone composing the Altyn thins out somewhere between the Yahk river and the Wigwam river, being replaced on the westward by the contemporaneous Creston quartzite. The great lenses of Dewdney conglomerate and Wolf grit similarly, but more rapidly, thin out to the eastward and are replaced by homogeneous Creston quartzite.

The lithological variations in the geosynclinal as a whole, when considered in transverse section, are relatively rapid, distances of only fifty or a hundred miles corresponding to profound differences of composition in contemporaneous strata. The persistence of the lithological units along the N.W.-S.E. axis of the geosynclinal seem to be much more pronounced than in the transverse section established on the Boundary line. Yet the work of McConnell and Dawson north of the line and of several American geologists, particularly Walcott, Willis, Lindgren, Ransome, Calkins, and MacDonald, all working in Idaho and Montana, shows that, even along the Cordilleran axis, there is considerable lithological variation among contemporaneous beds of the geosynclinal.

The maps and sections accompanying this report represent the outcrop and relations of lithological individuals. If sufficient paleontological evidence to date the strata of all the series in an actual time-scale ever be secured, and the same Boundary belt be mapped to show the outcrops of strictly contemporaneous formations, that map would have a very different look from the one here presented.

METAMORPHISM OF THE GEOSYNCLINAL PRISM.

One of the most notable features of the Monk formation (Summit series) as it crops out in the Monk creek section, is the pronounced increase of metamorphic effects over those witnessed in the overlying and similarly upturned sediments. Slaty cleavage and true schistosity are inconspicuous structures in the Wolf, Ripple, and Beehive formations but are regularly recurring structures at most horizons of the Monk, Irene Volcanic, and Irene conglomerate formations. The development of these secondary structures on so great a scale is doubtless related to the original depths of burial of the lowest three members of the Summit series. Before the series was flexed up, the beds of the Monk formation lay blanketed beneath at least 15,000 to 20,000 feet of the overlying conformable beds; it is very

2 GEORGE V., A. 1912

possible that several thousand feet of still younger rocks were piled upon the Lone Star formation. Assuming present normal temperature gradient of about 1° C. per 100 feet of ... the Monk and Irene formations must have had original temperature ... an 150 to 300 or more Centigrade.

It is evident that ... of the formation of new, metamorphic minerals under the tremen... ial stresses of mountain building, were greatly facilitated by these ... degrees of heating. One can hardly wonder that every member ... formation and almost every foot of the Irene formations, show some ... In the Monk formation the increase of -bearing with depth of origi... well ... conglomeratic bands, b, d, h and j. The conglom... is identical in appearance with the conglomerate ba... relatively unsheared Dowdney formation. On the other ... embers h and j show strong pressure-flattening or stret... aracterizing the basal Irene conglomerate.

Analogous relations were ... in the Purcell series. While the Moyie, Kitchener, and upper part of the Creston quartzite seldom showed schistosity even when strongly deformed, the lowest beds of the Creston often displayed a tendency toward the development of sericitic schists where deformed to about the same extent. This parallel behaviour of the Summit and Purcell series under similar dynamic stress favours, though of course not compelling, the correlation of the two series as parts of one large sedimentary prism.

In general, it may be stated that, from the summit of the Selkirks to the Great Plains on the Forty-ninth Parallel, the sediments of the Rocky Mountain Geosynclinal to a depth of about 20,000 feet below the top of the Carboniferous lime-stone, show few traces of what is ordinarily called dynamic metamorphism. This does not mean that the strata have not been strongly upturned, for at many points they approach or reach verticality. Within that 20,000-foot zone the sediments have been thoroughly indurated and very largely recrystallized, but almost entirely through static metamorphism (Belastungsmetamorphismus); the recrystallization seems to have been essentially completed before the prism was folded and faulted. Below the 20,000-foot zone the rocks were at such temperatures and pressures that, when deformation began, shearing and true dynamic metamorphism with the creation of schistose structures, were the rule.

SPECIFIC GRAVITY OF THE GEOSYNCLINAL PRISM.

During the laboratory study of the rock collections, specific gravity determinations of the stratified rocks were often found to be desirable. The purposes of the determinations were so various, and the number of specimens handled so considerable that it involved but little extra time and labour to make a fairly complete set of determinations for the typical, fresh specimens collected between Waterton lake and the summit of the Selkirk range. The result has been to give a tolerable idea of the average specific gravity of the different formations composing each series.

Weighting each value according to the thickness of the corresponding formation, the average specific gravity of each of the four great series has been calculated. The final result affords a fair estimate of the actual density of the Rocky Mountain, Purcell, and Selkirk ranges where these mountains, as at the Forty-ninth Parallel, are almost entirely carved out of the pre-Silurian geosynclinal rocks.

On account of the thorough induration and compactness of nearly all the specimens, the usual trouble arising from included air has not been seriously felt. Other sources of observational error were partly obviated by the use of selected, whole hand specimens (880 to 1,000 grams in weight), whereby rather reliable averages for the formations were secured.

The following tables embody the principal results:

TABLE V.—*Showing the calculated average densities of the four series, including the interbedded lavas.*

	Number of specimens measured.	Thickness of series in feet.	Average sp. gr. of series.
Lewis series	39	13,750	2.737
Galton series	73	12,160	2.711
Purcell series, western phase	57	20,300	2.701
Summit series	55	32,050	2.750
	262		

TABLE VI.—*Showing the calculated average densities of the four series, excluding the interbedded lavas.*

	Number of specimens measured.	Thickness of sediments in feet.	Average sp. gr. of sediments.
Lewis series	37	13,100	2.714
Galton series	67	12,150	2.708
Purcell series, western phase	35	20,300	2.701
Summit series	40	20,050	2.710
	179		

TABLE VII.—*Showing the calculated average densities of that part of the geosynclinal which corresponds to the full thickness of sediments in the Galton series.*

	Average sp. gr.
Lewis series	2.726
Galton series	2.708
Purcell series	2.702
Summit series	2.705

2 GEORGE V., A. 1912

Table V. gives actual densities in a part of the Cordilleran region and may possibly be of some value in discussions of pendulum observations or of other geophysical problems as they may be concerned with this region in the future.

The averages of Table VI. express the range of densities in a typical, thoroughly consolidated (statically metamorphosed) geosynclinal prism.

Table VII. indicates the approximate density relations of the Galton series, the least completely exposed series from the prism, to those of the equivalent strata of the other three series. The average densities of the two western equivalents are sensibly the same as that of the Galton series. On account of the extensive development of dolomite in the Lewis series, its average density is, as was to be expected, considerably higher than that of any other of the series.

CORRELATION OF THE FOUR BOUNDARY SERIES WITH THE CASTLE MOUNTAIN-BOW RIVER (CAMBRIAN) GROUP.

During the course of the field work in 1905 it gradually became suspected that the as yet unfossiliferous Siyeh limestone is the stratigraphic equivalent of the Cambrian Castle Mountain limestone of McConnell's well-known section on the main line of the Canadian Pacific railway. This suspicion was strengthened in the course of a brief examination of the rocks at and east of Mt. Stephen in the autumn of that year. The importance of the correlation prompted a second and longer field-study which might, to some extent, supplement McConnell's all-too-brief report on the great section. Toward the close of the season of 1906 the writer accordingly spent five days in working over the type sections on the northeast and southwest sides of the Bow river valley. The time available was too limited to secure a detailed columnar section of the group; yet the field evidence was clearly in favour of the correlation of the Siyeh and Castle Mountain formations.

The principal information was obtained from two partial sections, the one running northeastward from Eldon station to the 9,800-foot, unnamed summit northwest of Castle mountain; the second, running westward from Lake Louise chalet to the base of Popes Peak. Combining the results of the two traverses, the following succession was established:

Top, erosion surface.

3,500 feet.—Impure magnesian limestone with thin interbeds of shaly metargillite.
1,500 " 　 Quartzite in thin to thick beds.
1,200+ " 　 Fine-grained conglomerate, grit and quartzitic sandstone.

Base concealed.

Reference to the published report and, afterwards, personal consultation with Mr. McConnell, gave assurance that the limestone typically represented the Castle Mountain formation in its lower part, while the quartzites, conglomerate, and grit as typically represented the Bow River formation in its upper part.

PLATE 22.

Molar-tooth structure in Siyeh limestone (weathered), Clarke Range. Sunken parts calcium carbonate; remainder dolomitic. Two-thirds natural size.

Molar-tooth structure in Castle Mountain dolomite (unweathered, on main line of Canadian Pacific Railway. Light parts dolomitic; dark parts calcium carbonate. Two-thirds natural size.

PLATE 22.

Molar-tooth structure in Siyeh limestone (weathered), Clarke Range. Sunken parts calcium carbonate ; remainder dolomitic. Two-thirds natural size.

Molar-tooth structure in Castle Mountain dolomite (unweathered) on main line of Canadian Pacific Railway. Light parts dolomitic ; dark parts calcium carbonate. Two-thirds natural size.

25a--vol. ii p. 174.

In the section northeast of Eldon the Bow River rocks are not well exposed but the limestone shows about 3,500 feet of its thickness. The dip averages 25° northeast. At its base the limestone is rather massive and is composed of firmly knit, thin beds of alternating, gray-weathering and light brown to buff weathering, impure carbonate. The gray layers carry little magnesium carbonate, but the brown-buff layers are dolomitic. The thickness of these layers runs from a fraction of an inch to two inches. About 1,000 feet above the base a band, 100 feet or more in thickness, is exceptional in being thin-bedded and easily cleaved but it preserves the buff weather-tint and a dolomitic composition. Similar thin zones occur both above and below this band. In general, however, the limestone is not only thick-platy in structure as in the coursing of heavy masonry, but, like the Siyeh limestone, shows a rather uniform, buff to brown weathering tint and high content of magnesium carbonate. Both phases of the fresh limestone are normally rather dark-gray or bluish-gray. The rock is often arenaceous or argillaceous; the weathered surface is roughened very often, through the projection of the sand-grains. The unequal distribution of carbonate and impurity renders the surface characteristically pitted.

In a specially argillaceous bed at the top of the 100-foot, thin-bedded limestone band, fossils, chiefly trilobite fragments of apparently Middle Cambrian age were discovered. Middle Cambrian fossils were also found at the base of the whole limestone formation.

The likewise excellent section above Laggan, thirteen miles to the northwestward and across Bow river valley, disclosed practically identical features in the limestone. The lower beds, which were found to crop out at Lake Agnes afforded fragments of indeterminable trilobites and crustacean tracks.

Perhaps the most significant fact derived from the section is that the limestone very often possesses the typical molar-tooth structure so characteristic of the Siyeh limestone. (Plate 22).

This structure is not well developed in any part of the Eldon section but it was again seen in the limestone at Mt. Stephen and at several other points along the railroad, where the line cuts across outcrops of the Castle Mountain formation. As in the Siyeh limestone the molar-tooth structure seems to be best developed where the rock has been locally cleaved or cracked by orogenic stress.

The thick quartzite underlying the limestone is well exposed at the Laggan section. It is a thick-bedded formation, heavy plates of quartzite alternating with subordinate, thin, fissile intercalations of silicious metargillite. The general colour of the quartzite on a fresh fracture is pale reddish to reddish-gray; the colours of the weathered rock are in general, rusty-brown and red but vary through white, pale gray, pale red, pink, brown, purple, and, toward the top of the formation, deep maroon-red. The argillaceous interbeds are dark-gray or greenish, weathering greenish or grayish-brown. Cross-bedding, ripple-marks, annelide trails, and borings are all common. The lithological similarity of the lighter tinted and thicker beds to the Ripple quartzite

2 GEORGE V., A. 1912

of the Summit series was marked, and even more striking was the likeness of the reddish beds to standard phases of the Wigwam formation of the Galton series and the Grinnell formation of the Lewis series. No fossils were discovered in the quartzite.

The Bow River conglomerate and grit were also best seen in the Laggan section. Only 1,200 feet of these rocks appear, the base being here hidden beneath the Glacial gravels of the Bow valley. All stages of transition are represented between the conglomerate with well-rounded pebbles an inch or less in diameter, to a quartzitic sandstone of medium grain. The abundant and heavy beds of grit represent the rock of intermediate grain. These three sedimentary types occur in alternation through the whole 1,200 feet, though the conglomerate lenses seem most common toward the top. All the strata belong to one great lithological individual of heterogeneous grain but rather constant chemical composition. The conglomerate is made up of glassy, white, or bluish, often opalescent quartz pebbles with subordinate, large rounded grains of feldspar. These fragments are all cemented in a silicious matrix, itself feldspathic to some extent. The grits and sandstones are but finer grained phases of the same silicious, sedimentary material. In composition and the gray and greenish colours of fresh and weathered surfaces, the conglomerate and grit can hardly be distinguished from staple phases of the Wolf grit of the southern Selkirks. The similarity even extends to such a detail as the changeable tints of the opalescent quartz pebbles and grains. No fossils were found in this division of the Bow River formation, though it was apparently within this subdivision that Dawson found Lower Cambrian fossils at Vermilion Pass.

It would be highly desirable to have studied in the field the Bow River beds below the conglomerate-grit member and also the upper part of the Castle Mountain formation, but sufficient time for this could not be spared out of the field season. Yet it is believed—and Mr. McConnell, to whom the field data and typical specimens were submitted, agrees in the belief,—that sufficient evidence has already been secured to suggest the stratigraphic relation of the Castle Mountain-Bow River group to the old sedimentary prism traversed at the Forty-ninth Parallel.

The suggested correlation is as follows. The lower 4,000 feet or more of the Castle Mountain limestone is stratigraphically equivalent to the Siyeh formation and thus to the larger part of the Kitchener quartzite, and, again, to the larger part of the Beehive quartzite. The 1,500-foot quartzite immediately underlying the Castle Mountain limestone is the equivalent of the Grinnell and Wigwam formations, of the lowest beds of the Kitchener quartzite, and of the Ripple quartzite. The Bow River conglomerate-grit member at the base of the Laggan section is equivalent to the Dewdney quartzite and upper part of the Wolf grit formation in the southern Selkirks.

Since the foregoing paragraphs were written, Walcott has made a detailed study of the Castle Mountain group. His results corroborate McConnell's stratigraphy and show yet more precisely the range of the Upper, Middle, and

Lower Cambrian horizons in the great series.* Fossils were obtained in sufficient abundance to show that the base of the massive dolomitic limestone is the plane separating the Middle Cambrian from the Lower Cambrian in the region. McConnell's early view of the correlation is, therefore, finally established. It follows that, if the Siyeh formation and Castle Mountain dolomite are synchronous deposits, a critical horizon in the Forty-ninth Parallel series has been somewhat definitely fixed. Walcott also found Lower Cambrian fossils in the Lake Louise formation, at a zone about 3,000 feet below the top of the Bow River group.

Summary.—The evidences for the correlation may be restated in summary form.

In composition, in colours of fresh and weathered surfaces, in character of bedding and general influence on mountain forms, the Siyeh and Castle Mountain limestones are almost identical. The similarity is specially marked in the occurrence of the highly peculiar molar-tooth structure in both limestones. The correlation on these grounds is strengthened through the strong improbability that two magnesian limestones of such immense thickness and of similar characters should have been deposited so near together as these Bow River and Boundary line sections and yet be of widely different dates of formation. The discovery of fossiliferous Castle Mountain limestone in large development at Nyack creek, only ten or fifteen miles from Siyeh mountain itself, renders this improbability all the more convincing.§

The Siyeh limestone in the Galton, Clarke, and Lewis ranges is underlain by red quartzitic sandstones which correspond in essential features to the quartzite at the top of the Bow River formation. Certain whitish and massive beds in this quartzite also strongly recall the Ripple quartzite of the Summit series, a formation which, on independent grounds, has been correlated with the Wigwam and Grinnell formations.

Finally, the Bow River conglomerate is as strikingly similar to the Monk grit of the Summit series as the Castle Mountain limestone is like the Siyeh. Also on independent grounds the Wolf grit has been correlated with the Appekunny quartzite-metargillite which underlies the Grinnell and thus belongs to a stratigraphic horizon below the 1,500-foot Bow River quartzite at Laggan and Eldon.

Not only are there close similarities of lithological detail between the northern and southern rock-groups; the succession of formations is alike. The differences between the successive members of the two sections is due simply to the expected differences subsisting between contemporaneous sediments laid down in the one sea-basin. The Castle Mountain-Bow River group of strata is, in a sense, a composite of the entire pre-Silurian geosynclinal as exposed at the Forty-ninth Parallel. The Castle Mountain formation has its nearest lithological equivalent in the extreme eastern, Lewis series at the Boundary

* C. D. Walcott, Smithsonian Miscellaneous Collections, Vol. 53, No. 1804, 1908, p. 1 and No. 1812, 1908, p. 167.
§ Cf. C. D. Walcott, Bull. Geol. Soc., America, Vol. 17, 1906, p. 13.

25a—vol. ii—12

2 GEORGE V., A. 1912

line. The Bow River grit and conglomerate, where examined, have features identical with those of the Wolf formation in the extreme western, Summit series at the Boundary. The Bow River quartzite, where examined, has features like both the Ripple quartzite of the Selkirks and the equivalent Wigwam sandstone of the Galton range.

The systematic position of the Sheppard and Kintla formations, as of their respective equivalents, the Gateway, Phillips and Roosville formations of the Galton series, the Moyie formation of the Purcell series, and the Lone Star formation of the Summit series, is not apparent from their lithological comparison with the rocks of McConnell's section.

All of these Forty-ninth Parallel formations are unfossiliferous and their conditions of deposition (chiefly subaerial or in shallow water) were markedly different from those under which the upper beds of the Castle Mountain series (dolomitic limestones) were laid down.

A further clue to the correlation has been found in the fact that, in the Belt mountains and to the westward, the equivalent of the Siyeh formation (Marsh-Helena beds) is, over large areas, conformably overlain by the fossiliferous, Middle Cambrian Flathead sandstone. This sandstone is often coarse, little metamorphosed, and clearly shows its origin as a sandy deposit on the floor of a transgressing sea.§ This genetic feature seems to be well matched in the character of the massive, coarse sandstone beds occurring at the base of the Gateway formation, immediately above the (Purcell) lava-cap of the Siyeh. The lithological resemblance, coupled with the similar stratigraphic relations to the common (Siyeh, Marsh-Helena) horizon, suggests that the lower beds of the Lone Star, Moyie, Gateway, and Sheppard formations are, respectively, equivalents of the Flathead sandstone and are thus of Middle Cambrian age.

Since the Flathead horizon is well below the recognized top of the Middle Cambrian, since the succeeding Middle Cambrian time was long enough for the deposition of at least 1,500 feet of limestone in the area of the Canadian Pacific section, and since there is perfect conformity in all four of the Forty-ninth Parallel series above the Siyeh or its equivalent, it seems probable that the argillites and dolomites overlying the Siyeh or its equivalent are all or nearly all of Middle Cambrian age.

The foregoing tentative correlations are expressed in Cols. 1, 2, 7, 8, 10 and 12 of Table VIII.

§C. D. Walcott, Bull Geol. Soc., America, Vol. 10, 1879, p. 209.

Icosaic?

Lone Star. Metargillite, phyllite, and quartzite; dark tints of gray, green, etc. 2,000 + feet.	*Moyie.* Metargillite, shale, and quartzite; various tints of (deep) gray, red, green, etc.; shallow water features. 3,000 + feet.	*Striped Peak.* Shales and sandstones, red and green. 1,000 + feet.	*Striped Peak.* Shales and shaly sandstones, prevailingly dark red; rip ple marks, etc. 2,000 + feet.
Berkire. Quartzite and metargillite; gray and pale greenish, weathering brown of different tints. 7,000 feet.	*Kitchener,* upper part. Quartzite and metargillite; greenish and gray, weathering to browns. 6,000± feet.	*Wallace.* Shales, more or less calcareous, with thin limestone interbeds. Limestones and calcareous shales weather buff. 4,000 feet. *St Regis.* Shales and sandstones, purple and green. 1,000 feet.	*Blackfoot* (called *Newland* by Calkins). Lime stones, thin-bedded, si licious and ferruginous, interbedded with more or less calcareous shale. 5,000± feet.
Ripple. Quartzite; white, pale pink and yellow; very massive. 1,650 feet. *Dewdney.* Quartzite, with interbeds of conglomerate and metargillite; weathering gray and pale brown. 2,000 feet. *Wolf,* upper part. Grit, sandstone, and fine conglomerate; gray, fresh and weathered. 1,000± feet.	*Kitchener,* lower part. Quartzite, relatively massive; gray, green, and white, weathering light brown. 1,400± feet. *Creston,* upper part. Quartzite, in thick and thin beds; gray and greenish, weathering gray. 3,000± feet.	*Revett.* Quartzite; white, massive. 1,200 feet. *Burke.* Silicious metargillite, with quartzite, prevailingly gray green. 2,000 feet. *Prichard,* upper part. Metargillite and quartzitic sandstones, with shallow water features; gray to black. 1,500± feet.	*Ravalli,* upper part. Quartzites and metargil lites; gray, greenish, white, and purplish. 5,000± feet. *Ravalli,* lower part: lithology as above. 3,000± feet.
Wolf, lower part. Grit, conglomerate, and sandstone; gray. 1,900± feet. *Monk.* Quartzite, metargillite and conglomerate; gray tones. 5,500 feet. *Irene Volcanic formation.* 6,000± feet. *Irene Conglomerate.* 5,000 + feet.	*Creston,* lower part. Quartzite, with interbeds of metargillite; gray colours. 6,500± feet. Base concealed.	*Prichard,* lower part. Metargillite and quartzite; gray to black. 6,500± feet. Base concealed.	*Prichard.* Metargillite; dark bluish, banded. 2,000 feet. Sandstones; thick-bedded to shaly; gray. 10,000± feet. Base concealed.
Total—32,050 + feet.	Total—20,300 + feet.	Total—17,200 + feet.	Total—27,000 + feet.

Unconformity.

Priest River Terrane.

(1) F. C. Calkins. Bull. 384. U.S. Geol. Survey. 1909. p. 40.

GALTON SERIES, CLARKE ... ...
RANGE, 49° N. LAT. AND ... ...

9

10

SERIES IN ... ...
MOUNTAINS ...

CASTLE MOUNTAINS BOW
RIVER SERIES (?)

SERIES
F...

Erosion surface.   Erosion

...formity with upper Pa-   Conformity with upper Pa-   Confor-
leozoic.   leozoic.   Pale-

Sherbrooke. Limestone; St. Cha-
at ay. 1,375 feet.   with
Pentl. Gray limestone.   gray.
300± feet.
Bosworth. Gray limestone
and shale. 1,835+ feet.

...stone. Gateville. Metargillite Kintla. Argi-   ...alliatin. Limestone.
...lstones   and quartzite; gray   zite and d-   Flathead. Quartzitic
...stones   and greenish. 600 feet.   terbeds; ...   sandstone.
Phillips. Metargillite   Sheppard.   an   ...Mount
and quartzite; red. 350   Dolomite; ...   ...often
feet.   thering buff   ...Blamm
Gateway, upper part.   Aldon. Limestone; gray,   gray
Metargillite with quart-   weathering buff and   shale
zite; gray when fresh,   gray. 2,728 feet.   Blacka
gray and brown weath-   Stephen. Limestone and   stone
ered. 1,880 feet   shale; gray. 640 feet.   Ute. G
Cathedral. Dolomitic   shaly
bedded li   Gateway, lower part.   Sheppard. D-   limestone; gray, weath-   Spence
layers of   Quartzite, dolomite,   Buff weathe-   Marsh. Red shale. 80± feet.   ring buff and gray. 1,595   30 fe
buff wea-   and sandstone; various   ...te and reddish   Helena. Limestone with   feet.   Langst
feet.   colours. 125 feet.   stone. 100 feet   some shale; gray, wea-   stone
Siyeh. Dolomite lime-   Siyeh. Dolomite, lime-   thering buff. 2,400 feet.   Brigha
stone with much metar-   stone with ...   Empire. Argillite; green-   ian c
gillite; gray and green-   interbeds; ...   ish gray. 600 feet.   ing b
ish, weathering buff.   Galton series ... ...
1,000 feet.   feet.

rt. Quart-   Wigwam. Sand-stones and   Grinnell. Metargi-   Spokane. Argillite; deep   Mount White. Gray lime-   Reigat
urple and   metargillites; red colors   and sandstones gen-   red. 1,500 feet   stone and shale. 380 feet.   Tota
b.   dominant. 1,200 feet.   rally red. 1,600 feet   Greyson. Argillite and   St. Piran. Greenish and   fe
MacDonald. Chiefly   Ipokwana. Metargil-   quartzite; gray. 2,000±   gray sandstone. 2,705 feet.
metargillite; gray and   lite; gray or greenish   feet.   Lake Louise. Shale;
greenish, weathering   gray, weathering g-   gray. 105 feet.
gray and brown. 2,360   or brownish 2,640 feet   Fairview. Gray quartzite,
feet.   weathering brownish.
Hefty. Sandstones, with   600+ feet.
some shaly beds, some-
times calcareous; red
and reddish gray. 775
feet.

t. Quart-   Altyn. Siliceous dolomite,   Altyn. Siliceous ... ...   Greyson. lower part. Argil-   Continuation of Bow Ri-
gray.   thin-bedded; gray and   sandy dolomites; ...   lite; gray. 1,000+feet.   ver argillites, etc.
greenish-gray, weather-   gray, weathering buff.   Newland. Silicious lime-
ing buff. 650 feet.   3,500+feet.   stone; gray, weathering
Waterton. Dolomite;   buff. 2,200 feet.
gray. 300 feet.   Chamberlain. Argillite;
dark gray to black. 1,500
feet.
Neihart. Gray and green-
ish quartzite. 700 feet.

Base concealed.   Base concealed.   Base concealed.   Base c

feet.   Total—12,100 feet.   Total 15,120 feet   Total—14,000±feet.   Total—12,35 feet.   To

Unconformity.

Cherry Creek Beds.

SESSIONAL PAPER No. 25a

CORRELATION WITH THE BELT TERRANE.

Earlier Views on the Belt Terrane.—To one acquainted with the geological literature of the Cordillera, the writer's foregoing correlation is obviously quite different from that adopted by the United States Geological Survey for these old formations as exposed in the United States. By the present official views of that survey, (1) the whole of the Belt Mountains series which underlies the Flathead sandstone (the Belt terrane) is referred to the pre-Cambrian (pre-Olenellus) or latest 'Algonkian'; and (2) the whole of the Lewis series as described by Willis is considered as belonging to the same terrane and to the same age.

A brief history of the explorations on which these conclusions are based is given by Walcott in his paper on 'Pre-Cambrian Fossiliferous Formations.'[†] To facilitate the present discussion it is advisable to review the history at least so far as to show the divergence of views on the correlation.

In 1875 Dawson had incorrectly referred the Sheep formation to the Carboniferous, but later he stated the possibility that the Siyeh limestone is the equivalent of McConnell's Castle Mountain limestone. He referred all the underlying formations as far as the upper Altyn to the Cambrian.[‡]

In 1882 Davis made several sections through the Belt terrane rocks and referred them provisionally to the Lower Cambrian, though without fossil evidence.[§]

The next year Peale sectioned 2,300 feet of the terrane, then called the East Gallatin group, and referred the whole series to the Cambrian.[**]

In 1893 he published a second account of the terrane, calling it the 'Belt formation.' A significant paragraph may be quoted:

'There is no doubt that after the Belt formation was deposited there was an orographic movement by which the Archean area of nearly the entire region represented on our map south of the Gallatin and Three Forks was submerged just prior to the beginning of the Cambrian, before the Flathead quartzite was deposited. Whether this movement occurred immediately after the laying down of the Belt beds or after an interval is of course the question to be decided, and the decision cannot be positively reached with the meagre data now at hand. I am inclined to think that the subsidence of the Archean continent (or possibly islands) began with the first accumulation of the sediments that formed the lower portion of these beds and was coincident with their deposition throughout the

---

[*] Recently Walcott has used the adjective "Beltian," a systemic form, to designate the "Belt terrane" of earlier publications. (C. D. Walcott, Smithsonian Misc. Collections, Vol. 53, 1908, p. 169).
[†] C. D. Walcott, Bull. Geol. Soc. America, Vol. 10, 1899, p. 201.
[‡] G. M. Dawson, Report on the Geology and Resources of the Region in the vicinity of the Forty-ninth Parallel, 1875, p. 74; Bull. Geol. Soc. America, Vol. 12, 1901, p. 69; cf. Ann. Report. Geol. Surv. of Canada for 1885, p. 39B and 50-51B.
[§] W. M. Davis, Tenth Census report, Vol. 15 on Mining Industries, 1886, pp. 697-702.
[**] A. C. Peale, 6th Ann. Report, U.S. Geol. Surv., 1885, p. 56.

2 GEORGE V., A. 1912

entire period. It may have been succeeded by an emergence of the land area for a brief period, but the probability is that the interruption to the downward movement, if it occurred, was slight. Next, the widespread pre-Cambrian subsidence preceding the formation of the Flathead quartzite took place, and the Cambrian sea covered large areas that had hitherto been above the sea level. There is a marked difference in the character of the beds of the two groups. Little, if any, induration is seen in the Flathead formation, while the Belt beds are so altered in most cases as to resemble the metamorphic rocks which underlie them, and from the breaking down of which they were derived. Notwithstanding the metamorphism, there is no mistaking their sedimentary character.'[+]

In 1896 Peale summarized his net conclusion regarding the correlation in the following words:

'It is possible that further investigation may result in the reference of this formation to the lower part of the Cambrian. At present, however, it is referred provisionally to the Algonkian.'[*]

Weed, Iddings, and Pirsson in several publications issued between the years 1894 and 1899, refer the terrane to the Algonkian, though Weed and Pirsson, after close study of the Castle Mountain (Montana) mining district, wrote:

'Both the character of the sediments and their position beneath the beds of Middle Cambrian age indicate their similarity to the Bow River beds of the Canadian geologists, in which Lower Cambrian fossils are found. It has, however, been decided to class the beds as Algonkian.'[§]

In 1898 Walcott made a general study of the terrane as exposed in the Big Belt and Little Belt mountains and in the Helena district. He writes:—

'The results of my investigation were the discovery of a great stratigraphic unconformity between the Cambrian and the Belt formations; that the Belt terrane was divisible into several formations, and that fossils occurred in the Greyson shales nearly 7,000 feet beneath the highest bed of the Belt terrane.'[‡]

Walcott's columnar section of the terrane is that given in Col. 9, of Table VIII.

In 1902 Willis stated in the following words his correlation of the Lewis series as exposed in the Clarke and Lewis ranges:

'The oldest formation of the series, the Altyn limestone, is assigned to the Algonkian period on the basis of fossils discovered by Weller in its characteristic occurrence at the foot of Appekunny mountain near Altyn, Montana. These fossils are fragments of very thin shells of crustaceans [chiefly *Beltina danai*].............The fossiliferous strata of the Belt

[+] A. C. Peale, Bull. U.S. Geol. Survey, No. 110, 1893, p. 19.
[*] Three Forks Folio, U.S. Geol. Surv., 1896, p. 2,
[§] Bull. U.S. Geol. Surv., No. 139, 1896, p. 139.
[‡] C. D. Walcott, Bull. Geol. Soc. America, Vol. 10. 1899, p. 204.

formation in the Belt range are separated from the Cambrian by 7,700 feet of sediments and an extensive unconformity. In the Front range of the Rockies 10,700 feet of apparently conformable strata overlie the fossiliferous bed, and it is possible that the plane of division between the Algonkian and Cambrian as determined by paleontologic evidence will be found in this great series. In the upper part of the Siyeh limestone near the head of Mineral creek, Weller found some indistinct forms which he considers as possibly to be parts of crustaceans. Walcott expresses a similar view, saying:

'" Mr. Weller's suggestion that the fragments possibly represent crustacean remains appears to be the most plausible. If from a Devonian horizon they would suggest the genus *Licas*, or some of its subgenera. It is a case where more material is needed in order to arrive at any definite conclusion." '

In his paper on ' Algonkian Formations of Northwestern Montana ' Walcott refers the entire Lewis series and its equivalents to the pre-Cambrian (Algonkian) system.† In this conclusion he has been followed by Calkins, Ransome, MacDonald, and Lindgren, all working on the western phase of the Belt terrane in Idaho.§ The same view has governed the compiling of the geological map of North America which was prepared for the session of the International Geological Congress held at Mexico City in 1906; the large area of ' Neo-Algonkian ' shown in the States of Montana and Idaho represents the Belt terrane.

As one of Walcott's last (1906) papers on this subject shows the trend of opinion among the United States geologists, the more important parts of his table of equivalents has been reproduced in Table IX. of the present report.

---

* B. Willis, Bull. Geol. Soc. America, Vol. 13, 1902, p. 317.
† C. D. Walcott, Bull. Geol. Soc. America, Vol. 17, 1906, p. 17.
§ W. Lindgren, U.S. Geol. Surv., Prof. Paper, No. 27, 1904, p. 16.
F. L. Ransome, U.S. Geol. Surv., Bull. No. 260, 1904, p. 277.
D. F. MacDonald, U.S. Geol. Surv., Bull. 285, 1906. p. 43.
F. C. Calkins, U.S. Geol. Surv. Bull. 384, 1909, p. 27

TABLE IX.—*Showing Walcott's correlations in the 'Belt Terrane.'*

BELT MTS., MONTANA.	LEWIS AND CLARKE RANGES, MONTANA.	CAMP CREEK, MISSION RANGE, MONTANA.	CŒUR D'ALENE DISTRICT, IDAHO.	PURCELL RANGE, IDAHO-BRIT. COLUMBIA.
	No superjacent strata.	*Cambrian,* Unconformity.		
		1a, arenaceous gray, 1,762'		
Cambrian, (Flathead ss) Unconformity.	*C. da,* *Sheppard,* Quartzites, 1,200'	2a, calcareous and arenaceous, 1,566'		
Marsh, 800'				
Helena, calcareous, 2,400'	*Siyeh,* Limestone, 4,000'	3a. to 3g. Arenaceous, mostly reddish, 4,491'		
Empire, 800'+ *Spokane,* 1,500'+ *Gregson,* 3,000'+			No superjacent strata.	
Arenaceous, 5,100'	*Grinnell, Appekunny,* Silicious, 3,800'	4. to 7e. Arenaceous, red and gray, 3,887' (198' of li. near summit.)	*Striped Peak,* 2,000'	
Newland, Calcareous,2,200'+	*Altyn,* Calcareous and silicious, 700'			
	Base concealed.			
Chamberlain, Silicious, 1,500'	Total, 9,700'	*Blackfoot,* Calcareous and silicious, 4,800'	*Wallace,* Calcareous and silicious, 5,000'+	
				No superjacent strata.
Neihart, 700'				
Unconformity.		*Ravalli,* silicious and arenaceous, purple, greenish and gray beds, 8,255'	*Burke–St. Regis,* Silicious and arenaceous, purple, greenish and gray beds, 8,000'	*Monir,* Metargillite and quartzite, 3,500'
Archean.				*Kitchener,* Quartzite, 7,400'
Total, 12,000'				
		Base concealed. Total, 24,770'	*Prichard,* Banded, dark blue gray, blue black and gray, silicious series, 10,000'	*Creston,* Quartzite, 9,500'+
		Base concealed. Total, 25,000'		Base concealed. Total, 20,400'

A more recent table of correlation has been published by Calkins, who has similarly equated the Kitchener with the Ravalli and Burke horizons; Wallace and Blackfoot with Newland and Altyn.* The present writer has not been able to agree with these correlations. As stated in his summary report for 1905, the Kitchener quartzite passes into, and is the equivalent of the Siyeh limestone; and part of the Creston quartzite is equivalent to the Appekunny metargillite. In 1907 the writer came to suspect that the Blackfoot limestone is the equivalent of the Siyeh limestone, and the next year Walcott proved this to be the case.† It seems necessary, therefore, to make significant alterations in Calkins' table. His 'Newland' limestone in the Philipsburg and Cabinet Range districts, if the equivalent of the Blackfoot limestone, as he states, must belong to the Siyeh horizon. The other members of each stratigraphic column must be correspondingly shifted. These changes are noted in the general table VIII. Though it may be at fault in details, that table is believed to express the main relations subsisting among the different sections in the 'Belt terrane.' Walcott's discovery of the equivalence of the Blackfoot and Siyeh limestones has clearly simplified the whole stratigraphic situation in Montana and Idaho.

Evidence of Fossils.—The principal difficulty in the correlation with recognized systems is, of course, the rarity of fossils which can in any sense determine horizons. The only well characterized fossil horizons yet found in the Belt terrane as defined by Walcott, occur in the lower part of the Greyson shales of the Belt mountains and in the upper part of the Altyn formation in the Lewis range. These two horizons may well be practically contemporaneous, as they occur in similar stratigraphic relations and both carry the abundant species, *Beltina danai*. Neither that species nor any of the associated obscure organisms can directly date the horizon, which has never been found in undoubted association with the Olenellus zone or other general horizon of the Cambrian. So far as purely paleontological evidence is concerned, it is quite within the bounds of possibility that the *Beltina* horizon is really a lower phase of the Lower Cambrian, Olenellus zone, or is but slightly older than that zone.

No organic remains giving decisive indications of age have been found in the overlying Spokane, Empire, Helena, and Marsh formations of the Belt mountains or in the equivalents of these, either at the Forty-ninth Parallel or in the thick deposits of Idaho and western Montana.

Walcott recognizes the Cambrian-Ordovician equivalent of McConnell's Castle Mountain group as occurring near Belton, Montana, and at Nyack creek, Montana.§ At these localities, massive bluish and greenish limestones bearing a species of Raphistoma and a Stromatoporoid form, were found in great development. As shown by Plate 6 of Walcott's paper, the field-habit of these limestones is extremely similar to that of the Siyeh limestone at Mt. Siyeh.

---

* F. C. Calkins, Bull. No. 384 and Professional Paper No. 62, U.S. Geological Survey, 1909 and 1908.
† Information supplied by letter.
§ Bull. Geol. Soc. America, Vol 17, 1906, pp. 12, 19, 22.

2 GEORGE V., A. 1912

which is less than 15 miles distant from the Nyack creek locality. It is difficult to avoid the suspicion that these Castle Mountain limestones are, in truth, identical with the Siyeh limestone, in which, therefore, Middle Cambrian fossils may at some future time be discovered. Walcott has, however, come to a quite different view. He writes:

'The series of limestones at the head of Nyack creek, illustrated by plate 6, are of Cambrian or Ordovician age, as indicated by fragments of fossils that I found in them. I do not think the Siyeh limestone is to be correlated with them, nor with the Castle Mountain limestones of McConnell.'‡

Walcott's latest correlation paper for this region contains a section on the Dearborn river, which carries both Middle and Lower Cambrian fossils.†† The fossiliferous rocks are quite conformable and show a thickness of 2,205 feet; they are chiefly massive or thin-bedded limestones (described as weathering yellow to buff at some horizons), with interbedded shales. The description of the rocks is of essentially the same quality as that which must be applied to the Siyeh formation. Only a few miles away another section of somewhat similar beds, with, however, dominant argillites, had been measured by Walcott, and the whole referred to the Algonkian as part of the 'Belt terrane.'* No statement is given as to the precise relation of these two sections except the following (page 203 of the 1908 paper):

'Beneath the Cambrian sandstone the Empire shales of the Belt Terrane of the Algonkian occur with apparently the same strike and dip as the base of the sandstone. Traced on the strike, however, they appear to be unconformably beneath the sandstone.'

If the apparent unconformity should be explained in the manner suggested (on a later page) by the present writer, it follows from these studies at Dearborn river that the Empire shale is either Lower Cambrian or is not significantly older than the Lower Cambrian. In favour of the writer's view is the fact that these Dearborn river sections occur at points more than 100 miles distant from the old shore-line zone of the Belt mountains, that is, at such a part of the geosynclinal downwarp where the Lower Cambrian beds should be expected to appear in fairly full development.

We have, therefore, at Nyack creek, Belton, and the Dearborn river, three localities where fossiliferous Cambrian formations lie, respectively, side by side with typical members of the 'Belt terrane.' At one or more of these points some geologist may be fortunate enough to find the paleontological evidence which will, at no distant day, fix the position of the Belt terrane among the standard geological systems.†

‡Bull. Geol. Soc. America, Vol. 17, 1906, p. 19.
†† C. D. Walcott: Smithsonian Miscellaneous Collections, Vol. 53, 1908, p. 200.
* C. D. Walcott, Bull. Geol. Soc. America, Vol. 17, 1906, p. 8.
† In this connection it is of interest to add that, as reported by Wood in 1892, 'in the vicinity of Missoula, a few fossils were obtained in the siliceous limestone (dolomite) and identified by Mr. Charles Schuchert as Obolella.' The relation of this formation to the Belt terrane not stated. Herbert Wood, Amer. Jour., Sci., 3rd ser., Vol. 44, 1892, p. 404.

The argument that the Lewis, Galton, and equivalent series should be referred to the pre-Cambrian because they are almost or quite unfossiliferous is dangerous one. Most of the known Cambrian strata of the world are quite unfossiliferous, as far as present knowledge goes. They have been assigned to the Cambrian because a few, generally very thin interbeds bear determinable fossils. In the Bow and Bosworth section of British Columbia, one of the strongly fossiliferous among Cambrian series, Walcott found no traces in some 1,355 feet of continuous strata.[*] The famous Ogygopsis shale at Mt. Stephen is clearly a lens. It peters out rather rapidly and is not represented in either the Mt. Bosworth section or at Castle mountain. Except for a few worm burrows, a massive dolomitic limestone totalling 1,680 feet in thickness, at Mt. Stephen, is unfossiliferous.[†] A thousand feet of the calcareous Nounan formation of Walcott' Blacksmith Fork section in Utah is as poor in organic remains. Why interbeds similar to the Ogygopsis shale fail to appear in the Forty-ninth Parallel section is not apparent. Barrell has suggested a continental origin for much of the Belt terrane sediment, but we have seen that this is true of probably but a small part of the series. That chitinous fossils are relatively abundant at Mount Stephen and very rare in contemporaneous marine sediment its one hundred mile away is not more difficult to understand than that continuous fossils often occur in the Cambrian and generally do not occur in equally unmetamorphosed pre-Olenellus strata. The one contrast means conditions different in space, the other, conditions different in time. In each case explanation is needed. While awaiting complete explanation we must regard this negative character of the 'Belt terrane' as of little direct value in correlation.

Relative Induration and Metamorphism of the Belt Terrane and Flathead Formations.—One of Professor's arguments for the Algonkian age of the Belt terrane is noted in the first of the foregoing quotations from his writings. The point consists in the recognition of a much greater degree of metamorphism in the Belt terrane rocks as compared with the 'little, if any, induration' of the Flathead sandstone.

The weight of this argument is considerably lessened by reason of the fact that the Belt terrane where exposed in other regions, is often little folded or sheared and is scarcely at all affected by dynamic metamorphism. Its rocks have truly been well indurated and largely recrystallized under deep-burial conditions, but such alteration by static metamorphism is not of itself evidence of great difference of age between older underlying beds and the younger beds of a rock group. Messrs. Cantwell and Walcott have proved that the Cambrian period was long enough for the accumulation of 11,500 feet of strata, chiefly the slowly deposited limestone, in the Mt. Bosworth district of British Columbia.[‡] In the same period time shales, sandstones, and subordinate limestones might have elsewhere accumulated to even greater thicknesses. It is reasonable

---

[*] C. D. Walcott, Smithsonian Misc. Collections, Vol. 53, No. 1812, 1908, p. 208.
[†] C. D. Walcott, Canadian Alpine Journal, Vol. 1, 1908, p. 232.
[‡] C. D. Walcott, Smithsonian Misc. Collections, Vol. 53, No. 1804, 1908, p. 2.

to expect that the lower part of such a colossal deposit would show more induration than the upper part.

Microscopic examination of many specimens has, in fact, showed that in each of the Forty-ninth Parallel series, static metamorphism has operated much more strongly in the older formations than in the younger, quite conformable ones. This rule is, however, not absolute. Many of the Grinnell beds still largely preserve their original clastic structure, while the overlying argillaceous beds of the Siyeh are now typical metargillite. In the field these Grinnell strata look as young as the Carboniferous shales in the mountains farther north.

The writer was much struck with the relatively slight induration of the sandstones at the base of the Moyie and Gateway formations. Yet there can be little doubt that they are, respectively, thoroughly conformable to the Kitchener and Siyeh formations and, with these, make a mass of continuous sedimentation. These particular sandstones are just those which the writer has, on other grounds, correlated with the Flathead sandstone. In all these cases the relative lack of metamorphism is to be attributed more to the peculiar nature of the sandstone than to any great difference of age between each sandstone lens and the immediately underlying beds.

It appears fair to conclude that the criterion of relative induration does not imply a great erosion-gap between the Belt terrane and the Middle Cambrian sandstone.

Evidence of Unconformity.—The one controlling principle used in referring the Belt terrane to the pre-Cambrian consists in postulating a strong unconformity between the Middle Cambrian Flathead sandstone and the entire series below the top of the Marsh shale, the uppermost member of the terrane at the original localities. The unconformity is believed by Walcott and his colleagues to be similar to that found between the Middle Cambrian Tonto sandstone and the tilted Chuar series in the Grand canyon of the Colorado. In his original announcement of the westward extension of the unconformity beyond the region where Peale had first suspected its existence, Walcott wrote as follows:—

'The unconformity now known proves that in late Algonkian time an orographic movement raised the indurated sediments of the Belt terrane above sea-level, that folding of the Belt rocks formed ridges of considerable elevation, and that areal (sic) erosion and the Cambrian sea cut away in places from 3,000 to 4,000 feet of the upper formations of the Belt terrane before the sands that now form the middle Cambrian sandstones were deposited.'[*]

In one of his later publications Walcott states that:—

'One hundred miles farther north the section appears to be conformable from the Ordovician down through the Middle Cambrian and the Lower Cambrian of the Bow River series, and not to reach down to the Algonkian

---

[*] C. D. Walcott, Bull. Geol. Soc. America, Vol. 10, 1899, p. 213.

SESSIONA PAPER No. 25a

as it occurs in Montana, the Bow River series being the sediment deposited, in part, at least, in the erosion interval between the Algonkian and the Middle Cambrian.'†

In another place he writes:—

'Absence of Lower Cambrian rocks and fauna is accounted for by the fact that that portion of the continent now covered by the Belt and associated middle and upper Cambrian rocks was a land surface during lower Cambrian time.'‡

The detailed work of Weed and Pirsson resulted in the definite conclusion that during the deposition of the Belt terrane, there was a land area covering the region north of Neihart in the Little Belt mountain district. While the Belt beds were being laid down the pre-Belt rocks were reduced to a nearly level plain. In Flathead time there was a submergence of the old peneplained surface, with a resulting overlap of the sandstone and shale upon the pre-Belt formations.* Similarly, during most or all of the Belt terrane period there seems to have been land in the southern and eastern parts of the Livingston folio area in Montana (see folio); in the area covered by the Yellowstone National Park**; in the Absaroka quadrangle (see folio); in the Black Hills area of South Dakota and Wyoming††; in the Bighorn mountains and vicinity.§

There thus seems to be little doubt that the Belt terrane sediments were in part supplied by the erosion of a large land area covering South Dakota, Wyoming, and eastern Montana. In part they were supplied from the mountainous pre-Cambrian land of western Idaho, Washington, and Oregon. In other words, the rocks of the Belt terrane were laid down in the relation of a typical geosynclinal prism elongated in a meridional sense between the two land areas. The unconformity postulated by Walcott and his colleagues has been deduced from a study of the eastern shore-zone of the ancient gulf or sea.

The irregularities of such a coast line, coupled, it may be, with minor oscillations of level, would necessarily involve maximum variations of thickness in the different sedimentary lenses of the geosynclinal. The lenses must thin to nothing either at the actual shore-line, on the rims of off-shore depressions, or at the outer edge of the coastal shelf which was swept by waves and currents during a long period of stationary sealevel. The resulting irregularities of deposition are homologous to those observed in the section of a river delta which has grown out into sea or lake; those irregularities are necessarily most pronounced near the shore. The failure of individual members of the Belt terrane to appear beneath the Flathead sandstones cannot, therefore, be directly

† Bull. Geol. Soc. America, Vol. 17, 1906, p. 16.
‡ Bull. Geol. Soc. America, Vol. 10, 1899, p. 210.
* U.S. Geol. Surv., Little Belt Mountains folio, 1899, and Fort Benton folio, 1899.
** See Yellowstone Park folio and U.S. Geol. Surv., Monograph 32, part 2, 1899.
†† See the various United States government publications on the Black Hills.
§ See the Hartville, Aladdin, and Sundance folios (1903-5) of the U.S. Geol. Survey also N. H. Darton's Geology of the Bighorn Mountains, Prof. Paper, No. 51, 1906.

2 GEORGE V., A. 1912

taken to mean an erosion unconformity or structural unconformity of the Belt and Flathead beds by the amount of missing strata at any one or more localities.

The section which, according to Walcott, most clearly shows the extent of the unconformity is that running eastward through the Spokane Hills. In the diagram illustrating the relationships there, the Middle Cambrian beds are represented by Walcott as conformably overlying the Helena formation both at Helena and near White's canyon in the Belt mountains uplift.[*] These two localities are twenty-four miles apart. Nearly midway between them, at the Spokane Hills, the Middle Cambrian is represented as again conformable on the Belt beds but this time resting on the Spokane shales. Thus fully 3,000 feet of strata, the thickness of the Helena and Empire formations, are considered as lacking beneath the Middle Cambrian at the Spokane Hills.

The text accompanying the diagram does not state whether the fossiliferous Cambrian at the Spokane Hills is the stratigraphic equivalent of the Flathead sandstone. Mr. Walcott has, by letter, very kindly informed the writer that the Middle Cambrian beds at the Spokane Hills are not only faunally but lithologically the equivalent of the Flathead. There thus seems to be an actual failure of at least 3,000 feet of Belt beds at the Spokane Hills. Whether the failure is due to a lack of original deposition or to the erosion of a local upwarp of the Belt beds is apparently not an easy question to decide. Walcott has taken the latter view for this locality. On the other hand, he himself writes, concerning somewhat similar relations about the town of Neihart:—

'Whether the shore-line conditions, which are known to have existed near Neihart during the period when the Belt terrane was formed, causing a wedging out of the beds to the north, so that the Cambrian rests on different horizons at this locality, or whether pre-Cambrian erosion was extensive enough to pare down the exposed edges of the beds, is not certain from the evidence, though the latter view seems improbable.'[†]

All workers on these Montana rocks have observed that, wherever the Flathead sandstone is seen in contact with Belt formations, no important angular discordance of dip can be seen. Such slight discordances as have been described and figured by Walcott in his 1899 paper, can be explained either by slight, perhaps submarine, erosion of the older surface, or by local and very gentle warping of the surface just before the Flathead subsidence.

As early pointed out by Peale, Flathead time saw a general, rapid, but not very pronounced subsidence of the western Montana region. A large supply of quartzose debris was thus brought from the drowning land and deposited alike over Archean schists and the various lenses of the geosynclinal. In this way a fairly homogeneous formation was spread over a sedimentary mass which, in the nature of the case, must have been composed of many and varied lenses all of which petered out toward mainland, island, or shallow.

---

[*] See C. D. Walcott. Bull. Geol. Soc. America, Vol. 10, 1899, p. 211.
[†] Ibid, p. 210.

In brief, it seems to the writer that the facts . . . .
a sedimentary overlap of the Flathead but not a great structural unconformity
or even erosion unconformity, which is general at the base of the sand---
Considering the size of the area, the observed minor discordances of dip
not be used safely as positive evidence. The observed rollers of beds at various
points can be explained by original wedging-out or by the quite moderate
erosion of local upwarps just before the Flathead subsidence. It must be
remembered that Middle Cambrian time was exceedingly long. It sufficed for
the deposition of 5,000 feet of limestone in the Canadian Rocky mountains
at the localities recently studied by Walcott.[*] During the deposition of such
a slowly accumulated sediment, there was evidently plenty of time for local
upheavals, considerable erosion, and renewed subsidence along the border of
the Cambrian sea. It took only a portion of Pliocene time (next to the Plei-
tocene, probably one shortest of all the major divisions of geological time) to
form 5,800 feet of sediments represented in the Merced series of California,
a series which itself rests on a Pliocene land surface.[+]

Summary of Conclusions.—In view of the foregoing conclusions the writer
does not believe that the pre-Cambrian age of the upper part of the Belt terrane
as defined by Walcott, is proved; and regards the Helena Sixeh formation as
probably Middle Cambrian, somewhat older than the Flathead sandstone. On
the supposition that the Lewis series and the original Belt terrane have been
correctly correlated, that terrane as far down as the upper part of the Greyson
shale is tentatively considered to be of Middle and Lower Cambrian age. From
the lower part of the Greyson shale to the base of the Neihart formation the
beds are correlated as pre-Cambrian (pre-Olenellus) but conformable to beds
equivalent to the Olenellus zone elsewhere. The name 'Belt terrane' (or
Walcott's 'Beltian'), for the remainder of this report, is restricted to this
pre-Cambrian portion of the great geosynclinal prism.

Table VIII. presents a résumé of the writer's tentative correlation of the
Forty-ninth Parallel series with the formations described to the south of the
Boundary line.

The Cœur D'Alene series has been tied on to the Purcell and Summit
series through lithological resemblances. Calkins has traced the Prichard
formation northward, where he found it to pass into the Creston quartzite;
we have seen that the Creston is the off-shore equivalent of the Wolf and Mont
formations. The special white colour and massive appearance of the Revet
quartzite are duplicated in the Ripple quartzite, the two doubtless representing another definite common horizon for the two series. Both series are,
in the table, tied on to the Lewis series and thus indirectly to the more fossiliferous series on the line of the Canadian Pacific railway.

The table embodies, with some modifications, the correlation of the Lewis
series, Belt series, and the Camp Creek-Blackfoot-Ravalli group, as suggested

[*] C. D. Walcott, Smithsonian Misc. Collections, Vol. 53, No. 1804, 1908, p. 2.
[+] Cf., A. C. Lawson, Bull. Department of Geology, University of California, Vol. 1,
1894, p. 112.

2 GEORGE V, A. 1912

by Walcott.[*] The whole Purcell series as shown in Table IX. is certainly placed by Walcott much too low in the geological scale. His reason for making that particular correlation was probably in part due to the following statement of the present writer's in his summary report for 1904 (P. 97):—

'The nearest relatives of the Creston and Kitchener quartzites in the Rockies are respectively the two thick members of the Altyn limestone delimited by Mr. Bailey Willis, who, in the year 1901, carried out a reconnaissance survey of the Boundary belt on the Montana side.'

The expression 'the nearest relatives' should have been 'the nearest lithological relatives,' as the intention was to note the lithological relations of the Purcell and Lewis series, rather than to imply equivalence of age amongst individual members. As a matter of fact the Altyn formation is believed to be the stratigraphic equivalent of a part of the Creston quartzite. On the other hand, we have seen that the facts point towards the correlation of the Kitchener quartzite with the Siyeh and Grinnell formations of the Lewis series.

The somewhat elaborate correlation Table VIII. is intended to illustrate a suggestion rather than a proof. The unfossiliferous rocks of the Forty-ninth Parallel were approached by Dawson, McEvoy and others from the north, where lithologically similar formations bear Cambrian fossils; and, somewhat naturally, regarded the thick quartzites, etc., to the south as probably Cambrian. The United States geologists have as naturally refused to place the nearly unfossiliferous Belt terrane in the same part of the geological column as the formations of Utah and Nevada, where Cambrian fossils are not rare. The present writer has had to rely chiefly on lithological characters in making correlation and his tentative conclusion may be ultimately proved to illustrate once again the danger of using this criterion. It is certain, however, that the pre-Cambrian age of the Belt terrane is not proved, and we are yet at the stage where all reasonable correlations should be fully stated and carefully examined.

By the writer's suggested view the Eastern half of the Cordillera carries a simple Paleozoic-Beltian geosynclinal prism which is only locally interrupted by unconformities. The pre-Ordovician thickness of this prism has an observed maximum of about 30,000 feet. According to the view of Walcott and his former colleagues in the United States Geological Survey, the Eastern belt of the Cordillera carries what may be called a compound geosynclinal prism, made up of a pre-Cambrian series reaching observed thickness of about 30,000 feet, separated by a strong erosion unconformity from a Cambrian series reaching a maximum observed thickness of at least 20,000 feet. The pre-Ordovician sedimentaries, excluding such huge series as those represented in the Priest River terrane, the Cherry Creek beds of the Belt mountains, the Red Creek quartzite of the Uinta mountains, etc., are thus credited with some 50,000 feet of maximum thickness.

By the writer's view the Eastern Cordilleran belt (including the Great Basin), from the Yukon boundary to northern Arizona, was the scene of

[*] Bull. Geol. Soc. America, Vol. 17, 1906, p. 18.

generally uninterrupted sedimentation through Cambrian time. Walcott's correlation involves the conclusion that a very large area included within southern British Columbia and Alberta, much of Idaho and of western Montana, represents more or less continuous land (Belt terrane), separating the Cambrian basin of the Canadian Rockies from the Cambrian basin of Utah and Nevada. These fundamentally different conceptions are important not merely in stratigraphy; they should be in the mind of anyone who attempts to decipher the conditions under which orogenic forces built the ranges of the Great Basin and the Front ranges of Montana and Alberta.

CORRELATION WITH DAWSON'S SELKIRK AND ADAMS LAKE SERIES.

Shortly before his death George M. Dawson read before the Geological Society of America a paper summarizing his views regarding the geology of the Canadian Cordillera.* It is fortunate for the science that he was enabled to complete this able review of his discoveries during a quarter of a century of nearly continuous exploration in the mountains. In this delicate and principal matter of correlation no other person could have so authoritatively digested Dawson's numerous reports along with the others published before the year 1900. The reader of his summary will note how Dawson used his accustomed scientific caution in making correlations among the older rocks of British Columbia. In so brief a review of a vast area it was inevitable that all of his doubts and qualifications could not be expressed. Still more in his original government reports he shows how other interpretations might be deduced as field work progressed. Somewhat different correlations are, in fact, suggested by the field data at the Forty-ninth Parallel. The present writer believes that the lithology of these sections as described, is sufficiently similar to that of the Forty-ninth Parallel formations to warrant certain tentative correlations within the Selkirk mountain-system. Where all is so difficult in the study of these unfossiliferous groups of strata, it is well to entertain all possible views of the relations until accumulating facts shall narrow down the alternatives.

Dawson had found in the Selkirk range and the 'Gold ranges' (the Columbia system of the present report) three thick groups of rocks, which he named the 'Nisconlith series,' the 'Selkirk series,' and the 'Adams Lake series.' All three were referred to the Cambrian and each series was regarded by Dawson as a stratigraphic equivalent of some part of the Castle Mountain-Bow River group of the Rocky Mountain range. For the purposes of the present discussion no briefer, more accurate way of presenting Dawson's salient conclusions concerning these series can be devised than to quote his own summary in full. He wrote:—

'Passing now to the next mountain-system, to the southwest of the Laramide range and parallel with it, the Gold ranges, we find in the

* G. M. Dawson, Bull. Geol. Soc. America, Vol. 12, 1901, pp. 57-92.

MICROCOPY RESOLUTION TEST CHART

(ANSI and ISO TEST CHART No. 2)

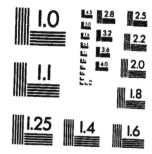

APPLIED IMAGE Inc

1653 East Main Street
Rochester, New York 14609   USA
(716) 482 - 0300 - Phone
(716) 288 - 5989 - Fax

2 GEORGE V., A. 1912

Selkirk mountains a great thickness of rocks that have not yet yielded any fossils, but appear to represent, more or less exactly, the Cambrian of our typical section. Resting on the Archean rocks of the Shuswap series is an estimated volume of 15,000 feet of dark gray or blackish argillite schists or phyllites, usually calcareous, and toward the base with one or more beds of nearly pure limestone and a considerable thickness of gray flaggy quartzites. To these, where first defined in the vicinity of the Shuswap lakes, the name Nisconlith series has been applied. The rocks vary a good deal in different areas, and on Great Shuswap lake are often locally represented by a considerable thickness of blackish flaggy limestone. In other portions of their extent dark-gray quartzites or graywackes are notably abundant. Their colour is almost everywhere due to carbonaceous matter, probably often graphitic, and the abundance of carbon in them must be regarded as a somewhat notable and characteristic feature. These beds have also been recognized in the southern part of the West Kootenay district and in the western portion of the Interior plateau of British Columbia.

'The Nisconlith series is believed, from its stratigraphical position and because of its lithological similarity, to represent in a general way the Bow River series of the adjacent and parallel Laramide range, but there is reason to think that its upper limit is somewhat below that assigned on lithological grounds to the Bow River series.

'Conformably overlying the Nisconlith in the Selkirk mountains, and blending with it at the junction to some extent, is the Selkirk series, with an estimated thickness of 25,000 feet, consisting, where not rendered micaceous by pressure, of gray and greenish-gray schists and quartzites, sometimes with conglomerates and occasional intercalations of blackish argillites like those of the Nisconlith. These rocks are evidently in the main equivalent to the Castle Mountain group, representing that group as affected by the further and nearly complete substitution of clastic materials for the limestones of its eastern development.

'In the vicinity of Shuswap lakes and on the western border of the Interior plateau, the beds overlying the Nisconlith and there occupying the place of the Selkirk series are found to still further change their character. These rocks have been named the Adams Lake series. They consist chiefly of green and gray chloritic, feldspathic, sericitic, and sometimes nacreous schists, greenish colours preponderating in the lower and gray in the upper parts of the section. Silicious conglomerates are but rarely seen, and on following the series beyond the flexures of the mountain region it is found to be represented by volcanic agglomerates and ash-beds, with diabases and other effusive rocks, into which the passage may be traced by easy gradations. The best sections are found where these materials have been almost completely foliated and much altered by dynamic metamorphism, but the approximate thickness of this series is again about 25,000 feet.

' The upper part of the Cambrian system, above the Bow River and Xi-conlith series, may thus be said to be represented chiefly by limestones in the eastern part of the Laramide range, calc-schists in the western part of the same range, quartzites, graywackes, and conglomerates in the Selkirk mountains, and by volcanic materials still further to the west. It is believed that a gradual passage exists from one to another of these zones, and that the finer ashy materials of volcanic origin have extended in appreciable quantity eastward to what is now the continental watershed in the Laramide range. No contemporaneous volcanic materials have, however, been observed in the underlying Bow River or Nisconlith series.'*

The writer has studied Dawson's original reports with a view to understand the grounds of the correlations mentioned in the foregoing quotation. Unfortunately the arduous and rapid nature of his reconnaissance surveys prevented Dawson from constructing columnar sections in detail sufficient to make intensive lithological comparisons possible. Nevertheless, the more detailed facts certainly seem to warrant the belief that the Selkirk series is, in the main, equivalent to the Summit series of the Forty-ninth Parallel section and to the Castle Mountain-Bow River group of McConnell's section.

On the other hand, any satisfactory conclusion as to the relation of the Nisconlith-Adams Lake terrane to the formations mapped at the International Boundary could not be reached without further field-work. Since the forwarding of the original manuscript of this report for publication, the writer has spent a season in the principal area, along the main line of the Canadian Pacific railway, where Dawson studied these old rocks. At the time of the present writing (November, 1911), the results of that season's work are not fully compiled, but certain of them, bearing on the question of correlation, are already in shape for definite statement.

The writer has been forced to differ from Dawson in several important conclusions. The evidences in each case are necessarily too detailed to be stated in the present report, wherein the writer's relevant conclusions only will be briefly noted, as follows:—

1. The ' Nisconlith ' series of the Selkirks, as sectioned by Dawson between Albert Canyon station and Glacier House, represents the northern continuation of the Beltian (Belt terrane) rocks at the Forty-ninth Parallel, and conformably underlies the thick quartzites of the Selkirk series, which are probably of Cambrian age. The writer believes that these ' Nisconlith ' rocks of the Selkirk mountains should logically be included in the Selkirk series.

2. The ' Nisconlith ' series of the Shuswap lakes area is an entirely different, pre-Cambrian and pre-Beltian, group of sediments, which underlie the ' Nisconlith ' of the Selkirks unconformably.

* Ibid. pp. 66-7. In the second volume of the same bulletins (1891), p. 165. Dawson treats the Selkirk section at greater length, giving a structure-section and table of correlations. He attributes nearly 40,000 feet of thickness to the Cambrian alone.

3. The Adams Lake volcanic series conformably overlies the thick limestones of the 'Nisconlith' series in the Shuswap lakes area and is likewise of pre-Beltian age.

4. The 'Shuswap series' of the Shuswap lakes region is not a distinct gneissic group unconformably underlying the 'Nisconlith' series, but represents a facies of the 'Nisconlith' series of the same region, where the latter has been specially metamorphosed. This metamorphism is thermal and is largely due to batholithic intrusion. The batholiths are pre-Beltian in age.

5. In many essential respects the lithology of the Priest River terrane corresponds with that of the 'Nisconlith'-Adams Lake group of the Shuswap lakes. In a general way those two pre-Beltian groups may be tentatively correlated.

The correlations suggested by the new facts are summarized in Table X.

TABLE X.—*Correlation with Canadian Pacific Railway Section.*

SELKIRK RANGE, 49TH PARALLEL.	WESTERN PART OF COLUMBIA RANGE; INTERIOR PLATEAUS.	SELKIRK RANGE AT MAIN LINE OF CAN. PAC. RY.	ROCKY MOUNTAIN RANGE; BOW RIVER SECTION.	AGE.
Summit Series:				
Lone Star	....	*Selkirk Series, upper*	*Castle Mountain*	Middle Cambrian.
Beehive	....	*part.*	*Series, lower part.*	
Ripple............	....	*Selkirk Series, middle*	*Bow River Series,*	Lower Cambrian.
Dewdney............	....	*part.*	*upper part.*	
Wolf, upper part....	....			
Wolf, lower part....	....	*Selkirk Series, lower*	*Bow River Series,*	Age of Belt terrane as defined in present report.
Monk	....	*part ("Nisconlith"*	*lower part.*	
Irene Volcanics. ..	....	*of Dawson).*		
Irene Conglomerate.	....			
Unconformity	Unconformity ..	Unconformity	?	
Priest River Terrane	*Adams Lake Series; Nisconlith Series* (of Dawson);*Shuswap Series* (of Dawson).	*Granite batholith* cutting schists of, Dawson's "Nisconlith of Shuswap Lakes area".	?	Pre-Beltian.

SESSIONAL PAPER No. 25a

EASTERN GEOSYNCLINAL BELT OF THE CORDILLERA.

In the final generalization regarding a mountain-chain, namely, the theory of its origin, it is of first importance to include a definite conception of the geosynclinal sedimentary prism or prisms out of which the rock folds of that chain have been made. For the eastern half of the North American Cordillera the complex orogenic history must be discussed in terms of at least three periods of specially important geosynclinal sedimentation. As Dana long ago pointed out, the principal period is that of the deposition of the stratified series from the Cambrian (and conformable pre-Olenellus) system to the Mississippian system inclusive. For this huge accumulation of clastic and chemical deposits the present writer has proposed the name ' Rocky Mountain Geosynclinal '; the down-warped surface of the pre-Cambrian on which the prism rests may be sald, for distinction, to form the ' Rocky Mountain Geosyncline.'

In northern Alaska, northeastern Alaska, eastern Yukon, eastern British Columbia, Alberta, Montana and central Utah, the Rocky Mountains, in the common and narrower sense of the term, are chiefly or largely composed of rock-forming part of this prism. So far the proposed name is appropriate. In Colorado the Rocky Mountains are principally composed of other terranes, so that the folded and faulted rocks of the prism constitute the ranges of the Great Basin, all of which lie well back of the front range of the Rockies proper. For this part of the Cordillera the proposed name is not fitting, except as the prism is, by its name, located alongside the local range of the true Rocky Mountains. However, the fact that by far the greatest part of the Rocky Mountain chain (proper) is actually made of the rocks of this prism, has impelled the writer to suggest the name chosen. Dana has offered the name ' Rocky Mountain geosynclines ' for the post-Cretaceous down-warps affecting a local part of the Cordillera, namely, that in the Wasatch-Green river region.[*] For the student of continental geology this name seems hardly appropriate; the larger part of the Rocky Mountain group has not been affected by down-warps of this date, at least to the extent demanding the formation of thick prisms of sediment. In any case the main uplift of the Rockies proper has not been due to the generation of Tertiary geosynclinals but has rather been one of the causes of their subsequent formation.

During the other two periods of heavy sedimentation, the resulting geosynclinals were incomparably smaller and all of more local nature than the enormous mass of strata upon the back of which, and from the substance of which, these younger prisms were made. The latter include the Cretaceous geosynclinal of the Crowsnest district in Canada as well as that in Colorado; also the Eocene geosynclinals of the great down-warps north and south of the Uinta mountains.

With the Cretaceous and Eocene geosynclinals we are not now engaged, but they are mentioned in this place in order to indicate once more the advisability of having a convenient name for the eastern half of the Cordillera which

---

[*] J. D. Dana, Manual of Geology, 4th edition, 1895, p. 365.

25a—vol. ii—13½

has been built so largely of the rocks laid down in these three periods. For use in the present report this part of the Cordillera will be called the ' Eastern Geosynclinal Belt.' In a later chapter details will be given which tend to corroborate the prevailing view that the western part of the Cordillera, from Alaska to southern California at least, is a second vast unit deeply contrasted in composition and history with the Eastern Belt. One result of the correlations so far made is to give some indication of the approximate line which may be taken as separating the Eastern belt from the ' Western Geosynclinal Belt.'

AXIS OF THE ROCKY MOUNTAIN GEOSYNCLINAL.

If the foregoing correlations of the formations in the Forty-ninth Parallel section be justified, it seems possible to determine, in a very general way, the thickness and extent of the geosynclinal which was accumulated during the time elapsing between the deposition of the oldest beds of the Belt terrane and the deposition of the Upper Cambrian formations. Since these older rocks, where developed in the heart of the geo-syncline, rival or surpass in thickness the whole of later Paleozoic formations in the same area, the delimitation of the pre-Upper Cambrian sediments effectively locates the main axis of the Rocky Mountain geosyncline. (Figure 12, page 202.)

Needless to say, the field evidence is far too incomplete to permit of anything like an accurate picture of the ancient down-warp or of its sedimentary filling. Nevertheless, the materials are already in hand to warrant a substantial corroboration of the view of J. D. Dana, G. M. Dawson, and others, that the Rocky Mountain system has been built up through the upturning of a vast geosynclinal lens whose main axis lay to the eastward of an Archean protaxis in the Cordillera; and, secondly, that the geosynclinal axis lay parallel to the general axis of the present Cordillera.

In the eastern part of the Selkirk range at the Forty-ninth Parallel the thickness of all the conformable pre-Upper Cambrian beds, excluding the 6,000 feet of Irene volcanics is about 26,000 feet. The character of these sediments show that their material was in largest part derived from the rapidly eroded lands lying to the westward. The old shore-line, or rather zone of shore-lines, was probably located not far from the crossing of the Columbia river at the International Boundary. As yet the only other columnar section of these Cambrian-Belt rocks which includes their base, has been constructed from outcrops observed in the Belt mountains 350 miles to the eastward, where the whole thickness is 12,000 feet. Not far to the eastward of this section there was land during the deposition of the Belt, Lower Cambrian, and some of the Middle Cambrian beds; during the Middle Cambrian much of this eastern land area was itself transgressed by a wide shallow sea. Between the Belt Mountains section and the Selkirk (Boundary) section, a great thickness of Belt-Cambrian beds, considerably excelling 20,000 feet, was laid down in apparently perfect continuity .

SESSIONAL PAPER No. 25a

The observations of Dawson, McConnell, and McEvoy serve to warrant the belief that the western rim of the geosynclinal may be traced through the whole length of British Columbia to the Sixtieth Parallel of latitude. The study of British Columbia geology impresses one, however, with the difficulty of locating this rim with precision. For hundreds of square miles together the beds of the geosynclinal are either buried out of sight by younger formations or have been replaced by batholithic intrusions on a gigantic scale. Even where, in many places, the Belt-Cambrian rocks are exposed, they have been so metamorphosed by crushing and by thermal action that the true nature and relation of the beds is very obscure. In each one of the following cases, therefore, the location of the rim of the geosynclinal is to be considered as only approximate. Future investigation may show that errors as great as fifty miles in longitude may have been made in these locations. The scale of the geosyncline is, however, so great that the main conclusions regarding the position and extent of the huge down-warp and of its sedimentary filling are considered as approximately correct.

At the Canadian Pacific railway section Dawson himself placed the western rim of the geosynclinal within the area occupied by the present Columbia system.

'In the earlier series of deposits assigned to the Cambrian, we discover evidence of a more or less continuous land area occupying the position of the Gold ranges and their northern representatives and aligned in a general northwesterly direction. The Archean rocks were here undergoing denudation, and it is along this axis that they are still chiefly exposed, for although they may at more than one time have been entirely buried beneath accumulating strata, they have been brought to the surface again by succeeding uplifts and renewed denudation. We find here, in effect, an Archean axis or genanticline that constitutes, I believe, the key to the structure of this entire region of the Cordillera. To the east of it lies the Laramide geosyncline (with the conception of which Dana has familiarized us), on the west another and wider geosyncline, to which more detailed allusion will be made later.

'Conglomerates in the Bow River series indicate sea margins on the east side of this old land, but these are not a marked feature in the Nisconlith, or corresponding series on its western side. Fossils have so far been discovered only in the upper part of the Bow River series, but the prevalence of carbonaceous and calcareous material (particularly in the Nisconlith) appears to indicate the abundant presence of organisms of some kind at this time.

'Although no evidence has been found of any great physical break, the conditions indicated by the upper half of the Cambrian are very different from those of the lower. Volcanic materials, due to local eruptions, were accumulated in great mass in the region bordering on the Archean axis to the west, while on the east materials of this kind appear to be mingled with the preponderant shore deposits of that side of the

2 GEORGE V., A. 1912

Archean land, and to enter sparingly into the composition of the generally calcareous sediments lying still farther eastward. Where these sediments now appear, in the eastern part of the Laramide range they are chiefly limestone, indicating marine deposition at a considerable distance from any land.'*

McEvoy has described the Bow River series as occurring in the mountains just east of the Rocky Mountain Trench at the divide between the Canoe and Fraser rivers, latitude 53° north.** He maps the Shuswap (Archean) series on the west of the great trench, showing a spatial relation between the Cambrian-Belt rocks to the Archean which is similar to that which Dawson had discerned farther south. On this ground and allowing for some overlap to the westward, the rim of the geosyncline may be provisionally placed some distance to the west of the Rocky Mountain Trench in latitude 53° north.

According to Dawson a parallel relation exists between the Bow River (Misinchinca schists) and the Archean on the Parsnip river, which also flows in the Rocky Mountain Trench.† Again, the zone of old shore-lines off which these Cambrian-Belt sediments were deposited, may be placed, here at 55° north latitude, to the westward of the trench. How far to the westward of the Parsnip river it should be placed it is now impossible to state but probably not more than fifty miles.

McConnell found the Bow River-Castle Mountain rocks on the east side of the Rocky Mountain Trench at the Finlay river 57° 30′ north latitude. He also discovered crystalline rocks, referred to the Archean on the west side of the trench. The relatively small thickness of the Bow River (4,000 feet) and its conglomeratic character point once more to proximity to the old shore-line zone. The zone probably lay not many miles to the westward of the trench.‡ Dawson and McConnell have followed the continuation of the Castle Mountain series to the Kachika and Liard rivers at 60° north latitude. They also describe a large area of Archean to the westward and it is not improbable that the rim of the geosyncline here lay west of the Kachika river.§

The enormous length and singular straightness of the Rocky Mountain Trench suggests that it is a line of dislocation. Detailed study at the Forty-ninth Parallel and at a few other points to the north corroborate this idea. It is, therefore, probable that the occurrence of Archean and Bow River-Castle Mountain rocks, respectively on the west and east sides of the trench, may simply show that the uplift has been greater on the west side of the line of dislocation and that erosion has removed the sedimentary veneer on that uplifted side, while it has not been able to destroy the veneer on the eastern, down-thrown side. This conclusion is undoubtedly just, and it is certain that

* G. M. Dawson, Bull. Geol. Soc. America, Vol. 12, 1901, p. 84.
** J. McEvoy, Ann. Report, Geol. Surv. Canada, Vol. 11, 1900, Part D.
† Report of Progress, Geol. Sur. of Can., 1879-80, Part B. p. 108.
‡ R. G. McConnell, Ann. Report Geol. Surv. Canada, Vol. 5, 1894, Pt. C.
§ R. G. McConnell, Ann. Report, Geol. Surv. Canada, Vol. 4, 1888-9, Pt. D, pp. 13-14; and G. M. Dawson, Ann. Report Geol. Surv. Canada, Vol. 3, 1887-8, Pt. B. pp. 31-4.

the present line of contact between the Archean and the later terrane is not itself the old shore-line. A careful study of the reports cited, has, however, caused the writer to believe that, in general, the zone of shore-lines probably lay not more than two or three score of miles to the southwestward of the trench.

From at least 60 north latitude to about 52 north latitude the western rim of the geosyncline ran roughly parallel to the course of the present Rocky Mountain Trench. It is a question worthy of investigation whether there is a genetic connection between the trench and this zone of shore-lines. Was the line of dislocation established where it is because of the contrast of rigidity between the strong rocks of the pre-Beltian and the weaker rocks of the geosyncline? At the Forty-ninth Parallel the trench is at least 100 miles from the zone of old shore-lines; it is possible that the specially thick and rigid Creston and Kitchener quartzites functioned in the same way as the pre-Beltian rocks in locating this main line of dislocation at the western edge of the weaker rocks of the geosyncline, thus controlling the divergence of the shore-line zone and the trench near the great bend of the Columbia river.

Not far north of the Sixtieth Parallel the Castle Mountain-Bow River group of rocks disappears under newer formations and, as yet, the geosyncline cannot be traced farther northwestward.

Nowhere on the Canadian side has the eastern rim of the geosyncline been discovered. The cover of Upper Paleozoic and Mesozoic formations will apparently always forbid its discovery in Alberta and farther north. In the accompanying map (Figure 12) the eastern rim of the Belt-Cambrian portion of the Rocky Mountain geosyncline is sketched in hypothetically. Its position is marked almost wholly on the supposition that the width of the geosyncline remains fairly constant from the Forty-ninth Parallel northward. The notable constancy in the lithological character and the great total thickness of the geosynclinal beds where studied in the mountain belt from Montana to Yukon Territory, lends some colour to the supposition. That the geosynclinal holds its width to 62° 30′ north latitude is rendered almost certain by McConnell's discovery of Castle Mountain dolomites and limestones on the Mackenzie river, seventy miles below Fort Simpson."

In borings made by Baron von Hammerstein at the Athabaska river near Fort McMurray, granitic and gneissic rocks, probably referable to the Archean, were encountered at the depth of 1,000 feet, the overlying rock being Devonian limestone. The Belt-Cambrian rocks seem thus to be lacking at this point, where their absence is possibly due to non-deposition.

Southward from the International Boundary the geosyncline can be traced with greater confidence. Like the Summit series of the southern Selkirks, the Cœur d'Alene series, described by Messrs. Calkins, Ransome, and MacDonald, and the equivalent Lolo series studied by Lindgren in sections farther south

* R. G. McConnell, Ann. Report Geol. Surv. Canada, Vol. 4, 1888.9, Pt. D, p. 89 and map.

in Idaho, both include thick members which were deposited at no great distance from shore.[**]

The land in both cases lay to the westward and in both districts, rocks of Archean habit are well developed to the westward of the areas occupied by the Belt-Cambrian rocks. The western rim of the geosynclinal as representing the zone of old shore-lines may here be tentatively fixed at about 118° west longitude.

The heavy lava blanket of southern Idaho and southeastern Oregon effectually precludes the discovery of either the rocks or the relations of the early Paleozoic or pre-Paleozoic terranes. There is little doubt, however, that the great geosynclinal once stretched far to the southward and probably without serious interruption, into Nevada. The geologists of the Fortieth Parallel survey showed that a great thickness of conformable Cambrian and pre-Cambrian beds were deposited over the area of what is now the Great Basin of Utah and Nevada. Hague agrees with King that the bulk of the detrital material in these deposits was washed out from a zone of shore-lines, located on the Fortieth Parallel near the meridian of 117° 30′, west longitude. For the Belt-Cambrian rocks King places the shore-line zone at 116° 30′, west longitude.[†] In the Eureka district Hague found 6,250 feet of pre-Upper Cambrian strata which represent only the upper part of the geosynclinal, as the base was everywhere concealed.[‡]

In the Wasatch, Walcott describes more than 11,000 feet of beds conformable to overlying strata bearing the Olenellus zone.[§] The Uinta quartzite formation, 12,000 + feet thick, underlies the Middle Cambrian Ophore shales quite conformably. Being unfossiliferous, the quartzite is referred to the pre-Cambrian. The lithology is very similar to that of the Purcell series and it is noteworthy that sheets of contemporary lava, analogous to the Purcell Lava, occur in the Uinta quartzite.[§]

In all of these standard sections of the Great Basin geosyncline, the lithological character of the sediments corresponds well with that of the many sections now run through the geosyncline near the Forty-ninth Parallel. There is every probability that these northern and southern sections include different parts of the same great unit, the Rocky Mountain Geosynclinal, which thus extended, without sensible interruption, from the Fortieth Parallel to and beyond the Sixtieth Parallel of latitude.

[**] U.S. Geol. Surv. Bull. No. 260, 1905, p. 274; Bull. No. 285, 1906, p. 41; and Prof. Paper No. 27, 1904, p. 16.

[†] C. King—Geological Exploration of the Fortieth Parallel—Systematic Geology, Vol. 1, 1878, p. 534, and map, p. 127; A. Hague, Geology of the Eureka District, Monograph 20, U.S. Geol. Survey, 1892, p. 175.—The more recent discovery of Lower Cambrian formations in the White Mountain range of eastern California, shows that, for at least part of Belt-Cambrian time, the shore-line must have been situated west of the limit set by King.—cf. C. D. Walcott, Amer. Jour. of Science, 3rd ser., Vol. 19, 1895, p. 111.

[‡] Op. cit. p. 13.

[§] C. D. Walcott, Tenth Annual Report, U.S. Geol. Survey, 1890, p. 550.

[§] F. B. Weeks, Bull. Geol. Soc. America, Vol. 18, 1907, p. 431.

How much farther southward the geosyncline stretched is not easy to declare, even in the tentative way held advisable for the extent of the geosyncline as just outlined. There is something to be said for the correlation of the Chuar series of Arizona with the lower part of the Belt-Cambrian group, for in that southern region the history of the geosyncline was complicated by orogenic upturning, erosion and subsidence, all of these affecting the strata older than the Middle Cambrian Tonto sandstone. Just before Tonto time, therefore, the geosynclinal sedimentation seems to have been, for a certain period, largely or wholly interrupted in the latitude of northern Arizona. During the earlier period we must believe, on the hypothesis that the Chuar series and Belt terranes are, at least in part, stratigraphic equivalents, that the geosynclinal extended still farther southward, perhaps into Mexico.

In southwestern Colorado, Cross and his colleagues of the United States Geological Survey have discovered a remarkable series of sections in 'Algonkian' rocks, unconformably underlying the apparently Middle Cambrian Ignacio quartzite of the region.* As now understood, the older rock series consists of at least 8,000 feet of exceptionally massive quartzite with argillitic interbeds (the whole called the Uncompahgre formation), conformably overlying the Vallecito conglomerate, 1,000 feet thick, which in turn rests on the Irving greenstone, believed to be over 10,000 feet thick. The relation of the greenstone and conglomerate are obscure but an unconformity is postulated by the authors of the reports on the Needle Mountains district. The evidence for this unconformity largely consists in the fact that the conglomerate is composed of pebbles derived from the greenstone. Neither that fact nor any other of those in favour of the unconformity's existence can be regarded as showing a great period of time as elapsing between the effusion of the lavas now represented in the Irving greenstone, nor do the authors of the Needle Mountains folio state that there has been any considerable time-gap at this horizon.

The importance of the series in the present connection is that it seems to correspond well with the basalt members of the Belt terrane as represented in the Summit series of the Selkirk mountains. The Irving greenstone is certainly lithologically very similar to the Irene volcanics of the Boundary section, and it bears the same relation to the Priest River terrane as the Irving greenstone bears to the Archean schists of Colorado, except that the equivalent of the Irene conglomerate is not directly apparent in the Colorado section. The Vallecito conglomerate and Uncompahgre formation match well with the Monk formation, and as with the lower part of the massive Creston quartzite of the Purcell range. The question arises as to whether we have in this Colorado section the southern part of the great Belt-terrane geosynclinal and, in fact, the base of it near its thickest section. The relation of the Uncompahgre formation to the Middle Cambrian Ignacio quartzite is like that of the Chuar-Unkar series to the Middle Cambrian Tonto sandstone of the Grand Canyon section. The correlation of all of these with the conformable series at

* W. Cross, E. Howe, J. D. Irving, and W. H. Emmons in the Needle Mountains folio, 1905; and W. Cross, E. Howe and F. L. Ransome in the Silverton folio, 1905.

DEPARTMENT OF THE INTERIOR

2 GEORGE V., A. 1912

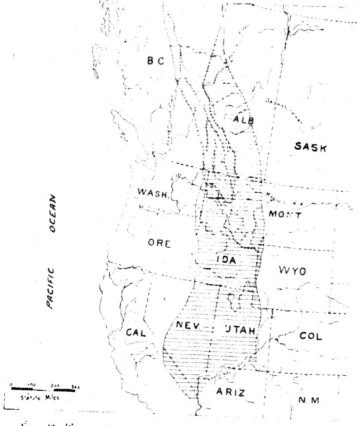

Figure 12. Diagrammatic map showing approximate position of the Rocky Mountain Geosynclinal basin in Cambrian (Beltian, Lower Cambrian, and lower Middle Cambrian pre-Flathead) times.

SESSIONAL PAPER No. 25a

the Forty-ninth Parallel means, in the writer's view, that the post-lunaic and pre-lunaic deformation affected the larger part of the southern end of the Rocky Mountain Geosyncline, and while the greater, northern part was not essentially affected by this phase of deformation.

King's sections and context in the Fortieth Parallel survey reports clearly show that the Belt-Cambrian geosyncline was bounded on the east by land or by marine shallows in the vicinity of the meridian of 110° west longitude at the Uinta mountains.[*] This eastern rim of the geosynclinal seems to be the western edge of an extensive land mass stretching from the Belt mountains southward, as already described.

In the latitude of the Uinta mountains the width of the Belt-Cambrian geosyncline was about 375 miles; in the latitude of the Belt mountains it about 300 miles; its average width in the United States seems to have be about 350 miles. In southern British Columbia and Alberta the width of the exposed part of the geosynclinal is not more than 150 miles; at the Mackenzie river the exposed part is about 225 miles wide. At both ends of the Canadian portion of the geosynclinal and in all the stretch between, the actual width was doubtless everywhere over 200 miles and, as noted above, is provisionally assigned a magnitude similar to that observed in the United States. The observed length of the geosyncline is 1,500 miles, and there are reasons for believing that this huge sedimentary prism was yet longer, extending, at the north, into Yukon Territory and at the south, into Arizona. The map, Fig. 12, illustrates the fact, important to the theory of mountain-building, that the axis of this old geosyncline ran faithfully parallel to the general axis of the present Cordillera.

The foregoing summary of many facts recently determined in Montana, Idaho, and at the International Boundary, thus serves to confirm the view of Dana, Dawson, King, and others concerning the existence of thick sedimentary prisms, of which the Rocky Mountains of Canada and the United States, as well as the ranges of the Great Basin, are largely composed. The present compilation is intended principally to enforce the writer's belief that the Canadian geosynclinal and the Fortieth Parallel geosynclinal are but parts of the same thing. The great 'Belt terrane' of Walcott is regarded by the writer as an integral part of this immense sedimentary unit, being the stratigraphic equivalent of the Bow River-Castle Mountain series in the north, and of the conformable series below the Upper Cambrian in the Wasatch, Eureka, and other districts of the Great Basin.

UPPER PALÆOZOIC PORTION OF THE ROCKY MOUNTAIN GEOSYNCLINAL.

We have seen that formations younger than the Middle Cambrian compose but an insignificant fraction of the mountains crossed by the International Boundary between the Great Plains and the summit of the Selkirk ranges.

---

[*]See especially analytical map and section facing p. 127 in King's Systematic Geology, 1878.

2 GEORGE V., A. 1912

There can be little doubt, however, that the Devonian and Carboniferous beds once covered the Cambrian rocks through all, or nearly all, of this distance. The minimum thickness given to the younger formations—2,000 ± feet—is such that we may well believe that the original thickness of the Devonian and Carboniferous combined was, at the Forty-ninth Parallel, of the same order as that determined by McConnell for the contemporaneous strata on the Canadian Pacific railway (main line) section. In the vicinity of Banff he found excellent exposures, giving a total thickness of 6,600 feet. From that section northward through all British Columbia and Yukon, and on to northern Alaska, this wonderfully persistent group of rocks may be followed; such breaks as occur in the outcrops through the long traverse are nowhere sufficient to make us doubt that these latter Palæozoic strata retain much of their great thickness all the way to Arctic waters.

In the Little Belt mountains 2,425 feet of beds referred to the Devonian and Mississippian are recorded in the text of the Little Belt Mountains folio (by W. H. Weed). At Mt. Dearborn, Montana, Walcott found more than 2,000 feet of Carboniferous limestone.* The Eureka district affords 11,000 feet of contemporaneous rocks, largely limestone.§ In the Bisbee district of Arizona, Ransome found about 1,000 feet of such rocks.†

In all of the sections above-mentioned there seems to be perfect conformity between the Devonian and Carboniferous, except possibly in parts of Alaska. In the Grand Canyon (Arizona) section about 1,000 feet of Devonian and Mississippian are represented, with an unconformity between them, just as the same region shows unconformity between the Middle Cambrian Tonto formation and the underlying Chuar series, both of which are conceivably of Cambrian age.

In the Black Hills of Dakota and in Wyoming the Devonian is wanting and the Mississippian is very thin, though its occurrence there is significant.

Without going into the details of the many other measured sections on the American side of the Boundary line, the writer will state his belief that the facts of Cordilleran geology show the Devonian and Mississippian formations to form an organic part of the Rocky Mountain Geosynclinal from one end of it to the other, thus once covering practically the entire area of the Eastern Geosynclinal Belt. The Rocky Mountain Geosynclinal was somewhat wider during the Devonian than in the long period represented by the Lower Cambrian and the Belt terrane. The Mississippian represents a still wider transgression of the sea beyond the earlier limits of the down-warp. This early Carboniferous transgression was analogous to that of the Middle Cambrian (Flathead time). The former was so extensive as to make it very difficult, if not impossible, to draw even a rough map of the geosynclinal area for the period. For orogenic theory this partial and irregular drowning of the old

---

* C. D. Walcott, Smithsonian Misc. Collections, No. 1812, 1908, p. 200.
§ A. Hague, Geology of the Eureka District, Monograph 20, U.S. Geol. Survey, 1892, p. 13.
† F. L. Ransome, Bisbee folio, U.S. Geol. Survey.

SESSIONAL PAPER No. 25a

lands to east and west during the Devonian or Mississippian periods, is not of primary importance. The fact seems certain that the heaviest sedimentation of those periods took place in the axial region of the ancient Cambrian downwarp. The late Paleozoic (pre-Pennsylvanian) deposition, irregular as it may have been, thus tended to complete the one massive prism out of which the Rocky Mountains and the Great Basin ranges were later to be formed. The southern part of the geosynclinal, that sectioned at the Fortieth Parallel of latitude. for example, shews that the down-warping persisted into Pennsylvanian time, but for the most part the Eastern Geosynclinal Belt of the Cordillera seems to have been out of water during the Pennsylvanian.

The records of the east and west transgressions of the sea during the Devonian and Carboniferous periods tend, therefore, in a measure to obscure the real situation of the sedimentary prism which was the essential antecedent to the building of the Rocky Mountains of Alaska, Canada, and Montana, as well as the ranges of the Great Basin. The western limit of the pre-Devonian members of that prism is approximately the zone of shore-lines which has been traced from southern California to the Yukon boundary. The zone may be considered as including the rather indefinite line or limit separating the Eastern Geosynclinal Belt of the Cordillera from the Western Geosynclinal Belt. That line was, of course, neither straight nor smoothly curved. Bays of the Cambrian sea must have reached well into the western land on the west and we have already seen that that land was extensively transgressed in the time when the Rocky Mountain Geosynclinal, the essential sedimentary member of the Eastern Belt, was being completed.* On the other hand, when the conditions were reversed and the Eastern Belt, after upheaval, furnished detritus out of which the geosynclinals of the Western Belt were constructed there were deep bays running eastward into the land, and on the Fortieth Parallel, the Eastern Belt was entirely covered by the sea. In spite of all these complications the division of the Cordillera into the two great belts tends to aid one in the attempt to understand the true history of the Cordillera north of the Mexican boundary. The division is made at the behest of dynamic geology, not at that of paleogeography nor paleontology; in those groups of studies the suggested division and nomenclature would probably have little value and might even lead to confusion. In a word, the division is warranted only for the geologist who is bent on locating geosynclinals, not shore-lines. The full conception of the profound contrasts otherwise existing between the two belts is not possible until a review is made of the diastrophic, igneous, and erosional history of the Western Belt.

---

* The discovery of Silurian and Devonian sediments in the Taylorsville district of California, and again at one point in southwestern Alaska suggests that the Early Paleozoic land mass of the Western Cordilleran Belt may have been locally interrupted by straits connecting the marine area of the Rocky Mountain Geosyncline with the open Pacific.

CHAPTER IX.

PURCELL LAVA FORMATION AND ASSOCIATED INTRUSIVES.

INTRODUCTION.

Many of the higher peaks in the four ranges of the Rocky Mountain system, as well as in the McGillivray range, owe their special heights to the strength of the Purcell Lava, which is even more resistant to the forces of weathering than the massive Siyeh formation underlying. As above noted, this lava formation is the faithful friend of the stratigrapher throughout eighty miles of the transmontane section at the Forty-ninth Parallel. Its discovery in the McGillivray range is a fact of the first importance, since its presence and relations have removed the last doubt as to the equivalence of the Siyeh formation with the main body of the Kitchener. Therein we have a main link in the correlation of the staple sedimentary rocks occurring in the eastern third (150 miles) of the whole structure-section from the Pacific to the Great Plains. The lava formation thus deserves a somewhat detailed description. For convenience certain associated dikes and flows will be treated in the present section, which is to deal with the stratigraphy and petrography of the Purcell formation proper.

PURCELL LAVA OF THE MCGILLIVRAY RANGE.

The formation is displayed with unusual perfection in three different areas within the McGillivray range. The most westerly of these occurs at the strong meridional ridge situated about six miles east of the main Yahk river valley at the Boundary line. The great sheet there dips east-northeast at angles varying from 42° to 50°. It also caps the ridge, three to five miles farther eastward, where it reappears in the eastern limb of the broken and pitching syncline at the summit of the range. Here the dip is from 20° to 28° northwesterly. The third area of the lava as mapped is a small one, situated at the edge of the drift-covered flat of the Kootenay river valley, where the dip is 35° to the northeast and represents the attitude appropriate to the eastern limb of the broad anticline that forms the main structural element in the Kootenay slope of the McGillivray range.

These localities were those at which the writer first encountered the formation. Since it has its maximum known thickness in the McGillivray division of the Purcell mountain system, the formation has been called the 'Purcell Lava.'

One or more of the important vents must have been situated not many miles from the western line of outcrops in this range. The lava seems never

2 GEORGE V., A. 1912

to have extended as much as nine miles to the westward of these outcrops, for at that distance the Moyie formation (equivalent to the Gateway) rests directly upon the Kitchener (equivalent, in its upper part, to the Siyeh). Such relations, coupled with the fact that the lava thickens between the western summits of the Galton range and the summit of the McGillivray range, seems to indicate pretty clearly that the greater flows were supplied from vents located near the present summit of the McGillivray range and not far from the Forty-ninth Parallel. One vent seems to be represented in a long, 50-foot dike which cuts the upper beds of the Kitchener formation in a meridional direction, at a point just north of the Boundary line and about two yards west of the most westerly band of the lava. In the Galton, Clarke, and Lewis ranges the Purcell lava seems to have issued, in like manner, from local vents, some of which are dikes cutting the underlying Siyeh formation and can be examined in the mountain-walls of all three ranges. Everywhere the lava was emitted in true fissure-eruptions, which were vigorous and wide-spread while they lasted but were not of long duration. The immediate association of dolomites and metargillites both above and below the lava in the Rocky Mountains and the perfect conformity of these sediments with the lava flows suggest that the eruptions took place on the sea-floor.

One of the best sections of the formation as exposed in the McGillivray range, occurs at the summit of the 6,583-foot mountain, situated 2,000 yards north of the Boundary line and five miles east of the main fork of the Yahk river. The total thickness is there 465 feet. At the base forty feet of mottled, brecciated lava (zone *a*) lies directly on the Kitchener (Siyeh) metargillites. This member has, in field appearance, much resemblance to a true tuff or ash-bed and so it was described in the field notes. The microscope has shown, however, that the apparently fragmental masses of porphyrite are cemented in part by an altered glass, bearing feldspar microlites in rough fluidal arrangement. In other parts the cement has the composition of metargillite. The writer has concluded that this lowest member is not a product of volcanic explosion but the thick lower shell of a heavy mass of lava which flowed out over the old muds; as it ran, the mass froze and decrepitated, incorporating some of the mud in its progress.

The zone of overridden block lava is covered by a ten-foot layer (zone *b*) of compact, slightly vesicular lava of similar composition and of a texture like that of ordinary basalt. This zone also belongs to the chilled, though here not brecciated, lower part of the main flow and passes gradually upward into a massive, eminently porphyritic, non-vesicular phase, 200 feet thick (zone *c*). Zone *c* is similarly transitional into the fourth phase, which consists of 220 feet of massive, highly amygdaloidal lava devoid of macroscopic phenocrysts (zone *d*).

The lava of zones *a* and *b* is a dark gray-green rock, originally a basic glass charged with numerous microphenocrysts of labradorite near Ab₁ An₁. These are usually between 0.5 mm. and 0.8 mm. in length. Octahedra of magnetite represent the only other primary constituent, unless some of the

PLATE 25.

Porphyritic phase of the Purcell Lava; from summit of the McGillivray
Range. Three-fourths natural size.

Quartz amygdule in the Purcell Lava. The amygdule, tubular in form and here six
inches long, is partly weathered out of its rocky matrix. A part of it, of unknown
length, has been lost during the weathering of the lava. About one-half natural size.

isotropic base is glass. Otherwise the rock is composed of very abundant chlorite and limonite, with some calcite and secondary quartz. The last is always in surprisingly small amount in the base, though the decomposition of the rock is profound. The pores are filled with quartz, chlorite, and opal.

The non-vesicular zone c is also of a gray-green colour. It is conspicuous by reason of the relatively great size of its abundant feldspar phenocrysts. (Plate 23). These range from one to three centimetres in length, by one to two millimetres in width. In the freshest specimens the phenocrysts have a dull lustre and brownish or greenish colour, both being due to the advanced alteration of the mineral. The feldspar is a plagioclase twinned polysynthetically after the albite law; it proved to be a labradorite near $Ab_4 An_5$. Under the microscope the crystals were seen to be filled with swarms of minute, secondary foils of sericitic habit but indeterminable (hydrargillite?). These large crystals are embedded in a base which again shows evidence of thorough decomposition, with the formation of much chlorite, much leucoxene, and the same colourless to pale greenish micaceous mineral found in the altered phenocrysts. In this mass there occur fairly numerous microlites of labradorite (also near Ab_5An_5), one millimetre or less in length. The specific gravities of two of the freshest and most typical specimens are 2·835 and 2·792. Notwithstanding the profound alteration of the rock it was thought that chemical analysis would throw light on its original character. Professor Dittrich has accordingly analyzed the freshest of the collected specimens (No. 1202). It was obtained on the high eastern ridge of the McGillivray range at a point about one mile south of the Boundary line. His results are as follows:--

Analysis of Purcell Lava (Zone c.)

SiO_2	41·50
TiO_2	3·33
Al_2O_3	17·09
Fe_2O_3	3·31
FeO	10·08
MnO	trace.
MgO	12·74*
CaO	0·97*
Na_2O	2·84
K_2O	0·22
H_2O at 110°C.	0·21
H_2O above 110°C.	6·99
CO_2	none.
P_2O_5	1·08
	·36
Sp. gr.	·92

The analysis evidently does not lend itself to profitable calculation. In spite of the very great alteration, however, the rock is pretty clearly a basalt.

* A second determination of CaO gave 0·89 per cent; a third gave 1·01 per cent, with MgO 12·57 per cent.

25a--vol. ii--14

2 GEORGE V., A. 1912

The rock of zone d (150 feet thick) shows an occasional large phenocryst of labradorite, but usually it is a blackish green, compact, homogeneous mass, bearing numerous amygdules of all sizes up to 8 centimetres in length. The amygdules, often orientated roughly parallel to the surface of the lava flow. are composed of infiltrated quartz or chlorite, or both; sometimes green biotite replaces some of the chlorite. In thin section the only original constituents are octahedra of magnetite and a plagioclase. The latter in individuals ranging from 0·5 mm. or less to 1 mm. in length, has a maximum extinction of 19° or 20° and seems to be acid labradorite, as in the underlying rocks. The characteristic arrangement of the abundant plagioclase crystals is that of a typical diabase. The interspaces are entirely filled with pale green chlorite and the original grains of magnetite, along with leucoxene and a little limonite. It seems impossible to say whether the chlorite has been derived from a pyroxene or from a glass. The habit of the rock is that of an ordinary basalt. Its specific gravity varies from 2·909 to 3·078; the average of three specimens is 3·000.

In zone d numerous, though small, angular fragments of quartzite and metamorphosed argillite, studded with numerous conspicuous octahedra of magnetite, were observed.

At the top of zone d is a conformable bed of argillite a few inches thick. The lava (65 feet thick) overlying this sediment belongs to a second period of extrusion closely following the former one.

In view of all the facts it seems certain that the whole 465 feet of lava represented in the section represent a single chemical type. It is highly probable that the lower 400 feet belong to one great flow and that the high vesicularity of zone d, the conspicuously porphyritic character of zone c, and the special features of zones a and b are all the results of different conditions of cooling in that thick flow.

The columnar section of the formation in this section is, therefore, as follows:—

Top, conformable base of Gateway (Moyie) formation

Second lava-flow: f. 65 feet—amygdaloidal lava poor in phenocrysts.

Inter-bed: e. 4 inches—argillite.

First lava-flow:
d. 150 feet—amygdaloidal lava poor in phenocrysts.
c. 200 " highly porphyritic, non-vesicular lava.
b. 10 " compact lava.
a. 40 " brecciated 'aa' lava.

Total lava........ 465 feet.

Base, conformable top of Kitchener (Siyeh) formation.

On examining the sections of the formation further east it was found that the four lava phases just described were not regularly represented. On the summit twelve miles west of Gateway the striking porphyritic phase is almost entirely replaced by the amygdaloid of zone d in the type section, while zone *a* is only about twenty feet thick near the summit monument. Zone *a* includes blocks of quartzite and metargillite, these rocks being torn and slivered as if the sediments were scarcely consolidated when they were over ridden by the flood of lava. At this section, with the exception of the twenty feet of brecciated lava, the whole formation, again nearly 500 feet thick, is made up of the deep gray-green amygdaloid. The reason for the non-appearance of the porphyritic phase in this well-exposed section is not apparent.

On the east-west ridge two and a half miles south-southeast of the monument, the porphyritic zone reappears at its proper place in the section though it is not so thick as in the type section. No argillitic beds here break the continuity of the lava. The most easterly exposure of the Purcell Lava in the McGillivray range is that at the Kootenay River flats. There the section showed three members with approximate thickness as follows:—

Top, erosion surface.

300 ± feet		blackish-green amygdaloid.
110	"	porphyritic lava with large phenocrysts of labradorite.
15	"	brecciated lower-contact zone.

Total,,455+ feet.

Base, conformable Siyeh metargillite.

About 220 t below the base a second sheet of highly scoriaceous amygdaloid, fifty feet thick, is conformably intercalated in the Siyeh strata. This lava corresponds in all respects with the uppermost member of the Purcell Lava proper. It occurs nowhere else in the Boundary sections and must have been a quite local flow.

To north and south of the summit monument a twenty-foot flow of *rhyolite* lies interbedded with the Gateway metargillites. Its base is at a horizon about fifty feet above the top of the Purcell amygdaloid. This occurrence of acid lava is unique in the range and has no known parallel in the Galton or Clarke ranges. It can be easily studied at the 6,400-foot contour on the main Commission trail, a half-mile north of the monument. The rock is a greenish-gray, slightly vesicular lava, bearing abundant phenocrysts of quartz and feldspar, from 1 mm. to 5 mm. in diameter; no dark-coloured mineral is macroscopically visible.

The thin section shows that the phenocrystic feldspar includes orthoclase (often microperthitic in look) and acid oligoclase. Like the quartz these are idiomorphic. A few small, deep yellow crystals of allanite are accessory. The feldspars are greatly kaolinized. The ground-mass was probably once mostly glass but is now completely devitrified. It is a very pale greenish mass of secondary material enclosing minute feldspar crystals and rounded quartzes, with apatite and altered ilmenite (leucoxene). Rutile needles have developed in the alteration of the ore. The main part of the ground-mass always polarizes, at least faintly. Most of it consists of quartz and a secondary, micaceous mineral, probably sericite, whose abundance seems to explain the relatively high specific gravity of the rock (2.735). The small steam-vesicles are filled with quartz and calcite.

25a—vol. ii—14½

2 GEORGE V., A. 1912

At the lower contact this lava flow, like the great basic flow, has ruptured, shredded, and balled up the underlying argillite which was clearly unconsolidated at the time of the eruption.

DIKES AND SILLS IN THE MCGILLIVRAY RANGE.

The fifty-foot dike already noted as cutting the Kitchener beds on the 6,583-foot summit merits description, since it is regarded as probably one feeder of the fissure eruption. The dike is vertical and strikes north 10° east. It has a marked zone of chilling on each wall.

In the chilled zones, acid labradorite, arranged as in diabase, is the only primary essential present. The interspaces are filled with a confused mass of chlorite, calcite, yellow epidote, kaolin, muscovite (the last occasionally in large, distinctive foils), and limonite, with a little secondary quartz. Abundant ilmenite or titaniferous magnetite is the one original accessory. The specific gravity of the freshest-looking specimen, taken three feet from the dike contact, is 2.840.

From its general habit, mineralogical composition, and mode of alteration, this chilled, fine-grained phase of the dike is almost certainly a much changed diabase. Except for the size of grain and lack of vesicularity it is lithologically identical with the diabase phase of the Purcell Lava.

The main body of the dike is composed of a medium-grained gabbroid rock which is similar to the chilled phase in all essential respects except in its coarseness of grain and in the occurrence of chlorite pseudomorphs with the forms of long prisms of amphibole. The latter mineral was an original constituent but has been completely altered. Other chlorite has resulted from the likewise complete alteration of interstitial pyroxene which seems to have accompanied the amphibole and labradorite in the list of primary constituents. The main body of this dike had thus originally the composition and structure of a hornblende gabbro, transitional to hornblende diabase. The specific gravity of one specimen is 2.853.

Two sills, respectively three and four feet thick, cut the sediments immediately east of the dike, from which they are probably offshoots.

At the head of the broad gulch, a mile farther west, the Commission trail crosses a second, thirty-foot, vertical dike, striking N. 30° W. The microscope has corroborated the impression won from the macroscopic appearance that this dike is of essentially the same composition as the first and may also represent the filling of a fissure whence issued part of the Purcell Lava flood.

PURCELL LAVA IN THE GALTON RANGE.

The Purcell formation reappears on the eastern side of the Kootenay valley and, as shown in the map sheet, crops out very liberally. Complete sections were made at ten different points, at each of which the thickness was found to be close to 400 feet. The section most favourable for the analysis of the formation was seen across the north slope of Phillips creek valley, about three miles above the cascade. The field study gave the following result:—

	Top, conformable base of Gateway formation.
d. 60 feet—	greenish-black amygdaloid.
c. 40 "	coarse basic breccia.
b. 200 "	greenish-black amygdaloid with occasional large phenocrysts of labradorite.
a 90 "	porphyritic, non-vesicular, with abundant large phenocrysts of labradorite.
380 feet.	Base, conformable top of Siyeh formation.

Mineralogically and chemically these rocks are similar to the corresponding phases of the formation in the McGillivray range. Zone c appears to be a true explosion-breccia but is apparently of quite local extent. In the other sections it is replaced by an approximately equal thickness of the black amygdaloid. The conspicuous porphyritic phase is also replaced by the amygdaloid in several sections made on the Kootenay valley slope, north and south of Phillips creek. In each of these latter sections the formation is very homogeneous and massive, as if formed of a single great flow. The intercalation of tuffaceous rock in zone c seems to show that zone d belongs to a later flow distinct from that represented in zone b. There is no plane of separation between zones a and b, which merge gradually into each other, being probably phases of one erupted mass.

At the summit of the Galton range the formation is cut off by a master fault. To the east of the fault the lava has been completely eroded away and it does not appear on the map of the belt covering the eastern half of the Galton range and the whole of the MacDonald range.

PURCELL LAVA IN THE CLARKE RANGE.

The most westerly outcrop of the lava in the Clarke range occurs at the head of Starvation creek, twenty-seven miles east of the summit fault of the Galton range. From that point to the lake at the eastern extremity of the Commission map the formation forms a conspicuous feature of the cliffs. From a commanding point it can be seen contouring the mountains through several miles of continuous exposure. In all, the Boundary map has twenty-five miles of this outcrop. It rigidly preserves its conformable position between the Siyeh and Sheppard formations and steadily holds a thickness of about 260 feet. As in the western ranges it is, on account of its hardness, a strong cliff-maker, often forming unscaleable precipices at cirque or canyon.

Wherever examined the whole formation is a homogeneous, dark greenish-gray to blackish amygdaloid, scoriaceous and of typical ropy structure at the upper contact. White amygdules of quartz and calcite are there abundant and often reach great size, even to six or eight inches in length. (Plate 23.) The porphyritic phase and breccia of the western ranges are not associated with the

2 GEORGE V., A. 1912

amygdaloid; in one field section a phase suggesting rolled-in lava-crust forms a local variation.

As in the other ranges, care was taken to secure the freshest possible specimens but here also the microscope displayed profound alteration in them all. The dominant constituent is again labradorite (Ab An₂), with the usual diabasic arrangement. Abundant chlorite, calcite, kaolin, and limonite, with magnetite in laths and octahedra and many narrow prisms of apatite as the two surviving original accessories, fill the spaces between the idiomorphic feldspars. The rock is almost certainly a greatly altered basalt with diabasic structure. The specific gravities of two typical specimens are 2.828 and 2.846.

Lithologically similar lavas have been described in the accounts of the Grinnell, Sheppard, and Kintla formations, in which flows have been locally interbedded

DIKES AND SILLS IN THE CLARKE RANGE.

At the western end of the Sawtooth ridge, north of Lower Kintla lake and at the 7,000-foot contour, the Appekunny and Grinnell beds are respectively cut by two vertical dikes running northwest-southeast. Each dike is about twenty feet wide. Lithologically, even to microscopic details, these intrusives are not to be distinguished from the diabasic phase of the lava just described; the dikes were, most probably, feeders of the extrusive mass.

To north, south, and east of Upper Kintla lake a persistent intrusive sill, averaging forty feet in thickness, cuts the Sixth formation at a nearly constant horizon, about 1,200 feet below the base of the Purcell Lava. Macroscopically the rock of the sill is a deep greenish-gray, fine-grained trap like that of the two dikes. The thin section shows the rock is relatively fresh. Its essential constituents are diopsidic augite, labradorite, and green hornblende. The original accessories are micropegmatite (of quartz and orthoclase); much magnetite in octahedra, laths, and skeleton crystals; apatite, titanite, pyrite, and interstitial quartz. Yellow epidote, chlorite, zoisite, limonite, and a little calcite are secondary products. The feldspar is decidedly subordinate to the bisilicates in amount. Like the hornblende it is idiomorphic. The augite apparently occurs in two generations; a small proportion of it is crystallized in stout idiomorphic prisms up to 0.6 mm. in length, while most of it is in anhedral grains 0.1 mm. or less in diameter. The feldspar- and hornblende prisms average about 0.2 mm. in length or less, and are enclosed in a mesostasis of granular augite, micropegmatite, and quartz. The structure is transitional between that of a diabase and a gabbro with a stronger tendency to the gabbroid. The specific gravity of a type specimen is 3.057.

In chemical composition, in the dominance of the bisilicates, in structure, in the character of accessories, including the micropegmatite intergrowth, and in specific gravity, this rock closely resembles the staple phase of the much greater sills in the Moyie and Yahk ranges. The principal mineralogical difference consists in the fact that here the bisilicate is mostly augite, while, in the thick western sills, it seems to be entirely amphibole of the same habit as in this sill.

SESSIONAL PAPER No. 25a

At the same time there are many points of lithological resemblance between the Kintla canyon sill and the amygdaloid of the Purcell formation. It is clear that the extremely abundant chlorite of the amygdaloid could have been derived from a dominant original pyroxene identical with that in the underlying sill. The feldspars of sill and lava seem to be of exactly the same species, while the list of important accessories, excepting the micropegmatite and quartz, is common to both. The existing differences in mineralogical and chemical composition are to be explained by the contrasted conditions of crystallization, as well as by a slight acidification of the sill magma. The latter was thrust into a zone of silicious metargillites; a relatively slight resorption of the invaded rock would lead to the generation of interstitial quartz and micropegmatites as in the Moyie and other of the western sills. The significance of these parallels will be noted in the discussion of the latter intrusives. At present it may suffice to observe that the Kintla canyon sill seems to belong to the same eruptive period as the Purcell Lava and that both are probably contemporaneous with the great sills west of the Yahk river.

Another sill, fifty feet thick, cuts the Siyeh formation on the eastern slope of the Clarke range. It is well exposed on both sides of Oil creek, about two miles upstream from the derricks at Oil City. The intrusive has split a zone of silicious metargillites at a horizon roughly estimated to be 1,000 feet above the base of the Siyeh.

The rock is essentially a fine-grained duplicate of the Kintla canyon sill-rock but there is here a considerably greater amount of freely crystallized sodiferous, microperthitic orthoclase, which replaces some of the labradorite and becomes a major constituent. Micropegmatite is an abundant interstitial accessory. The rock is badly altered with the generation of epidote, chlorite, kaolin, sericite, saussurite, and limonite, but it is certain that at least half the volume of the rock was originally composed of bisilicate. Through most of the sill the same green, idiomorphically developed hornblende which was found in the Kintla canyon sill, is an abundant essential along with the colourless augite.

A specimen taken at a point five feet from the upper contact and thus representing the contact-zones, bears no hornblende, but the bisilicate is entirely augite, crystallized, as usual, in apparently two generations. The hornblende, here, as in the other sill, has every evidence of being a primary constituent. It seems to have been able to crystallize only in the interior part of the sill, while augite monopolized the contact zones. These contrasted, augitic and hornblendic, phases of the sill are homologous to the similar phases found in the fifty-foot dike near the summit of the McGillivray range. This dike has been noted as most probably one feeder of the Purcell Lava flood. The specific gravity of the augitic phase is 3.005; that of the normal hornblende-bearing phase, 3.048. These values further show the similarity of this sill to the Kintla canyon sill (sp. gr., 3.057).

2 GEORGE V., A. 1912

PURCELL LAVA AND ASSOCIATED INTRUSIVES IN THE LEWIS RANGE.

The most easterly exposures of the lava, yet described, are those found in the Lewis range by Willis and Finlay.* Finlay's account of the formation shows the close parallel between the relations of extrusive and intrusive phases of the rock in this range and their relations in the Clarke and McGillivray ranges. His descriptive note may be quoted in full:—

'The igneous rocks of the Siyeh limestone are two—an intrusive diorite and an extrusive diabase.

'*Diorite.*—On Mount Gould and on Mounts Grinnell, Wilbur, and Robertson there is found a band of diorite 60 to 100 feet thick. Near the upper and lower surfaces this intrusive sheet was chilled and is fine-grained. In the center the texture is medium or fine-grained. Several dikes which have acted as conduits for the molten rock are exposed in the region near Swift Current pass. One of these extends across the cirque occupied by the Siyeh glacier and runs vertically up the amphitheatral walls. It is 150 feet in width. A second dike, vertical and 30 feet wide, comes in beside the Sheppard glacier. Along the trail to the east of Swift Current pass the diorite sheet breaks across the Siyeh argillite and runs upward as a dike for 500 feet. It then resumes its horizontal position as an intercalated sheet between the beds of argillite. As a dike it skips for 600 feet across the strata on Mount Cleveland.

'Under the microscope the diorite is found to contain abundant plagioclase, with small amounts of another feldspar, much weathered, which does not show twinning. This mineral is closely intergrown with quartz. Brown hornblende is the principal dark silicate. The plagioclase has an extinction angle high enough for labradorite, but it gives no definite clue as to its exact basicity. No section of a fresh piece twinned on the albite and Carlsbad laws at the same time could be observed. The quartz is not present in sufficient amounts to make advisable the name quartz-diorite for the rock. The small patches of biotite originally present are entirely altered to chlorite. Pyrrhotite is occasionally met with, apatite occurs in crystals of unusual length, and magnetite in lath-shaped pieces is common.

'*Diabase.*—In the field this rock is always much weathered, presenting a dull green colour by reason of the secondary chlorite which it contains. It is a typical altered diabase. Exposures are found near the top of Mount Grinnell, where the thickness of the sheet is 42 feet, and on Sheppard mountain opposite Mount Flattop. Here the extrusive character of the flow is well shown, for its upper surface is ropy and vesicular, with amygdaloidal cavities containing calcite. Its place is at the top of the Siyeh formation, 600 feet above the sheet of diorite, with heavy bedded ferruginous sandstone and green argillite immediately below and above it

---

* G. I. Finlay, Bull. Geol. Soc. America, Vol. 13, 1902, p. 349.

SESSIONAL PAPER No. 25a

respectively. The argillite has filled in the irregularities of the upper surface of the diabase. Five dikes of the same rock, genetically connected with it, were observed on Flattop. They contain inclusions of the argillite, and range from an inch to six feet in width. They are nearly vertical.

'Under the microscope the rock is seen to be made up principally of augite and plagioclase, arranged in such a manner as to give the normal diabase structure. The plagioclase is idiomorphic in long, slender laths. It has the habit of labradorite, but no material was studied which offered data for its accurate determination. The extinction angle is high. The augite is much more abundant than the feldspar. It is an allotriomorphic mineral, red-brown when fresh, but frequently entirely gone over to chlorite. The small amount of olivine originally present in the rock is now altered to serpentine and chlorite. Besides the chlorite, which is the chief alteration product, resulting from the plagioclase as well as from the augite and olivine, much secondary calcite has been derived from the feldspar. Apatite is found and titaniferous magnetite, in grains and definite crystals, is abundant. The medium texture of the diabase is fairly uniform throughout the flow.'

The present writer had an opportunity of studying both the intrusive and extrusive types as they occur near the summit on the Swift Current Pass trail. At the upper edge of 'Granite Park,' on the western side of the divide (see Chief Mountain Quadrangle sheet, U. S. Geological Survey), the Purcell formation is represented by two contiguous flows resting on the Siyeh metargillite and overlain by typical Sheppard beds. The lower flow is forty feet thick. Its upper surface, as noted by Finlay, is ropy. This structure passes beneath into a pronounced pillowy structure, which, in place, characterizes most of the thickness of the lava sheet. The pillows are generally quite round and of spheroidal form. They range from a foot or less to two or three feet in greatest diameter. No sign of the variolitic composition so common in pillow-lavas could be detected. The interstices between the pillows are filled either with chert, or, more commonly, with an obscure, breccia-like mass of aphanitic material whose microscopic characters are those of palagonite. This material bears a few minute crystals of feldspar but is chiefly composed of finely divided chlorite, quartz, calcite, and abundant yellowish-brown isotropic substance like sideromelane. The whole seems to form a greatly altered basaltic glass.

The pillows themselves and the non-pillowy parts of the flow are composed of the same type of vesicular microporphyritic, occasionally diabasic basalt that makes up the upper flow. This is eighteen feet thick and lies immediately upon the ropy surface of the forty-foot flow. The latter is without the pillow structure but is massive like the normal Purcell amygdaloid. Microscopic evidence shows that the rock is of the chemical type recognized in all the occurrences of the lava in the western ranges.

The writer's examination of the sill (50-70 feet thick) and dike (50 feet thick) noted by Finlay as outcropping to the east of the Swift Current Pass,

2 GEORGE V., A. 1912

led to similar results except that the untwinned feldspar, which is present in large amount, has been determined as orthoclase, probably bearing soda. The other constituents, both primary and secondary, are the same as in the Oil creek and Kintla canyon sills. Augite is as important as the hornblende and micropegmatite is again a prominent accessory.

On account of the striking predominance of the bisilicates compared to the feldspar, this rock can scarcely be called a true diorite. Its systematic position is better recognized by calling it a somewhat acidified, abnormal gabbro. It constitutes both the sill and the dike at the Swift Current Pass. The specific gravity of a typical specimen from the dike is 3.055, a value almost identical with those found for the Kintla canyon and Oil creek sills.

RELATION OF THE SILLS AND DIKES TO THE PURCELL EXTRUSIVE.

The Kintla canyon sill and dikes crop out twelve miles or more to the west of the Oil creek sill, while the Swift Current Pass locality is about twenty miles from either of the other two. Thus, at each of three widely distributed localities, we have a constant association of an extrusive basaltic lava resting on the top bed of the Siyeh formation and an intrusive gabbroid sill-rock thrust in' the Siyeh itself. Though the vertical dikes, either feeding the visible sills or apparently independent of them, are relatively numerous in the Siyeh, no dike or sill has yet been observed in the admirably exposed Sheppard formation. These facts, of themselves, afford good presumptive evidence that the Purcell Lava proper is genetically connected with the sills and dikes. This conclusion is amply corroborated by microscopic study, which, even in face of the great alteration of all the rocks, goes to show an essential identity of the principal minerals respectively occurring in intrusive and extrusive.

The main difference of chemical composition consists in the presence of the silica and potash represented in the primary orthoclase, micropegmatite, and quartz which are so abundant in the sills while entirely absent in the surface flows. The marked rarity of secondary quartz in the altered lava seems to indicate that these acid materials were not originally dissolved in the glassy base of the amygdaloids. Neither quartz nor orthoclase were appreciable constituents of the holocrystalline phases of the lava. It appears, therefore, highly probable that they enter into the composition of the intrusives because of a special modification of the magma when in purely intrusive relation. The simplest cause for the appearance of the acid constituents is to be found in the absorption of a small amount of the invaded metargillites along the contact-surfaces; and this the writer believes to be the true cause. If the sills had been considerably thicker, their greater heat-supply would have led to yet more pronounced acidification; as in the case of the Moyie sills (described in the following chapter), a facies of granitic acidity might have been develop d, preferably at the top of such a sill.

On the other hand, the amygdaloid was not so acidified because of the manifest speed with which the extrusive magma must have passed through the dike fissures to form the highly fluid and hence widespread floods of lava. In its rapid mounting there was not time enought for the basic magma to dissolve an appreciable amount from the walls of the fissures. The Purcell Lava is, in this view, to be considered as representing the pure, original magma that, at the end of Siyeh time, underlay the Rocky Mountain and Purcell mountain system at the Forty-ninth Parallel. Some of the feeding dikes are composed of the same material, chemically considered, while others, like the sills, are made of the magma which has been enriched in silica and potash by slight but appreciable assimilation of the invaded quartzitic and metargillitic strata.* The uniformity of the extrusive lava through the ninety miles of distance between the Swift Current Pass and summit of the McGillivray range, is matched by the uniformity in the lithology of the intrusive bodies wherever discovered in the Clarke and Lewis ranges.

The further correlation of these sills, dikes and flows with the huge sills of the Moyie and Yahk ranges will be discussed in chapter X.

SUMMARY.

The variations in thickness, field-habit and associations of the Purcell Lava may be conveniently shown in the form of the following table:

Purcell Range.	Galton Range.	Clarke Range.	Lewis Range (Swift Current Pass).
Local 20-foot flow of rhyolite about 50 feet above f. f. 65', amygdaloid. e. 4' argillite. d. 150', amygdaloid. c. 200', porphyry. b. 10', compact lava. a. 40', "aa" lava.	d. 60', amygdaloid. c. 40', coarse breccia. b. 200', amygdaloid with phenocrysts. a. 90', non-vesicular porphyry.	260', amygdaloid.	35' of amygdaloid in Sheppard and 40' of amygdaloid in Kintla. 18', massive, amygdaloidal flow. 40', ropy flow passing below into pillow lava
Total lava, 465'.	Total lava, 390'.	Total lava, 260'.	Total lava, 58'.
Dikes and sills cutting Kitchener (Siyeh) immediately below. Locally, 220' below a, a 50-foot flow of amygdaloid.		Dikes and sills cutting Siyeh immediately below.	Dikes and sill cutting Siyeh immediately below.

The lavas are everywhere thoroughly conformable to the overlying and underlying sediments.

Excepting the local rhyolite the lavas are petrographically similar and belong to the basaltic family. The more unusual characters include the local

___

* It is conceivable that the local rhyolite flow overlying the main sheet of Purcell lava, represents a product of differentiation following the acidification of a large, though invisible body of the gabbroid magma.

2 GEORGE V., A. 1912

pillow-structure observed in the Lewis range, and the extraordinary size of the feldspar phenocrysts in the porphyritic phases. The individual flows were generally of great thickness, reaching as much as 400 feet.

From the close association with basaltic dikes and sills cutting the Siyeh (Kitchener) formation, it is believed that the feeders of the fissure eruptions can be actually seen. The eruptions began at the close of Siyeh time but were intermittently continued through the Sheppard and early Kintla times. Following the correlation of the preceding chapter, all of this vulcanism falls within the Middle Cambrian period; the Purcell Lava proper, underlying the equivalent of the Flathead sandstone, seems to belong to a period of crustal fissuring which accompanied the widespread Middle Cambrian subsidence.

CHAPTER X.

INTRUSIVE SILLS OF THE PURCELL MOUNTAIN SYSTEM

INTRODUCTION.

Within the area of the Boundary belt where it crosses the Yahk and Moyie ranges, no extrusive lava was anywhere observed, but intrusive basic masses were found in large development. On the map they form twenty-four bands, covering in all about one-sixth of the area between the Kootenay river at Porthill and the main fork of the Yahk. One of them is a true dike; a second is either a dike or a sill; the other twenty-two bands are all more or less certainly to be classified as sills. All the bodies are intrusive into either the Kitchener or Creston quartzites, and, as noted below, are referred to the Middle Cambrian period. No one of this group of intrusives, so far as known, cuts the Moyie formation.

The exposure of the igneous bodies has become possible through extensive block faulting and upturning, followed by erosion. The faulting has repeated the outcrops at several points, so that the number of different bodies is less than the number of igneous bands shown on the map. On account of the unusual continuity of the forest cover, obscuring the field relations, and also because of the general lack of easily recognized horizon-markers in the invaded quartzites, it has been impossible to determine the actual amount of this repetition of outcrops through faulting. Its occurrence in certain localities is hypothetically indicated in the general structure section. The rarity of dikes is probably only apparent. If the overwhelming forest-cap were removed from these ranges, dikes in considerable number might be displayed. The sills are, on the average, so large that they were discovered even under the peculiarly difficult field conditions of these mountains. At the same time, the mapping of several of the igneous bands, especially in the eastern half of the Yahk range, must be regarded as merely approximate.

The sills vary in thickness from fifty feet to about 1,000 feet. The thickest of these is one of a genetically related group of five adjacent sills which are distinguished by peculiar composition and history and may be conveniently referred to as the Moyie sills. Several of the bodies are from 200 to 500 feet in thickness. The dip varies from about 5° to 90°, averaging perhaps 40°. In general, it is a simple matter to locate both top and bottom of each intrusive sheet. On reference to the maps it will be seen that most of the bands hold their respective widths for several miles. In no case was it possible, owing to the conditions of field work, to follow a sill far beyond the limits of the Boundary belt, but, from the fact that the bodies hold nearly uniform thickness

2 GEORGE V., A. 1912

... across the belt, it is believed that the true sill form, rather than the cushion form of the laccolith, is characteristic of all the intrusions which follow bedding planes. The large irregular igneous mass whose western contact crosses the Boundary line at a point seven miles east of the Movie river, is, in part, a crosscutting body; its north-south arms are in sill relation, while the east-west band seems to be in the form of a huge dike.

USUAL COMPOSITION OF THE INTRUSIVES.

Throughout both mountain ranges the main mass of each intrusive body is composed of a notably uniform type of rock. Macroscopically, the type has the habit of a dark greenish-gray hornblende gabbro of medium grain. Already in the hand-specimen it can be seen that hornblende and feldspar are the essential constituents and that the former dominates in quantity. Occasional glints of light from accessory pyrrhotite may be observed. The hornblende forms elongated prisms from 1 mm. to 3 mm. or more in length. They generally lack the usually high lustre of the amphiboles occurring in plutonic rocks. The whitish feldspars and the accessories together form a kind of cement for these prisms. The principal variations in macroscopic character are due either to the local coarsening of grain, as so commonly seen in gabbros, or to a likewise frequent, local development of a phase richer in hornblende and poorer in feldspar than the type. In the latter case the rock becomes almost peridotitic in look.

On examination of the rock in thin sections the list of constituents is enlarged by the addition of titanite, ilmenite or titaniferous magnetite, pyrrhotite, apatite, rare zircons, and never-failing, though variable amounts of accessory, interstitial quartz. Accessory biotite and orthoclase were found in many specimens.

The amphibole was found to have characters which changed rather regularly with the freshness of the rock. In the freshest specimens it was a compact, strongly pleochroic mineral with the following scheme of absorption:—

Parallel to **a**—light yellowish-green.
      "   **b**—strong olive green.
      "   **c**—deep bluish-green.

b > c > a

In specimens which appear to have been slightly altered, the hornblende is still compact but the colours are considerably paler, so as to give the mineral the look of actinolite. A further stage of alteration is represented in a fibrous phase of the amphibole, suggesting uralite in colour and other essential respects. This fibrous amphibole is so common in the slides that it was at first believed that it might be secondary after a pyroxene. A close study of a large number of thin sections has, however, led to the conclusion that the fibrous amphibole is really secondary after the compact form. All stages of transition can be found between the two, and the fibrous type has demonstrably grown at the expense of the other, which has been simultaneously decolourized. No trace of any pyroxene or of pseudomorphs of pyroxene has been discovered in any slide. Many of the sliced specimens are so fresh, as shown by the preservation of the essential minerals as well as of biotite, that the pyroxene must be discoverable if it had ever entered into the composition of the rock at the time of crystallization from the magma. Another hypothesis, that some of the fibrous hornblende has resulted from the speedy alteration of originally crystallized pyroxene, through the influence of magmatic vapours which acted long before the rock was exposed to ordinary weathering, cannot be excluded. So far, however, the positive microscopic evidence declares in favour of the first view. Similar cases of the derivation of fibrous amphibole from compact amphibole through metasomatic changes are described by Zirkel.*

The hornblende is, in the prismatic zone, idiomorphic against the feldspar; it fails to show good terminal planes. The ends of the crystals characteristically run out into narrow forked blades. The extinction on (010) averages about 13° 30'; that on (110), about 14°. In phases of the rock where quartz is an abundant accessory, the amphibole is often highly poikilitic, the prisms being charged with swarms of minute droplets of quartz. For the purpose of finding the optical orientation the attempt was made to produce etch-figures on the more likely looking specimens of the amphibole but, on account of the poikilitic and bladly character of the amphibole, the attempt was not successful.

From the chemical and quantitative analyses of the type rock, a rough calculation of the chemical composition of the hornblende gave the following proportions:—

SiO_2	49·8
Al_2O_3	5·2
Fe_2O_3	5·2
FeO	12·1
MnO	·2
MgO	15·3
CaO	11·9
	99·7

The estimate is crude but it shows that the amphibole is a common hornblende high in silica, iron, magnesia, and lime, but low in alumina.

The feldspar is plagioclase, always well twinned on the albite law and often on the Carlsbad law. Many individuals extinguish with angles referring them to labradorite, $Ab_1 An_1$; some have the extinction angles appropriate to basic bytownite; a very few others are zoned, with anorthite in the cores and andesine in the outermost shell. The average composition of the plagioclase in the normal rock is near that of the basic labradorite, $Ab_1 An_2$.

Magnetite, titanite, pyrrhotite, and apatite are all present but are strikingly rare in most of the slides. Their forms and relations are those normal

* F. Zirkel, Lehrbuch der Petrographie, Vol. 1, 1893, p. 325.

2 GEORGE V., A. 1912

to gabbros. The quartz often bears many fluid inclusions. Chlorite, epidote, leucoxene, and a little calcite are rare secondary minerals.

Professor Dittrich has analyzed a specimen of the fresh sill-rock from a point situated about nine miles east of the Moyie river and 1·5 miles north of the Boundary line. This specimen (No. 1153) represents the principal rock type of most of the sills. The analysis resulted as follows:—

Analysis of dominant gabbroid type in the Purcell sills.

		Mol.
SiO_2	51·92	·865
TiO_2	·83	·010
Al_2O_3	14·13	·137
Fe_2O_3	2·97	·019
FeO	6·92	·096
MnO	·14	·001
MgO	3·22	·205
CaO	11·53	·205
Na_2O	1·38	·023
K_2O	·47	·005
H_2O at 110°C	·10	....
H_2O above 110°C	1·07	....
P_2O_5	·04	....
CO_2	·06	....
	99·78	
Sp. gr..........(corrected value)	2·990	

A fairly accurate optical determination of the weight percentages among the principal mineral constituents (Rosiwal method) gave the result:

Hornblende	58·7
Labradorite	34·8
Quartz	4·0
Titanite and magnetite	1·4
Biotite	·9
Apatite	·2
	100·0

The comparative poverty in alumina and the high acidity are evidently related to the composition of the hornblende, which has been estimated as above. In some respects the analysis recalls the diorites but both the magnesia and lime, as well as the amount of femic material in the rock, are too high for that class. It seems best, for the present, to place this type among the hornblende gabbros, although it is to be regarded as an abnormal variety in that class.

The standard mineral composition or 'norm' of the Norm classification was calculated to be:—

Quartz..	6·78
Orthoclase..	2·79
Albite..	11·53
Anorthite..	50·96
Diopside..	21·07
Hypersthene..	19·44
Ilmenite..	1·52
Magnetite..	4·41
H_2O and CO_2..	1·23
	99·62

Accordingly, in this method of classification, the type belongs to the presodic subrang of the percalcic rang, in the order, vaalare, of the salfemane class. The ratio of **Q** to **F** in the norm is very close to that which would place the rock in the order, gallare.

VARIATIONS FROM THE NORMAL COMPOSITION.

Variations from this gabbro type are very common in most of the sills. These generally consist in an increase of quartz and biotite, along with the appearance of orthoclase, which is crystallized either independently or in the form of micrographic intergrowth with quartz. As these constituents increase in amount, the hornblende seems to preserve its usual characters, but the plagioclase shows a strong tendency toward assuming the zoned structure; the cores average basic labradorite, Ab₁ An₂, and the outermost shells average andesine, near Ab₄ An₃. When the quartz and micropegmatite become especially abundant, the plagioclase averages acid andesine or basic oligoclase. In several thin sections the plagioclase is seen to be mostly replaced by orthoclase and quartz, which, with the still dominant hornblendes, form the essential substance of the rock.

These changes in composition, indicating that the sill-rock has become more acid, are always most notable along the contacts and especially along the upper contacts. A good illustration of the acidification along the upper contact occurs in a well exposed sill outcropping in the band that runs south from the Boundary line at a point nine miles east of the Moyie river. This sill is about 500 feet thick. A specimen (No. 1) taken twenty feet from the lower contact is unusually rich in hornblende but bears much quartz and orthoclase along with the subordinate essential, acid andesine. It carries no biotite nor micropegmatite but orthoclase dominates over the plagioclase. Specimen No. 2, taken seventy-five feet from the lower contact, is a very similar rock in which the plagioclase is an unzoned labradorite somewhat subordinate to the orthoclase in amount. Specimen No. 3, taken fifteen feet from the upper contact, is gabbroid in look, though lighter in colour than either No. 1 or No. 2. It is essentially composed of hornblende, quartz, orthoclase, and basic andesine, named in the order of decreasing abundance. The accessories include micropegmatite and much biotite, the latter in small, disseminated foils. The essentials are all poikilitic with mutual interpenetrations and enclosures. The structure is quite confused.

25a—vol. ii—15

2 GEORGE V., A. 1912

All three specimens are very fresh and their densities clearly indicate the acidification along the upper contact. The respective specific gravities are:—

No. 3, 13 feet below upper contact..	.. ..	.. ..	.. ..	.. ..	2·983
No. 2, 425 " " "	.. ..	.. ..	.. ..	.. ..	3·001
No. 1, 480 " " " "	.. ..	.. ..	.. ..	.. ..	3·072

Lower contact, 500 feet below upper contact.

MOYIE SILLS.

Of all the intrusions these outcropping on the isolated 'Moyie Mountain,' immediately west of the Moyie river at the Boundary line, show the most remarkable variations in composition. (Figures 13 and 14; Plate 25.) Some years ago the writer published two papers detailing the petrography of the more important phases of these sills.* The description was based on field work during only a few days in the season of 1904. The importance of this particular section was not fully apparent until the field season was over and the rock collection had been microscopically studied. If time could have been spared during the continued reconnaissance of the Boundary belt, the writer would have early made a second visit to the Moyie sills to test the conclusions of the 1904 season regarding field relations. Unfortunately, no such opportunity for additional personal field work became available. In 1905 an untrained assistant was sent to the locality, and he collected new petrographic material at points along the Boundary slash, as designated by the writer. The character of the specimens thus added, to the material in hand seemed to corroborate the general conclusions of the writer and the two publications above mentioned were issued.

Thus, in 1905 and 1906, the writer believed that the intrusive rocks occurring on the western slope of Moyie mountain together form a single sill about 2,600 feet in thickness. Such was his belief at the time when the present report was sent in for publication. In 1910, Mr. Stuart J. Schofield was commissioned by the Director of the Dominion Geological Survey to make a geological study of the Purcell range. At the writer's request, Mr. Schofield examined the section of Moyie mountain at the Boundary slash. He found that the igneous rocks of the western slope really compose three sills, separated by Kitchener quartzite. He also found two thinner sills on the eastern slope of the mountain and in the same Boundary-line section, an area which the writer was not able to traverse in 1904. In 1911 Mr. Schofield guided the writer to his various contacts, all of which were seen to be correctly located in his profile of the mountain. Recent forest fires had cleared the exposures somewhat since 1904 and there can now be little doubt as to the structural relations hereafter described. The writer's sincere thanks are due to Mr. Schofield for his careful, efficient field-work on this problem.

The relations are, therefore, more complex than was formerly believed by the writer. However, it may be stated well in advance that the theoretical con-

---

* American Journal of Science, Vol. 20, 1905, p. 185; Festschrift zum siebzigsten Geburtstage von Harry Rosenbusch, Stuttgart, 1906, p. 203.

clusions published in 1905 and 1906 as a result of a study of the ' Moyie Sill' are *essentially strengthened* by the new facts of structure. Gravitative differentiation is illustrated not once but thrice, that is, in each of the sills occurring on the western slope of the mountain. It is illustrated a fourth time in the more important of the two sills on the eastern slope.

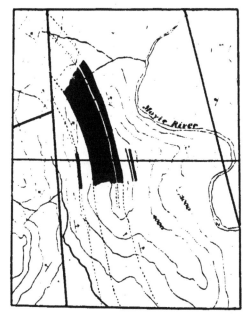

Figure 13. Locality map of the Moyie sills, showing in solid black the parts best exposed. The straight line in the middle of the map represents the Boundary line; the other straight lines represent the approximate outcrops of major faults. The block between the faults includes Moyie mountain.' Contour interval is 500 feet Scale, 1 : 68,000.

For convenience, the five sills of Moyie mountain will be distinguished by the letters, A, B, C, D, and E, named in stratigraphic order, with A the highest, E the lowest in the series. (See Figures 13, 14 and 15). Of these sills, C, D, and E correspond to the whole ' Moyie sill ' of the 1905 and 1906 papers.

Of the five sills, B is the only one with a sensibly homogeneous composition. Each of the other four presents phasal variations of notable character.

2 GEORGE V., A. 1912

With the addition of one type, the list of rock varieties recognised in the early publications will serve for all the bodies to be described. The description of the individual sills may be anticipated by an account of all the phases,

Figure 11.- Section of Moyie mountain and the Moyie sills, along the International Boundary line. Sills in solid black. Bedding-planes of the Quartzite and fault-planes shown.

beginning with the most acid one, a granite, and ending with the most basic and ferromagnesian one, a metagabbro or abnormal hornblende gabbro. The following account of petrography and theory will largely consist of a revised edition of that contained in the 1905 and 1906 papers.

ABNORMAL BIOTITE GRANITE.

In sills A, C, and D the intrusive rock forms distinctly acid zones. The chief constituent is a biotite granite. This is a gray rock, much lighter in tint than the deep green gabbro (Plate 24). The grain varies from quite fine to medium. Very often roundish grains of bluish, opalescent quartz interrupt the continuity of the rock. These are considered to be of exotic origin as they were seen to graduate in size into larger blocks of quartzite (xenoliths) shattered from the sill-contacts.

To show the average composition of the granite, and the approximate limits of its lithological variation, fresh specimens, taken from sill C at three points in the section following the wagon-road, west of the mountain, will be described. They were collected at respectively 15, 40, and 50 feet from the upper contact with the quartzite.

The specimen taken at a point 15 feet from the contact, and representing what may be called Phase 1, has the macroscopic appearance of a finely granular gray granite. In thin section it is seen to be a micropegmatite with a hypidiomorphic granular structure sporadically developed in many parts of the section. The crystallization is confused and does not show the regular sequence of true granites. The essential constituents are quartz, micropegmatite, microperthite, orthoclase, oligoclase-andesine and biotite; the accessories include titaniferous magnetite, a little titanite, and minute acicular crystals of apatite and rarer zircons. The characters of all these minerals are those normally belonging to

PLATE 24.

Secondary granite of the Movie sill C, fifty feet from upper contact.   natal size.

common granite. The chemical analysis of the rock shows the mica to be magnesian.

A striking feature of this, as of the other phases of the acid rock, is the advanced alteration of the feldspars which are usually filled with dust-like aggregates of epidote, kaolin and muscovite. This alteration is believed to be due to magmatic after-action, probably the result of the expulsion of vapours during the solidification of the underlying gabbro.

The calculation of the quantitative mineralogical composition of the rock has been attempted by the Rosiwal method. In the process the secondary products were neglected and the feldspars were arbitrarily regarded as fresh. The inaccuracy of the result is manifest but it does not affect the value of the comparison among all the phases of the sill. Especially between the gabbro and the acid zone the contrasts of quality emerge with the same clearness and certainty as characterize the related contrasts established in the chemical analyses.

The total chemical analysis by Prof. Dittrich of this Phase 1 (specimen No. 1137) is here given:

Analysis of Granite (Phase 1) of Monic Sills.

		Mol.
SiO_2	71·69	1·195
TiO_2	·59	·005
Al_2O_3	13·29	·130
Fe_2O_3	·83	·005
FeO	4·23	·058
MnO	·09	·001
MgO	1·28	·032
CaO	1·66	·030
Na_2O	2·48	·040
K_2O	2·37	·025
H_2O at 110° C	·14	....
H_2O above 110°C	1·31	....
P_2O_5	·07	....
CO_2	·13	....
	100·16	

Sp. gr. ....(corrected value). 2·733

This rock is clearly an unusual type of biotite granite. The most evident peculiarity is the low total for the alkalies; it accords with the relatively small proportion of feldspar present. Notwithstanding the abundance of free quartz, the silica percentage is kept low by the comparative richness in biotite and by the magmatic alteration of the rock. The estimate of the mineralogical composition gave the following result in weight percentages:—

Quartz	41·6
Sodiferous orthoclase	32·5
Biotite	15·2
Muscovite	4·6
Microperthite	3·9
Oligoclase	1·0
Magnetite	1·0
Apatite	·2
	100·0

2 GEORGE V., A. 1912

In the Norm classification the rock enters the sodipotassic subrang, teha-nose, of the domalkalic rang, alsbachase, of the order, columbare, and the persalane class. The norm has been calculated as follows:—

Quartz..	40·14
Orthoclase..	13·90
Albite..	20·96
Anorthite..	7·23
Corundum..	3·94
Hypersthene..	9·27
Magnetite..	1·16
Imenite..	1·21
H₂O and CO₂..	1·53
	99·43

The second analyzed specimen of the biotite granite, Phase 2, is that collected at the point 40 feet from the upper contact of sill C. It is closely allied in composition to the phase just described and is chiefly distinguished from the latter by a coarser grain and a different structure. Microscopic examination shows this rock to be eugranitic (hypidiomorphic-granular), with small isolated areas of the micrographic intergrowth of quartz and feldspar. The constituents are nearly the same as in Phase 1. Here, however, muscovite is an accessory so rare as not to enter the table of quantitative mineral proportions. True soda-orthoclase replaces nearly all the micropertite of the micropegmatitic facies. Calcite enters the list of constituents; it may be in part of primary origin.

The chemical analysis by Prof. Dittrich of Phase 2 (specimen No. 1133) is as follows:—

Analysis of Granite (Phase 2) of Moyie Sills.

		Mol.
SiO₂..	72·42	1·207
TiO₂..	·68	·009
Al₂O₃..	10·47	·103
Fe₂O₃..	·83	·005
FeO..	5·50	·076
MnO..	·16	·002
MgO..	·41	·010
CaO..	2·53	·045
Na₂O..	1·93	·031
K₂O..	2·94	·031
H₂O at 110°C..	·06	....
H₂O above 110°C..	1·11	....
P₂O₅..	·11	·001
CO₂..	·61	·014
	99·76	
Sp. gr..	2·728	

The corresponding mineral composition in weight percentages was roughly determined by optical means, thus:—

Quartz..	16·0
Soda-orthoclase..	29·1
Biotite..	22·0
Oligoclase..	1·5
Magnetite..	·5
Apatite..	·5
Calcite..	·4
	100·0

The high proportion of quartz, the very low percentages of the alkalies, yet lower than in Phase 1, and the low percentage of alumina indicate that we have here again, as in Phase 1, a quite abnormal kind of granite.

In the Norm classification this rock belongs to the sodipotassic subrang of the domalkalic rang of the order, hispanare, in the dosalane class, with the following norm:—

Quartz..	42·30
Orthoclase..	17·24
Albite..	16·24
Anorthite..	7·78
Corundum..	1·30
Hypersthene..	9·45
Ilmenite..	1·36
Magnetite..	1·16
Apatite..	·31
Calcite..	1·40
H·O..	1·17
	99·71

N · · · · rang nor subrang has yet received a distinct name in the system. . collected at the point 50 feet from the upper sill-contact, is unusi rtzose. It has nearly the same qualitative composition as Phase 2 but structure is more like that of Phase 1, being essentially that of a rather coarse-grained micropegmatite. The optical method gave the following weight percentages for the different constituents:—

Quartz..	57·1
Sodiferous orthoclase..	24·9
Biotite..	8·9
Muscovite..	3·2
Calcite..	2·5
Magnetite..	1·9
Oligoclase..	1·5
	100·0

It is clear that there is notable variation in the composition of the biotite-granite zone as represented in Phases 1, 2, and 3. The apparently regular increase in acidity in the zone from above downwards is fortuitous. The zone is in reality irregularly streaked in many such phases, carrying variable proportions of the mineral and oxide constituents. Whatever the cause, the

2 GEORGE V., A. 1912

magma was not homogeneous at the time of its solidification. To that fact is doubtless to be related the confused, rapid crystallization of the essential mineral constituents.

ABNORMAL HORNBLENDE-BIOTITE GRANITE.

The biotite granite of sill C graduates downward into a rock of similar habit, with hornblende added to the list of essential constituents (Phase 4). The amphibole resembles that of the unaltered gabbro of the Purcell sills. The structure of this hornblende-biotite granite changes rapidly and apparently irregularly from the micrographic to the hypidiomorphic-granular. The top zone of sill D is composed of the same rock type. No chemical analysis has been made of it. The specific gravity of a specimen from sill C is 2·765, being slightly greater than the average for the overlying biotite granite.

INTERMEDIATE ROCK TYPE.

Underlying the hornblende-biotite granite in both sill C and sill D, and underlying the biotite granite in sill A, are zones composed of a rock which combines features of granite and gabbro (Phase 5). It is, in fact, a rock directly transitional into the dominant gabbro of the Purcell sills. A specimen illustrating this intermediate rock was collected at a point 200 feet below the upper contact of sill C, and has been analyzed.

Macroscopically this phase is much like the usual gabbro of the sills. It is a dark, green-gray, granular rock of basic habit. Its essential minerals are hornblende, biotite, and andesine; the accessories, quartz, orthoclase, titanite, titaniferous magnetite and apatite. The secondary minerals are zoisite, kaolin and epidote. The structure of the rock is in general the hypidiomorphic-granular, but local areas of micropegmatite are common in the section.

The total analysis of this phase (specimen No. 1140) by Prof. Dittrich gave the following result:—

Analysis of Intermediate Rock (Phase 5) of Moyie Sills.

		Mol.
SiO_2	52·63	·877
TiO_2	·62	·008
Al_2O_3	16·76	·165
Fe_2O_3	2·86	·018
FeO	10·74	·149
MnO	·38	·006
MgO	4·33	·108
CaO	6·17	·110
Na_2O	1·41	·023
K_2O	2·29	·024
H_2O at 110°C	·12	....
H_2O above 110°C	1·17	....
P_2O_5	·33	·002
CO_2	·10	....
	99·91	
Sp. gr. .... (corrected value).	2·954	

The quantitative mineral composition by weight percentages was determined (orthoclase not separately estimated but included in the andesine) thus:—

Hornblende	49.4
Biotite	22.0
Andesine	16.5
Quartz	11.7
Apatite	.3
Magnetite	.1
	100.0

The abundant biotite and quartz go far to explain the differences between the chemical analysis here and that of the normal gabbro. It also appears from the analysis that the hornblende is here unusually aluminous. Chemically considered this intermediate rock has its nearest relatives among the diorites; yet the low feldspar content forbids our placing this rock variety in that family. Like both the gabbro and the granite it is an anomalous type.

In the Norm classification the intermediate rock appears in the as yet unnamed sodipotassic subrang of bandose, the docalcic rang of the dosalane order, austrare, with the following norm:—

Quartz	9.72
Orthoclase	13.34
Albite	12.05
Anorthite	28.63
Corundum	1.53
Hypersthene	26.51
Ilmenite	1.22
Magnetite	4.18
Apatite	.62
H_2O and CO_2	1.39
	99.19

At the perpendicular distance of 200 feet from the lower contact of sill E, another specimen of the intermediate rock was collected. It gave the following weight percentages (mode):—

Hornblende	42.9
Quartz	22.8
Andesine	18.5
Biotite	6.6
Sodiferous orthoclase	5.5
Titanite	3.7
	100.0

ABNORMAL HORNBLENDE GABBRO.

The whole of sill B, and the lower part of each of sills C, D, and E are all constituted of dark, heavy gabbro (Phase 6), which is either sensibly like the usual gabbro of the thinner Purcell sills, or differs from it in unessential

2 GEORGE V., A. 1912

details. The foregoing description of the usual gabbro will suffice, also, for most of the femic rock in these Moyie sills.

Yet microscopic and chemical study of the lower internal zone of contact of sill E, shows that here the rock is not quite the same as the usual gabbro. This Phase 7 was collected at a point 30 feet perpendicularly from the lower surface of contact. In macroscopic appearance and internal structure it is not markedly different from the usual gabbro. The essential minerals are hornblende and labradorite; the accessories, quartz, potash feldspar, titanite, magnetite. Zoisite, kaolin, and much chlorite are the secondary constituents.

Chemical analysis of Phase 7 (specimen No. 1143) by Prof. Dittrich gave the following result:—

Analysis of Gabbro (Phase 7) of Moyie Sills.

		Mol.
SiO_2	52·94	·882
TiO_2	·73	·009
Al_2O_3	14·22	·139
Fe_2O_3	2·08	·013
FeO	8·11	·113
MnO	·35	·005
MgO	6·99	·175
CaO	10·92	·195
Na_2O	1·40	·023
K_2O	·49	·005
H_2O at 110°C	·12	....
H_2O above 110°C	1·56	....
P_2O_5	·08	·001
	99·99	
Sp. gr.	2·980	

The corresponding mineral composition in weight percentages is roughly as follows:

Hornblende	54·8
Labradorite	25·6
Chlorite	11·0
Quartz	6·3
Titanite	2·0
Magnetite	·3
	100·0

On account of some alteration in the rock, it was found difficult to distinguish with certainty the small amount of alkaline feldspar; which has, accordingly, been entered in the total for labradorite.

Phases of the Moyie sills: specimens one half natural size.
Upper left : average gabbro.
Lower left : tonic phase of gabbro.
Upper right : granite fifteen feet from upper contact of sill C.
Lower right : granite fifty feet from upper contact of sill C.

The calculated norm is:

Quartz	8.40
Orthoclase	2.78
Albite	12.05
Anorthite	30.86
Hypersthene	21.14
Diopside	18.40
Magnetite	3.02
Ilmenite	1.36
Apatite	.31
Water	1.68
	100.04

This rock belongs to the prosodic suborang of the as yet unnamed docaleiorang of the salfemane order, vaalare.

RÉSUMÉ OF PETROGRAPHY.

As a convenient summary, the mineralogical and chemical analyses of the different phases have been assembled in Tables XI and XII.

TABLE XI.—*Weight percentages of minerals as determined by the Rosiwal method.*

	Usual Gabbro of Pur cell sills.	Gabbro, 30 feet above lower contact in sill E.	Intermediate rock, 280 feet below upper contact in sill C.	Intermediate rock, 280 feet above lower contact in sill E.	Hornblende- biotite granite, 100 feet below upper contact in sill C.	Biotite granite, 50 feet below upper contact in sill C.	Biotite granite, 40 feet below upper contact in sill C.	Biotite granite, 15 feet below upper contact in sill C.
Hornblende	58·7	54·8	49·4	42·9	16·0			
Biotite	·9		22·0	6·6	17.3	8·9	22·0	15·2
Labradorite	34·8	25·6						
Andesine			16·5	18·5				
Oligoclase						1·5	1·5	1·0
Soda-bearing orthoclase				5·5	27·8	24·9	29·1	32·5
Microperthite								3·9
Quartz	4·0	6·3	11·7	22·8	37·2	57·1	16·0	41·6
Muscovite						3·2		4·6
Apatite	·2		·3				·5	·2
Titanite	1·4	2·0		3·7	·5			
Magnetite or ilmenite		·5	·1		1·2	1·9	·5	1·0
Chlorite		11·0						
Calcite						2·5	·4	

The total is 100·0 in each case.

Specific gravity	2·980	2·980	2·954	2·942		2·751	2·728	2·733

2 GEORGE V., A. 1912

Table XII.--Chemical analyses of Phases of the Moyie Sills.

	Coral Gabbroid of cell sills	Gabbro, 30 feet from lower contact in sill E	Intermediate rock, 290 feet below upper contact in sill C	Biotite granite, 40 feet below upper contact in sill C	Biotite granite, 15 feet below upper contact in sill C
SiO_2	51 92	52 94	52 64	72 42	71 69
TiO_2	83	73	62	68	59
Al_2O_3	14 14	14 22	16 76	10 47	13 29
Fe_2O_3	2 97	2 08	2 86	83	83
FeO	6 92	8 11	10 74	5 50	4 23
MnO	14	35	38	16	09
MgO	8 22	6 99	4 33	11	1 28
CaO	11 53	10 92	6 17	2 53	1 66
Na_2O	1 38	1 40	1 41	1 93	2 48
K_2O	47	49	2 29	2 94	2 37
H_2O at 110°C	10	12	12	06	14
H_2O above 110°C	1 07	1 56	1 17	1 11	1 31
P_2O_5	04	08	33	11	07
CO_2	06	.....	10	61	13
	99 78	99 99	99 91	99 3	100 16

The two tables illustrate the abnormal character of every one of the rock types occurring in these sills. The tables also show the great range of rock variation. The changes in mineralogical and chemical composition and in density are clearly systematic in the series from gabbro, through intermediate rock, to hornblende-biotite granite, and then to biotite granite. It now remains to indicate that the same serial arrangement characterizes the rock-zones in each of four of the sills; and that there is an analogous series in passing upward from sill E, through sill D to the top of sill C.

ESSENTIAL FEATURES OF THE DIFFERENT SILLS.

The reader will readily seize the situation by a glance over the following stratigraphic column (Table XIII) and the corresponding diagram (Figure 15). At the top of the column is the recently discovered sill A with its cap of quartzite; at the bottom is the quartzite underlying sill E at the valley floor west of Moyie mountain. It should be noted that thicknesses of sills and zones, and the positions of type specimens have been determined only with approximate accuracy. The section described occurs almost exactly in the line of the Boundary slash.

SESSIONAL PAPER No. 25a

TABLE XIII.—*Showing Columnar Section through the Moyie Sills.*

Sills, thicknesses in feet.	Rock zones. thicknesses in feet.	Character of rock	Average specific gravities of igneous rocks
		Quartzite of great thickness.	
A. 135	25	Acidified gabbro, 2 specimens, sp. gr. 2·89 and 2·97.	2·93
	85	Biotite granite	2·76
	25	Slightly acidified gabbro	2·97
	100	Quartzite	
B, 30	30	Usual gabbro of Purcell sills	2·99
	500	Quartzite	
C, 530	80	Biotite granite, 5 specimens, sp. gr. 2·72—2·794	2·73
	110	Hornblende-biotite granite, 3 specimens, sp. gr. 2·74—2·84.	2·78
	60	Intermediate rock, 2 specimens, sp. gr. 2·95 and 3·00	2·97
	280	Gabbro	3·00
	75	Quartzite	
D, 1050	?	Hornblende-biotite granite	2·85
	?	Intermediate rock	2·92
	950	Gabbro	2·99
	750—	Quartzite	
E, 200	50	Intermediate rock	2·94
	150	Gabbro (somewhat weathered)	2·97
		Quartzite of great thickness.	

Sill A is well exposed on the eastern slope of the mountain at the contour 300 feet lower than Monument No. 211. The overlying quartzite dips 51° in a northeasterly direction. The uppermost 25-foot zone of this sill is acidified gabbro. That rock shows rapid transition into an underlying, 80-foot zone of biotite granite, which similarly graduates with some rapidity into a nearly quartz-free gabbro approximating the usual rock of the thinner Purcell sills. This is the only one of the sills which has been seen to have a gabbroid zone overlying a granitic one. The upper gabbroid zone seems to represent a layer of magma which was rapidly chilled against the cool roof of quartzite. The lower part of the mass had a longer period of fluidity and became stratified through gravitative differentiation.

Between sills A and B is a 100-foot layer of quartzite, with strike N. 25° W. and dip 65° E.N.E.

Sill B, 30 feet thick, is a fine-grained gabbro of the usual type in the Purcell sills; it is apparently of quite homogeneous composition throughout.

2 GEORGE V., A. 1912

Below sill B is a 550-foot band of quartzite with strike N. 25° W. and dip 65° E.N.E.

Sill C, about 530 feet thick, is well exposed in the Boundary-line section as well as in that on the wagon road north-northwest of the summit of the mountain, where titre of the analyzed specimens were collected. This body is the most striking of all in its evidence of gravitative differentiation. The 80-foot zone of biotite granite at the top passes gradually into the underlying 110-foot zone of hornblende-biotite granite, which, in turn, merges into the 60-foot zone of intermediate rock overlying the 280-foot zone of the usual Purcell gabbro at the bottom of the sill.

Between sills C and D comes a band of quartzite with strike N. 30° W. dip 60° E.N.E.

Sill D is poorly exposed but seems to be largely composed of the usual gabbro, overlain by successive zones of intermediate rock and hornblende-biotite granite. The outcrops do not suffice to show the exact thickness of any of these zones, but it seemed clear in the field that the total thickness of the two more acid zones was little more than 100 feet.

Between sills D and E is a band of quartzite, estimated as about 750 feet in thickness; its strike is N. 30° W. and dip 60° E.N.E.

Sill E. The lower part of sill E is well exposed a few hundred yards south of the Boundary slash, but its contact with the overlying quartzite was nowhere discovered. As already noted, the existence of that layer of quartzite was not even suspected in 1904, as it was entirely covered by talus along the line of traverse then followed by the writer. It then seemed most probable that the gabbro masses exposed at the top and bottom of the great talus slope formed parts of a single sill. For 100 feet or more from its lower contact the rock of sill E is practically the usual gabbro of the Purcell sills. That zone is overlain by a zone of intermediate rock, the top of which has not been discovered. The two zones show a gradual transition into each other.

ORIGIN OF THE ACID PHASES.

Preferred Explanation.– Among all the conceived hypotheses as to the origin of the acid zones, the writer has been forced to retain one as the best qualified to elucidate the facts concerning the Moyie sills. More important still, this hypothesis, better than any of the others, affords a coherent, fruitful, and, it seems, satisfactory explanation of similar occurrences in other parts of the world. It will be presented in some detail, since it is believed that these sills, and similar ones in Minnesota and Ontario represent gigantic natural experiments bearing on the genetic problem of granites and allied rocks in general. The view adopted includes what has been called 'the assimilation-differentiation theory.' The acid zone is thereby conceived as due to the digestion and assimilation of the acid sediments, together with the segregation of most of the assimilated material along the upper contact.

SESSIONAL PAPER No. 25a

Flat Position of Quartzite at Epoch of Intrusion.—Since the granophyre-granite zone of sill C is known to have a tolerably constant thickness throughout an exposure of at least three miles along the outcrop, the hypothesis involves the assumption that that sill and the adjacent ones lay much more nearly horizontal at the time of intrusion than they do now. This assumption is favoured by all the pertinent facts determined during field work, though it cannot be claimed that they furnish absolute proof.

In the first place, it is probable that the majority of the faulting and upturning suffered by the Purcell sedimentary series was brought about at one orogenic period. The intrusive sills are themselves profoundly faulted and their outcrops are repeated by faulting in such a way as to indicate throws of thousands of feet. If this extensive disturbance of the sills had followed their intrusion, which itself followed earlier important dislocations of the intruded sediments, we might reasonably expect that the detailed structures of the twice-faulted sediments would show some evidence of the history. As a matter of fact, the cleavage often fully developed in the quartzites apparently belongs to one orogenic period and to one only. It was developed after the intrusion of the sills, for the gabbro itself is occasionally cleaved with its planes of cleavage parallel to those in the adjacent quartzites. The sediments must, of course, have been slightly disturbed as the intrusive bodies were injected, but true mountain building seems to have been postponed until long after the solidification of the magmas. The repetition of sill outcrops by faulting is most easily understood if it be believed that the dips have been greatly increased by the relatively late disturbance. If the strata had been well faulted, tilted, and cleaved before the intrusions took place, the injected bodies should show much greater irregularity of form than they now actually show; most of them would be in the relation of dikes or chonoliths (injected bodies of irregular form), following faults and other secondary planes of weakness, rather than in the relation of sills following bedding-planes.

A second argument is to be derived from the fact that sills and dikes of hornblendic gabbro, mineralogically and chemically very similar to these sills of the Yahk and Moyie ranges, cut the Kitchener formation and equivalent Siyeh formation of the McGillivray, Clarke, and Lewis ranges at horizons immediately below the Purcell Lava, and, almost without question, represent feeders or offshoots of the magma represented in the widespread lava flood. That these eastern dikes and sills do thus represent the contemporaneous intrusive facies of the lava is suggested, as above remarked, not only by the lithological consanguinity, but also by the fact that none of the formations overlying the Purcell Lava horizon, i.e., the Sheppard, Kintla, Gateway, Phillips, Roosville, or Moyie, is known to have been cut by dikes or sills which are younger than the older beds of the Kintla formation. Granting, further, the contemporaneity of *all* these Purcell mountain-system sills with the (Middle Cambrian?) Purcell Lava, which is rigidly conformable to the geosynclinal sediments, it follows that the intrusions took place when the strata lay flat and, in the eastern ranges, were covered at the end of Kitchener (Siyeh) time, by the great flows of the extrusive, post-Siyeh lava.

2 GEORGE V., A. 1912

Evidently neither of these two arguments is quite conclusive, but the balance of probabilities is certainly on the side of the belief that the strata cut by the Purcell sills lay nearly horizontal as the thick bodies were injected. In view of the perfect conformity of the Moyie and Kitchener formations (both laid down in shallow water) it appears probable that the surface of the Kitchener formation was not elevated through the full 2,000 feet represented in the total thickness of the Moyie sills; it seems more likely that the beds underlying the sill-horizon were down-warped nearly or quite 2,000 feet, so as to make room for the sill magma.

Superfusion of Sill Magma.—The hypothesis carries the second assumption that the gabbroid magma was, at the time of intrusion, hot enough and fluid enough to permit of the solution of a considerable body of quartzite and the diffusion of the dissolved material to the upper contact. The assumption is supported by the discovery of the great horizontal extent and uniform thickness of the intrusive bodies; if the magmatic viscosity had been high, each body would have probably assumed the true cushion shape of the typical laccolith. The extreme fluidity of the Purcell Lava is proved by the great distances to which its flows ran before solidifying. If the Moyie and other sills were but the contemporaneous intrusive facies of the same lava, the intrusive magma must have been highly fluid. Its temperature was at least slightly higher than that of the extrusive and therefore somewhat chilled lava; and, secondly, the pressure of the few thousand feet of overlying Kitchener beds could raise the solidifying point only to an insignificant extent (probably less than 5° C). From a study of the grain in the Moyie sills, Lane has calculated that the magma, when injected, must have been considerably superheated, and therefore quite fluid.[*]

Finally, whatever theory of the acid zones be adopted—whether that of pure differentiation, of assimilation, or of both—the fact is clear, from the foregoing lithological description, that the diffusion of silicious material through the gabbro actually occurred on a large scale and that this diffusion could not have taken place unless the original magma were possessed of a high degree of fluidity.

Chemical Comparison of Granite and Intruded Sediment.—A third, even more clearly indispensable condition of the hypothesis relates to the composition of the invaded sediments. One of the most noteworthy features of the huge series of conformable strata in the Creston-Kitchener series in this particular district is the marvellous homogeneity of the whole group. As already indicated, even the division into the two great subgroups, Creston and Kitchener, is founded on merely subordinate details of composition. Hence it is that the study of comparatively few type specimens can give a very tolerable idea of the average constitution of the quartzites. For convenience a brief description of both Creston and Kitchener specimens analyzed will be here repeated. Single beds typical of the Creston occur interleaved in the

[*] A. C. Lane. Jour. Canadian Mining Institute, Vol. 9, 1906.

Kitchener and occasionally rusty beds are intercalated in the Creston series. In both series the average rock is a quartzite, always micaceous and often decidedly feldspathic. Many of the strata above and below the Moyie sills have a composition essentially identical with that of typical Creston quartzite. Hence the chemical analysis of this latter rock partly shows the constitution of the sedimentary group invaded by the gabbro. From Mr. Schofield's description, the underlying Aldridge quartzite seems to be like the Kitchener.

Professor Dittrich has analyzed such a type specimen collected several miles to the westward of the Moyie river. It is very hard, light gray, fine-grained to compact, and breaks with a subconchoidal fracture and sonorous ring under the hammer. The hand-specimen shows glints of light reflected from the cleavage-faces of minute feldspars scattered through the dominant quartz. A faint greenish hue is given to the rock by the disseminated mica. This rock occurs in great thick-platy outcrops, the individual beds running from a metre to three metres or more in thickness. Occasionally a notable increase in dark mica and iron ore is seen in thin, darker-coloured intercalations of silicious metargillite.

In thin section this characteristic Creston quartzite is found to be chiefly composed of quartz, feldspar, and mica, all interlocking in the manner usual with such old sandstones. The clastic form of the mineral grains has been largely lost through static metamorphism. The feldspars are orthoclase, microcline, microperthite, oligoclase, and probably albite. The mica is biotite and muscovite (possibly paragonite), the latter either well developed in plates or occurring with shreddy, sericitic habit. The biotite is the more abundant of the two micas. Subordinate constituents are titanite in anhedra, with less abundant titaniferous magnetite and a few grains of epidote and zoisite.

The chemical analysis (Table XIV., Col. 1) shows a notably high proportion of alkalies, and therewith the importance of the feldspathic constituents, especially of the albite molecule, which alone holds about 15 per cent of the silica in combination.

TABLE XIV.—*Analyses of Sill Granite and Invaded Sediments.*

	1.	2.	3.	4.	5.
SiO_2..	82·10	76·90	74·23	79·50	72·05
TiO_2..	·40	·35	·58	·38	·63
Al_2O_3..	8·86	11·25	13·23	10·13	11·88
Fe_2O_3..	·49	·69	·84	·59	·83
FeO..	1·38	3·04	2·65	2·21	4·87
MnO..	·03	·02	·07	·02	·12
MgO..	·56	1·01	1·02	·78	·85
CaO..	·82	·88	1·13	·85	2·10
SrO..	....	....	tr.	....	....
Na_2O..	2·51	3·28	2·78	2·89	2·20
K_2O..	2·41	1·36	2·66	1·89	2·66
H_2O at 110°C..	·05	·20	·06	·12	·10
H_2O above 110°C..	·37	1·20	·81	·78	1·21
CO_2..	..	tr.	·08	....	·37
P_2O_5..	·04	·15	....	·09	·09
	100·02	100·33	100·16	100·23	99·96
Sp. gr. (corrected)	2·681	2·680	2·722	2·680	2·730

25a—vol. ii—16

2 GEORGE V., A. 1912

1. Type specimen of Creston quartzite. Analyst: Prof. Dittrich.
2. Type specimen of Kitchener quartzite. Analyst: Mr. Connor.
3. Specimen of Kitchener quartzite from contact zone, Moyie sill C.
   Analyst: Prof. Dittrich.
4. Average of analyses 1 and 2.
5. Average of two analyses of biotite granite in the Moyie sills.

Mr. Connor has analyzed a specimen collected as a type of the Kitchener quartzite itself. It was taken from a point about 400 feet measured perpendicularly from the upper contact of the Moyie sill C, and this specimen represents what appears to be the average quartzite above, below, and between the sills. It is rather thin-bedded, the thin individual strata being grouped in strong, thick plates sometimes rivalling in massiveness the beds of the Creston quartzite.

The thin section discloses a fine-grained interlocking aggregate of quartz grains cemented with abundant grains of feldspar and mica. The feldspar is so far altered to kaolin and other secondary products that it is most difficult of accurate determination. Only one or two small grains exhibit polysynthetic twinning and the preliminary study referred practically all the feldspar to the potash group. Mr. Connor's analysis shows conclusively, however, that soda feldspar is really dominant. The analysis was most carefully performed, the second complete determination of the alkalies agreeing very closely with the first. Supplementary optical study of the rock has pointed to the probability that pure albite, as well as highly sodiferous orthoclase, is present. Quartz makes up, by weight, 50 to 60 per cent of the rock, and feldspar from 25 to 40 per cent. Biotite both fresh and chloritized is the chief mica; sericite is here quite rare. Colourless epidote is the principal accessory; titanite. magnetite, apatite, a few zircons and pyrite crystals are the remaining constituents

The analysis is given in Table XIV, Col. 2. Column 4 of the same table shows the average of Cols. 1 and 2 and may be taken as nearly representing the average chemical composition of the quartzite invaded by the Moyie sills. This average is to be compared with that of the two analyses of the biotite granite of the sills, represented in Col. 5. The general similarity of the two averages is manifest. There is clear chemical proof that the greater proportion of the elements in the granite could have been derived directly by fusion of the quartzite.

The conviction as to such a secondary origin for the granite has been enforced by an examination of the exomorphic contact-zone at the upper limit of sill C. For the perpendicular distance of at least 60 feet from the upper surface of contact, the quartzite has been intensely metamorphosed. The rock is here vitreous, lightened in colour-tint, and exceedingly hard. Under the microscope the clastic structure is seen to have totally disappeared. Recrystallization is the rule. It takes the form of poikilitic or micrographic intergrowth of quartz with various feldspars, along with the development of abundant well crystallized biotite and (less) muscovite. The feldspar is chiefly microperthite and orthoclase, the latter often, perhaps always, sodiferous. Albite in independent, twinned grains of small size seems certainly determined by various optical tests. Innumerable, minute grains of zoisite and epidote occur as dust clouding the feldspars, micropegmatitic intergrowths, and even the quartz. Scattered anhedra of magnetite and small crystals of anatase and apatite are rather rare constituents.

The chemical analysis of this highly metamorphosed quartzite is entered in Col. 3, Table XIV. In the preliminary study of the sill it was considered as probable that the quartzite had been somewhat feldspar   during the metamorphism, but the critical analyses seem hardly to bear out any certain conclusion on that point. The analysis shows that in several respects the metamorphosed rock is intermediate in composition between the granite of the sill and the unaltered quartzite. However, there is a perfectly sharp line of contact between the granite and this metamorphosed zone of the quartzite. The former rock has been in complete fusion; the latter rock still preserves its bedded structure.

The net result of the foregoing mineralogical and chemical comparisons affords good grounds for believing that the striking similarity of granite and quartzite is really due to a kind of consanguinity: that the igneous rock is due to the fusion of the sediment.

Comparison with Other Sills in the Purcell Range.—The assimilation theory assumes sufficient heat to perform the work of fusion. It is, hence, an indication of great value that there is some acidification of the respective upper-contact zones in all of eight different sill-outcrops optically studied in the 60-mile stretch from Porthill to Gateway; yet that this acidification is, in general, in a direct proportion to the thickness of the sills. The closely associated Moyie sills A, B, C, and D together have about three times the thickness of any other of the intrusive bodies. Presumably, therefore, the total store of heat in the Moyie group was a local maximum and the capacity for energetic contact-action was there much the largest. As a matter of fact, the Moyie sills are the only sills bearing the truly granitic phase. The other sills are also somewhat more acid at their upper contacts than at their respective lower contacts, but the rock throughout is of gabbroid habit. The acidification in these cases has, as we have seen, led to the development of abundant interstitial and poikilitic quartz, abundant biotite, and less abundant alkaline feldspar in the hornblende-plagioclase rock. The rock of the acidified zones is here very similar to, if not identical with, the intermediate rock of the Moyie sills. The acidification is relatively slight because these sills have been more rapidly chilled than the huge Moyie complex. This point is based on deductive reasoning but it is no less positively in favour of the assimilation theory than the testimony of chemical comparison between the acid zone and the sediments.

Evidence of Xenoliths.—There is, finally, direct field evidence that the gabbro has actually digested some of the quartzite. Along both the lower and upper contacts and, less often, within the main mass of the sill, fragments of the quartzite are to be found. These rocks have, as a rule, sharp contacts with

2 GEORGE V., A. 1912

the gabbro, but, none the less, they have the appearance of having suffered loss of volume through the solvent action of the magma.

In some of the other sills the blocks are yet more numerous and many of them are surrounded by shells of mixed material such as would result from the solution of the quartzite in the basic magma. Since the blocks were suspended in a magma of different density and since the product of solution was not diffused away, the viscosity of the magma must have been high. Under these special circumstances, it is not surprising that a limited amount of solution was possible, even though the viscosity of the pure gabbro was relatively high. On the one hand, the very strong contrast in the ionic composition of solvent and substance dissolved, implies a specially great lowering of the melting point. On the other hand, the original water of the sedimentary rock would facilitate solution even at the comparatively low temperature of 1000° C. or less, at which the nearly anhydrous gabbro became toughly viscous in cooling.

Hybrid Rock.—A special instance was studied, optically and chemically, in connection with material collected in one of the sills at a point on the main Commission trail which is six miles up stream from the Boundary slash on the west fork of the Yahk river and two miles north of the Boundary line. The sill rock there forms low knobs on each side of the trail. Scores of gray, angular quartzite blocks, surrounded by the gabbro, can be seen on the glaciated ledges.

One of these, measuring perhaps 100 cubic feet in volume, is enclosed in a shell a foot or two thick, composed of the solutional mixture. The quartzite has been completely recrystallized, with the development of large, poikilitic quartzes, as in the case of the quartzite metamorphosed on the main contact of the Moyie sill C (analysis in Col. 3, Table XIV.). Abundant, minute granules of epidote were also developed. Recrystallized orthoclase (probably sodiferous) and a little oligoclase are accessory constituents; no biotite could be found. The original sediment must have been composed of nearly pure quartz and seems to have been far less feldspathic than the strata cut by the Moyie sills.

A certain amount of osmotic action has taken place, for the quartzite is shot through with narrow, greenish-black prisms of hornblende, 10 to 20 mm. in length. This exotic hornblende has the optical properties of that in the normal gabbro. It is specially abundant near the surface of the block which is, however, sharply marked off from the shell of mixed material. Titanite and apatite in notable amounts have also been introduced into the body of the inclusion.

The shell of mixed material consists of a coarse aggregate of deep green hornblende in prisms 10 to 40 mm. long, and poikilitic quartz, which encloses much granular epidote, titanite, apatite, a little ilmenite, and abundant, minute prisms of the amphibole. No feldspar whatever is apparent in thin section.

An analysis of this mixed material (specimen No. 1164) gave Professor Dittrich the following result:—

Analysis of hybrid rock in gabbro sill.

SiO₂	54·02
TiO₂	1·95
Al₂O₃	12·08
Fe₂O₃	6·85
FeO	5·61
MnO	·09
MgO	2·82
CaO	14·63
SrO	tr.
Na₂O	·60
K₂O	·14
H₂O at 110°C	·06
H₂O above 110°C	·62
P₂O₅	·21
CO₂	·19
	99·87
Sp. gr.	3·111

The alkalies appear to belong, wholly or in largest part, to the hornblende; the alumina, ferric iron, and lime to the epidote and hornblende. The epidote has all the appearance of a primary mineral. In any case it has not been derived through ordinary weathering, for the rock is strikingly fresh.

The composition of the shell is evidently anomalous and represents a double effect. On the one side, the abundant quartz and probably part of the alkaline constituent in the hornblende represent material dissolved from the block; on the other side, the special abundance of the amphibole, to the apparently entire exclusion of soda-lime feldspar, shows that the block formed a centre around which the amphibole, as one of the earliest minerals in the magma to crystallize, segregated. As the amphibole substance was osmotically transferred into the quartzite block, so the quartz substance was diffused outward into the magma. The shell has clearly not the composition expected through the mere solution of the quartzite; the actual composition has also been controlled by the concentration of the basic hornblendic material around a foreign body. The latter may have acted after the manner of the crystal introduced by the chemist into a saturated solution so as to produce crystallization through 'inoculation.'

In other words, magmatic assimilation and differentiation are both illustrated in the history of this shell of mixed material about the quartzite-block. It is, nevertheless, certain that the sill magma as a whole was acidified by the solution of this block and still more by the solution of others now invisible because completely dissolved.

The phenomenon of the partial digestion of xenoliths is quite familiar at intrusive contacts; its significance is only properly appreciated if one remembers that the visible effects of digestion have but a small ratio to the total solvent effects wrought by the magma in its earlier, more energetic, because hotter, condition. It is not a violent assumption to consider that many quartzite blocks have thus been completely digested in the original gabbro

2 GEORGE V., A. 1912

magma. The product of this digestion is not now evenly disseminated through the crystallized gabbro, which, except near its upper and lower contacts, is very nearly identical in composition with the unacidified gabbro occurring elsewhere in the district. No conclusion seems more probable than that the material of the dissolved blocks is now for the most part resident in the acid zone at the upper contact. The same view holds for the perhaps much more voluminous material dissolved by the magma at the main contacts themselves. The excess of acid material at the lower contact was held there because of the viscosity of the magma in its final, cooling stage. For the greater bulk of the digested material there has been, it appears, a vertical transfer upwards, a continuous cleansing of the foreign material from the basic magma.

Assimilation at Deeper Levels.—Another cause of acidification is to be sought in the conditions of sill-injection. In the Purcell range, as generally throughout the world, channels (dike-fissures) through which the magmas have been forced into the greater sill chambers, are relatively narrow as compared with the thickness of the respective sills. In most cases the feeding fissures seem also to be few in number for each sill. The magma must pass through such a fissure during a considerable period of time in order to form the enormous bulk of a first-class sill. At that stage the magma is at its hottest and it is being moved rapidly past the country rock. The effect is analogous to that of stirring a mixture of salt and water: solution is stimulated by the movement. The original magma is thus converted into a syntectic magma, with greater or less chemical contrast to the original.

Such a case may be represented in the great sill in New Jersey, which outcrops for a distance of more than a hundred miles. Lewis has shown that its rock is chiefly a quartz diabase.* Since the sill-rock shows chilled contacts, it appears probable that its magma after reaching the sill-chamber, was too cool to accomplish much solution on roof or floor, though some xenolithic material (sandstone) may have been dissolved. The special composition and structure of the New Jersey sill can be explained as that of a syntectic of primary basaltic magma, which dissolved a small proportion of the acid rocks (sand-stones and pre-Cambrian crystallines) forming the walls of the feeding fissures. Though the temperature of this sill was too low for much evident solution of the sill's floor and roof, it was high enough, and the sill magma therefore fluid enough, to permit of the remarkable gravitative differentiation described by Lewis.

Some acidification of the Purcell gabbro may, thus, have occurred in its long passage through the thick lower quartzites and other sediments of the Purcell series. Nevertheless, the great chemical similarity between the biotite granite and the average quartzite strongly suggests that the assimilation in the Moyie-sill magma chiefly occurred in the quartzite formation, and not in the underlying pre-Cambrian formations of differing composition.

---

*J. V. Lewis, Annual Report, State Geologist of New Jersey, 1907, p. 99.

Assimilation through Magmatic Vapours. Again, the influence of magmatic water and other vapours must be given due weight. The quartzites to-day are not entirely dry rocks. They must have been moister in that early time when the intrusions occurred. From heated roof and floor of each sill, and from each heated xenolith, water vapour must have been injected and forced into the sill magma. The volatile matter contained in assimilated sediment must similarly enter the magma. A large part of such vapour would rise to the roof, and there aid in the solution of the quartzite. Such *resurgent* vapour must not only lower the solution-point (of temperature) for the roof-rock; it must also specially metamorphose the sediment outside of the magma chamber. It is a fact that the quartzite above each sill seems to be more thoroughly crystallized than the quartzite below the sill.

Summary of the Arguments for Assimilation.—The facts and deductions bearing on the subject are so numerous that it will be convenient to review them in brief statement. The writer's belief in the principle of assimilation as a partial explanation for the acid zones in the Moyie sills is founded on the following considerations:—

1. The strong mineralogical and chemical similarity between the biotite granite and the invaded quartzite.

2. The existence of solution aureoles about the visible xenoliths of quartzite.

3. The field evidences of superfusion in the sill gabbro.

4. The relation between sill-thickness (heat supply) and degree of acidification.

5. The necessary recognition of various loci of solution in the sill, namely, at roof and floor, at xenolith contacts, and in the feeding channels below the sills. Resurgent and juvenile vapours, collected at the roof, must tend to hasten solution in that place specially.

6. The fact that differentiation may partially mask the direct evidence of assimilation.

7. The existence of many other sills and sill-like intrusions showing similar or analogous relations of gabbroid magma to sediments. Some of these cases will be listed after the nature of the differentiating process at the Moyie sills has been sketched.

8. The inadequacy of the hypothesis that the various phases of the sills are due only to the pure and simple differentiation of a primary earth-magma. This point is implied in the foregoing argument; it was briefly discussed in the writer's 1905 paper.

Gravitative Differentiation.—Inspection of Table XIII. and of Figure 15 will lead to the conviction that, in sills C, D, and E, the igneous rock is stratified. In each of these instances the specific gravity increases from top to bottom of the sill. The same is true of sill A, with the exception of the shell of gabbroid rock next the roof and overlying the biotite granite. An explanation for this exceptional arrangement of zones has been given in the preceding descrip-

2 GEORGE V., A. 1912

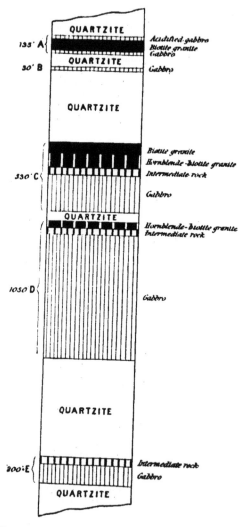

Figure 15. Diagram showing the petrographic nature of each of

tion of sill A. The zonal character of the four individual sills is clearly due to gravitative adjustment.

The same principle has probably controlled the rough system implied in the succession of average rock-densities of sills D, C, and A, as illustrated in the following table:—

Sill.	Approximate mean densities.
A	2·83
C	2·91
D	2·98

The layer of quartzite separating sills C and D is only 75 feet thick. It is entirely possible that this layer is wedge-shaped, or else was penetrated by one or more connecting dikes. By such means B and C might have been in magmatic communication. We may imagine a partial differentiation within this larger chamber, whereby sill C became more acid, on the average, than sill D. Continued differentiation by gravity, within the partially separated masses C and D, led to the observable stratification of each. It is not impossible that all four sills were similarly connected in a common (sill-like) magma chamber, from which each visible sill was a kind of great, flat apophysis.

On the other hand, these sills may not have been of exactly contemporaneous intrusion. A large part of the magma now represented in the rock of sill A may have formerly rested in the chamber of sill B. Since then, after partial differentiation, that part of the magma may have broken through the roof of chamber B to the new horizon now occupied by sill A, where continued differentiation produced the actual zonal arrangement. Similarly, sills A, B and C may have been apophysal from the great sill D. One cause for such successive injections may be found in the enormous gas-pressure generated by the assimilation of moist quartzite—a tension amply sufficient, under certain conditions, to fissure the roof of the slightly older chamber and cause the rise of the magma to the higher horizons of the existing upper sills. (See Figure 16.)

In spite of the relative complexity of the whole system, we may conclude that gravitative differentiation is clearly " 'ominant process in developing the zonal structures of the Moyie sill gro....

It is hardly necessary to dwell on the chemical side of the differentiation. It was probably founded on the limited miscibility of gabbro and secondary magma at the low temperature immediately preceding crystallization. The magma was not quite the chemical equivalent of the invaded sediments. Each of the two granite types contains more ferrous iron and lime than the average quartzite. The hornblende-bearing granite is clearly more ferromagnesian and calcic than the sediment. However, the total volume of the gabbro in the sill system is so great that its average original composition was not essentially affected through the transfer of the extra lime, iron oxides, and magnesia to the granite zones.

Similar and Analogous Cases.—The writer's explanation of the Moyie sills has been greatly strengthened by the discovery of similar features in other

2 GEORGE V., A. 1912

thick basic sills cutting siliceous sediments. More indirect corroboration is offered in certain cases where large basic injections have become differentiated by gravity, apparently after the absorption of considerable limestone.

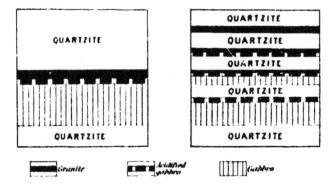

Figure 16. Diagram illustrating the hypothesis that the partially differentiated syntectic magma of a thick sill may break through the roof and form, at stratigraphically higher horizons, several thinner sills differing in composition among themselves. Some later differentiation in the derived sills is assumed. The original sill is shown in the drawing on the left; the derived sills and the remnant of the original sill are shown in the drawing on the right. The channels (dikes) connecting the sills are not indicated.

Direct parallels to the Moyie sills have already been noted as occurring in the Purcell range to the east of the Moyie river. Schofield has found many other stratified sills in the same range north of the Boundary belt.* At Sudbury, Ontario; at Pigeon Point, at Governor's Island, and at Spar, Jarvis, and Victoria islands, and other localities in Minnesota, and among the Logan sills on Lake Superior the same general association of gabbroid-granitic magmas and quartzose sediments occurs. These instances were cited in the writer's 1905 paper, where a rather full summary of the facts concerning Pigeon Point and Sudbury intrusives was given. There, and still better in the original memoirs of Bayley, Barlow, and Coleman the reader will find evidence of the extreme similarity of these cases to that of the Moyie sills. Many other examples have been described in late years, but it is hardly appropriate or necessary to note them individually in the present report.

---

* S. J. Schofield, Summary Report of the Director, Geol. Survey of Canada, 1909, p. 136, and 1910, p. 131.

SESSIONAL PAPER No. 25a

Likewise significant is the analogy of the Moyie sills to several igneous masses which have cut thick lime-tones and have then undergone differentiation by gravity.

The well known laccoliths of Square Butte and Shonkin Sag, in Montana, have been ably described by Weed and Pirsson.[*] In a later, independent publication, Pirsson has described the differentiation as due to the combined effect of fractional crystallization, convection currents, and gravity.[†] In the present writer's opinion, thermal convection must be of infinitesimal strength in such bodies and he cannot find adequate explanation of the shonkinite and other basic phases of these sills in fractional crystallization. On the other hand, the writer finds most satisfaction in the view that the leucite-basalt porphyry of Shonkin Sag, occurring at top and bottom of the laccolith, represents the quickly chilled magma originally injected into the chamber. The syenite and the shonkinite are the two poles of a gravitative differentiation of the remaining leucite-basalt magma, which, in the heart of the mass, remained fluid long enough for splitting. The segregation of the two polar magmas is of a kind suggesting limited miscibility between them. The leucite-basalt can be explained as itself a differentiate from basaltic magma which had dissolved a moderate amount of the thick pre-Tertiary limestones traversed by the magmatic feeder of this laccolith. A similar explanation may be applied to Square Butte.

Tyrrell has described another noteworthy analogy in the Lugar sill of Scotland, where the alkaline pole is teschenite overlying the femic pole, a picrite. This sill is injected into the Mill-stone Grit; its feeder doubtless traversed the underlying Carboniferous lime-tones and perhaps absorbed them in some measure. Tyrrell explains the differentiation of the Lugar sill in essentially the same way as that outlined by the present writer for the Shonkin Sag laccolith.[‡]

Shand has recently described a large laccolith near Loch Borolan, Scotland, in which quartz syenite (specific gravity 2.635) overlies quartz-free syenite (specific gravity 2.65), which in turn overlies nephelite syenite (specific gravity 2.67), and 'ledmorite' (specific gravity 2.74 — 2.78). This mass clearly cuts thick Cambrian limestone and other sediments. Shand attributes the layered condition of the laccolith to differentiation under gravity.[§] He makes no statement as to the origin of the magma thus differentiated. On account of its 'desilicated' character, the present writer is inclined to suspect its derivation from a basalt-limestone syntectic.

Finally, the thick sill described by Noble as cutting the shales in the Colorado canyon is worthy of special emphasis in the present connection.[**] The

[*] W. H. Weed and L. V. Pirsson, American Journal of Science, Vol. 11, 1901, p. 1. and Fort Benton Folio, U.S. Geological Survey, 1899.
[†] L. V. Pirsson, Bulletin 237, U. S. Geological Survey, 1905, p. 42.
[‡] G. W. Tyrrell, Transactions of the Geol. Society of Glasgow. Vol. 13, part 3, 1909, p. 298.
[§] S. J. Shand, Transactions of the Geol. Society of Edinburgh, Vol. 9, 1910, p. 376.
[**] L. F. Noble, American Journal of Science, Vol. 29, 1910, p. 517,

2 GEORGE V., A. 1912

main mass of this intrusive is olivine diabase. Towards its upper contact the diabase appears to grade into a pink rock, which proved to be a hornblende syenite. The syenite makes a sharp contact with the argillites at the roof of the sill. The floor rock is also an argillite. The feeding channel or channels must have traversed the calcareous shale, limestone, and jasper of the lower Unkar series. We need not discuss the genesis of the syenite. Is this rock not a granitic differentiate from a syntectic of the various sediments dissolved in diabase magma? Since the argillites are dominant, one must expect for the different differentiates a degree of acidity intermediate between that of the lighter differentiate in the Moyie sill which has assimilated quartzite, and that of the lighter differentiate in a sill which assimilated limestone. This expectation seems to be justified in fact.

In the way illustration that might be cited, the chemical composition of each lighter differentiated sills varies with the chemical composition of the diabase. Herein rests a powerful argument favouring the secondary origin of respective silic or alkalic magmas. As indicated in each instance, the main mass is stratified in the way demanded by the theory of gravitative differentiation in a syntectic.

GENERAL CONCLUSION AND APPLICATION.

The remaining paragraphs of this chapter are devoted to the broader bearing of the main conclusions regarding the Moyie sills. The statement is almost identical with that already published in the writer's 1895 and 1906 papers, but a few changes have been made in the form of presentation.

Sooner or later experience must teach every careful field student of igneous rocks the truth of the principle of magmatic differentiation. That principle is, indeed, so generally accepted by petrologists that it may be considered as a permanent acquisition in the theory of their science. Yet it is a long step from the recognition of the doctrine to its application to the origin of igneous rocks as actually found in the earth's crust. The principle becomes really fruitful, in fact becomes first completely understood and realized, when certain chief problems have been solved.

Among those problems there are naturally three that are fundamental. Only after they are solved has petrology done that which it has set out to do, namely, determine, under the difficult conditions of earth study, the true nature and genesis of rocks. The first insistent question is, in every case, what was the magmatic mixture or matrix from which the material of the existing rock-mass or rock-masses was produced through differentiation? The second question is, how far did the differentiating process operate? The third insistent question is, what was the process of differentiation itself?

All three problems are interdependent and involve a study in structural geology. They cannot be solved simply by acquiring even the fullest information to be derived from single plutonic contacts, nor, as a rule, from such as may be derived from entire ground-plan contact lines. On the other hand, it is necessary that, more or less completely, the petrologist shall know his magma chamber as the chemist or metallurgist knows his crucible. No student of fused slags ear obtain safe results from the profoundest examination of merely one surface or one section of the fused product. He must think in three dimensions. In the same way, the petrologist attempting to unravel the complex history of a magma chamber, should, ideally, know its general shape, size, and contents, as well as the method by which the chamber has been opened within the earth's crust. Until these conditions are fulfilled his problem of rock-genesis through magmatic differentiation remains wholly or in part unsolved.

The geologist knows how hard those conditions are. He is dependent upon erosion's rendering his contact accessible; yet erosion destroys surfaces of contact. He can find no bottom to the chamber of stock or of batholith, though large-scale differentiation is most commonly evinced in stocks and batholiths. It is not to be wondered at that, notwithstanding the great number of described instances of magmatic differentiation, the phenomenon itself is so little understood or that the origin of the igneous rocks is still shrouded in the mists of hypothesis. In view of the difficulties surrounding the study, the discovery of single cases where the requisite field conditions are tolerably well fulfilled, merits special statement. Descriptions of bodies differentiated in chambers of known form are in the highest degree rare. Nevertheless, it is precisely in the light of these rare cases that the laws of differentiation can be most intelligently discussed.

Such instances are discussed in this chapter, in which have been described exceptionally clear examples of differentiation within magmatic chambers, the crystallized contents of which can now be examined from top to bottom. The form and geological relations of the chambers are sufficiently well determined to serve for the discussion of the magmatic problem. The general nature of the magma whence differentiation has evolved the existing igneous rocks is believed to be deducible from the field and chemical relations in each case. The compound magmas were themselves derived, owing their composition to the digestion or solution of acid sedimentary rock- in original gabbro magmas. Finally, the facts seem indisputable as to the nature of the method by which the differentiation took place. The actual segregation of the salt-magmas appears to have been directed by gravity, producing simple stratification in the chambers. In each sill the less dense sub-magma of splitting overlies the denser sub-magma of splitting.

In almost every case the opponents of the assimilation theory have treated of the assimilation as essentially a static phenomenon. Each interpretation of field facts has been phrased in terms of magmatic differentiation *versus* magmatic assimilation as explaining the eruptive rocks actually seen on the contacts discussed. Nothing seems more probable, however, than that such rocks are often to be referred to the compound process of assimilation accompanied and followed by differentiation. The chemical composition of an intrusive rock at a contact of magmatic assimilation is thus not simply the direct

2 GEORGE V., A. 1912

product of digestion. It is a net result of rearrangements brought about in the compound magma of assimilation. In the magma, intrusion currents and the currents set up by the sinking or rising of xenoliths must take a part in destroying any simple relation between the chemical constitutions of the intrusive and invaded formation. Still more effective may be the laws of differentiation in a magma made heterogeneous by the absorption of foreign material which is itself generally heterogeneous. The formation of eutectic mixtures, the development of density stratification, and other causes for the chemical and physical resorting of materials in the new magma ought certainly to be regarded as of powerful effect in the same sense.

A second fundamental principle has, as a rule, been disregarded in the discussions on magmatic assimilation. If differentiation of the compound magma has taken place so as to produce within the magma chamber layers of magma of different density, the lightest at the top, the heaviest at the bottom, the actual chemical composition of the resulting rock at any contact will depend directly on the magmatic stratum rather than on the composition of the adjacent country-rocks.

In the foregoing discussion the secondary origin of some granites has been deduced from the study of intrusive sills or sheets; but it is evidently by no means necessary that the igneous rock body should have the sill form. The wider and more important question is immediately at hand—does the assimilation-differentiation theory apply to truly abyssal contacts? Do the granites of stocks and batholiths sometimes originate in a manner similar or analogous to that just outlined for the sills?

General reasons affording affirmative answers to these questions are noted in chapter XXVI.

Gabbro and granophyre are often characteristically associated at various localities in the British Islands as in other parts of the world.† The field relations are there not so simple as in the case of the Moyie sills, for example, but otherwise the recurrence of many common features among all these rock-associations suggests the possibility of extending the assimilation-differentiation theory to all the granophyres. Harker's excellent memoir on the gabbro and granophyre of the Carrockfell District, England, shows remarkable parallels between his ' laccolite ' rocks and those of Minnesota and Ontario‡

At Carrock Fell there is again a commonly occurring transition from the granophyre to true granite, and again the granophyre is a peripheral phase. Still larger bodies of gabbro, digesting acid sediments yet more energetically than in the intrusive sheets, and at still greater depth, would yield a thoroughly granular acid rock as the product of that absorption with the consequent differentiation.

The difficulty of discussing these questions is largely owing to the absence of accessible lower contacts in the average granite body. All the more valuable

† See A. Geikie, Ancient Volcanoes of Great Britain, 1897.
‡ Quart. Journal Geol. Soc., Vol. 50, 1894, p. 311 and Vol. 51, 1895, p. 125.

must be the information derived from intrusive sills. The comparative rarity of such rock-relations as are described in this chapter does not at all indicate the exceptional nature of the petrogenic events signalized in the Moyie, Pigeon Point, or Sudbury intrusives. It is manifest that extensive assimilation and differentiation can only take place in sills when the sills are thick, well buried, and originally of high temperature. All these conditions apply to each case cited in this chapter. The phenomena described are relatively rare largely because *thick* basic sills cutting acid sediments are comparatively rare.

On the other hand, there are good reasons for believing that a subcrustal gabbroid magma, actually or potentially fluid, is general all around the earth; and secondly, that the overlying solid rocks are, on the average, gneisses and other crystalline schists, and sediments more acid than gabbro. Through local, though wide-spread and profound, assimilation of these acid terranes by the gabbro, accompanied and followed by differentiation, the batholithic granites may in large part have been derived. True batholiths of gabbro are rare, perhaps because batholithic intrusion is always dependent on assimilation.

The argument necessarily extends still farther. It is not logical to restrict the assimilation-differentiation theory to the granites. For example, the preparation of the magmas from which the alkaline rocks have crystallized, may have been similarly affected by the local assimilation of special rock-formations. See chapter XXVIII.

The officers of the Minnesota Geological Survey have shown that the same magma represented in the soda granite and granophyre of Pigeon Point forms both dikes and amygdaloidal surface flows.* The assimilation-differentiation theory is evidently as applicable to lavas as to intrusive bodies. But demonstration of the truth or error of the theory will doubtless be found in the study of intrusive igneous bodies rather than in the study of volcanoes either ancient or modern.

Finally, the fact of 'consanguinity' among the igneous rocks of a petrographical province may be due as much to assimilation as to differentiation.

* N. H. Winchell, Final Rep. Minn. Geol. Surv., Vol. 1, 1899, pp. 519-22. The Duluth gabbro and the broad fringe of red rock (partly extrusive) on the southeast, together seem to form a *gigantic* replica of the Pigeon Point intrusive!

CHAPTER XI.

STRATIGRAPHY AND STRUCTURE OF THE SELKIRK MOUNTAIN SYSTEM (RESUMED).

Between the Purcell Trench and the Selkirk Valley (Columbia river) the ten-mile belt includes stratified rocks belonging to four groups in addition to those forming the Summit series. (Maps No. 6, 7, and 8). These other groups have been named the Priest River terrane, the Pend D'Oreille group, the Kitchener quartzite, and the Beaver Mountain group. The first two groups rival the Summit series in areal importance within the Boundary belt. The Kitchener quartzite and the Beaver Mountain group cover but small patches and their description can be given in few words. The Beaver Mountain sediments are intimately associated with basic volcanic rocks which in turn are involved with the Rossland Volcanic group. Their description is best postponed to chapter XIII. in which the igneous rocks of the Rossland mountains are discussed.

KITCHENER FORMATION.

Along the western edge of the Kootenay river alluvium and north of the Rykert granite opposite Porthill, the foot-hills are composed of unfossiliferous quartzite and interbedded metargillite, which in lithological characters are essentially like the Kitchener strata across the river. These beds are apparently not metamorphosed in any sense different from that which is true of the unfolded Kitchener quartzite of the Purcell mountains; that is, one misses in them the evidences of great dynamic metamorphism, intense mashing, and recrystallization observed in the neighbouring Priest River terrane and the evidences of likewise intense contact metamorphism which has affected the Priest River rocks in the batholithic aureoles farther west. The relative lack of dynamic metamorphism is quite striking and largely on that account the writer has separated these rocks from the Priest River terrane, postulating a master fault of great throw on the west side of the Purcell Trench. This fault is thus considered as bringing into contact a very old member of the Priest River terrane (Belt G) and the quartzite which is tentatively correlated with the Kitchener formation. The down-throw is on the east (see map), and may be as much as 30,000 feet.

On account of the great structural importance of this correlation a detailed study of the sediments west of the alluvial flat of the Kootenay is imperative. While in the field the writer was not entirely conscious of the importance of the lithological comparison, for at that time the existence of the Kitchener forma-

2 GEORGE V., A. 1912

tion itself was unknown and was not determined until the camp had been moved many miles to the eastward. Since then no favourable opportunity has arisen by which the study of this quartzite could be continued in the field. It is now only known that, throughout most of the meridional belt of the Kitchener quartzite as mapped on the west side of the Kootenay, the rocks are indistinguishable from types of the Kitchener strata collected at the Moyie river. The staple rock is a greenish gray quartzite, weathering brownish. Under the microscope the dominant quartz is seen to be regularly associated with small grains of microperthite and orthoclase, with generally a little plagioclase, a few zircons, and pyrite crystals. There is always mica present, generally colourless and sericitic, though minute biotites are seemingly never absent. Where the quartzite is cleaved, as it is at certain points north of Corn creek, the micas are specially developed in the cleavage planes. The metargillitic interbeds have not been microscopically examined but they appear to be composed of the same materials as the metargillites of the Kitchener formation.

It is equally true that this local quartzite-metargillite series is lithologically similar to the Beehive formation as developed on the summit of the range. This is, of course, natural if the writer is correct in correlating the Kitchener and Beehive quartzites.

At Summit creek and north of it for a half-mile the quartzite is extremely massive and of a gray colour when fresh, and very often grayish to light brownish-gray when weathered; it is possible that here we have a large outcrop of the Creston formation underlying the Kitchener. There is so little certainty of this, however, that the colour representing the Kitchener on the map has been extended northward across Summit creek.

North of Summit creek the strike averages about N. 16° E., and the dip is about vertical. The same strike (dip observed at 60° E.) is preserved fairly well for a couple of miles south of the creek when it abruptly changes to N. 22° W. then to N. 90° E., becoming highly variable in place of structural turmoil. A half-mile farther south the strike is N. 45° E., and the average dip about 55° S. E. This general attitude of the beds was observed at several points south of Corn creek. On the whole it must be said that the strike of the quartzite is distinctly transverse to the trend of the Purcell Trench.

The western limit of the quartzite is shown on the map only approximately. For the reason already noted, the amount of structural and areal work done in the field was insufficient to show that limit and therewith the exact place of the postulated master fault. Few points in the structure section along the Forty-ninth Parallel are more important than this one and it is especially here that further and more detailed work is needed.

PRIEST RIVER TERRANE.

It has already been noted that the basal conglomerate of the Summit series rests unconformably on older rocks outcropping at, and to the eastward of, the head-waters of Priest river. The name 'Priest River terrane' may be appro-

PLATE 25.

Looking eastward over the heavily wooded mountains composed of the Priest River Terrane; Nelson Range. Glacial lakes (rock-basins) in Irene Conglomerate formation.

priately given to this whole group as exposed in the southern Selkirks at the Boundary. It appears to be the oldest series anywhere exposed on the Forty-ninth Parallel. The group is of sedimentary origin but has been largely recrystallized. It is as yet entirely unfossiliferous. Its stratigraphic relation to the Summit series leaves no room for doubt that the Priest River terrane is both pre-Cambrian and pre-Beltian in age.

Exposures and Conditions of Study.—Within the 10-mile Boundary belt where it crosses the Selkirk range, this old terrane covers about one hundred square miles. Such an area would seem sufficient to afford leading data as to the composition and structure of the series. Yet a comparatively long and certainly arduous field attack on the area has been exceedingly unsatisfactory in its results. The difficulties of geological exploration in this area are unsurpassed in the entire Boundary section. The intense metamorphism of the series in almost every part, and its structural complexity would alone render the solution of the main geological problems as difficult as in most typical Archean terranes. The strong relief of the country and, above all, the heavy and continuous forest cap add special physical troubles in a field where the geologist's mental troubles in interpretation are already of the first order. (Plate 26.) With wearisome repetition outcrops failed at critical localities. For a mile or two together the sections were often left quite blank where fallen timber, deep moss, or humus effectually covered the rock ledges; so complete was this cover of vegetation that even the 'wash,' frost-riven from the ledges, was invisible for long stretches.

Under these conditions it has proved impossible to treat the Priest River terrane in anything like as satisfactory a manner as would be desirable. Though its rocks are almost entirely of clearly sedimentary origin, not the slightest clue was discovered as to the succession of beds. Neither top nor bottom, nor certain indication of relative ages among individual members has yet been determined. Four, more or less complete, traverses, besides several shorter ones, were run across the area, and a tolerable idea of the lithological nature of the series was obtained. The map and section as well as the following description of the series, indicate that the characters of the rock-members and the attitudes of the beds are not favourable to the discernment of stratigraphic sequence. It has thus seemed best to map the series on a purely lithological basis.

Compiling the data won from the several traverses it appears that the rocks of the terrane may be grouped into seven irregular belts which will be henceforth referred to by the letters *A* to *G.* In general they run meridionally and follow, more or less faithfully, the strike of the bedding planes, which appear usually to lie parallel to the planes of schistosity. Belts *A, B, C, D,* and *E* have been most fully investigated. The relative inaccessibility of the area covered by belts *F* and *G* has caused the information concerning them to be very scant. Along the northern edge of the Boundary belt all the belts exposed show specially complicated features as a result of the intrusion of the great

2 GEORGE V., A. 1912

Bayonne batholith. Peripheral schistosity and cleavage and a very intense degree of recrystallization have been developed about that batholith.

Belts *F* and *G* are also much disturbed and altered in the vicinity of the Rykert granite batholith in the southeastern corner of the area. The eastern limit of belt *G* occurs at a master fault, along which quartzites referred to the Kitchener formation have been dropped down into contact with the pre-Cambrian schists.

Petrography of Belt A.—South of Summit creek the Irene conglomerate directly overlies belt *A*. This is a heterogeneous group of rocks, including biotite, chlorite, and sericite schists; sheared, compact quartzites; and dolomites. The micaceous schists occupy most of the belt; sericitic quartzites are next most abundant; the dolomites occur as thin bands intercalated in schist and quartzite.

The schists vary in colour from light to dark greenish gray, according to the nature and abundance of the essential micaceous mineral, sericite, biotite or chlorite. They are often interrupted by veinlets of quartz and of dolomite lying in the schistosity planes. In certain phases crystals of dolomite occur in individuals or groups disseminated through the schist. Rock types transitional between the true schists and impure dolomite are found. On the west side of the trail at Copper Camp the dark phyllitic schist is abundantly charged with single crystals and small clumps of a light brown ferruginous carbonate which is probably ankerite. The rock has, in consequence, a pseudo-porphyritic appearance.

. The quartzitic bands sometimes run over a hundred feet in thickness. They are always sheared, with an abundant development of sericite in the shearing planes. At several localities the quartzites, like the schists, are magnesian to some extent. They thus pass over into the dolomites which have the habit of compact, more or less silicious, marbles. On fresh fractures the dolomites range in colour from white to a delicate pinkish-brown, weathering to a light though decided buff tint. The exposures of the dolomites in belt *A* are very poor but it appears that no one bed measures much over fifty feet in thickness.

Throughout most of the belt the strike of both bedding and schistosity averages a few degrees west of north and seems to cut the plane of unconformity with the Summit series at angles varying from 10° to 25°. The dip is generally nearly vertical but angles of 75° to 80° to the eastward are not uncommon. About one mile south of the Dewdney trail the belt is broken by a strong transverse fault along which, as shown in the map, the block to the south has been displaced westward with respect to the block on the north. Within the northern block the belt rapidly narrows down as if there it had been cut away during the erosion preceding the deposition of the Irene conglomerate. In this short tongue of belt *A* the strike averages about N. 30° E.; the dip about 75° northwest.

Large quartz veins, usually lying in the planes of schistosity are common in the schists. One of these veins, from 15 to 20 feet in thickness, and well exposed in a high cliff occurs at a meadow on the divide between Priest river and a small fork of Summit creek. Fifty feet to the eastward of this vein are two narrow sill-like injections of minette. This association of vein and eruptive prompted the assay of the quartz for values in the precious metals. The result was negative.

The dolomites of the belt characteristically bear isolated crystals and small pockets of galena and chalcopyrite, and some active prospecting of these rocks has taken place at the forks of Priest river. The sulphides are reported to carry both silver and gold, but so far no workable lode has been discovered. The pockets of galena form the principal 'ore' of the prospect-dumps but the small size and rarity of the pockets—clumps of crystals only a few inches in diameter at most have led to the abandonment of the claims, which certainly seem to have no commercial value.

The intrusive rocks occurring in belt *A* will be described in the section on the igneous bodies of the Selkirks.

Petrography of Belt B.—The next belt to the east is, so far as lithological types are concerned, very similar to belt *A*; the chief contrast between the two lies in the different proportions of these types in the belts. Belt *B* bears thick and persistent bands of dolomite alternating with quartzites and phyllitic and coarser mica schists. The best exposures were seen on the divide between Priest river and Summit creek, to the northwestward of North Star mountain. A tolerably complete section of the belt was there made.

At the northwest end of this section the western limit of belt *B* occurs at a bed of silicious dolomite, one hundred feet in thickness. This dolomite is white to bluish white on the fresh fracture but weathers buff-yellow. Though generally massive, it is greatly cracked and shattered, the cracks being filled with vein-quartz which ramifies in all directions through the rock. The strike is N. 9° E.; the dip is practically vertical.

That limestone is followed on the east by 110 feet of biotite schist, which in turn is succeeded by about 300 feet of thinly laminated, schistose silicious dolomite of colour and composition like the first limestone. This rock too is highly charged with narrow, irregular veinlets of white quartz. The strike is here north and south; the dip, about 65° E. This second limestone is succeeded on the southeast by a 150-foot band of dark, glossy biotite-sericite schist with its planes of schistosity striking north and south and dipping 70° E. It is followed by 95 feet of white dolomitic quartzite (weathering yellowish) with conformable attitude. The quartzite is succeeded by a thick band of light to dark greenish gray phyllitic mica-schist. The observed width of this band was 1,400 feet across the strike, which runs N. 10° E. The dip is 85° E. On its eastern limit this schist is in contact with a band of dolomitic quartzite of which the thickness measured 340 feet. Here too this rock type is white on the fresh fracture and weathers buff-yellow. The staple dolomitic quartzite is interlaminated with thin beds of nearly pure dolomite and others of nearly pure white-weathering quartzite. The strike is N. 5° E.; the dip, 85° E. Next to that band, on the east, comes a conformable band of phyllite, followed

2 GEORGE V., A. 1912

by another band of dolomite, which is very similar to the first dolomite occurring at the western end of the section. The dolomite here is about 150 feet thick.

The specimen of this dolomite (No. 886) is fairly typical not only of the whole band but also of the whole group of carbonate bands occurring in the Priest River terrane. It has, accordingly, been selected for chemical analysis. On the fresh fracture the rock varies in colour from white to pale blue and weathers rather uniformly brownish-yellow or buff. It is transected by numerous veinlets of white quartz and by others of very compact dolomite. Otherwise the rock is a very homogeneous, fine-grained, marble-like mass of carbonate, which in the ledge shows no appreciable impurity. The specific gravity is 2·822, corresponding to normal dolomite pretty closely.

The analysis by Professor Dittrich, afforded the following results:

Analysis of dolomite, Priest River terrane.

		Mol.
SiO_2	5·84	·097
Al_2O_3	·80	·008
Fe_2O_3	·79	·005
FeO	·16	·002
MgO	19·38	·485
CaO	28·31	·506
Na_2O	·27	·004
K_2O	·09	·001
H_2O at 110°C	·03	
H_2O above 110°C	·63	·035
CO_2	43·55	·990
	99·85	
Sp. gr.	2·822	

Portion insoluble in hydrochloric acid, 5·96%.

Under the microscope the carbonate is seen to occur in the form of a granular aggregate, the grains being of rather uniform size and averaging about 0·08 mm. in diameter. They never show the rhombohedral outlines so common in the dolomites of the Lewis and Galton series. This difference may be easily explained by the fact that all of the Priest River dolomites have been thoroughly recrystallized and now have the structure of true marble, while the younger dolomites seem to have preserved their original sedimentary structure more or less perfectly. The granular dolomite of the thin section is interrupted by a few small grains of glass-clear quartz and feldspar. The visible quantity of these impurities matches well the portion of the rock found to be insoluble in hydrochloric acid. About 94 per cent of the rock by weight is made up of the carbonate, which, as shown by the ratio, CaO : MgO (1·46:1), is almost ideal dolomite.

It happens that a small veinlet of carbonate, cross-cutting the main mass of the rock, appears in the thin section studied. This veinlet is about 1 mm. in diameter. Throughout its visible extent its grains average about 0·02 mm.

in diameter or sensibly equal to the average diameter of the grains in the Waterton, Altyn, Siyeh, and Sheppard dolomites. Here as there we have a steady persistence in the size of grain which characterizes the chemically precipitated carbonate.

The strike and dip of the Gateway band of dolomite was, on account of the massiveness of the rock, not readily determined but, as usual in the zone, the former was a few degrees east of north, while the dip seemed to be nearly vertical.

East of the analyzed dolomite, outcrops were few for about 400 feet of cross-section but that stretch seems to be underlain by dolomitic chlorite schist and phyllitic mica schist. Immediately to the eastward and just at the western base of North Star peak, a 200-foot, nearly vertical, band of sheared dirty-white dolomite, weathering yellow, forms the most easterly part of belt B. The strike of the band and of its schistosity planes is about N. 5° E.; the dip averages 80° E.

A review of the field-notes suggests that belt B may constitute a closely appressed fold, the erosion of which has produced a duplication of the three dolomitic bands on the two sides of the belt. However, the very considerable differences of thickness between the respective bands thus supposed to be duplicated, are so great that one cannot be sure of the postulated repetition. In any case, there is no evidence in this section as to whether the fold is an anticline or a syncline. In no other part of belt B could this point be settled. In the general structure section, therefore, no attempt is made to show the true relations in the great monocline. It has seemed better to illustrate simply the empirical facts of field observation rather than to attempt the projection of folds which, under the circumstances, could be nothing else than fanciful.

Belt B is, thus, composed of both mica schists and dolomites. In that belt the carbonate rocks are relatively more abundant than in any other belt in the Priest River terrane. The persistence of the dolomites along the strike, their nearly vertical dip, the notable straightness of each bed and of the entire belt across North Star mountain, and the general parallelism of belts A and B to the band of Irene conglomerate, would, at first sight, suggest that at least the upper part of the Priest River series is conformable to the overlying Summit series. It is believed, however, as noted elsewhere, that this general parallelism of belts A and B to the band of Irene conglomerate is partly an incidental result of the strong upturning and mashing which have forced the two unconformable series into positions of apparent conformity. Six miles north of the Boundary, belt B has been broken by the same fault which offset belt A a mile or more south of Summit creek. At the creek all the dolomites and schists of belt B are entirely cut off by a second fault (see map) so that these rocks are replaced, north of the Dewdney trail, by the sheared quartzites characteristic of belt C.

The dolomitic bands of belt B, like those of belt A, carry small bunches of galena and occasional crystals of copper pyrites. Neither of these ores where they have been actually prospected, as at the claims of 'Copper Camp,'

2 GEORGE V., A. 1912

occurs in masses of workable size. No reliable information was obtained on the ground as to the values found in assayed specimens of the sulphides, but the material collected from the prospect-dumps nowhere suggested the possibility of a high-grade property. On the other hand, the small size and comparative rarity of the bunches of ore shows that no known claim in the 'camp' can prove successful as a low grade mine.

Petrography of Belt C.—In width, length, and axial trend, belt C is very similar to belts A and B. In composition C is, in some respects, like A but does not seem to bear any dolomitic bands. The most complete section across belt C was made at the summit of North Star mountain. Elsewhere in the belt, exposures are very poor and it is very possible that the boundary lines, especially that on the eastern side, are drawn too straight. This third belt is composed essentially of well and thinly foliated phyllites, chlorite-sericite schists, and phyllitic biotite-sericite schists, all tending toward a dark greenish gray colour. Within these staple rocks there occur strong bands of a very dark gray intensely sheared quartzite. The quartzite bears abundant little foils of sericite and biotite, disseminated in planes of schistosity. The interlocking, metamorphic quartz grains are full of opaque black dust which may be driven off before the blow-pipe and is probably carbon in graphitic or other form. This carbonaceous matter is abundant and explains the dark colour of the rock in ledge or hand-specimen. A few sheared quartz pebbles were found in the phyllite on North Star mountain near the western limit of the belt.

Wherever outcrops were found in the belt the attitude of the planes of schistosity corresponds well to the average attitude in belts A and B. Through most of the belt the strike varies from N. 7° E. to N. 10° W.; the dip averages about 75° E. At one locality near the summit of North Star mountain, the dip of the schistosity plane was 75° E. Such discordance appears, however, to be local and, in general, the planes of bedding and schistosity may be nearly coincident. The schists do not extend beyond the Dewdney trail and seem to be cut off by the same transverse fault which has been postulated to explain the failure of belt B north of the trail and so marked on the map.

Petrography of Belt D.—The fourth belt is dominantly quartzitic. The quartzite is normally more or less sheared. Both biotite and sericite are largely developed, in fact never failing entirely in this metamorphosed sedimentary. Within the quartzite beds are numerous, though thin intercalations of sericitic and chloritic schists along with beds of dolomite. The quartzites are of compact texture and vary in colour from white to pale greenish-gray, weathering white or buff. They are often charged with accessory grains of carbonate, which qualitative analysis shows to be probably typical dolomite. The same mineral is also an abundant accessory in the chlorite and biotitic schists. The study of thin sections seems to show that much, perhaps all, of the chlorite found in the schists is secondary after biotite and after the rather rare garnets which sometimes appear among the accessories.

At Summit creek and north of it, the rocks of the belt have been profoundly metamorphosed by the Bayonne granodiorite intrusion. The effects are most notable in the schistose bands. In them the small sitted and foil of sericite, chlorite and biotite are replaced by felted aggregates of large biotite and muscovite foils. The resulting coarse-grained mica-schist be a most striking contrast to the more phyllitic schists far from the batholithic contact. Though the recrystallization by contact-action is so pronounced, the original banding or bedding is as fully read as in the staple phases of these old sedimentaries. The thin bands of coarse schist are sharply marked off from the enclosing quartzite, which, though it bears disseminated plates of biotite and muscovite of relatively large size is still a true hard quartzite. Occasionally minute, deeply coloured tourmalines are seen under the microscope to be distributed through the quartzose matrix. Feldspars are characteristically absent, or at least, are indetectable in the normal schist and quartzite, but both plagioclase and orthoclase are recognizable in considerable amounts in the schists and quartzites of the specially metamorphosed part of the belt. It is not possible to attribute their presence with certainty to feldspathization by the granitic magma, however probable it may seem from the field relations of the feldspar-bearing phase.

Strong contact-metamorphism is visible for at least two miles from the granite contact. The great width of the metamorphic collar as illustrated in belts *D*, *E* and *G* indicate the probability that the contact-surface of the granite body plunges under the rocks at and south of Summit creek. The vertical distance between the granite and the rocks exposed in the depths of the canyon at the creek is probably less than two miles. (See Figure 19.)

The best exposures of the belt were found on the ridge running south from North Star mountain. There the strike of the bedding, the plan of which usually coincide with the schistosity planes, varied from N. 25° W. N.—and—S., the dip varies from 75° E. to 70° W., with the average about E. The nearly vertical dip and meridional strike persist for a distance of six miles north of the Boundary line; but along Summit creek the strike has swung around, so as to run, on the average, about N. 40° E. near belt *C* and gradually approaching N. 65° E. as the eastern limit of quartzites on the Dewdney trail is approached. Throughout the whole width of the belt on the Dewdney trail the strike thus follows very closely the general contact-line of the Bayonne granite; the relation affords an excellent illustration of the development of peripheral cleavage about a batholith. The dip of the banding (bedding) in the schist-quartzite along this contact collar seems to coincide generally with the dip of the schistosity. It averages about 60° to the northwest, but is highly variable, as expected in a belt of rocks energetically displaced and mashed during batholithic intrusion.

Petrography of Belt E.—Belt *E* is composed of a group of acid sediments even more intensely metamorphosed than those of belt *D*. The dominant type is a highly sericitic schist in which large biotite foils have been extensively

2 GEORGE V., A. 1912

developed along the planes of schistosity. (Plate 27, Fig. B.) The general ground-mass of the rock is, as a rule, a light to medium-tinted greenish-gray, silvery, glittering felt of quartz and abundant sericite. Sprinkled through the felt are the round or hexagonal biotite plates, which range from 1 mm. to 3 mm. in diameter. The biotite is highly lustrous, and, on account of its darker colour, stands out prominently on the surface of the rock. This special pseudo-phenocrystic development of biotite is characteristic of the whole belt and, while occasionally seen in narrow bands of belt *D*, is not an essential feature of any other than belt *E*. For this reason it may be called the belt of 'spangled schists.'

Along with the biotite spangles there are often many pale-reddish anhedral garnets also developed in the planes of schistosity. The sericite is commonly replaced by well characterized muscovite of the ordinary type, though it never takes on the size of the biotite spangles. Around the large biotites and the garnets the small shreds of sericite and quartz grains are often seen under the microscope to be arranged in concentric layers; this relation is the familiar one to be observed so often in garnetiferous schists. A little magnetite, a few zircons and needles of rutile form the accessories of the schist.

There are all stages of transition between the typical spangled schist and sheared quartzite, which is always sericitic and commonly speckled with minute dots of dark biotite. These quartzites are similar to those characteristic of both belt *D* and belt *E*.

Near the divide between Summit creek and the north fork of Corn creek, the spangled schists enclose a band of common amphibolite about one hundred feet in apparent thickness. This is evidently a sheared and highly metamorphosed basic igneous rock, probably of intrusive origin. A second sill-like intrusion of much altered basic rock (now a hornblende-chlorite schist), with an exposed width of ten feet, was found on the ridge about a mile and a half E. N. E. of North Star mountain. With these exceptions belt *E* is a fairly homogeneous body of acid, sedimentary rock wholly metamorphosed.

The schistosity planes usually strike parallel to the boundaries of the belt as laid down on the map; the dip is always very high, varying from 90° to 75° E. It is apparently more characteristic of this belt than of any of the others that the attitude of the bedding is highly discordant with that of the schistosity. The two planes were often seen, in the same ledge, to cut each other at angles of from 60 to 80 degrees. Unfortunately the exposures were not sufficiently numerous to enable the writer to determine even the main facts concerning the true position of the bedding planes throughout the belt. It is only known that these rocks are often greatly crumpled and that the folds and crinkles are crossed indifferently by the master-structure. Considering the intense metamorphism, the bedding is well preserved and is represented by good contrasts of colour between the lighter tinted, more quartzitic layers and the darker, more micaceous layers once rich in argillaceous material. (Plate 27, *B*).

The spangled schists were followed from the Boundary line to the ridge

A. Contrast of normal sericite schist of Monk formation (left) and contact metamorphosed equivalent in aureole of summit granite stock, a coarse-grained, glittering muscovite schist (right). The sericite schist specimen shows dark patches of surface stain. One-half natural size.

B. Spangled, garnetiferous schist characteristic of Belt of Priest River Terrane. Banding represents original bedding of a siliceous argillite. Three-fourths natural size.

just south of Summit creek. There belt *E* has already passed into the collar of contact metamorphism belonging to the Bayonne batholith. On the Dewdney trail all trace of the normal spangled schist is lost and the rocks which appear to represent it are relatively very coarse-grained, crinkled muscovite-biotite schist, alternating with micaceous quartzite. So complete is the recrystallization that it has proved impossible to separate the contact-metamorphosed part of belt *E* from the similarly altered schists of belt *G*. For this reason belt *E* is, in the map, represented as ending in an arbitrary line drawn to indicate the northernmost limit of the schist which actually shows the spangling with biotite. From near the Kootenay river flat to a point four miles up Summit creek, the coarse, glittering mica schists with their quartzitic intercalations represent the utmost crystallinity and a very striking parallel to typical mica schists in the great pre-Cambrian field of eastern Canada. This spectacular exomorphic collar is more than two miles wide as measured outward from the Bayonne granite. Within the collar the schists are powerfully crumpled and the strike of both schistosity and bedding has been forced around so as to be sensibly parallel to the contact-line of the Bayonne granite. The dip averaged about 75° to the north but is quite variable.

Petrography of Belt F.—East of the zone of spangled schist good outcrops are specially rare for several miles. These sections were run across belt *F* but, on account of the heavy forest cover, the information was but meagre. The net result of these traverses went to show that the zone is, like belt *D*, composed of sheared quartzite with subordinate interbeds of mica schist. The quartzite is here usually much more schistose than that in belt *D* and is chiefly a true quartz schist. Sericite or well developed muscovite, biotite, and chlorite, all in minute foils giving by reflexion point-like scintles of light from the planes of schistosity, are the micaceous minerals formed by the dynamic metamorphism. The intercalated mica schists are much like those of belts *A* to *D*, but almost never show the biotite spangles characteristic of the rocks of belt *E*. In both the quartz schists and the mica schists there is, close to the Rykert granite, an increase in the size of the mica foils and usually some development of reddish garnets. These features are regarded as due to special contact-metamorphism. A band of garnet-bearing amphibolite, 125 feet wide, and apparently following the bedding-planes of the schistose quartzite near the Rykert granite, is another example of greatly metamorphosed basic intrusives in the Priest River terrane.

Peripheral schistosity was developed in the belt by the Rykert granite intrusion. On the north slope of Boundary creek near the contact, the strike of bedding and schistosity was observed to run from N. 25° E. to N. 45° E., with dips varying from 30° to 45° N. W.

On the top of the ridge and a mile from the contact the average strike is about N. 30° W. and the dip varies from 75° E. N. E. to 75° W. S. W. Farther west the dip is northerly and flattens to 20° or less. Toward the western limit of belt *F*, on the same ridge, the strike is about N. 25° E. and the dip nearly vertical. In all these cases the strike and dip refer to the banding of the

2 GEORGE V., A. 1912

quartz-schist-mica schist series: this banding seems undoubtedly to represent original bedding. The schistosity for the most part apparently coincident with it

It looks as if the rocks of this belt lying to the west of the Rykert granite form an appressed and greatly crumpled syncline but, in view of the scanty field data, no great confidence can be felt in this interpretation.

Petrography of Belt G.—The most easterly of the seven belts is even more obscure as to its detailed structure than the other belts. Belt G lies between the Bayonne and Rykert granite batholiths which have conspired to perfect the metamorphism begun by the crush of earlier mountain-building pressures. Half-way between the two batholiths and from four to five miles from either, the rocks have the peculiar habit of micaceous contact-hornfelses. Intense crumpling of the sedimentaries in the zone has been brought about by a combination of the strong orogenic pressure which has affected all the belts, and of the outward pressures exerted during the forceful intrusion of the batholiths. The structural problem of the belt is further rendered difficult by the rarity of good bed-rock exposures.

The belt is essentially composed of glittering coarse- to medium-grained mica schists. These vary in colour from light to dark greenish-gray and dark rusty brown. The average phase is distinctly more ferruginous than the staple schists of any of the other six belts. As a rule the schists are well banded, much after the fashion of the spangled schists of belt E.

It is believed that the bands represent the true bedding. The original sediments were doubtless chiefly argillites more or less rich in silica, with subordinate thin interbeds of sandstone. Their existing metamorphic equivalents are muscovite-biotite schists carrying variable and often important amounts of red garnet, yellow epidote, and tourmaline. The muscovite is sometimes sericite but generally occurs in the form of the usual foils of relatively large size.

As already noted, the northern part of the belt along Summit creek includes schists which form the probable extension of belt E into the exomorphic collar of the Bayonne batholith. All across belt G, at the creek, the strike is a little north of east and thus roughly parallel to the contact-line of the batholith. It is possible that similar peripheral schistosity was developed in belt G north of the Rykert granite but this point could not be determined in the time that could be allotted to the area. Elsewhere in belt G the average strike of the banding varies from N. 25° E. to N. 45° E. The dips are exceedingly variable, those observed ranging from 70° N.W., through verticality, to 50° S.E.

Thicknesses and Structure in the Priest River Terrane.—With the exception of a few relatively unimportant bands of amphibolite, the whole of the Priest River terrane is composed of originally sedimentary rocks. The list of these include argillites, argillaceous sandstones, highly silicious sandstones, dolomitic sandstones and argillites, and dolomites. All these rocks are tremendously altered and metamorphosed so that not a single ledge observed in the field nor a single one of about one hundred specimens, more closely studied in the laboratory, is without abundant signs of crushing or, at least, recrystallization.

The carbonate rocks occur in belts *A*, *B*, and *C*, but are chiefly concentrated it belt *B*. In the section crossing that belt, northwest of North Star mountain, the six great beds of the dolomitic marbles aggregate about 1,500 feet in thickness. If the three beds outcropping on the western side of the belt are but duplications of the three beds outcropping on the eastern side, and, if half the mean of the thickness be assumed as indicating the real thickness of the three beds, this would total 750 feet. In belts *A* and *C* there must be at least 250 feet of highly magnesian rock additional. The writer believes, in fact, that 1,000 feet represents the minimum thickness of the total dolomitic rock as exposed in the area of the Priest River terrane.

Most of belts *A* and *C* and a large part of *B* are composed of rather homogeneous mica schists, including great masses of phyllite and chloritic schist. It is possible that belt *B* represents a duplication of *A*; with this assumption a very rough estimate of the minimum total thickness of the argillaceous strata corresponding to these schists is 5,000 feet.

The thicknesses of the dominant quartzites of belts *D* and *F*, which are lithologically very similar, are extremely difficult to estimate but it is believed that at least 6,000 feet of different beds must be represented. The total apparent thickness of the spangled schist is over 6,000 feet and an estimate of 3,000 feet based on the possibility that belt *E* coincides with a simple closed vertical fold, seems to be a safe minimum estimate for the thickness of the spangled schist. Belt *G* consists of mica schists which in several respects are very similar to the schists of belts *A*, *B* and *C*, yet no dolomites have been found in belt *G* and it would be unsafe to correlate the strata of belt *G* with those of any of the western zones. In any case, it appears that at least 3,000 feet of recrystallized strata, not appearing in any of the members estimated above, must be added to complete the total of strata exposed in the area.

It thus seems likely that this total is, at the minimum, 18,000 feet. Even that estimate is large in absolute measure but the total number of feet is but a relatively small fraction of the apparent thickness of the whole series. Rough as the estimate is, it indicates the fact, amply demonstrated by the field observations, that this old series is of great thickness even when compared with the more certain minimum totals for the neighbouring Summit and Purcell series.

At one stage in the work of interpreting the terrane it was postulated that at least part of belts *D*, *E*, *F*, and *G* really form part of the Cambrian-Beltian series, being truly equivalent to the quartzitic and argillaceous phases of the Summit and Purcell groups. A careful study of the field data and of the collected specimens, however, led the writer to believe that this supposition is inadmissible. Quite apart from the thermal action of the Rykert and Bayonne batholiths, the whole Priest River terrane is intensely metamorphosed, to a degree not or seldom in the Purcell series and only rarely, and then but locally, observed in the Summit series. Moreover, the detailed composition of none of the belts agrees with any similarly thick portion of the Summit or Purcell series. Lastly, it may be noted, as good evidence, that the huge basal conglomerate of the Summit series contains myriads of pebbles manifestly derived from ledges quite similar in composition to those of belts *A, B, C, D, F,* and *G*. Since the strikes and dips of the Irene conglomerate are nearly or quite parallel to those in belt *A* of the older terrane, one might doubt the existence of the unconformity at the base of the conglomerate, were it not especially for the similarity of the dolomitic pebbles in the conglomerate to the dolomitic bands in the Priest River series. Largely for this reason a pre-Beltian age is ascribed to all the schistose rocks (not intrusive) situated within the Boundary belt, between the Irene conglomerate and the down-faulted Kitchener quartzite at the western edge of the Kootenay River alluvium. This great Priest River group presents a structural problem as yet quite unsolved.

Correlation.—It is, of course, too early to attempt a fixed correlation of the Priest River terrane with the other pre-Cambrian terranes of the Cordillera, but it is not without interest to observe that in various regions there are thick masses of ancient sedimentaries which appear to correspond both lithologically and in stratigraphic relations to the Priest River terrane as exposed along the Boundary line. A few references to typical sections in the Belt mountains of Montana, the Black Hills of South Dakota and adjoining portions of Wyoming, the Fortieth Parallel region, and the sections worked out by Dawson on the main line of the Canadian Pacific railway, may be useful as showing the places where possible equivalents of the Priest River terrane may be sought.

In the Three Forks, Montana, folio of the United States Geological Survey (1896) Peale describes the 'Cherry Creek beds' as a series of mica-schists, quartzites, gneisses, and marbles or crystalline limestones. These beds are highly inclined, apparently conformable to one another, and, notwithstanding the obscurity of the folding, are known to total thousands of feet in thickness (at least 7,000 feet shown in columnar section). The series is lying 'probably' unconformably upon 'Archean gneisses' and is unconformably underlain by the Belt terrane, i.e., by equivalents of the lower members of the Rocky Mountain Geosynclinal prism as just described in this report.

In the Hartville, Wyoming, folio (1903), W. S. T. Smith and N. H. Darton describe, under the name of the "Whalen group" a series of schists, gneisses, quartzites, and limestones, which are said to resemble closely the 'Algonkian' rocks of the Black Hills. These rocks have high or even vertical dips. They appear to resemble also the pre-Cambrian schists of the area covered by the Sundance, Wyoming, folio, which are unconformably overlain by the Middle Cambrian Deadwood formation. The Algonkian rocks of the Black Hills have not been adequately described but include garnetiferous and other mica schists, graphitic schist, ferruginous quartzite and amphibolite.[*] These metamorphosed rocks, with high dips, lie unconformably beneath the Middle Cambrian overlapping strata.

[*] T. A. Jaggar, jr., Prof. Paper No. 26, U.S. Geol. Survey, 1904, p. 31.
　　See also Newton and Jenney's Report on the Geology and Resources of the Black Hills of Dakota, Washington, 1880, p. 50.

SESSIONAL PAPER No. 25a

MacDonald mentions an important group of metamorphosed and highly crystalline sediments, now schists, outcropping along the west shore of Cœur d'Alene lake.§ This locality is about 120 miles due south of the area of the Priest River terrane as mapped for the present report. It seems possible that the one terrane is a continuation of the other.

King recognized a greatly deformed series of slates, quartzites, limestones, dolomites, mica schists, and hornblende schists in the 'Archean' division of the rocks encountered during the Fortieth Parallel survey.† Farther south the quartzites and micaceous schists of the Vishnu group in the Grand Canyon section represent other pré-Cambrian sediments which have suffered, apparently, about the same measure of deformation and metamorphism as those characterizing the Priest River terrane.

In British Columbia, north of the Boundary belt, it is fully as difficult as in the cases already noted, to correlate with confidence. Among the described rock-groups, the nearest approach, lithologically, to the Priest River terrane is the Nisconlith series of Dawson, as exposed around the Shuswap lakes.‡ This series is made up of calcareous or graphitic mica schists, flaggy, often dark-coloured limestones, gray and blackish quartzites in apparent conformity. The series appears to lie conformably beneath the Adams Lake series and both are placed in the Cambrian, the Nisconlith overlying the truly Archean Shuswap series of gneisses, etc. All three series are quite unfossiliferous and the present writer suspects that the correlation of the Nisconlith with the Priest River terrane is at least as justifiable as that with the Cambrian of the Front ranges.

The foregoing brief statement of the constitution and relations of the various groups indicates lines of thought in the future correlation of the ancient formations of the Cordillera, rather than any definite view as to the correlation. One thing is certain, however; the Cordillera is at many points underlain by very thick and important groups of sediments which are not only pre-Cambrian but also pre-Beltian in age. It is possible if not, indeed, probable that the total thicknesses of these stratified rocks rival those of the pre-Cambrian terranes in the Great Lakes region of Canada and the United States, as well as those of the vast formations of Finland.

PEND D'OREILLE GROUP.

General Description.—Between the western limit of the Summit series monocline and the southeastern edge of the great central volcanic field, an area of about sixty square miles of the ten-mile belt is underlain by a thick group of unfossiliferous, heavily metamorphosed sediments. A considerable

§ D. F. MacDonald, Bull. 285, U.S. Geol. Survey, 1906, p. 42.
† Report, Vol. 1, Systematic Geology, 1878, p. 532.
‡ G. M. Dawson, Explanatory notes to Shuswap sheet, Geological Survey of Canada, 1898. For further references see Bull. Geol. Soc. America, Vol. 12, 1901, p. 66. Since this report went to press, the writer has proved the pre-Beltian age of the Nisconlith of the Shuswap district.

2 GEORGE V., A. 1912

part of the season of 1902 was spent in their study but the results were, in many essential respects, very meagre. These rocks occur in one of the Cordilleran zones of maximum orogenic shearing and mashing, with complete recrystallization. Numberless cramplings, overturnings and faulting characterize the region, which, as already noted, has been the scene of repeated igneous injections in the form of dikes, sills, stocks, and batholiths. Again and again the region has been buried deeply in volcanic ejectamenta. In spite of prolonged erosion these volcanics still cover hundreds of square miles and conceal many desired facts concerning the sedimentary rocks. To such principal difficulties in analyzing the complex assemblage of strata along the Pend D'Oreille river there was added that common disadvantage of the geologist on the Forty-ninth Parallel, the dense evergreen forest with its deep mat of brush and fallen timber. (Plate 28.)

At the time of the writer's exploration, the Commission trail on the south side of the Pend D'Oreille river, had not been cut. The crossing of this dangerous river above Waneta was effected only once; hence relatively little is known of the rocks and structures on the left bank of the river. In that part of the belt, outcrops of rock are relatively few. Attention was therefore concentrated on the strip of altered sediments lying between the river and the Rossland-Volcanic terrane to the north.

The ancient Priest River rocks themselves are scarcely more baffling in structural analysis than are these much younger schists along the Pend D'Oreille. Their clean-cut mapping, their order of superposition and the determination of thickness could not be thoroughly worked out. Nearly all that is possible, as a result of the reconnaissance in 1902, is to give a general qualitative description of the metamorphosed sediments. They are conveniently referred to in the present report, under the name, Pend D'Oreille Group; the wild canyon of the Pend D'Oreille river in the lower twenty miles of its course has been excavated in the rocks of this group. Their distribution in the Boundary belt is shown on the map though not with entire accuracy, for it is extremely difficult if not impossible with existing exposures, to separate, in several areas, the rocks of the group from the younger members of the Summit series or from the old, schistose phases of the Rossland volcanics.

The group may be divided into two parts, the Pend D'Oreille schists (including greenstone and amphibolite, as well as phyllite and quartzite), and the Pend D'Oreille marbles. They are primarily not stratigraphic subdivisions so much as purely lithological ones. It was found impracticable to use the limestones as definite horizon-markers and equally impossible to be sure of the relative ages of the limestones and their non-calcareous associates. Several of the larger bodies of limestone seem to form gigantic, isolated pods which have been squeezed, like a truly plastic substance, through the schists, to accumulate locally and with exaggerated thickness in these great masses. On this view the limestone pods are, in part, exotic—they might be called non-igneous intrusives—with reference to the enclosing schists. In any case, it has appeared unsafe to use the few legible records of original

Typical scene in Bonnington Pass Dekaille Mountains of the Selkirk system.   Looking down Fifteen mile Creek toward the Pend Oreille River.

bedding in schist or marble, as indicating the real stratigraphic relations in any detail. A columnar section is as yet impossible. The brief description of a few typical traverses may suffice to show the general characters of the rocks.

Area East of Salmon River.—One of the most continuous exposures of the group was found on the top of the broad ridge running westward from Lone Star mountain. East of that peak the schist is in contact with the Lone Star formation, but, as indicated above, the relation between the two formations is very obscure. A special reason for the uncertainty as to the true relation is found in the existence of the wide break in the section, caused by the intrusion of the Lost creek granite.

From the bottom of the col between Beehive and Lone Star mountains westward to the Salmon river, the dominant rock is a typical carbonaceous phyllite. It is a very dark gray or greenish-gray to black rock, highly schistose, and generally with few certain traces of the original bedding. For hundreds of yards together along the ridge this greatly crumpled schist shows marked homogeneity, but, in places, it passes into an abundant schistose, likewise carbonaceous quartzite. Both these phases may be calcareous and carry necessary tremolite and epidote as metamorphic products, about with the quartz, sericite, and carbon dust. The schists are often pyritized to some extent and in many parts, bear numerous veins of mineralized quartz. Biotite is very often developed as an abundant accessory constituent of the schists. Strain-slip schistosity with the resulting crinkly rock surface is well developed at many points.

On the top of Lone Star mountain a pod of banded, white and bluish marble is intercalated in the phyllite-quartzite. The limestone is enormously crumpled and mashed, so that it is impossible to determine its thickness. Its average dip seems to be 30° to the east. A mile west of Lone Star peak a much larger intercalation of banded, dark gray and bluish-white marble crosses the ridge. It can be followed continuously on a band of fairly constant width from Lost creek to, and beyond, Sheep creek on the north. The continuity suggests that this band represents a sedimentary member which retains nearly its original thickness and has not been seriously thinned, or thickened by orogenic shearing. In this view, the thickness must be at least 2,000 feet, for the true dip is 70° and is against the mountain slope, while the width of the band is nearly half a mile.

This limestone, like all the others found in the area now described, is a true marble, fine-grained and completely recrystallized. None of the limestones of the group seems to be magnesian to any great extent; all the specimens collected effervesce violently with cold, dilute acid. Occasionally flattened grains of quartz appear in thin sections and, more rarely, minute crystals of basic plagioclase, probably anorthite, lie scattered through the thin section. Chert nodules or beds were never seen in any of the marbles. In one bed on the south side of Salmon river, concretions of finely granular quartz of the size of large peas, are embedded in calcite. Excepting these accidental ingredients, the marbles are to be regarded as composed of notably pure calcium carbonate.

MICROCOPY RESOLUTION TEST CHART

(ANSI and ISO TEST CHART No. 2)

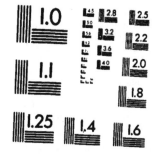

APPLIED IMAGE Inc

1653 East Main Street
Rochester, New York 14609 USA
(716) 482 - 0300 - Phone
(716) 288 - 5989 - Fax

2 GEORGE V, A. 1912

A strong cataclastic structure was microscopically observed to be a general feature of the marble.

It is impossible effectively to distinguish the true stratigraphic positions of all the marble bands in the area, or to be sure of their correlation among themselves. They are, therefore, mapped under the common name, 'Pend D'Oreille limestone.'

Between the mapped monzonite stock and the Salmon river flat the quartzite-phyllite bears one or more strong intercalations of amphibolite, composed of dark olive-green hornblende, quartz, and highly granulated residual individuals of basic plagioclase. Other intercalations of amphibolite and hornblende schist were observed on Lost creek just above its confluence with Lime creek, and on the Salmon river below Roseleaf creek.

Throughout the four-mile section the dips of the schistosity planes are generally high (40° to 75°) to the eastward, though they are, of course, often reversed in the numerous crumples affecting the schists. The banding of the limestones and their planes of contact with the schist were usually seen to dip eastward at similarly high angles. The attitude of the bedding-planes cannot be taken as directly indicating the succession from older to younger in this sedimentary monocline; there is every possibility that the whole group has been overturned along with the apparently conformable Lone Star and Beehive members of the Summit series. In favour of this conception is the fact that a massive limestone of great thickness, of similar lithological characters, and lying nearly flat, overlies a thick series of phyllitic and quartzitic rocks between Roseleaf creek and the Pend D'Oreille river. This limestone covers at least five square miles and dips from 10° to 30° south; it is highly improbable that so large a mass has been overturned. The underlying schists are in the main like those exposed in Lone Star mountain. The tentative conclusion has thus been reached that the schistose rocks composing the Lone Star section from the western contact of the Lone Star schist to the eastern contact of the great limestone band all underlie that limestone and, with it, have been overthrown so as now apparently to overlie the limestone. On the same tentative basis these older schistose sediments may be set down as totalling at least 3,500 feet in thickness.

The large body of marble situated at the confluence of Lost creek and the south fork of the Salmon river is probably the down faulted equivalent of the 2,000-foot band of limestone above described. If so, the phyllites and quartzites lying to the westward of that band may be wholly or in part of the same age as the schists lying to the eastward of the band. In this Lone Star-Salmon river section, therefore, one cannot be sure that there are any sediments younger than the great limestone. Unfortunately, no other area in the Boundary belt has afforded any more certain help in carrying the stratigraphic succession higher or completing the columnar section for this region. It is probable that the micaceous schists exposed in Sheep Creek valley for three miles from its intersection with Salmon river, are younger than the great limestone, but the exposures are much too imperfect to warrant a definite conclusion on the point.

Area West of Salmon River.—Dark greenish, or dark gray to black phyllite, alternating with blackish quartzite, is the dominant rock on both banks of the Pend D'Oreille, from its confluence with the Salmon to its mouth at Waneta. The schists enclose lenses of white to gray marble, varying from ten feet or less to 200 feet or more in thickness. Near the Columbia and especially on the west side of that river, the phyllites and quartzites are associated with very abundant, thick masses of greenstone and altered basic breccias, so that it there becomes very difficult to separate the Pend D'Oreille group from the younger Rossland Volcanic group.

Lithologically, there is a great similarity between the respectively dominant rock types on both sides of the Salmon but it is also clearly impossible to develop a useful columnar section of these metamorphosed sediments along the lower Pend D'Oreille. The great limestone is not represented. It is, however, probable that most of the phyllite and quartzite is the equivalent of the rocks tentatively regarded as stratigraphically underlying the great limestone on the Lone Star ridge. They are unconformably overlain by the Rossland lavas as developed east of Sayward. Between Nine-Mile and Twelve-Mile creeks a strong and persistent band of silicious limestone is intercalated in the phyllites; it is truncated at each end by the overlapping lavas in such a way as to illustrate the unconformity. (See map.)

The structure of this area is fully as complex as that east of the Salmon. The schists are well exposed for miles in the canyon of the Pend D'Oreille, where the dips and strikes of the schistosity planes were seen to shift every few hundred feet. On the average the strike runs a little north of east, so that the river section is not favourable to the discernment of the field relations. Numerous acid and basic intrusions have also affected the structural relations.

As a negative result of the field work among the schists it may be stated that the leading problems regarding their age, their subdivision into recognizable members, and their thickness must apparently be solved outside the ten-mile belt. It is most probable that, if ever found, the key to these secrets will be disclosed on the United States side of the Boundary line. In the present report the whole assemblage of phyllites, quartzites, traps, and limestones is included under the name, Pend D'Oreille group. Its minimum thickness is believed to be 5,500 feet.

Correlation.—The marbles and schists themselves carry no fossils, so far as known, but a hint as to their age is found in the fact that lithologically similar marbles bearing a Carboniferous species were found by McConnell and by the writer in the Rossland district.[*] In central Idaho, eighty to one hundred miles to the southeast of the Boundary section at the Pend D'Oreille, Lindgren has found closely allied rocks at several, rather widely separated localities, and at most of them some of the rocks bear Carboniferous fossils.[†] His description

---

[*] Cf. R. G. McConnell. Explanatory notes to Trail sheet issued by the Geological Survey of Canada; 1897.

[†] W. Lindgren, 20th. Annual Report, U.S. Geol. Survey. Part 3, pp. 86-90, 1900.

2 GEORGE V, A. 1912

of the Wood River series in his report will be found to match fairly well with the account of the Pend D'Oreille group just given.

About one hundred miles to the northward and north-northwestward of the Boundary section at the Pend D'Oreille are considerable areas of stratified rocks referred by Brock to the Cache Creek series or to the Slocan series which he regards as probably equivalent to the Cache Creek.‡ In that region the Cache Creek series is made up of 'dark argillites, greywackes, quartzites and limestone, with some eruptive material"; the description of the Slocan series is in similar terms. In the Kamloops district, still farther northwestward, the Cache Creek beds are well exposed and there they have been studied in some detail by Dawson. His summary statement of their succession is as follows:—

'The lower division consists of argillites, generally as slates or schists, cherty quartzites or hornstones, volcanic materials with serpentine and interstratified limestones. The volcanic materials are most abundant in the upper part of this division, largely constituting it. The minimum volume of the strata is about 6,500 feet. The upper division, or Marble Canyon limestones, consists almost entirely of massive limestones, but with occasional intercalations of rocks similar to those characterizing the lower part. Its volume is about 3,000 feet.

'The total thickness of the group in this region would therefore be about 9,500 feet, and this is regarded as a minimum. The argillites are generally dark, often black, and the so-called cherty quartzites are probably often silicified argillites; The volcanic members are usually much decomposed diabases or diabase-porphyrites, both effusive and fragmental, and have frequently been rendered more or less schistose by pressure.'*

Much of the Cache Creek series is fossiliferous and definitely Carboniferous (Pennsylvanian), but Dawson points out that the lower beds may include formations somewhat older than the Carboniferous. He emphasizes, after many years of experience, the great constancy of the series from the Yukon boundary of British Columbia southward throughout the length of British Columbia.

In view of these various facts it seems to be the best working hypothesis that these greatly metamorphosed rocks of the Pend D'Oreille group roughly correspond to the Cache Creek series and that they are in large part of Carboniferous age. The lower schists may include sediments of any age from the Carboniferous to the Silurian inclusive. There is no evidence of unconformity with the Summit series; the Pend D'Oreille schists seem, on the other hand, to pass gradually into the underlying Lone Star schists. Because of the special local intensity of ther   nd dynamic metamorphism it must be long before the correlation of the 1 ...d D'Oreille group is anything other than hypothesis. Yet, as in so many cases, the tentative correlation seems to be better than none.

---

‡ R. W. Brock, Explanatory notes to West Kootenay sheet, issued by the Geological Survey of Canada, 1902.
* G. M. Dawson, Bull. Geol. Soc. America, Vol. 12, 1901. p. 70.

PLATE 29.

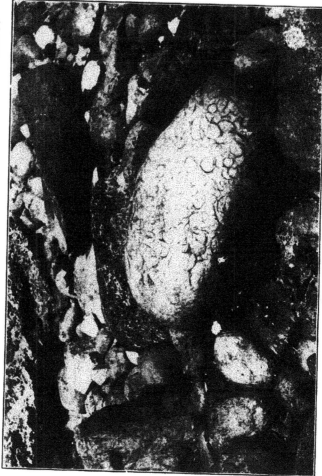

Percussion marks on quartzite boulder in bed of Pend D'Oreille River. Boulder about two feet long.

for even in its upsetting, the future observer's eye will be sharpened for the essential facts.

SUMMARY ON THE STRUCTURE OF THE NELSON RANGE.

The structural geology of the Nelson range where crossed by the ten-mile belt naturally involves a study of three different types of areal geology, corresponding respectively to the Priest River terrane, the rocks of the Summit series, and the large bodies of batholithic granite.

The obscurity of relations among the old sediments of the Priest River terrane has been described in the account of the different zones (belts) of the terrane. Schistosity and bedding often coincide. Both sets of planes are highly inclined, with dips averaging about 75° to the eastward. Quite vertical dips are very common in the southern half of the belt. In the northern half the Priest River rocks have been intensely crumpled by the intrusion of the Bayonne batholith, giving local dips at all angles and in all directions, with average northwesterly to north-northwesterly dips of about 70°. The original dips due to tangential pressure have likewise been greatly modified by the intrusion of the Rykert granite batholith. The failure to find recognizable folds in the terrane has already been sufficiently noted. South of Summit creek, zones *A*, *B* and *C* have been affected by a strong horizontal shift (a fault in which there has been horizontal movement of one block past the other). At the creek the three zones appear to be cut off entirely by a fault which is entered on the map. A less important break cuts off zone *B* near the Boundary line. With these exceptions the writer has failed to find structural elements which can be definitely mapped.

On Map No. 6 a long band of Kitchener quartzite is shown along the western edge of the Kootenay river delta between the Rykert mountain granite and the mouth of Summit Creek canyon. The quartzite is referred to the Kitchener formation on lithological grounds and there are many points of resemblance to the Beehive quartzite. The microscope shows that microperthite is a relatively abundant constituent of all three quartzites while the feldspathic material of the quartzites belonging to the Priest River terrane is quite different. In other respects also this quartzite along the river alluvium corresponds well with the Kitchener formation in essential features. Though the brushy slopes to the westward have not been thoroughly explored it appears safe to postulate a great north-northwest fault on which these Kitchener beds have been dropped down into contact with the Priest River terrane. This fault is shown on the map. Its exact course is represented only approximately; further field-work is imperative before greater precision may be attained. The fact remains, however, that this quartzite, which has thus been correlated with the Kitchener and the equivalent Beehive quartzite, has been downthrown through a vertical distance equal to the whole thickness of the Summit series below the Beehive formation plus an unknown thickness of the Priest River terrane. The downthrow may measure 20,000 to 30,000 feet.

2 GEORGE V, A: 1912

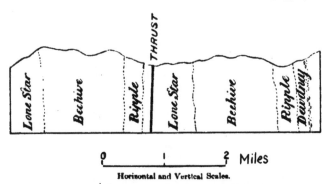

Figure 17.—East-west section on ridge north of Lost Creek, Nelson Range; showing duplication of beds by thrust, the plane of which has been rotated.

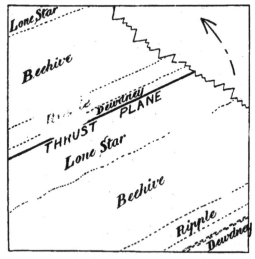

Figure 18.—Diagram showing stage of development of the thrust illustrated in Figure 17.

SESSIONAL PAPER No. 25a

On both sides of the Purcell Trench, therefore, we have evidence of huge displacements which have given this part of the Kootenay valley the character of a fault-trough. The down-faulted block or blocks have, of course, lost much substance through erosion but it seems most probable that the trench was located in a constructional depression due to faulting.

Another of the primary structural features of the Nelson range is the unconformity at the base of the Irene conglomerate. The existence of the unconformity is not conspicuously shown by contrasts of attitude between the conglomerate and the older sediments. In fact, as above noted, the strike and dip of the conglomerate and of zone *A* are often closely similar. The evidence is more fully derived from (1) the much stronger metamorphism of the Priest River terrane; (2) the abundant pebbles of Priest River rocks in the conglomerate; and (3) the truncation of zones *A*, *B* and *C* by the lower-contact plane of the conglomerate. The Nelson range covers the only part of the Boundary belt where the Rocky Mountain Geosynclinal is sounded to its full depth.

Within the great Summit series monocline itself one of the most conspicuous structural complications is the horizontal shift mapped as crossing the divide between Monk creek and the south fork of the Salmon river. In the field the effects of the shift are spectacularly clear. The almost vertical formations have been dislocated by a movement of about a mile along the vertical west-northwest plane of shifting. The relative displacement is that which would have been produced if the southern block had moved westward through that distance. The outcrop of the shift-plane could be readily followed for four miles; its continuation westward across the southern slope of Lost mountain is less evident in the field but seems competent to explain the relations of the Pend D'Oreille limestones and schists to the quartzites on Lost mountain. A second horizontal shift, not so evident, is mapped just south of Summit creek.

About three miles west of the main divide of the range the upper beds of the Summit series are duplicated for a great thickness by a powerful thrust. This thrust is among the most remarkable elements in the anatomy of the range. (Figures 17 and 18.) The plane of the thrust is stratigraphically located in or very near the 225-foot band of conglomerate in the Dewdney formation. The conglomerate has apparently acted as a local plane of weakness. Along that plane the entire overlying part of the Summit series has been driven eastward and has then been pushed up on the back of the Lone Star schists. Either during the thrusting, or, less probably, afterwards, the overridden and overriding blocks together with the thrust-plane have been rotated so as now to stand almost perfectly vertical or to show a slight overturning to the westward. As a result the observer traversing the ridges on either side of Summit creek will, on going westward, pass over the Dewdney beds, then the Ripple, Beehive, and Lone Star in regular order, and will then, after crossing the thrust-plane, pass over the upper Dewdney, the Ripple, Beehive, and Lone Star formations once more. These relations are illustrated in Map 7 and in Figure 17. They are specially clear on the high, nearly treeless ridges north and south of Summit creek. The extreme northern and southern extensions of the thrust-plane are not so well exposed and the mapping is there somewhat tentative.

2 GEORGE V, A. 1912

It is scarcely necessary to remark that the straightness of the bands of colour corresponding on the map to the Summit series formations, is controlled by a structural necessity, namely, the nearly or quite vertical dip which is general throughout the greater part of the monocline. The thrust-plane just described must similarly be nearly vertical. Deep as the canyons are, the out-crops of the different formations deviate but little from the straight line where the bands cross the canyons.

West of Beehive mountain the Pend D'Oreille series is so thoroughly disordered that the structure section in this part could be represented only in a schematic way. The same procedure is necessary for the continuation of these rocks across the Salmon river.

Finally, in the Nelson range section we find the outposts of the army of granitic intrusives which cut the stratified rocks of the Cordillera at intervals all the way from the Purcell Trench to the Pacific ocean. In general the sediments of that greater part of the Boundary belt are much younger than the rocks of the Rocky Mountain Geosynclinal; but, because of the inherent weakness of the younger sediments, because of the intrusion of many batholiths, and probably also because of a greater intensity of the orogenic forces in the western half of the Cordillera, these sediments are generally more metamorphosed than those of the older prism. Just east of the Salmon river the Summit series plunges under the Pend D'Oreille group of schists and marbles and the still younger Rossland volcanics, never to reappear in the sections farther west.

The Nelson range is the greatly worn product of the mightiest crustal upturning on the Forty-ninth Parallel; beside the range is the Purcell Trench, the eroded representative of one of the deepest structural depressions of the Cordillera.

CHAPTER XII.

INTRUSIVE ROCKS OF THE SELKIRK MOUNTAIN SYSTEM.

From the Great Plains to the Purcell Trench the igneous-rock geology of the Boundary section shows relative simplicity. It has centred principally around the discussion of the Purcell Lava formation and the basic sills and dikes of the Purcell mountain system. Crossing the trench westward we enter a region where igneous rocks become areally important and, because of their petrographical variety and complicated relations, deserve considerable attention. All the rest of the Boundary belt, from the Purcell Trench to the Fraser flats at the Pacific may be described as an igneous-rock field. It is not always possible to treat of the many intrusive and extrusive bodies in groups corresponding to the various mountain ranges crossed by that long belt. In some cases the igneous-rock bodies are crossed by the master valleys which have been taken as the convenient lines of separation between the ranges. This is true of several of the igneous-rock units which, in part, make up the Selkirk system at the Forty-ninth Parallel. It happens that the larger areas covered by these bodies occur in the adjacent Rossland mountain group of the Columbia system, and it is appropriate to discuss such areas in the following chapter devoted to the geology of the Rossland mountains. In that chapter will, then, be described the formations which have been mapped under the names 'Rossland Volcanic group,' 'Beaver Mountain Volcanic group,' 'Trail batholith,' 'Sheppard granite,' and 'Porphyritic olivine syenite.' (Maps No. 7 and 8.)

In the present chapter there will be described two granitic bodies named the Rykert and Bayonne batholiths; rocks which appear to be satellitic to the Bayonne batholith; a sill or dike very abnormal hornblende granite which cuts the Kitchener quartzite; creek; sills and dikes of metamorphosed basic intrusives cutting the Priest tetr... numerous lamprophyric dikes and sills and other basic in ... with a few acid dikes and sills cutting the younger sedimentary well as the Priest River terrane; and a boss of monzonite near the ... the Salmon river. No attempt will be made to describe these bodies ... their order of age or geographical arrangement, though the usual proc... ...ing them up in the order from east to west will be followed. The diffi... ...oblem of their succession in geological time will be discussed in a following ...

The Irene Volcanic formation has already been de... ... in its natural place as a member of the Summit series. Further reference ... ...essary except in the general summary on the igneous rocks of the S...

2 GEORGE V, A. 1912

METAMORPHOSED BASIC INTRUSIVES IN THE PRIEST RIVER TERRANE.

Various belts of the Priest River sedimentaries enclose narrow dikes and sills of basic igneous rocks and one or two basic bodies of larger size. With few exceptions these are poorly exposed and the intrusives are enormously altered. It is, therefore, impossible to give a satisfactory account of the intrusives either as to the field relations of several of the bodies or as to the original nature of the magmas whence they have been derived.

The largest of the bodies outcrops for a distance of several hundred yards on the trail running from Boundary lake to Summit creek and at a distance of about 2,000 yards west of the top of North Star mountain. The body is at least 500 feet broad. Whether it is a great dike or sill or an irregular intrusion could not be determined. The rock is a dark green, fine-grained, highly schistose trap. Labradorite in small broken individuals is apparently the only primary mineral remaining after the profound metamorphism that the rock has undergone. Most of it is now a mass of chlorite, uralite, epidote, secondary quartz, leucoxene, and pyrite. The original structure seems to have been the hypidiomorphic-granular. The rock was doubtless a gabbro, now altered to a chlorite-uralite-labradorite schist or greenstone.

A ten-foot sill-like intrusion of a somewhat similar rock was found in belt E where it crosses the main fork of Corn creek.

Just below the lower contact of the Irene conglomerate on Summit creek, belt A of the Priest River terrane is cut by a relatively uncrushed hornblendite, occurring as a sill three feet in thickness. The essential amphibole has nearly the same optical properties as the hornblende of the Purcell sill gabbros. Feldspar is absent. Magnetite and apatite are the observed accessories. Chlorite, quartz, and a little carbonate are secondary products. A larger sill-like body, at least 100 feet thick, cuts the quartzites of belt D at the junction of the North Fork and main fork of Summit creek. This rock bears much quartz, orthoclase, and some indeterminable plagioclase, along with the dominant green hornblende. Its composition and habit recall the acidified hornblende gabbro of the Purcell sills.

A quarter of a mile from the Rykert granite contact the schists of belt F are cut by a 125-foot sill of originally basic igneous rock which is now a dark green amphibolite, composed essentially of green hornblende, quartz, and basic plagioclase (labradorite to bytownite) along with much accessory orthoclase and red garnet. This sill has been squeezed to a highly schistose condition and thoroughly metamorphosed during the intrusion of the Rykert granite.

Beyond the fact that these intrusives cut the Priest River sedimentaries, there is little direct evidence as to their age. The thoroughness of their dynamic metamorphism indicates a pre-Tertiary age, while the lithological similarity of the gabbroid bodies to the gabbro of the Purcell sills suggests the possibility that the former may also be as old as the Middle Cambrian, to which the Purcell sills have been tentatively referred. Some of these intrusives may represent the deep-seated phase of the yet older Irene volcanics. In any case the impression won in the field was that the chlorite-uralite schist, the amphibolite, and the sheared hornblende gabbro are of much older date than any other of the igneous bodies occurring in this part of the ..?hirk.. .?ne i dated peridotitic sill, hornblendite, may be of the same general date as ..?e semi to se derivatives of the gabbro or may be younger.

ABNORMAL GRANITE INTRUSIVE INTO THE KITCHENER QUARTZITE.

At the edge of the Kootenay river alluvial flat and 2,000 yards south of Corn creek, the down-faulted Kitchener quartzite is cut by a peculiar granular rock exp.. ..l in the form of a band about 600 feet wide and elongated in the strike .. ..e invaded quartzite. The igne.. ..mass seems to be in sill-relation to the ..?..entaries, although the exposures are not sufficient to cause certainty on tha ..?nt. The dip of the adjacent quartzite is 60° to the southeastward; if the intrusive body is a sill its thickness is nearly 500 feet.

The igneous rock is dark bluish-gray, medium-grained, and has the habit of a quartz diorite. In the hand-specimen idiomorphic, lustrous black prisms of hornblende up to 5 mm. in length are very abundant; these are often arranged with a rough fluidal alignment. Quartz is easily recognized as a dominant constituent; feldspar is as clearly subordinate.

Under the microscope the rock is seen to be very fresh, though slightly strained, with possibly some granulation in places. The observed amount of deformation is not sufficient to explain the rough parallelism of the hornblende prisms, which is apparently a primary feature established during the crystallization of the magma. The essential and necessary constituents are here listed in their order of quantitative importance (by weight) as determined by the Rosiwal method:—

Quartz	41.3
Hornblende	33.4
Orthoclase	19.2
Garnet	2.8
Magnetite	2.1
Epidote	.6
Apatite	.5
Zircon	.1
	100.0

The hornblende is highly pleochroic, with unusually beautiful tints:

   a—Light yellowish green.

   b—Very deep olive green.

   c—Bottle green with pronounced bluish tinge.

   Absorption very strong: b>c>a.

The extinction on (010) is about 11° 15'; that on (110), about 13', as average of eight measurements on cleavage pieces. Etch-figures on (110) show that c lies in the obtuse angle β in Tschermak's orientation of amphibole, and also that the hornblende is rich in alumina. The hornblende is quite idiomorphic in the prismatic zone but the prisms are seldom, if ever, terminated by crystal faces. They lie in a mesostasis of quartz and feldspar and have suffered

2 GEORGE V, A. 1912

somewhat by resorption carried on by this acid matrix in the late magmatic period. The hornblende seems also to be truly poikilitic, through the
inclusion of minute droplets or microlites of quartz and feldspar. The feldspar
is either a sodiferous orthoclase or its chemical equivalent, a poorly developed
microperthite. Not a certain trace of soda-lime feldspar could be seen. The
surprisingly abundant quartz occurs as glassy-clear, granular aggregates. The
garnet is, in thin section, of a very pale pink colour and is probably a common
iron-lime variety.

The garnet is idiomorphic against the hornblende. The order of crystallization seems to be: zircon, apatite, and magnetite; followed by garnet; then,
in order, hornblende, orthoclase-microperthite, and quartz.

Calculation shows that the rock must carry about 68 per cent of silica,
not more than 8 or 9 per cent of alumina, and not more than about 5 per cent
of alkalies. The specific gravity of a typical hand-specimen is 2·894.

The presence of essential quartz and orthoclase would place this rock
among the granites, but it is clearly an aberrant type in that family. It
may be questioned that it is advisable to risk overweighting the granite family
by including this rock within it, but no other place is offered to it in the prevailing Mode classification. Its abnormal composition may be due to some assimilation of the quartzite. There are many points of similarity between this
rock and certain phases of the Purcell sills across the river and it is quite
possible that the abnormal granite is the result of the solution ot the quartzite
in an original hornblende gabbro magma. The quartzite is here very poor in
feldspathic and micaceous constituents; hence, possibly, the absence of biotite,
which is so universal a constituent in the acidified phases of the Purcell sills.

This abnormal hornblende granite is tentatively correlated with the
Purcell sills. Though little more crushed than those sills, it may also be
possible to credit a correlation with the sheared basic intrusives found in the
Priest River terrane; for the deformation of the latter must have taken place
at a depth several miles greater than that at which the intrusives cutting the
much younger Kitchener formation began to feel the post-Paleozoic orogenic
stresses. The higher temperatures and pressures of the more deeply buried
massive rocks at the time of deformation would seem to be amply sufficient
to explain such differential metamorphic effect.

RYKERT GRANITE BATHOLITH.

This granite, as shown on the map, covers some fifteen square miles
of the Boundary belt north of the line; it extends in a broad band southward for
an unknown distance into Washington and Idaho and the whole body is, doubtless, of batholithic size. It has intrusive relations to the Priest River terrane,
as shown by numerous apophyses, and by the development of a metamorphic
aureole about the granite. On the eastern side of the batholith the Kootenay
river alluvium conceals the bed-rock relations, but the granite is probably there

PLATE 20

Sheared phase of the Rykert granite, showing concentration of the femic elements of the rock (middle zone). Natural size.

Massive phase of the Rykert granite, showing large phenocrysts of alkaline feldspar.
25a—vol. ii—p. 284.

in contact with the Kitchener (?) quartzite which have been faulted down against it. This faulting is believed to be of later date than the intrusion of the granite; no apophyses were found in the quartzite.

Lithologically and structurally the batholith is unique in the whole Boundary belt, although in both respects this granite is paralleled by many intrusive bodies both in Idaho and in British Columbia. The rock is distinguished by a very coarse grain and commonly by an unusually perfect gneissic structure due to crush-metamorphism. (Plate 30.) The colour is a light gray to a light pinkish-gray. In the ledge and hand specimen the most conspicuous elements are large phenocrysts of alkaline feld-par, and, less commonly, of acid plagioclase; these are embedded in a coarse matrix of quartz, feldspar, and biotite. The phenocrysts range from 2 cm to 8 cm. in length. In the less crushed rock they are subidiomorphic and lie with their longer axes parallel, recalling a true fluidal structure. Such phenocrysts lie sensibly parallel to the planes of crush-schistosity. Generally, however, the crushing has been so intense that the phenocrysts are now lenticular and more or less rounded. In this case they stand out as ' eyes ' and, while the core of each crystal still holds its glassy lustre and recognizable cleavages, the outer shell of the crystal, for a depth of one to two millimetres, is opaque-white and lustreless, owing to the peripheral granulation of the phenocrysts. A third and very common phase consists of zones from a few inches to fifty feet or more wide, in which the crushing has developed a medium to coarse grained, equigranular biotite-gneiss or muscovite-biotite gneiss. This gneiss is devoid of phenocrysts, probably for the reason that these have been completely merged with their ground-mass through excessive granulation in zones of maximum shear. Of the three phases the augengneiss is the most abundant.

The planes of schistosity of the granite have a fairly constant attitude with a strike varying from N. 30° W. to N. 10° W., and dips varying from 60° W. to 75° E. The average attitude is about: strike, N. 15° W., and dip, 80° W. The gneissic bands are very seldom, if ever, crumpled, but continue nearly vertical through thousands of feet of depth in the mass.

The apophyses of the batholith are often coarsely pegmatitic. They are often greatly faulted, distorted or pulled out into discontinuous pods, showing that the country-rock about the intrusive has shared in the energetic deformation of the batholithic body. It is possible that the deforming stresses were at work before the granite had thoroughly solidified, thus explaining the apparent flow-structure in certain phases of the batholith; but most of the deformation must have followed the crystallization of the ground-mass, the minerals of which are so greatly strained or granulated.

Under the microscope the phenocrysts are seen to be chiefly orthoclase or microcline, more rarely acid oligoclase, near Ab_2An_1. The ground-mass is composed of quartz, orthoclase, oligoclase, microcline, microperthite, biotite, and muscovite with a little accessory magnetite, apatite and titanite. All of these minerals are more or less bent or fractured. The crushing has been so intense that it is now impossible to state the original diameters of the ground-mass essentials, though the average for the quartz and feldspars

2 GEORGE V, A. 1912

must have been several millimetres. It is likewise difficult to be certain of the exact nature of the original feldspars. Microcline, microperthite, and musco- vite are all more abundant in the phase of greatest crushing, and are probably in the main of metamorphic origin. Some part of their volume may thus represent the product of changes wrought in the somewhat sodiferous orthoclase. The soda-potash intergrowths of the microperthite have not, as a rule, the regularity of form characteristic of this feldspar when crystallized directly from an alkaline magma. In the present case the albitic material has been segregated in irregular lenticules and stringers which seem to represent fractures in the original ortho- clase. A little of the muscovite may be a primary accessory of the rock, for it then occurs in parallel intergrowths with the undoubtedly primary biotite.

It may be noted that the accessories, apatite, magnetite, and titanite, are either entirely absent or exceedingly rare in the zones of specially intense shear- ing and crushing; their removal seems to be one of the results of the meta- morphism. In several thin sections prisms of allanite were noted and, in one slide, a little fluorite; these minerals should, probably, be added to the list of primary accessories.

All phases are generally very fresh and the secondary products, kaolin, chlorite, and sericite, are unimportant. The observation was made in the field that the rock of the zones of maximum shearing and crushing is very consider- ably tougher under the hammer than the coarser porphyritic granite and augen- gneiss alongside. In the bed of Boundary creek the former rock stands out in long ridges or riffles, between which the softer granite has been eroded by the sluicing waters of the creek. This contrast of strength shows that the batholith lay deeply buried at the time of its shearing so that the crush-zones underwent cementation, which made them actually stronger than the rock more closely resembling the original granite.

The specific gravities of typical specimens from the batholith vary from 2.640 to 2.677, with an average for five specimens of 2.658.

A large type specimen, collected at a point on the Boundary creek wagon- road, about two miles from the ferry at the eastern end of the road, has been analyzed by Mr. Connor. The large phenocrysts are here generally micro- cline, although a few, twinned on the Carlsbad and albite laws, are acid oligo- clase near $Ab_4 An_1$. The essentials of the coarse ground-mass are quartz, microcline, orthoclase (sometimes obscurely microperthitic), oligoclase averag- ing apparently $Ab_3 An_1$, muscovite and biotite. The accessories are the same as those noted in the foregoing description of the average rock. A trace of secondary calcite was observed in the thin section.

The analysis of this specimen (No. 962) resulted as follows (Table XVII., Col. 1):—

Table XVII.—Analyses of Rykert granite and related rock.

	1.	1a.	2.
		Mol.	
SiO_2	70·78	1·180	72·07
TiO_2	·20	·003	·16
Al_2O_3	15·72	·154	15·51
Fe_2O_3	·36	·003	·31
FeO	1·61	·022	1·01
MnO	·03	....	tr.
MgO	·46	·011	·35
CaO	1·92	·034	1·93
SrO	tr.	....	....
BaO	·01	....	....
Na_2O	3·48	·056	4·02
K_2O	5·23	·055	4·09
H_2O at $110°C$	·10	....	·03
H_2O above $110°C$	·25	....	·30
P_2O_5	·26	·002	·11
	100·41		99·89
Sp. gr	2·654		

The calculated norm is:—

Quartz	25·38
Orthoclase	30·58
Albite	29·34
Anorthite	8·90
Corundum	1·12
Hypersthene	3·21
Ilmenite	·46
Magnetite	·70
Apatite	·62
Water	·35
	100·66

The mode (Rosiwal method) is approximately:—

Quartz	35·5
Orthoclase and microcline of phenocrysts	15·0
Orthoclase and microcline of ground-mass	17·5
Oligoclase of phenocrysts	3·0
Oligoclase of ground-mass	17·7
Muscovite	6·5
Biotite	3·0
Magnetite	·6
Apatite	·4
Zircon	·3
Kaolin	·5
	100·0

According to the Norm classification the rock enters the sodipotassic subrang, toscanose of the domalkalic rang. toscanase, in the persalane order **brittanare.**

2 GEORGE V., A. 1912

About 200 miles to the south-southeastward a huge batholith of a somewhat similar granite has been described by Lindgren. The two batholiths have been tentatively correlated by the present writer. The correlation is based entirely on petrographical likenesses; it is thus important to review in actual quotation the principal facts and conclusions reached by Lindgren:—.

'Granitic rocks prevail in the Bitterroot range and in the Clearwater mountains, and form a central mass of vast extent, bounded in the four corners of the region covered by this reconnaissance by smaller areas of different sedimentary series. To the north of this region the extent of the granite is not well known. But as the granite is absent in the Cœur d'Alene section it is probable that the main area does not continue far north of Lolo ridge except as detached masses. Southward this granite continues through all of central Idaho as a broad belt, and finally disappears below the sediments of Snake River valley. It does not reach Snake river at any place between Huntington and Lewiston. It forms on the whole an elongated area 300 miles from north to south and 50 to 100 miles from east to west, constituting one of the largest granitic batholiths of this continent.

'On the whole, this extensive area of granite shows great constancy in its petrographic character. It is a normal granular rock sometimes roughly porphyritic by the development of large orthoclase crystals up to 3 cm. in diameter. The colour is almost always light gray, the outcrops assuming a yellowish-gray colour, which in glaciated districts changes to a brilliant white or light-gray tone. Biotite is always present in small foils, and over large areas muscovite also enters into the composition; quartz is abundant in medium-size grains, while the feldspars are represented by both orthoclase and oligoclase, the latter usually in large quantities. Perthite is also frequently encountered, and more rarely microcline. The rock contains far too much oligoclase to be classed as a normal granite and should be rather characterized as a quartz-monzonite. Modifications more closely allied to granodiorite, diorite, and granite occur in subordinate quantity.

'The granite is typically developed near the head of Mill creek, Bitterroot range, where it is a light-gray, medium-grained rock, with small foils of biotite and a little muscovite. A few larger crystals of orthoclase reach one-half inch in length. Under the microscope the rock shows much quartz, a little normal orthoclase, and many large grains of microperthite. An acidic oligoclase with very narrow striations is very abundant. Biotite and muscovite occur in scattered straight foils. Few accessories except zircon and apatite were noted, though titanite occurs abundantly in basic concretions in the same granite. The structure is typically granitic; the oligoclase is in part idiomorphic and sometimes included in the perthite.'[*]

---

[*] W. Lindgren, Prof. Paper No. 27, U.S. Geol. Survey, 1904, p. 17.

In Table XVII., Col. 2, is entered a typical analysis of the Bitterroot granite, called by Lindgren a quartz monzonite. The chemical resemblance to the British Columbia rock, and a corresponding similarity in macroscopic habit, mineralogical composition (more plagioclase in the Bitterroot granite), structure, and dynamic history suggest the possibility that, in the future, the Rykert granite may be found to be an offshoot of the vast Bitterroot batholith.

Lindgren states that:—

'The age cannot be determined with certainty on account of the absence of fossils in the surrounding formations. In the southern part of the batholith, near Hailey, on Wood river, it has been shown that the intrusion is certainly post-Carboniferous.' As it has been shown that the sedimentary series on the South fork of the Clearwater, near Harpster, is very probably Triassic, a post-Triassic age may, with the same degree of certainty, be attributed to the great granitic batholith.'

Assuming that the Rykert and Bitterroot granite intrusions were contemporaneous, we see some reason for referring the Rykert granite to the Jurassic or to a yet later date. However, until further field-work is done in northern Idaho, this correlation must be regarded as quite hypothetical. The Rykert granite may, indeed, be of pre-Cambrian age, though, of course, younger than the Priest River terrane.

The fact that the Rykert granite is, on the whole, more schistose than the Bitterroot granite is not an argument against their correlation, for it is highly probable that the deformation of the Rykert granite took place when that body was under an exceptionally thick cover. This cover almost surely included the entire thickness of the Summit series, so that this granite lay at a depth of six miles or more before its final shearing, with uplift, began. Under those conditions the development of perfect crush-schistosity might be expected even in a Jurassic batholith.

The thermal or contact metamorphism produced by the Rykert batholith was studied only on its west side. There the effects are noticeable for many hundreds of feet from the contact. They consist chiefly in the development of plentiful garnets and of much muscovite and biotite in the schists of the Priest River terrane. The micas form much larger foils than are usual in the schists far from the contact. The metamorphic effects, thus, consist in recrystallization without the formation of rare minerals.

BAYONNE BATHOLITH AND ITS SATELLITES.

North of Summit creek an area of some ten square miles in the ten-mile belt is covered by intrusive basic granodiorite. This mass belongs to the southern extremity of a large batholith which extends northward far down Kootenay lake, and covers a total area of at least 350 square miles. The batholith has the form of a rude ellipse about 20 miles long from north to south and 16 miles in greatest width. The Bayonne gold mine is located well within the granitic mass which may, for convenience, be distinguished as the Bayonne batholith.

25a—vol. ii—19

2 GEORGE V, A. 1912

PETROGRAPHY OF THE BATHOLITH.

Within the part of the batholith covered by the ten-mile Boundary map (the only part investigated), the granodiorite is a notably homogeneous rock of a light-gray to pinkish-gray colour and a medium to fairly coarse grain. It is essentially composed of quartz, microperthite, orthoclase, microcline, hornblende, augite, and biotite. Crystals and aggregates of magnetite, well crystallized titanite and apatite, a few small zircons, and rare idiomorphic crystals of allanite are accessory constituents. Microperthite is the dominant feldspar; it often has the double lamellation of microcline-microperthite. The orthoclase is probably sodiferous. The soda-lime feldspar is of somewhat variable composition. Some crystals (in Carlsbad-albite twins) have the extinction angles of andesine, $Ab_4 An_1$; others are acid labradorite. Many of them are zoned, with cores of labradorite, $Ab_3 An_2$, and outer rims of oligoclase, $Ab_3 An_1$. The average plagioclase has about the composition of basic andesine; $Ab_3 An_2$.

Next to the feldspars and quartz, hornblende is the most important constituent. It forms idiomorphic crystals, bounded by planes at the extremities as well as in the prismatic zone. The colour scheme is:—

Parallel to **a**—Strong yellowish green.
    "   **b**—Deep olive green.
    "   **c**—Deep sea-green with bluish tinge.

The absorption is strong: $b>c>a$. In sections parallel to (010) the extinction is 16° 30'; in sections parallel to (110), 20° 15'. These values show that the optical angle is unusually small and near 50°.[*] The hornblende has properties somewhat similar to those of the variety 'philipstadite.'[†]

The biotite is deep brown with powerful pleochroism; it is sensibly uniaxial. The diopsidic augite is colourless to pale greenish in thin section and is not noticeably pleochroic. It is quantitatively subordinate to the biotite but in all the specimens collected must be ranked among the essentials.

The other constituents need no special note. Though the rock is unusually strong and fresh, a little secondary kaolin and yellow epidote may occasionally be seen.

The specific gravity of the rock varies from 2.743 to 2.785; the average for five fresh specimens is 2.757.

Mr. Connor has analyzed a typical specimen (No. 858) from the vicinity of the Bayonne mine, with the following result:—

[*] Cf R. A. Daly, Proc. Amer. Academy of Arts and Science, Vol. 34, 1899, p. 311.
[†] Proc. Amer. Academy of Arts and Science, Vol. 34, 1899, p. 433.

Analysis of basic granodiorite, Bayonne batholith.

		Mol.
SiO₂ SiO_2	60.27	1.005
TiO₂ TiO_2	.4	.008
Al₂O₃ Al_2O_3	17.17	.169
Fe₂O₃ Fe_2O_3	2.36	.015
FeO FeO	3.67	.051
MnO MnO	.14	.0
MgO MgO	2.45	.061
CaO CaO	6.49	.116
SrO SrO	.04	
BaO BaO	.04	
Na₂O Na_2O	2.92	.047
K₂O K_2O	3.25	.035
H₂O at 110°C H_2O	.15	
H₂O above 110°C	.23	
P₂O₅ P_2O_5	.20	.001
	100.01	
Sp. gr.	2.785	

The calculated norm is:

Quartz	13.32
Orthoclase	19.46
Albite	24.63
Anorthite	24.19
Hypersthene	7.17
Diopside	5.90
Magnetite	3.48
Ilmenite	1.21
Apatite	.31
Water	.38
	100.05

The mode (Rosiwal method) is approximately:

Quartz	19.5
Microperthite	17.4
Orthoclase	3.7
Andesine	23.6
Hornblende	16.2
Biotite	11.8
Augite	4.4
Titanite	.6
Magnetite	.9
Apatite	.5
Epidote and kaolin	1.2
	100.0

In the Norm classification the rock enters the sodipotassic subrang, harzose, of the alkalicalcic rang, tonalose, in the dosalane order, austrare. In the older classification the rock must be regarded as a basic granodiorite. It is quite possible that the batholith has more acid phases in the region north of the Boundary belt and thus nearer the centre of the mass.

25a—vol. ii—19½

2 GEORGE V, A. 1912

Basic segregations, in the form of deep green to black ellipsoids from five centimetres or less to ten or fifteen centimetres in diameter, are quite common. These small bodies are of two classes. In the one class the essential components are hornblende, labradorite (Ab, An), biotite, and augite, named in the order of decreasing abundance. A little quartz and orthoclase, with much crystallized titanite, magnetite, and apatite are accessory. Microperth and microcline seem to be entirely absent. The specific gravity of a typical specimen is 2.924. In the other class of segregations the colour is yet deeper and is explained by a complete lack of feldspar. The essentials are hornblende, biotite, and augite, also named in the order of decreasing importance. Quartz is accessory but is considerably more abundant than in the first-mentioned class of segregations. The other accessories are titanite, apatite, and specially abundant magnetite in crystals and rounded grains. The specific gravity of a typical sample is 3.214.

There can be little doubt that all these bodies are indigenous and that the segregation of the material, if not its actual crystallization, took place in the early stage of the magma's solidification.

The granodiorite is generally massive and uncrushed. Straining and granulation through pressure were not observed in any of seven thin sections cut from the specimens collected. Sometimes, though rarely, thin partings in the granodiorite carry much biotite, which is arranged with its lustrous foils lying in the planes of parting, as if there developed as a result of shearing in the crystallized batholith. At the Bayonne mine the rock is sheeted and locally sheared. On the whole, however, the batholith is notably free from evidences of dynamic disturbances and appears never to have suffered the stresses incidental to an important orogenic movement in the region.

As regards its influence on the intruded formations the Bayonne granodiorite has typical batholithic relations. A glance at the map suffices to convince one that this huge mass is a cross-cutting body. Four of the thickest members of the Summit series are sharply truncated by the main southern contact. For distances ranging from one to two miles from that contact the rocks of the Wolf, Monk, Irene Volcanic, and Irene Conglomerate formations are greatly crushed, fractured, and metamorphosed by the energetic intrusion. Farther to the eastward, for a distance of ten miles down the Dewdney trail, the schists and interbedded quartzites of the Priest River terrane, though likewise truncated, have been almost completely driven out of their regional strike and a well developed schistosity peripheral to the batholith has been found in these recrystallized rocks. For the lower twelve miles the east-west Summit creek canyon has been excavated along the strike of the schists, which have been forced out of their originally meridional trends by the force of the intrusion. Abundant apophyses of the batholith sometimes 300 or 400 yards in width, cut these various invaded formations. The main contact is sinuous but clean-cut. Inclusions of the invaded rocks are not common in the batholithic mass as studied in the ten-mile belt.

CONTACT METAMORPHISM.

The recrystallization of the rocks of the Priest River terrane through the influence of the intruded magma, is most conspicuously shown along the Dewdney trail. This trail threads the floor of the deep Summit creek canyon as it rises from the 2,000-foot level near the Kootenay river to the 3,000-foot level, about nine miles farther up Summit creek. The main contact of the

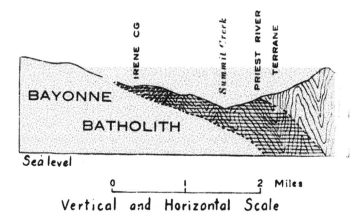

Sea level

0 _____ 1 _____ 2 Miles

Vertical   and   Horizontal   Scale

Figure 19. North-south section illustrating probable explanation of the great in⸍   ⸍y
and extent of the contact metamorphism at Summit Creek. Au   ⸍ of
contact metamorphism shown by cross-lining. Folds shown in   ⸍⸍t
River Terrane purely diagrammatic.

batholith runs nearly parallel to trail and creek and at an average distance of about 2,200 yards from both. The aureole of contact metamorphism is here two to three miles wide. The metamorphic effects seen along the trail are, however, greater than they would be at the same distance from the exposed igneous contact and on the same level as the nearest contact. The line of contact runs generally from 2,000 to 3,200 feet higher than the trail at the bottom of the canyon. The extraordinary intensity of the metamorphism along the trail is, thus, in part explained by the depth to which Summit creek has excavated its canyon in the sloping roof of the batholith. In other words, the strength of the metamorphism suggests that the contact-surface of the batholith is not vertical but dips under the creek bed, and that, on this southern extremity at least, the batholith has the section of a body enlarging downwardly. (Figure 19.)

In these Priest River rocks the thermal metamorphism has not developed new types of minerals to any notable extent. The changes in the quartzitic beds consist chiefly in their becoming micaceous, with the liberal generation of both muscovite and biotite. The phyllites, metargillites, and quartz-sericite schists, interbedded with the quartzites, have been converted into coarse, glittering mica schists, in which the individual mica-plates average scores o times the size of the original micaceous elements in the equivalent bands farther south and not thermally metamorphosed. These metamorphosed schists are regularly composed of dominant quartz, muscovite, and biotite, in variable proportion, giving muscovite-quartz schist, muscovite-biotite-quartz schist, and biotite schist. Grains of plagioclase and orthoclase are accessory in variable amount. Here and there prisms of tourmaline are developed in abundance. In general, the metamorphic effects along the trail are of a nature leading to a higher crystallinity and coarser granularity in the ancient sediments rather than to the generation of new minerals. This effect is manifest for distances as great as three miles from the main contact of the granite. Since the exomorphic collar was not thoroughly studied in the part lying north of the Dewdney trail, it is possible that many variations on the described simple scheme of metamorphism would be discovered by one exploring the inner edge of the collar.

On the other hand, the mineralogical changes in the Summit series of rocks are often very marked. This is the case even at long distances from the granodiorite contact.

One of the most remarkable instances is shown in the band of basal Irene conglomerate. At the Dewdney trail, nearly two miles south of the batholithic contact, this rock is exposed on a large scale. As usual it is intensely sheared, with its quartzite, carbonate, and slate pebbles rolled out into flat lenses and ribbons. The thermal metamorphic effects are most pronounced in the cement, which is often abundant. In ledge and hand-specimen the cement is of a dark green colour and of silky lustre, evidently due to abundant biotite and muscovite crystallized in minute individuals. In the less metamorphosed beds the microscope shows that grains of quartz and carbonate are the other essential constituents. There is considerable effervescence with cold dilute acid, showing that the disseminated grains of carbonate are, in part, calcitic. The numerous pebbles of carbonate are true dolomite. Through their mashing the cement has become mechanically impregnated with grains and small, granular aggregates of dolomite. The calcite may be, in part at least, of secondary origin and, in any case, is subordinate to the magnesian carbonate. On the whole, the composition of these few, relatively unaffected bands of the conglomerate is like that described for the standard sections of the Irene formation.

For hundreds of feet of thickness the cement has been very notably altered through contact action. The chief effect consists in the extremely abundant generation of dark green, actinolitic amphibole, forming long straight or curved prisms. These often shoot irregularly through the quartz-mica ground-mass or form beautifully developed sheaves and rosettes, which are specially well exhibited on fractures parallel to the schistosity. The individual prisms run from 1 cm. or less to 3 or 4 cm. in length, with widths usually under 1 mm. The amphibole has the optical properties of actinolite.

The study of several thin sections has convinced the writer that the amphibole has been generated at the expense of the dolomitic grains disseminated through the cement, thus illustrating a familiar phase of the metamorphism of carbonate-bearing rocks. When the carbonate was abundant in the cement, the actinolite now forms as much as a third or a half of the rock. Considerable epidote and basic plagioclase were also formed in some beds. Such metamorphic effects are noteworthy in view of the distance of these outcrops from the main batholithic contact,—about 3,000 yards. A partial explanation of the metamorphic intensity is again to be found in the probable fact that the granodiorite lies beneath these outcrops and at a distance of considerably less than 3,000 yards downward.

Two specimens of the Irene lavas were collected at the Dewdney trail. These seem to be typical of the lavas of the exomorphic zone where, as a rule, they have been completely changed to fine-grained or medium-grained, highly fissile hornblende schists. Green hornblende and quartz are the principal components; grains of carbonate, apparently dolomite, and a little basic plagioclase are present in both thin sections.

The phyllitic schists of the Monk formation have been signally metamorphosed by the batholithic intrusion. For a distance of a half mile or more outward from the granite, these rocks have been converted into a schistose hornfels composed of quartz, muscovite, biotite, sillimanite, and red garnet, along with much untwinned feldspar, apparently all orthoclase. The muscovite foils either lie in the plane of schistosity or occur with random orientations through the rock. In the latter case they are spangles measuring from 0·5 mm. to 1·5 mm. in diameter and are in phenocrystic relation to other constituents. The sillimanite has the usual development in needles which are often aggregated in tufts or sheaves very conspicuous under the microscope. The orthoclase grains show a tendency to aggregate along with some grains of quartz in lenses 1 mm. to 2 mm. long, these lenses lying in the plane of schistosity. The abundance of the orthoclase in some of the specimens suggests that its substance has been introduced from the magma, but this is not certain. The garnet is pale reddish to nearly colourless in thin section and has the usual habit of the mineral in contact-zones.

On the top of the 6,600-foot ridge which overlooks Summit creek on the north and runs eastward directly from the peaks at the western head of the creek, a thick series of ferruginous schists are exposed for a distance of a mile measured along the ridge. These schists dip under the Wolf grit and overlie the 200-foot bed of breccia-conglomerate at the top of the Irene volcanic formation. There is little doubt that these ferruginous schists are the much metamorphosed equivalents of the rocks of the Monk formation. Four type specimens were collected at points about 1·5 miles from the contact of the Bayonne granite. All of them have been microscopically examined and prove to belong to the one species of staurolite-schist. The staurolites form subidiomorphic crystals and anhedra of all sizes up to 15 mm. in length. In transmitted light they are usually of a strong yellow colour. As usual, quartz inclusions are

2 GEORGE V, A. 1912

very numerous, so that hundreds of minute clear lenses or droplets of that mineral are contained in a single crystal of the staurolite. The inclusions are almost invariably arranged with their longer axes parallel to each other and, at the same time, parallel to the plane of schistosity of the rock. This orientation of the inclusions appears to indicate that they are residuals of the quartz grains composing the schist before it was thermally metamorphosed; the staurolite crystals grew quietly in the rock without causing mechanical disturbance of the pre-existing, schistose structure. Sericitic muscovite, biotite, and quartz form the matrix in which the abundant staurolite lies. These relations of the staurolites to the ground mass find full analogy in the rocks illustrated in figures 88 and 89 of Rosenbusch's Elemente der Gesteinslehre, 1898, p. 498. Abundant twinned crystals of disthene, which do not show inclusions of the ground-mass often accompany the staurolites.

Even from the foregoing brief account of the contact action of the Bayonne batholith, it is clear that the exomorphic collar is unusually broad and that the action was correspondingly powerful. To the future geologist who plans to make a thorough study of the collar, interesting results may be promised. The different beds which have been altered should be identified and followed, so as to determine the whole gamut of changes involved in the metamorphism of each, and to find the relation of these changes to distance from the granodiorite. This work would entail the expenditure of much more time than could be devoted to the study during the Boundary belt survey. The mountains are very rough; the work must, in any case, be time-consuming and arduous, but the result would amply repay the effort.

SATELLITIC STOCKS ON THE DIVIDE.

On the main water-parting of the range and just south of the Dewdney trail a granite stock, cutting the middle members of the upturned Summit series, is well exposed. In ground-plan this body is an ellipse with a north-south major axis of 2·5 miles and a width of one mile. One-half mile west of this stock there occurs a small intrusive mass of the same granite which sends a long dike-like tongue northeastward across the Dewdney trail, where the rock is easily studied.

Petrography.—This granite is medium-grained, of a light pinkish-gray tint, and is noticeably poor in dark-coloured constituents. Quartz, microperthite, orthoclase, a little microcline, and considerable oligoclase, $Ab_4 An_1$, with a quite subordinate amount of biotite are the essentials; titanite, magnetite, apatite, and zircon are sparingly present. Primary muscovite is accessory and is often regularly intergrown with the biotite. Along the western contact of the larger stock the muscovite becomes so important that the rock may be called a two-mica granite; its structure in this contact zone tends to the panidiomorphic. The average specific gravity of four fresh specimens of the granite is 2·628.

A typical specimen from this stock was studied quantitatively according to the Rosiwal method and the following weight percentages of the different constituents were found:

Quartz..	34·9
Microperthite..	31·3
Sodiferous orthoclase..	15·9
Oligoclase..	12·1
Biotite..	3·8
Muscovite..	·9
Magnetite..	·9
Apatite and zircon..	·2
	100·0
Sp. gr..	2·622

Silica must form about 75 per cent of the rock.

The larger stock is surrounded by an irregular fringe of strong apophyses, some of which follow the trend of master joints in the invaded quartzites and grits. A finely exposed 20-foot dike (mapped) of porphyritic biotite-granite cutting the Wolf grit on the summit about 1,300 yards north of the Dewdney trail, is probably an offshoot of the same magma. This dike runs east and west with remarkable straightness and can be followed with the eye for nearly two miles over the mountain slopes. It seems to lie in the prolongation of a strong vertical thrust-fault which is marked on the map.

A great number of other dikes occur in an unusually broad shatter-zone occurring on the eastern and southern sides of the larger stock, where the rocks of the Monk and Wolf formations are tremendously shattered for distances varying from 0·6 mile to 1·5 miles. Figure 20 illustrates the general form and relations of stock and shatter-zone, which on the southeast is actually broader than the stock itself. It is apparent in the figure that the bands of the various invaded formations are not seriously disturbed from their regional strike. The shattering has locally broken up each formation into a vast number of fragments, but neither the Monk schist band, the Wolf grit-band, nor the Dewdney quartzite band has been driven out of alignment with the unshattered portions lying to north and south of the intrusions. The granite of the main stock has evidently replaced an equal volume of the sediments. There is no hint in the field-relations that the intrusion is of the laccolithic or 'chonolithic' order and thus due to a mere parting of the strata which permitted of the 'hydrostatic' injection into the opening so provided. The fact that the granite was not intruded after the manner of a laccolith is further demonstrated by the exceedingly strong contact metamorphism in the invaded strata.

Contact Metamorphism.—This exomorphic action is signally illustrated throughout the shatter-zone to the southeast of the stock. (Figures 20 and 21.) For square miles together the Monk phyllites have, in that zone, been converted into greenish gray, medium-grained, hornfelsy rocks of quite different habit and composition. (Plate 27, A.)

Muscovite in foils running from 0.5 mm. to 2 mm. or more in diameter is so prominent a constituent of these altered rocks as to give them a highly lustrous and glittering look, quite similar to the mica schists in the Bayonne granite aureole. Biotite in foils from 0.05 mm. to 0.5 mm. in diameter is a second essential mica in most phases and quartz is invariably a third essential. Along with these minerals, cordierite, cyanite, andalusite, and tourmaline (Plate 31) are developed in varying amounts, giving the following principal types of rock: muscovite-cyanite-quartz schist, cordierite-muscovite-biotite-quartz schist, cordierite-andalusite-tourmaline-muscovite-biotite-quartz schist. Cordierite is especially abundant and seldom fails from any of the thin sections. The optical properties of all these minerals are typical of them as described in the standard text-books of petrography; their detailed description need not burden this report and is omitted. An exceptional, non-micaceous hornfels, composed of green hornblende, quartz, epidote, and zoisite with a little f   ar, probably represents a greatly metamorphosed dolomitic quartzite.

The gritty rocks of the Monk and Wolf formations have been but little altered, though the once-argillaceous cement of the conglomerates in the Dewdney formation has been completely recrystallized, with the generation of much cordierite, sillimanite, muscovite, and biotite. Microperthite is present in surprising amount and appears to have been in part introduced from the magma. This mineral also occurs in the rocks of the Bayonne batholith aureole. Its development in the phyllites at Ascutney mountain, Vermont, where again it has been transferred from an alkaline magma, is another example of the special ease with which this particular feldspathic substance migrates into contact aureoles.* Two specimens of the Dewdney quartzite taken from a point about 300 feet from the granite are very rich in microperthite, soda-orthoclase, and a feldspar which is almost certainly anorthoclase. In this case some feldspathization by the magma is probable but is not so certain, since grains of microperthite occur in the unmetamorphosed quartzite.

On the other hand, the composition of the intrusive has been affected by the incorporation of material from the walls. The granite of the larger stock is abundantly charged with fragments of quartzite, schist, and conglomerate. In many cases these show no direct evidence of having lost substance by solution in the magma but the included blocks of conglomerate afford conclusive proofs that even in the magmatic period immediately preceding solidification, the magma was able to absorb such material.

---

*R. A. Daly, Bull. 209, U.S. Geol. Survey, 1903, p. 34.

PLATE 31.

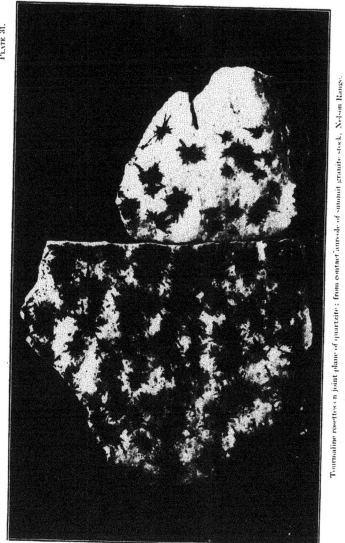

Tourmaline rosettes on joint plane of quartzite; from contact aureole of summit granite-stock, Nelson Range.

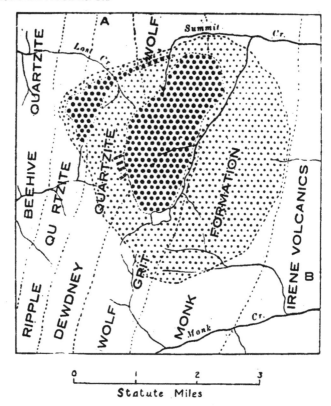

FIGURE 20.—Diagrammatic map of summit granite stocks (large dots) with wide aureole of contact metamorphism (smaller dots). The cross-cutting relation of the stocks is illustrated.

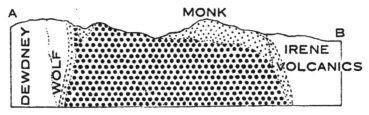

FIGURE 21.—Section along line A-B of figure 20, on the same scale ; illustrating explanation of great breadth of metamorphic aureole south-east of larger granite stock. Apophysal dikes in roof of stock not shown.

2 GEORGE V, A. 1912

On the main ridge-divide masses of conglomerate were found in a wide apophysis at the southwest side of the larger stock. The magma has eaten into the rock, dissolved out the cement in large amount, and has thus not only thoroughly impregnated the conglomerate with granitic material but has quite separated many of the larger quartzite pebbles which, still rounded, are now completely enclosed in granite. The cement was evidently more soluble in the magma than were the quartzite pebbles—a conclusion to be expected in view of the fact that the heterogeneous cement has a lower fusion-point and, in relation to the acid granite, a lower solution-point of temperature than the more highly silicious quartzite. This partial absorption of the conglomerate must have taken place when the magma was (because cooled down) sufficiently viscous to allow of the suspension of the blocks and pebbles. At an earlier period, when the cooling was less advanced, the quartzite pebbles themselves like the main quartzitic and schistose formations could have been dissolved. For reasons which will be stated in chapter XXVI., the absorption of foreign material in this earlier and more potent condition of the magma should not be directly demonstrable on the main contacts, but it is at least possible that the muscovite, which is concentrated in the endomorphic zone of the stocks, is a magmatic derivative from the sericitic and feldspathic country-rock dissolved by the main body of magma in a late stage of its history.

Quartz-diorite Apophyses.—On the 7,000-foot ridge a mile or more southeast of the larger stock, the shattering of the heavily metamorphosed schists is well displayed. Hundreds of irregular dikes and tongues of granular rock cut the schists in all directions. From one of these a typical specimen was collected and has been studied microscopically.

Its mineralogical composition differs widely from that of the stock granite. The dominant essential is andesine feldspar, near Ab, An, occurring in remarkably idiomorphic, twinned crystals averaging 1 mm. in diameter. Quartz, which is always interstitial, is next in importance. Biotite in foils from 1 mm. to 4 mm. in diameter is an abundant essential. The accessories include titanite, magnetite, apatite, zircon, and muscovite. The micas are often regularly intergrown, with common basal plane. They show some tendency to cluster in the rock and especially so where the muscovite is primary; the rock is quite fresh. Not a trace of alkaline feldspar was found, though the detection of any such would not be difficult in this case.

A quantitative estimate of the composition was made by the Rosiwal method. A high degree of accuracy was impossible on account of the leaf-shapes of the micas. Rough as it is, the estimate serves to show how widely divergent the rock is from the staple granite of the stocks. The proportions are as follows:—

Quartz..	25·0
Andesine..	60·0
Biotite..	.10·0
Muscovite..	4·0
Magnetite..	·7
Other accessories..	·3
	100·0

The calculated chemical composition (biotite assumed to have the average composition of average biotite in California and Montana quartz-monzonites) is approximately:—

SiO_2..	66·4
Al_2O..	18·0
Fe_2O_3..	1·0
FeO..	1·6
MgO..	1·2
CaO..	4·2
Na_2O..	4·8
K_2O..	1·4
Water, etc..	1·4
	100·0

The specific gravity of the specimens is 2·687. The rock is a muscovite-bearing quartz diorite.

Relation of the Stocks to the Bayonne Batholith.—The width of the shatter-belt and of the metamorphic aureole about these intrusions is out of all proportion to the visible size of the latter. Their explanation is simple if it be credited that granite underlies, at no great depth, these belts of profound mechanical and mineralogical changes or, in other words, that the altered schists belong to the roof of an intrusive body much larger than the whole area of plutonic rock actually exposed. Following this line of argument the writer believes that the granite of the larger stock represents the crystallized product of the uppermost part of a batholithic mass. Further, the evident similarity of the dominant salic components of this stock granite with those of the Bayonne granite, as well as the matching of the biotitic and minor accessory constituents in the two cases, leads to the hypothesis that the stocks are not only spatially but genetically satellites of the Bayonne batholith. (Figure 22.) Like the latter, the stocks are quite uncrushed and cut the upturned Summit series of sediments; the exposed stocks and the batholith are contemporaneous, so far as the field evidence can decide.

If it be assumed that the stocks and the exposed batholiths are really connected underground from a greater only partially unroofed, batholithic mass, an important question arises. Erosion has removed from the Bayonne body several thousand feet of granitic rock, measured vertically. We are thus in a position to determine the density of the mass crystallized at points thousands of feet below the roof of this part of the batholith. In the Summit stock we are able to determine the density of the granite crystallized very near the roof of an offshoot from the same batholith. The average specific gravities are, respectively, 2·757

2 GEORGE V, A. 1912

and 2.62×. These considerations suggest the possibility that this huge batholith, including the Bayonne granodiorite, the Summit stock granites and, as we shall see, the Lost Creek body, is stratified according to the law of density—biotite granite above and granodiorite below. On this view, similar contrasts of densities existed in the magmatic period and would find explanation in the

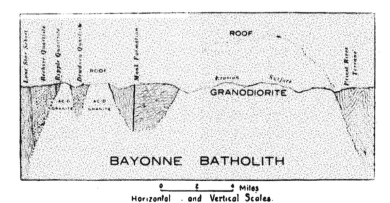

Horizontal and Vertical Scales.

FIGURE 22.—Diagrammatic section showing relation of the summit stocks of Nelson Range to the Bayonne batholith.

differentiation of the magma through gravitative adjustment. On the other hand, the smaller bodies may owe their lower density to their having been specially acidified by the solution of the invaded quartzites; or, thirdly, the more salic character of the satellitic stocks may be due to special concentration of magmatic fluids in the smaller chambers, facilitating more extreme differentiation in them than in the main Bayonne batholith. Probably all three causes have operated.

LOST CREEK GRANITE BODY.

The peculiarly shaped mass of granite over which Lost creek flows is, mineralogically, chemically, and genetically, akin to the granite of the Summit stocks. The staple rock is alkaline, with microperthite and orthoclase as the dominant feldspars. Oligoclase is the subordinate feldspar, biotite the only femic essential; primary muscovite, magnetite, apatite, and zircon are the accessories. In places the muscovite has the rank of a subordinate essential, so that the rock varies from biotite granite to biotite-muscovite granite. The average specific gravity is 2.617.

Along all observed contacts this granite, for a distance of several score of feet inward, is aplitic and poor in mica. The apophyses are generally composed of the same aplitic phase. At certain points numerous blocks and shreddy fragments of quartzite and schist were observed in the granite. These xenoliths, especially the schists, have undergone much metamorphism, with the generation of abundant andalusite in stout prisms, broad leaves of muscovite, and biotite in aggregates which mottle the rock in striking fashion. Again large amounts of microperthite are disseminated through the altered schist, as if introduced from the magma.

The northern arm of the body has the form of a huge irregular dike or sill which follows the strike of the invaded schists. The exposures do not favour the decision as to whether or not the mass here follows planes of bedding or schistosity. The other and larger arm of the mass is clearly in cross-cutting relations. The width of this hand is doubtless the greater because of the excavation of the deep canyon of Lost creek. If erosion should remove a few thousand feet more of the sedimentary cover at the head of the creek, the Lost creek body and the summit stocks would doubtless be found to form one continuous batholithic mass.

A small intrusion of the Lost creek granite occurs on the divide between Sheep and Lost creeks and 1·5 miles east of Salmon river. It cuts schists and limestone probably of Carboniferous age, and the youngest bedrock formations with which this whole group of granites, including the Bayonne batholith and its satellites, is known to make contact. The date of these intrusions will be further discussed in a following summary.

BUNKER HILL STOCK.

Within the ten-mile belt an igneous body which appears to be the most westerly satellite of the Bayonne batholith is a stock covering about eighteen square miles and lying almost wholly on the western side of the Salmon river. This stock is composed of a medium to rather coarse, alkaline biotite-granite (specific gravity, 2·610) which, in all essential respects, is identical with the granite forming the small summit stocks and the Lost Creek body. The Bunker Hill mine (now shut down) is situated in the metamorphic aureole of this stock and it may, for convenience be referred to as the Bunker Hill granite stock.

Being generally more weathered, this granite has a more reddish tint than the Lost creek and Summit granites. The stronger weathering effect may be partly due to the fact that the Bunker Hill granite has been much more strained and crushed than the more easterly bodies. A distinct schistosity has been thus produced at many points in the stock. The gneissic structure is most pronounced near the southeastern contact, at the confluence of Lost creek and Salmon river. For a distance of 500 feet or more from the contact the granite is specially basic and consists essentially of quartz, biotite, plagioclase (labradorite Ab, An, to basic oligoclase, Ab, An,), with very subordinate orthoclase, and abundant muscovite foils. The plates of the white mica lie in the planes of schistosity and are of metamorphic origin. This basic phase recalls the muscovite-bearing quartz diorite which forms the many apophyses in the shatter-zone about the summit stocks.

2 GEORGE V, A, 1912

This granite also has thermally metamorphosed its country rock, in this case the Pend D'Oreille schists. The metamorphic aureole is nowhere as wide as those about the summit stocks or the Bayonne batholith; it is thus probable that the contact surface of this stock dips under the invaded rocks at higher angles than those characteristic of the contacts in the eastern bodies. The Bunker Hill aureole has not been systematically studied with the microscope. Thin sections of two specimens collected, one at the southwestern contact, the other at a point about 1,000 feet from the contact, both showed the abundant generation of andalusite prisms in the characteristic micaceous hornfels. At Bunker Hill mine the andalusite schist is enormously crumpled and is cut by veins of gold-bearing quartz. On one of the veins the mine shaft has been sunk for free-milling ore.

SALMON RIVER MONZONITE.

Halfway between Sheep creek and Lost creek, and a mile east of the Salmon river, the Pend D'Oreille schist is cut by a small stock of plutonic rock, which, in chemical and mineralogical composition, is unique among the known intrusives of the Selkirks within the ten-mile belt. The stock has the subcircular ground plan of a typical granitic boss, measuring 700 yards in diameter. The rock is relatively prone to disintegration and it has weathered freely into huge bouldery masses, whose forms have been produced by exfoliation and concentric weathering on joint blocks. By the energetic intrusion the schists round about have been crumpled, hardened, and converted into hornfelsy, massive rock. This contact aureole is a few hundred feet in width; it has not been studied microscopically.

The igneous rock is dark greenish-gray and rather coarse-grained. It is massive and quite uncrushed. With the unaided eye, augite, biotite, and feldspar can be readily identified as the essential constituents. The first named mineral forms highly idiomorphic, stout prisms of varying lengths up to that ct 7 mm. or 8 mm. The biotite occurs in lustrous black, often idiomorphic foils which may be 2 mm. or more in diameter but average about 0·6 mm. Between these femic essentials the feldspar forms a kind of mesostasis, numerous individuals approaching 5 mm. in diameter. Many of the larger cr, stals schillerize in vivid sky-blue colours which are specially brilliant when the rock is wetted.

Under the microscope the augite shows the cleavages, the very pale green almost colourless tint, double refraction, and extinction angles of a diopside. One crystal in a thin section showed a narrow interrupted mantle of green hornblende about the pyroxene. The biotite is sensibly uniaxial and has powerful absorption. The feldspar belongs to the alkaline and soda-lime groups, which are represented in nearly equal proportions. The larger, schillerizing individuals have the optical properties of soda-orthoclase and microperthite. The same crystal often has the homogeneous structure of soda-orthoclase in one part and the familiar microperthitic intergrowth irregularly developed

In other parts; these two feldspar varieties are here clearly transitional into each other. The extinction-angle of the soda-orthoclase is 10° 30' on (010), showing a high content of soda. Its double refraction is markedly low. It is possible that some of this homogeneous feldspar is anorthoclase, but the extinction of flakes cleaved parallel to (001) was found, in three cases, to be parallel and thus corresponding to the monoclinic isomorphic mixture. The schillerizing effect, like the chemical composition, relates this feldspar to the dominant feldspar of Brøgger's original laurvikite.

The alkaline feldspar often encloses poikilitically idiomorphic to subidiomorphic plagioclase, which occurs always in relatively small crystals, averaging about 0·5 mm. in length. These are commonly twinned according to the Carlsbad and albite laws and are often irregularly zoned. The average plagioclase is labradorite, near Ab₂ An₃. Moderate amounts of apatite and magnetite are accessory, while very rare, interstitial grains of quartz are also found. The structure is the hypidiomorphic-granular. The order of crystallization appears to be: apatite and magnetite; augite; biotite; plagioclase; soda-orthoclase (and microperthite); quartz.

Mr. Connor's analysis of a fresh specimen (No. 671) gave the result: -

Analysis of Salmon River Monzonite.

		Mol.
SiO₂	50·66	·844
TiO₂	1·32	·016
Al₂O₃	16·91	·166
Fe₂O₃	1·71	·011
FeO	6·17	·086
MnO	·16	·002
MgO	5·50	·138
CaO	8·26	·147
SrO	·08	·001
BaO	·23	·001
Na₂O	2·89	·047
K₂O	4·45	·047
H₂O at 110°C	·14	
H₂O above 110°C	1·66	
P₂O₅	·91	·006
	100·45	
Sp. gr.	2·843	

The calculated norm is:

Orthoclase	26·13
Albite	16·77
Nephelite	4·26
Anorthite	20·02
Diopside	13·32
Olivine	11·19
Ilmenite	2·43
Magnetite	2·55
Apatite	1·86
Water	1·26
	99·73

2 GEORGE V, A. 1912

The mode (Rosiwal method) is appr imat 'v:

Alkaline feldspar (soda orthoclase)	32·6
Labradorite	27·9
Augite	20·1
Biotite	16·9
Magnetite	1·2
Apatite	·8
Quartz	·5
	100·0

(Specific gravity calculated from mode 2·818, closely agreeing with observed specific gravity).

According to the Norm classification the rock enters the sodipotassic subrang, kentallenose, of the alkalicalcic rang, camptonase, in the salfemane order, gallare; but it is also very close to the sodipotassic subrang, shoshonose, of the alkalicalcic rang, andase, in the dosalane order, germanare. According to the older classification the rock is evidently a typical (basic) augite-biotite monzonite.

LAMPROPHYRIC DIKES AND SILLS.

In the Selkirk mountain system many of the formations older than the Rossland volcanics are cut by lamprophyric dikes and thin sills which are sometimes very abundant. Both sills and dikes are generally highly inclined, approaching the vertical, and are bodies of quite moderate size; widths of either dikes or sills are seldom as much as 20 feet and average only a few feet. The larger number of these different ...ion products are minettes but there are also representatives of the kersantites, camptonites, and odinites. The dikes are specially numerous in the Pend D'Oreille schists, quartzites, and limestones where these rocks crop out in the canyon of the Pend D'Oreille river and on the west side of the Columbia. Others cut the large masses of Pend D'Oreille marble, the Wolf grit, Irene conglomerate, and doubtless other members of the Summit series. Still others transect the different belts of rock in the Priest River terrane.

The minettes, as the most abundant lamprophyres in the region, have merited most attention. On account of their fine grain and degree of alteration their diagnosis merely through microscopic study was not to be entirely trusted. For that reason as well as on account of their intrinsic interest a number of chemical analyses have been made of the minettes. With the help of the analyses and rather numerous thin sections the conclusion was reached that four different types of minette occur more or less abundantly in the Boundary belt. The types are augite minette, mica minette (biotite the only femic essential), augite-olivine minette, and hornblende-augite minette.

Porphyritic Mica Minette.—The type which for distinction may be called mica minette was found in the form of a three dikes cutting the Pend D'Oreille series near the railroad bridge over the Pend D'Oreille river. The dikes run from three to six feet in width and seem to be composed throughout of this one type, though they are associated with dikes of augite minette. The mica minette is a dark gray, fine-grained, highly micaceous rock usually showing phenocrysts of biotite up to 2 mm. or more in diameter. The ground-mass is the common hypidiomorphic aggregate of biotite, orthoclase, with little labradorite; the accessories are apatite, titanite, and magnetite, with a little interstitial quartz which may be secondary. The alteration products are the same as in the augite minette, from which this rock differs mineralogically only in the fact that the pyroxene is here absent. The mica minette represented in all of the collected specimens is rather badly altered—so much so as to discourage the idea of chemical analysis in their case—but there seems to be little question that both augite and olivine were absent from this rock in its original condition; in any case they were present in but accessory amounts. The specific gravity of a typical specimen was found to be 2·790.

Augite Minette.—A large proportion of the lamprophyric intrusives belong to the species, augite minette. This rock was found in dikes in the Pend D'Oreille group as exposed on both sides of the Columbia river and at many points along the walls of the Pend D'Oreille canyon. The freshest specimen collected was, however, taken from a 60-foot dike cutting biotite-spangled and garnetiferous mica schist in Belt F of the Priest River terrane on the summit of the ridge two miles E.N.E. from the peak of North Star mountain. The dike is nearly vertical and strikes north and south.

The rock is dark greenish to slate-gray and is porphyritic, with conspicuous, lustrous phenocrysts of brown biotite measuring 5 mm. or less across the foils. In thin section, idiomorphic prisms of a nearly colourless, diopsidic augite are seen to be yet more abundant phenocrysts than the mica. The prisms range from 0·5 mm. to 1·5 mm. or more in length. The ground-mass is a fine-grained hypidiomorphic-granular aggregate of minute augite and biotite crystals with abundant orthoclase. The last often encloses the femic minerals poikilitically. Apatite, magnetite, and a little quartz, which is interstial between the feldspars, are the primary accessories. The orthoclase is somewhat kaolinized, while chlorite, epidote, and calcite have been secondarily developed, but, for a minette, this rock must be regarded as unusually fresh.

Mr. Connor's analysis of the same specimen (No. 900) resulted as follows:—

Analysis of augite minette.

		Mol.
SiO$_2$	53·32	·889
TiO$_2$	·90	·011
Al$_2$O$_3$	14·16	·139
Fe$_2$O$_3$	2·15	·013
FeO	5·08	·071
MnO	·10	·001
MgO	7·90	·198
CaO	7·12	·127
SrO	·05	......
BaO	·12	·001

2 GEORGE V, A. 1912

Analysis of augite minette—Continued.

		Mol.
Na$_2$O	2·39	·039
K$_2$O	4·80	·051
H$_2$O at 110°C	·26	......
H$_2$O above 110°C	1·24	......
P$_2$O$_5$	·66	·005
	100·25	
Sp. gr.	2·831	

The calculated norm is:—

Orthoclase	28·36
Albite	29·44
Anorthite	13·62
Diopside	11·18
Olivine	9·47
Hypersthene	7·48
Magnetite	3·02
Ilmenite	1·67
Apatite	1·55
Water	1·50
	100·29

In the Norm classification the rock enters the sodipotassic subrang, non-zonose, of the domalkalic rang, monzonase, in the dosalane order, germanare.

Hornblende-augite Minette.—A somewhat allied type of minette, distinguished, however, by a notable and essential proportion of hornblende among the phenocrysts, occurs as a ten-foot dike outcropping on the western bank of the Columbia river, a few hundred feet south of the Boundary line. Though occurring just beyond the limit of the Selkirk system this dike may best be described here. Its relations are shown in Figure 23. Huge dike-like masses of an uncrushed, fresh biotite-hornblende granite porphyry cut the intensely crumpled Pend D'Oreille phyllites and one of the porphyry masses is itself cut by the dike in question, which is ten feet wide and strikes N. 10° E. with a dip of about 75° to the eastward. It truncates a seven-inch dike of augite minette cutting the phyllite, as shown in the figure.

The dike of hornblende-augite minette shows a very marked chilling along both walls. Its main mass is composed of a dark greenish-gray to dark ash-gray, fine-grained rock, macroscopically showing occasional phenocrystic foils of biotite up to 2 mm. in diameter and many minute prisms of augite and green hornblende varying from 0·5 mm. or less to 1 mm. in length. These three femic minerals are embedded in a very fine-grained paste of doubtless sodiferous orthoclase and oligoclase (in about equal proportions) accompanied by abundant titanite, apatite, interstitial quartz, and a small amount of ilmenite as the accessories. The rock is somewhat weathered, with calcite, quartz, kaolin, chlorite, and epidote as the secondary products. The hornblende and augite are present in nearly equal amounts and each rivals the biotite in abundance. The specific gravity of this phase is 2·740.

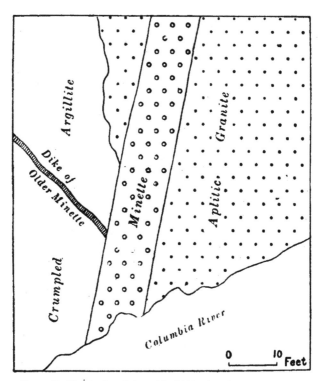

FIGURE 23.— Map showing relations of Pend D'Oreille argillite, aplitic granite, and two dikes of minette, just south of Boundary line on the shore of Columbia River.

2 GEORGE V, A. 1912

On account of its superior freshness the rock from the chilled zone (specimen No. 493) was selected for chemical analysis. Mineralogically it resembles the coarser phase except that hornblende is scarcely more than accessory and that here minute biotite and augite crystals with the dominant alkaline feldspar form the ground-mass. Its chemical analysis yielded Mr. Connor the following proportions:—

Analysis of hornblende-augite minette.

		Mol.
SiO_2	53·68	·895
TiO_2	·90	·011
Al_2O_3	16·89	·166
Fe_2O_3	1·28	·008
FeO	5·53	·076
MnO	·11	·001
MgO	3·70	·092
CaO	6·08	·109
SrO	·10	·001
BaO	·38	·003
Na_2O	4·03	·065
K_2O	4·32	·046
H_2O at 110°C	·10	......
H_2O above 110°C	1·85	......
P_2O_5	1·05	·007
	100·00	
Sp. gr.	2·723	

The calculated norm is:—

Orthoclase	25·58
Albite	32·49
Nephelite	·85
Anorthite	15·29
Diopside	8·47
Olivine	9·29
Ilmenite	1·67
Magnetite	1·86
Apatite	2·17
Water	1·95
	99·62

According to the Norm classification the rock enters the sodipotassic subrang, monzonose, of the domalkalic rang, monzonase, in the dosalane order, germanare. The analysis doubtless represents also the composition of the main, unchilled part of the dike, in which hornblende is more abundant.

Olivine-augite Minette.—A fourth variety of the porphyritic minettes forms a two-foot sill cutting the Wolf grit on the main summit of the Selkirks, and one mile north of the Dewdney trail. It appears itself to be cut off by the long east-west granite dike shown on the map as occurring at this locality. The granite is perhaps contemporaneous with the Bayonne batholith and it is thus possible that this minette injection antedates the batholithic intrusion.

The phenocrysts are pale green augite (up to 1·5 mm. in length), a few biotite foils, up to 1 mm. in diameter, and abundant round masses of serpentine, almost certainly derived from olivine. The latter measure 1 to 2 mm. in diameter. The ground-mass is composed of a multitude of idiomorphic deep brown biotites and a few microlites of orthoclase embedded in colourless glass. The rock (specimen No. 836) is relatively fresh but chlorite and calcite are secondary constituents, like the serpentine. It has the composition of an olivine-augite minette, as shown by Mr. Connor's analysis:—

Analysis of olivine-augite minette.

		Mol.
SiO_2..	48·33	·806
TiO_2..	·81	·010
Al_2O_3..	12·56	·124
Fe_2O_3..	1·87	·012
FeO..	5·26	·073
MnO..	·13	·001
MgO..	9·07	·227
CaO..	8·94	·160
SrO..	·05	......
BaO..	·24	·001
Na_2O..	1·81	·029
K_2O..	4·67	·050
H_2O at 110°C..	·97	......
H_2O above 110°C..	2·63	......
P_2O_5..	·78	·006
CO_2..	2·64	......
	100·76	
Sp. gr..	2·771	

The calculated norm is:—

Orthoclase..	27·80
Albite..	8·91
Nephelite..	3·41
Anorthite..	12·51
Diopside..	21·75
Olivine..	13·76
Ilmenite..	1·52
Magnetite..	2·78
Apatite..	1·86
Water and CO_2..	6·24
	100·54

According to the Norm classification the rock enters the dopotassic subrang, prowersose, of the domalkalic rang, kilauase, in the salfemane order, gallare; it is, however, near the sodipotassic subrang, lamarose, of the same rang.

Comparison of the Minettes with the World Average.—In Table XVIII. the three minette analyses are entered and, as well, their mean and the average of ten analyses recorded for the world in Osann's compilation.

2 GEORGE V, A. 1912

TABLE XVIII.

	1. Augite minette (No. 900).	2. Hornblende-augite minette (No. 493).	3. Olivine-augite minette (No. 836).	4. Mean of 1, 2 and 3.	5. World-average minette.
SiO_2	53·32	53·68	48·23	51·78	49·45
TiO_2	·50	·90	·81	·87	1·23
Al_2O_3	14·16	16·89	12·56	14·54	14·41
Fe_2O_3	2·15	1·28	1·87	1·77	3·39
FeO	5·08	5·53	5·26	5·29	5·01
MnO	·10	·11	·13	·11	·13
MgO	7·90	3·70	9·07	6·89	8·26
CaO	7·12	6·08	8·94	7·35	6·73
SrO	·05	·10	·05	·07	... ...
BaO	·12	·38	·24	·25	... ...
Na_2O	2·39	4·03	1·81	2·74	2·54
K_2O	4·80	4·32	4·67	4·60	4·69
H_2O-	·26	·10	·97	·44	2·43
H_2O+	1·24	1·85	2·63	1·91	
P_2O_5	·66	1·05	·78	·83	1·12
CO_2			2·64	·88	·61
	100·25	100·00	100·76	100·32	100·00

Considering the relatively great amount of alteration suffered by all these rocks the correspondence of the two averages is quite close. This essential equivalence of chemical types points clearly to the prevalence of a general law which underlies the generation of these lamprophyres wherever found. It should be noted that the world-average includes analyses of minettes that are mineralogically transitional to the kersantites. The analysis of the hornblende-augite minette (No. 493) corresponds to a similar transitional type.

Kersantite.—The long band of Pend D'Oreille limestone running from Lost creek to Sheep creek, parallel to Salmon river and two miles distant from it, is traversed by dikes of mica-lamprophyre. Two of these, each about four feet wide, crop out on the summit of the ridge dividing the waters of the two creeks. Petrographically, they are similar to other dikes, occurring along the Pend D'Oreille river. All of them are altered in varying degree, so that the microscopic diagnosis of these lamprophyres is difficult. For this reason, one (four feet wide) of the two dikes cutting the limestone, the freshest of all those encountered in the different traverses, has been selected for chemical analysis.

The rock is a dark, greenish-gray, fine-grained, non-porphyritic trap, evidently highly micaceous. Under the microscope it is seen to be essentially a panidiomorphic aggregate of brown biotite and an imperfectly twinned plagioclase. A little orthoclase is almost certainly present. Magnetite and apatite are accessory. Quartz, kaolin, chlorite, and especially calcite are abundant secondary products. Neither augite, hornblende, nor olivine could be found.

A peculiarity of this dike is the occurrence of numerous small spherical aggregates of the plagioclase crystals, often mixed with quartz or calcite or with both. These aggregates apparently characterize the whole dike, from wall to wall. The little balls, a millimetre or less in diameter, are wrapped about with mica foils, much as phenocrystic leucites, as they enlarged, have often displaced small crystals of biotite in other types of rocks. The microscopic evidence is not decisive in the present case, but seems to indicate that the feldspar balls were formed during the crystallization of the rock and are not due to amygdaloidal filling. An account of other 'Kugelkersantite' may be found on page 665, in the second volume of Rosenbusch's Mikroskopische Physiographie der Massigen Gesteine (1907). Pirsson has described in detail the 'variolitic' facies of a minette occurring as dikes and thin sheets in the Little Belt mountains of Montana. From his description it is clear that we have a very close structural parallel, in these Montana minettes, to the kersantite just described. Before reading Pirsson's report the present writer had independently come to the conclusion that the feldspathic 'varioles' of the kersantite are of primary origin. The fact that Pirsson had announced his view in connection with the closely related lamprophyre, has given the greater confidence in the truth of the explanation.*

The chemical analysis (specimen No. 666) by Mr. Connor resulted as follows:—

Analysis of kersantite.

		Mol.
SiO₂	47·12	·790
TiO₂	·70	·009
Al₂O₃	15·65	·154
Fe₂O₃	2·66	·017
FeO	4·05	·056
MnO	·1C	·001
MgO	4·90	·122
CaO	8·56	·153
SrO	·10	·001
BaO	·14	·001
Na₂O	2·60	·042
K₂O	4·10	·044
H₂O at 110°C	·90	......
H₂O above 110°C	2·55	......
P₂O₅	·	·004
CO₂	· ·	......
	100·66	
Sp. gr.	2·740	

* Cf. L. V. Pirsson, 20th Annual Report, U.S. Geol. Survey, part 3, 1899, p. 532.

2 GEORGE V, A. 1912

The calculated norm is:—

Orthoclase..	24·46
Albite,.	15·20
Nephelite..	3·69
Anorthite..	18·90
Diopside..	16·68
Olivine..	5·97
Ilmenite..	1·36
Magnetite..	3·94
Apatite..	1·24
Water and CO_2..	9·14
	100·58

According to the Norm classification the rock enters the sodipotassic sub-rang, shoshonose, of the alkalicalcic rang, andase, in the dosalane order, germanare. Chemically it is nearer minette than a typical kersantite, but, by the older classification, the character of the feldspar places the rock in the kersantites.

Camptonite.—Only one occurrence of camptonite is known as a result of field study in the Boundary belt across the Selkirks. This rock, which microscopic study shows to conform well with the type camptonite, forms a wide but very poorly exposed dike cutting the Pend D'Oreille phyllite on the south side of the Pend D'Oreille river about 1,900 yards east of Waneta.

Odinite.—A half mile farther up the river and on the same bank, the phyllite is cut by a six-inch dike of a rock which appears to represent another occurrence of typical odinite as described by Rosenbusch in his last edition of the Mikroskopische Physiographie der Massigen Gesteine. This lamprophyre is a dark greenish-gray, compact rock with conspicuous though small phenocrysts of augite and others of labradorite. The microscope shows these to be embedded in a microcrystalline ground-mass composed essentially of very many minute prisms of hornblende, feldspar microlites, and less abundant granules of augite. A detailed description of this one thin dike, though composed of a relatively rare species of lamprophyre, is scarcely warranted in the present report.

APLITIC AND ACID APOPHYSAL DIKES.

Practically all of the granitic bodies in the Selkirks where crossed by the Boundary belt have sent tongues or apophyses into their respective country-rocks. These dikes show the familiar variation from quartz-feldspar aplites to the aschistic porphyries corresponding to the different types of plutonics. Other sills and dikes occur at distances too great to be regarded as necessarily apophyses from any visible stock or batholith, and in some cases it is not possible to determine whether these detached acid eruptives represent distinct periods of eruption. None of the bodies seems to demand special description. One of the dikes is cut by augite minette and by the analyzed hornblende-augite minette which occur on the western bank of the Columbia river about

300 yards south of the Boundary slash. The acid dike is a typical biotite granite porphyry. It is between 200 and 300 feet wide and is paralleled by other great dikes of similar material outcropping at low water in the islets of the river channel. They may be acid apophyses from the extensive Trail batholith toward which they strike; they are, however, noted here because their relation to the younger minettes is very clear. (See Figure 23.)

A white aplitic sill cutting the Pend D'Oreille phyllitic schist on the right bank slope of the South Fork of the Salmon, about 2.5 miles S. 30° W. of the summit of Lost mountain, may be mentioned on account of the unusual structure of the rock. It is slightly porphyritic with phenocrysts of quartz and sodiferous orthoclase. The ground-mass is partly the common panidiomorphic aggregate of quartz and alkaline feldspar (much sericitized) but contains quite numerous, small spherulites of alkaline feldspar which is developed in rosettes. A few grains of magnetite represent the only other constituent. The relations of this sill to the other granitic rocks of the range are unknown.

DIKE PHASES OF THE ROSSLAND AND BEAVER MOUNTAIN VOLCANICS.

The formation older than the Rossland and Beaver Mountain lavas are, naturally, cut by dikes which indicate vents for the lavas or the fillings of fissures connected with those vents. A few of these dikes have been found in localities where erosion has stripped away the volcanic cover and some of them have been microscopically examined. Among these, four types may be listed but it should be understood that the list does not exhaust the different varieties of the dikes genetically connected with the volcanics.

Just east of the large boss of Sheppard granite mapped on the Pend D'Oreille river, the schists are traversed by a fifty-foot, nearly vertical, north-south dike of porphyritic monzonite. The phenocrysts are stout prisms of augite up to 8 mm. in length. The essentials of the hypidiomorphic-granular ground-mass are orthoclase, microperthite, labradorite (Ab₁An₁), augite and biotite; the essentials are magnetite, apatite, zircon and a little interstitial quartz. The plagioclase crystals are characteristically clumped in the orthoclase mesostasis.

About three-quarters of a mile north of Old Fort Sheppard, where the mountain-spur projects through the terrace sands and gravels to the Columbia river, there are large outcrops of slaty and quartzitic rock which have been mapped as part of the Pend D'Oreille group. The crumpled and mashed slate is here cut by a 25-foot vertical dike of dark-gray hornblende-biotite monzonite striking N. 8° E. (visible at low water). Some 300 yards south of the Boundary slash on the same side of the river and at the water's edge, three dikes from ten to thirty-five feet wide and of macroscopic appearance somewhat similar to the monzonite were found to consist of hornblende-augite gabbro. In this type the feldspar is basic labradorite (Ab₁An₁), and alkaline feldspar is entirely absent; a few foils of biotite are accessory.

2 GEORGE V, A. 1912

Finally, a three-foot, north-south, vertical dike of highly amygdaloidal basalt, cutting the Pend D'Oreille phyllite about fifty yards west of the mouth of Twelve-mile creek, may be noted.

RELATIVE AGES OF THE ERUPTIVE BODIES.

The entire lack of paleontological evidence within the ten-mile belt makes it impossible to form a full chronological column for the formations occurring in this part of the Selkirks. It may be recalled that the Priest River terrane unconformably underlies the great Summit series, with a part of which (the Beehive formation) the Kitchener quartzite is believed to be equivalent. The Pend D'Oreille group overlies, with apparent conformity, the Summit series and, as will be further indicated in the next chapter, unconformably underlies the Rossland and Beaver Mountain groups of sediments and volcanics. The relative ages of the igneous rocks can be partly indicated through their relations to these sedimentary groups as well as through their relations to each other. The observed facts may be briefly summarized.

The intensely crushed Rykert granite batholith cuts the Priest River terrane, including bodies of metamorphosed hornblende gabbro which themselves cut the schists and quartzites of the terrane. The uncrushed and very rarely sheared Bayonne batholith cuts formations belonging to the Priest River terrane and Summit series respectively. The satellitic stocks believed to be contemporaneous with the Bayonne batholith cut the Pend D'Oreille group and one of them—the Bunker Hill stock—seems to cut the older members of the Rossland volcanic group. The Salmon river monzonite stock cuts the Pend D'Oreille schists and limestone. The abnormal hornblende granite at Corn creek cuts the Kitchener quartzite and is tentatively correlated with the Purcell sills. The minettes, kersantites, camptonites, and odinites cut the Pend D'Oreille schists or limestones and probably also cut the Rossland volcanics, since similar lamprophyres cut the Rossland monzonite stock which is almost certainly of the same general age as many of the Rossland lava flows. The peculiar porphyritic olivine syenite cuts the Rossland volcanics; in the next chapter the correlation of this syenite with the minettes will be indicated. The Sheppard granite cuts the Trail granodiorite which itself cuts the older members of the Rossland volcanics. Since a half dozen of the principal formations in these Selkirk mountains are more directly associated with fossiliferous sediments in the mountains across the Columbia river, the discussion of the final correlation of the Selkirk rocks will be postponed to the chapter dealing with the geology of the Rossland mountains. At this point it will be sufficient to anticipate that discussion by tabulating the Selkirk formations in their probable order of age:—

SESSIONAL PAPER No. 25a

Salmon River monzonite stock . . . .
Bayonne batholith and its satellitic stocks. . . . . } *Post-Eocene (Miocene?*

Sheppard granite stocks and dikes. . . . .
Lamprophyres, minettes, kersantites, odinite and camptonite
Aplitic dikes. . . . . } *Post-Laramie (Eocene?)*
Trail granodiorite batholith. . . . .

Beaver Mountain group. . . . .
Rossland volcanic group, with interbedded sediments (in part). } *Mesozoic (Cretaceous?).*
Monzonite, gabbro, and basaltic dikes cutting Pend d'Oreille group

UNCONFORMITY.

Rykert granite batholith. . . . . *Late Jurassic.*

Pend d'Oreille group. . . . . *Carboniferous? (and older?).*

Abnormal hornblende granite sill cutting Kitchener quartzite at
Corn creek. . . . . } *Middle Cambrian?*
Metamorphosed gabbro sills and dikes cutting Priest River terrane

Kitchener formation. *Middle Cambrian?*

Summit series. . . . . *Cambrian and 'Beltian.'*

UNCONFORMITY.

Priest River Terrane . . . . *Pre-Cambrian and pre-Beltian.*

CHAPTER XIII.

FORMATIONS OF THE ROSSLAND MOUNTAIN GROUP.

It will be recalled that, in the chapter on the nomenclature of the mountain ranges, the Rossland mountain group where crossed by the ten-mile Boundary Belt, is bounded on the east by the Selkirk Valley (Columbia river) and on the west by the meridional valley occupied by Christina lake and the 1 ver Kettle river. On the east the formations of the Rossland mountain group in several instances extend over into the Selkirk system. Of these the Pend D'Oreille series has already been described, as well as a few of the dikes cutting that series along the western bank of the Columbia river. The Trail batholith, Sheppard granite, Rossland and Beaver Mountain volcanic groups, and small bodies of a peculiar porphyritic olivine syenite are represented on both sides of the Columbia and will be described in the present chapter. The western topographic limit of the Rossland mountain group is also, within the limits of the Boundary belt, a clean-cut and convenient line of division between the geological formations of the Rossland and Midway-Christina mountain groups. (See Maps No. 8 and 9.)

From the Columbia to Christina lake igneous-rock formations dominate very grea   Sedimentary rocks appear only in small patches, and are nearly always much deformed and metamorphosed. Though there are good reasons for believing that these rocks are chiefly if not altogether late Paleozoic or post-Paleozoic in age, fossils are almost as rare as they are in the formations of the Rocky Mountain Geosynclinal. The writer has been able to add but little to the stratigraphic information secured by McConnell, Brock, and others who have made studies in the region. However, the interpretation given the few scattered facts in hand differs somewhat from that adopted by these observers.

The older sedimentary formations will be describ- first. They include, besides the small area of the Pend D'Oreille slates, phyllites, quartzite, and limestone near the Columbia, a small patch of obscurely fossiliferous limestone associated with chert in Little Sheep creek valley; fossiliferous limestone occurring with the older traps north of Rossland; an intensely deformed series of limestones, quartzites, and schists sectioned by the railway line east of Christina lake and named, for convenience, the Sutherland schistose complex; and a few small outcrops of old-looking quartzite and argillitic rocks intimately associated with the Rossland volcanics.

A very limited exposure of fossiliferous (plant-bearing) argillite, probably of Mesozoic age, will then be described. The youngest sedimentaries observed in this part of the Boundary belt are conglomerates and sandstones which, again from very imperfect fossil evidence, seem to be of early Tertiary or mid-Tertiary age; these beds form four small patches at or near the Boundary line.

319

2 GEORGE V, A. 1912

The igneous formations to be treated include those which have been named by McConnell and Brock the Rossland and Beaver Mountain volcanic groups; and those which are referred to by the present writer as the Trail batholith; the Sheppard granite (stocks and dikes); the Coryell syenite batholith with its satellitic dikes, and a satellitic chonolith of syenite porphyry; the Rossland monzonite; several bodies of gabbroid and ultra-basic intrusives; and certain of the numerous dikes which have certain special petrographic interest.

At the time when the writer made his examination of the Rossland mountain group it was understood that the Geological Survey of Canada was planning a detailed study of the Rossland camp and its vicinity. Accordingly, very little work was done in the region of the town and, in fact, no attempt was made to plan an exhaustive report for the region between Sophie mountain and the Columbia. Specimens of the rocks were collected, but many of the field relations could not be decided in the limited time which it seemed advisable to devote to this part of the Boundary belt. Nearly all of McConnell's contacts, as published in the Trail sheet, were followed up and verified. For the rest the present chapter can claim to be no more than a report of progress on the geology of these unusually complicated mountains.

PALEOZOIC FORMATIONS.

Carboniferous Beds in Little Sheep Creek Valley.—In the bottom flat of Little Sheep creek valley, about 1,000 yards north of the Boundary line and on the west side of the creek, there is a low hill of limestone surrounded on all sides by alluvium. The limestone is of blue-gray to white colour and is much brecciated and highly crystalline. It contains cherty and quartz lenses and true quartz veins. The attitude of the bedding is obscure, observed strikes ranging from N. 55° E. to N. 80° E., with an average northerly dip of about 60 . The limestone contains numerous, poorly preserved crinoid stems which are of some value as pointing to the probability that the limestone is of Paleozoic age. Across the creek there are several large outcrops of cherty quartzite also greatly deformed, with average strike, N. 35° E. and dip, 90°. That rock extends 200 feet vertically up the steep eastern slope of the valley, where it is unconformably overlain by a coarse breccia (probably a volcanic explosion breccia) containing fragments of the same obscurely fossiliferous limestone and chert as that just described. The breccia is part of the Rossland volcanic formation, which has here an average strike, N-S. and dip, 35° E. From the composition of the breccia and from the stratigraphic relations the Rossland volcanics as represented are clearly unconformable to the Paleozoic strata. The latter seem, in fact, to be part of the foundation on which the volcanic mass was spread.

During his mapping of the Trail sheet McConnell found in the similar breccia outcropping on the opposite side of this valley, fragments of marble bearing the fossil remains of a species of *Lonsdalia* and the marble was referred by Dr. Whiteaves to the Carboniferous. It would seem simplest directly to

SESSIONAL PAPER No. 25a

correlate the limestone in place with the limestone fragments in the breccia on each side of the valley, and the formation, including the limestone and chert, is tentatively placed in the Carboniferous system.

Carboniferous Limestone in the Rossland Mining Camp. In 1905 Brock discovered, in a limestone band interbedded with andesitic greenstone at the O.K. mine, four miles north of the last mentioned locality, certain fossils which have been referred to Carboniferous species.

Sutherland Schistose Complex.—A group of metamorphic rocks, exposed in the railway cuttings between Cascade and Coryell stations, were sectioned during the season of 1902. Although nearly a week was spent on the section, the results of the structural study were meagre. The oldest rocks of the section consist of highly crystalline schists of sedimentary origin. With these are associated many irregular bands of gneissic, gabbroid rocks and amphibolites and sheared hornblende porphyrites, all of which represent greatly altered basic intrusives. The metamorphosed sedimentary rocks are now represented by garnetiferous schist, sericite schist or phyllite, biotite-epidote schist, actinolite-biotite schist and andalusite-biotite schist. Massive, often brecciated, greenish quartzite and at least two large pods of white to light gray marble are interbedded with the schists.

Structurally the complex is characterized by utter confusion. Neither bedding-planes nor planes of schistosity preserve a steady attitude for more than a few score or hundreds of feet together. The section is located in a zone of maximum dislocation, a zone now followed by the deep trough of Christina lake. The immense alteration of these formations is further due to the intrusion of numerous large bodies of acid and basic igneous rock, including various gabbros and peridotites as well as the great Coryell syenite batholith.

No trace of a fossil was found in the sedimentaries and it is still impossible to correlate them with known horizons. The quartzite and limestone associated with the schists are, in general, similar to the quartzite and crinoidal limestone of Little Sheep creek valley and to staple phases of the Pend D'Oreille group. All of them are possibly of Carboniferous age. The gabbroid and peridotitic masses cutting the schists are evidently of more recent date; some of them show neither crushing nor even appreciable straining under the microscope. Three of these basic intrusive bodies will be briefly described below; a microscopic description of the schists themselves is scarcely warranted by any special petrographic interest they possess.

Summary. In conclusion, it may be noted that some at least of these old looking metamorphosed sediments are almost certainly of Carboniferous age. Others may be either pre-Carboniferous or else Triassic, if not as late as Jurassic. For the present the writer follows the tradition of McConnell and Brock, in placing all of these formations in the Paleozoic. Whatever the age of the sediments, some of them seem to be contemporaneous with the massive greenstones and metamorphosed ash-beds of andesitic sort, and it is highly probable that the

2 GEORGE V, A, 1912

greenstones occurring in the Pend D'Oreille group (especially those near the Columbia river) are of the same age. The quartzites and slaty rocks of the Pend D'Oreille group are almost if not quite indistinguishable both in composition and in degree of metamorphism from the quartzites and slates interbedded with the greenstones of the Rossland mountains. The Pend D'Oreille marbles are lithologically identical with the obscurely fossiliferous limestones just described. As the best working hypothesis, therefore, the writer is inclined to believe that the western slope of the Selkirk range and the eastern half of the Columbia system are underlain by residuals of a very thick upper Paleozoic, probably Carboniferous, series which represents the oldest sedimentary rocks of those parts of the Boundary belt. It will be seen that the same series probably has similar fundamental relations in the Midway and more westerly mountain groups.

MESOZOIC SEDIMENTS AT LITTLE SHEEP CREEK.

At Monument 175 in Little Sheep creek valley, erosion has laid bare a considerable thickness of stratified rocks which are evidently much younger than the marbles and quartzites farther up the valley. The exposures are not good but, since these younger rocks are also obscurely fossiliferous, the field observations so far made may be detailed. At the Boundary monument the steep slope of Malde mountain is seen to be largely underlain by black and red argillite, enclosing thin beds of gray sandstone and of angular conglomerate, as well as a number of layers of sandstone which is described in the field notes as hard black quartzite. The quartzite is sulphide-bearing. These beds are greatly deformed, the argillite specially showing frequent changes of strike and dip in short distances both up the slope and along its foot. The more rigid sandstone beds tend to have a fairly steady strike of N. 0°–10° E., with an average dip of from 35° E. to 90°. The series, chiefly argillitic, continues eastward to a contour about 600 feet above Little Sheep creek, and there it appears to dip under the volcanic breccias of Malde mountain. This general eastward dip appears to characterize the series throughout its extent of 600 yards up the valley from the Boundary slash. The exposures south of the line did not promise useful results and the beds were not followed in that direction. The exposures are likewise very poor on the west side of the creek, but the shale-sandstone series seems to extend on the Sophie mountain slope at least 500 feet above the creek. The argillite is there greatly crumpled, but probably strikes in the average direction, N. 65° E., with dip high to the northwest.

The series seems thus to be at least 600 feet thick and to have the attitude of a broken and mashed anticline plunging to the north, carrying the sediments beneath the Malde mountain and Sophie mountain breccias and lavas. The field relations are, however, so obscure that this conception must be regarded as only suggestive and by no means proved to be correct.

At the rock-bluffs along the railway track and on each side of the Boundary slash, a number of very poorly preserved remains of plants were found in the shales. These fossils were submitted to Professor D. P. Penhallow, who identified them 'as the rachises of a fern, in all probability of *Gleichenia* (*gilbert*-

PLATE 32.

Eruptives composed of Rossland volcanics, Record Mountain ridge, west of Rossland.   Old Glory Mountain in the background.

thompsoni), and tentatively correlates the beds with the lower Cretaceous *Gleichenia*-bearing strata on the Pasayten river.* The only other information in hand on this question of age is that based on the condition of the stratified series. It is, apparently, too greatly deformed to be placed in a post-Eocene period.

"... on the other hand, the degree of metamorphism is too low to warrant our referring the series to the Paleozoic. Either a Mesozoic or Eocene date would be preferable to either of those alternatives. For the present, it seems best to consider the beds broadly as of Mesozoic age.

ROSSLAND VOLCANIC GROUP.

GENERAL DESCRIPTION.

From the Salmon river to the Kettle river at Cascade, a distance of forty miles, the ten-mile Boundary belt contains an irregular though continuous band of basic volcanic rocks. This band covers about 150 square miles of the belt and is part of a volcanic area in the West Kootenay district of British Columbia aggregating 500 square miles. West of the Columbia river the volcanics are developed on the United States side of the Boundary but how extensively is not known. (See Plate 32.)

The entire volcanic area is highly accented by basic and acid plutonic masses which, in general, are younger than the volcanics and cut them. Long continued erosion has revealed many of the dikes, stocks, and batholiths, so that the mapped contact-lines of the effusive rocks are extremely sinuous. Owing to severe orogenic stresses the lava flows, ash-beds and breccias usually have high dips and complicated structures. Most of these rocks are altered by crush-metamorphism and contact-metamorphism. They are often involved most obscurely with the Paleozoic sediments just described and also with younger strata which are generally unfossiliferous. The differentiation of the lavas on the ground of geological age cannot as yet be carried out systematically.

It is certain that the volcanics were erupted in at least two different periods. The oldest lavas, ash-beds, and agglomerates seem to have been extruded contemporaneously with the Carboniferous limestones, cherts, and slaty rocks, and have since, through regional metamorphism, been converted into massive and schistose greenstones which often keep their porphyritic structure more or less plainly preserved. No chemical study has been made of these older volcanics, and microscopic analysis is generally helpless in the attempt to refer them to definite types of lava. From their general habit and from the nature of the alteration and metamorphic products it appears probable that the whole series of Carboniferous extrusives should be classed with the common augite andesites and basalts. In his reconnaissance of the region during the preparation of material for the Trail sheet, McConnell recognized the Carboniferous age of these rocks and called the more massive, porphyritic

---

* D. P. Pen... w, Transactions, Royal Society of Canada, ser. iii, Vol 1, pp. 290 and 329, 1908.

2 GEORGE V, A. 1912

phase 'augite porphyrite.' One of the chief difficulties in mapping these rocks lies in the fact that the distinct and much younger augite latites are extremely difficult to distinguish in the field from the older augite andesites. There are, moreover, true augite andesites and basalts belonging to the younger series of lavas and the problem of differentiating them from the Carboniferous lavas is in many cases not to be solved.

Since, therefore, most of the volcanic belt has defied clear-cut division on the map, the writer has followed McConnell and Brock in colouring under one legend, the 'Rossland Volcanic Group,' most of the volcanic formations occurring in the Boundary belt between the Salmon river and Christina lake. Between the Columbia river and Christina lake the larger part of the volcanic masses have been found to belong to the family of latites, although there are some flows of true basalt and augite andesite associated with them. In the Beaver Mountain region there is a considerable area of relatively unaltered lavas and tuffs which nowhere seem to have any latitic phase. Chiefly because of their relatively fresh and recent appearance, Brock has already separated this series of volcanics, and he has given the series the name, 'Beaver Mountain Group.' The petrographic distinction just noted further justifies our following Brock in his mapping, and this part of the whole volcanic area will be separately described, as well as separately mapped in the accompanying sheet. If, in the future, the Rossland volcanic group can be analyzed with sufficient accuracy to permit of its subdivision on the map, it would be appropriate to reserve the name 'Rossland Volcanic Group' for the latitic lavas and associated pyroclastics, for these seem to be the dominant extrusives of the area.

PETROGRAPHY OF THE LAVAS AND PYROCLASTICS.

The writer has collected about one hundred specimens of the freshest and most typical rocks of the volcanic belt, and from them about eighty-five thin sections were cut. It was not until these had been microscopically examined that the lithological diversity of the lavas became fully apparent. Seven varieties of latite, olivine basalt, olivine-free basalt, augite andesite, and possibly picrite (corresponding to harzburgite among the plutonic rocks and described among the latter) have been recognized among the less altered lavas. The most abundant types are probably the augite latite and biotite-augite latite. These are respectively transitional into olivine-augite latite and biotite latite. Hornblende-biotite latite and hornblende (-augite) latite and a specially femic augite latite are of more local occurrence. The true basalts are far less common than one would suspect in the field, since so many of the latites have basaltic habit. True augite andesite is probably more abundant than the basalts.

Augite Latite.—Massive lava belonging to this variety was found at widely spaced localities, among which are specially noted the area between Castle mountain (southeast slope) and Record mountain ridge, the divide between Malde and Little Sheep creeks, and the bluffs on the west side of the Columbia river about four miles north of the line. The following brief description of a typical, relatively unaltered phase relates to one of the younger flows occurring on the unnamed conical peak west of the Murphy creek-Gladstone trail and about two miles north of Stony creek. The volcanic rocks are there exceptionally well exposed above tree-line, where thick sheets of highly porphyritic latite alternate with more basaltic sheets and with coarse agglomerates composed of these lavas. The latite when fresh is a deep greenish-gray to almost black rock bearing abundant phenocrysts of tabular plagioclase up to 3 mm. in greatest diameters and of smaller, stout prisms of greenish-black pyroxene.

Microscopic examination shows that the rock is uncrushed, the phenocrysts being unstrained and almost perfectly unaltered. The plagioclase is the more abundant. On (010) and in the zone of symmetrical extinctions for simultaneous Carlsbad-albite twins, individual crystals give extinction angles appropriate to the series from labradorite, $Ab_1 An_1$, to bytownite, $Ab_1 An_2$. Occasionally one of these basic individuals is surrounded with a narrow rim of orthoclase. The average plagioclase phenocryst has about the composition of labradorite, $Ab_1 An$. The pyroxene is a common, non-pleochroic, pale greenish augite of diopsidic habit.

The ground-mass has been somewhat altered, with the generation of uralite in small needles, zoisite in rather rare granules, chlorite, abundant biotite, and more sericitic mica in minute foils and shreds. Orthoclase was not certainly detected in the ground-mass, which was originally hyalopilitic, with plagioclase microlites embedded in glass. Magnetite and apatite occur in the usual well-formed crystals.

A specimen collected at this locality (No. 543) and answering to the foregoing description has been analyzed by Mr. Connor, with result as follows. (Table XIX., Col. 1.):—

Table XIX.—Analyses of augite latites, Rossland district and Sierra Nevada.

	1.	1a. Mol.	2.
SiO_2	54·54	·909	56·19
TiO_2	·96	·012	·69
Al_2O_3	18·10	·177	16·76
Fe_2O_3	1·14	·007	3·05
FeO	4·63	·064	4·18
MnO	·10	·001	·10
MgO	4·56	·114	3·79
CaO	5·85	·104	6·53
SrO	·15	·001	tr.
BaO	·21	·001	·19
Na_2O	3·38	·055	2·53
K_2O	5·44	·058	4·46
H_2O at 110°C.	·10	....	·34
H_2O above 110°C.	·50	....	·66
P_2O_5	·46	·004	·55
	100·12		100·02
Sp. gr.	2·745		

2 GEORGE V, A. 1912

The calculated norm is:—

Orthoclase	32·25
Albite	26·20
Nephelite	1·42
Anorthite	17·79
Diopside	6·87
Olivine	10·18
Ilmenite	1·82
Magnetite	1·62
Apatite	1·24
Water	·60
	99·99

According to the Norm classification the rock enters the sodipotassic sub-rang, monzonose, of the domalkalic rang, monzonase, in the dosalane order, germanare. The mineralogical and chemical composition and structure all perfectly match the typical augite latite of Table mountain, California, as originally described by Ransome.* The analysis of the more basic phase of the Table mountain flow is entered in Col. 2 of the foregoing table.

From the fresh rock just described all transitions to profoundly altered phases are represented in the area. The latite has often been transformed into a dark green, massive rock, still showing its porphyritic character by the presence of broken and altered feldspar phenocrysts or of uralitic pseudomorphs after the augite. For the rest the completely changed rock is, in thin section, seen to be a confused mass of epidote, calcite, quartz, chalcedony, chlorite, biotite, uralitic and actinolitic amphibole, zoisite, pyrite, etc., in ever varying proportion. Sometimes, though not often, an amygdaloidal structure is preserved. This is not so much because it has been obliterated by metamorphism as because these lavas were largely non-vesicular when first consolidated.

Augite-biotite Latite.—This type of massive lava is at least as important in the area as the augite latite. As above noted, the two varieties grade into each other, and the only noteworthy persistent difference is the absence or presence of biotite among the original phenocrysts. Biotite also often occurs in minute, shreddy foils in the ground-mass but it appears to be general of secondary origin. The phenocrystic biotite is of a deep, rich brown colour and has powerful absorption; its optical angle is probably under 2°. The other phenocrysts, the accessories, and the ground-mass have characters essentially identical with those of the augite latite.

No perfectly fresh specimen of the augite-biotite latite was secured. One of the least altered ones, collected on the ridge joining Record and Sophie mountains, at a point two miles north of the Dewdney trail (No. 456), has been analyzed by Mr. Connor. It is a compact, deep greenish-gray rock with numerous small phenocrysts of labradorite (averaging about Ab_1An_2), biotite, and uralitized augite. These minerals are embedded in an abundant, originally hyalopilitic, greenish base. The latter is chiefly devitrified glass. Its advanced

* F. L. Ransome, American Journal of Science, Ser. iv. Vol. 5, 1898, p. 359.

alteration has led to the formation of kaolin, uralite, sericite, epidote, zoisite, chlorite, carbonate, and a little quartz. Orthoclase was apparently never individualized.

Mr. Connor's analysis resulted as follows (Table XX., Col. 1):

Table XX. -Analyses of augite-biotite latite.

	1.	1a.	2.
		Mol.	
SiO₂	59·06	·984	62·33
TiO₂	1·08	·014	1·05
Al₂O₃	16·24	·159	17·35
Fe₂O₃	·43	·003	2·98
FeO	4·88	·068	1·63
MnO	·20	·003	·08
MgO	3·51	·088	1·05
CaO	5·59	·100	3·23
SrO	·12	·001	·05
BaO	·11	·001	·24
Na₂O	2·84	·046	4·21
K₂O	3·95	·042	4·46
H₂O at 110°C	·21	....	·44
H₂O above 110°C	·13	....	·75
P₂O₅	·21	·001	·29
CO₂	·70	....	....
FeS₂	....	....	·08
C	....	....	·11
	99·32		100·33
Sp. gr	2·796		

In the Norm classification the rock enters the sodipotassic subrang, shoshonose, of the alkalicalcic rang, andase, in the dosalane order germanare. The norm is as follows:—

Quartz	8·64
Orthoclase	23·35
Albite	24·10
Anorthite	19·74
Hypersthene	12·78
Diopside	6·40
Ilmenite	2·13
Magnetite	·70
Apatite	·31
H₂O and CO₂	1·10
	99·25

In the older classification this variety is clearly a biotite-augite latite. In Col. 2 of Table XX, the analysis of one of Ransome's types, that from near Clover Meadow, California, is entered. The alkalies are a little lower in the British Columbia rock, but the respective differences are too small to cause doubt as to the classification.

2 GEORGE V, A. 1912

Augite-olivine Latite.—This type has been identified at only two localities in the Boundary belt. On Record mountain ridge it is interbedded with the chemically analyzed biotite-augite latite; it also occurs on the top of the broad ridge west of Mable ridge at a point about a mile and a half north of the Boundary line. The specimens collected at these places are comparatively fresh and are uncrushed.

Macroscopically, there is little to distinguish these rocks from the more common augite latite. The colour, grain, and general habit is the same. The phenocrysts are augite, olivine, and labradorite (averaging $Ab_1 An_1$). The ground-mass may be cryptocrystalline, devitrified-glassy, or microcrystalline, with greater or less development of microlitic augite and labradorite. The accessories and secondary products are the same as those in the augite latite, except that a little phenocrystic biotite is developed in the specimen from Record mountain ridge.

That specimen (No. 465) has been analyzed by Mr. Connor. The microscope showed that the augite is here somewhat uralitized and the olivine partly serpentinized, while the plagioclase is very fresh. The hyalopilitic base bears microlites of labradorite, magnetite, apatite, and possibly orthoclase; most of the ground-mass is, however, a glass which is turbid through the very abundant generation of sericitic mica and other secondary products. The specific gravity of three specimens from this locality varies from 2·700 to 2·751; the higher value is the more reliable since it refers to the freshest specimen.

From the chemical analysis it is clear that this latite verges on augite-andesite.

Analysis of augite-olivine latite.

		Mol.
SiO_2	58·67	·978
TiO_2	1·00	·013
Al_2O_3	15·67	·154
Fe_2O_3	2·85	·018
FeO	3·28	·046
MnO	·11	·001
MgO	3·86	·097
CaO	5·33	·095
SrO	·09	·001
BaO	·11	·001
Na_2O	4·77	·077
K_2O	3·06	·033
H_2O at 110°C	·02	....
H_2O above 110°C	·54	....
P_2O_5	·16	·001
	99·54	
Sp. gr	2·751	

In the Norm classification the rock enters the dosodic subrang, akerose, of the domalkalic rang, monzonase, in the dosalane order, germanare. The norm is as follows:—

SESSIONAL PAPER No. 25a

Quartz..	3·90
Orthoclase..	18·35
Albite..	·0·35
Anorthite..	12·23
Diopside..	11·02
Hypersthene..	6·75
Magnetite..	4·18
Ilmenite..	1·98
Apatite..	·31
H₂O..	·56
	99·66

Hornblende-augite Latite.—A fourth type was collected at the 3,100-foot contour on the slope due east of Sayward railway station. It is a dark gray rock with conspicuous, lustrous, black prisms of phenocrystic hornblende in a gray-tinted ground-mass. The acicular hornblendes vary from 1 mm. to 4 mm. in length and are arranged in roughly fluidal fashion. They are accompanied by a subordinate number of idiomorphic augite prisms, also phenocrystic but first discovered in thin section. The ground-mass is a rather confused, microcrystalline aggregate of the same bisilicates and feldspar. In this case there can be no question that orthoclase forms a large proportion of the ground-mass feldspar microlites, which for the rest are probably labradorite. Magnetite, pyrite, pyrrhotite, and a little titanite are accessory minerals; calcite, chlorite, epidote, kaolin, and sericite are secondary products. This specimen (No. 557) is comparatively fresh. Its analysis, by Mr. Connor, resulted in the form shown in Table XXI., Col. 1.

Table XXI.—Analyses of hornblende-augite latite.

	1.	2.	2a. Mol.
SiO₂..	52·17	52·17	·870
TiO₂..	·80	·80	·010
Al₂O₃..	16·59	16·59	·163
Fe₂O₃..	8·32	1·86	·012
FeO..	not det.	3·74	·052
MnO..	·11	·11	·001
MgO..	3·87	3·37	·097
CaO..	8·25	8·25	·147
SrO..	·05	·05	....
BaO..	·15	·15	·001
Na₂O..	3·91	3·91	·063
K₂O..	4·00	4·00	·043
H₂O at 110°C..	·13	·13	....
H₂O above 110°C..	1·17	1·17	....
P₂O₅..	·24	·24	·001
CO₂..	·56	·56	....
S..	1·37		....
FeS₂ and Fe₇S₈..	....	2·31	....
	01·69	99·91	
Sp. gr..	2·852		

On account of the presence of pyrrhotite the ferrous oxide could not be directly determined. The proportion of this oxide was estimated, as shown in

2 GEORGE V, A. 1912

Col. 2. First, an amount of Fe_2O_3 representing sufficient Fe to satisfy the sulphur present, was apportioned. The sulphides of iron were arbitrarily considered as half pyrite and half pyrrhotite. The remaining Fe_2O_3 was calculated to represent the FeO and FeO, of this rock by assuming that these oxides occur in the average proportions which they have in other analyzed Rossland lavas (Nos. 456 and 543). The analysis, so recalculated, is entered in Col. 2; the corresponding molecular proportions are shown in Col. 2a.

The norm was calculated from the values given in Cols. 2 and 2a, with result as follows:—

Orthoclase	23·91
Albite	23·06
Nephelite	5·40
Anorthite	15·85
Diopside	19·71
Olivine	3·09
Ilmenite	1·52
Magnetite	2·78
Apatite	·31
Pyrite and pyrrhotite	2·31
H_2O and CO_2	1·86
	99·80

According to the Norm classification the rock enters the sodipotassic subrang, monzonose, of the domalkalic rang, monzonase, in the dosalane order, germanare.

According to the older classification the rock is both mineralogically and chemically a hornblende-augite latite. It was nowhere seen to be vesicular but, on account of its persistent fine grain, it is believed to belong to a massive flow rather than to an intrusive body.

A somewhat similar porphyritic rock, perhaps intrusive, crops out on the Dewdney trail where it crosses the low ridge between Sophie mountain and (the western) Sheep creek. Orthoclase is very abundant in the ground-mass of this rock.

Hornblende-biotite Latite.—A fifth type of latite was collected on the mountain spur running up from Bitter creek southeastward at a point four miles due east of the railroad station at Cascade. The rock crops out at the 3,300-foot contour as a massive, gray to greenish gray, porphyritic, non-vesicular trap, and seems to extend uninterruptedly along the ridge to the 4,300-foot contour, where it is interbedded with hard bands of fine basic ash. Continuing southeastward to the Boundary line, the same lava is seen interbedded with coarse quartz conglomerate. This type of lava was not identified at any other locality.

The phenocrysts are brown biotite and green hornblende, the former predominating. The determinable feldspar, averaging labradorite, $Ab_2 An_3$, is confined to the ground-mass where it forms minute, tabular, twinned crystals in great number. A green shreddy biotite of low absorptive power and evidently of different composition from the phenocrystic mica, is extensively developed among the plagioclase microlites. This green biotite also forms complete pseudomorphs after the hornblende phenocrysts, so that it is doubtful that any of the mica of the ground-mass is original. Orthoclase was not observed and it is practically impossible to determine the character of the original ground-mass, so great has been the alteration of the rock. The other secondary products, as well as the accessory minerals are like those in the augite-biot... latite, to which the hornblende-biotite latite must be chemically quite similar.

Biotite Latite.—A sixth type represents a lava which is macroscopically like a mica andesite, but under the microscope shows features relating it to the latites just described. In its present condition it is a greenish-gray rock of decidedly lighter tint than the great majority of the Rossland lavas. Biotite and labradorite (averaging about Ab_1An_1) are the only phenocrysts. The base was probably once largely glass in which microlites of labradorite and more irregular ones of (probably) orthoclase were embedded. The ground-mass is now abundantly charged with secondary sericitic mica a... some quartz which is doubtless also of secondary origin.

This rock has not been chemically analyzed but the analysis of a fresh specimen would correspond to many mica andesites which are rich in potash. In view of the intimate association of this type with the undoubted, analyzed latites, it seems best to regard the rock as a salic latite rather than a true andesite, though it must be on the border-line between the two species.

Femic Augite Latite.—Finally, an altered lava which seems to represent an opposite pole in the differentiation of the latitic magma, was found on the eastern slope of the hog-back ridge in (the western) Sheep creek valley. A second but more doubtful occurrence was noted on the ridge between Boundary monument No. 170 and the Coryell batholith. The chief mineralogical difference between this type and the analyzed augite latite consists in a great increase in the number of augite phenocrysts, a corresponding decrease in the abundance of labradorite phenocrysts (which may entirely fail in the thin section), and apparently a decrease in the relative amount of the ground-mass. The accessory and secondary minerals are the same as those noted for the augite latite; orthoclase was not observed in the ground-mass, which in all the collected specimens has largely gone over to green biotite and sericitic mica.

Comparison with Sierra Nevada Latite and with Average Monzonite.— Before noting the other types of lava in the Rossland group it will be instructive to review their classification in terms of the chemical constitution of the original latites as defined by Ransome. In Table XXII., Col. 1, the average of the Rossland latites is given, and in Col. 2 the average of six typical latites from California. Col. 3 shows the average of all ten latites and Col. 4, the average of the twelve monzonites recorded in Osann's compilation of chemical analyses throughout the world. The last three averages have been reduced to 100 per cent. In making the average of the Rossland latites the augite latite and augite-biotite latite were considered as of equal weight and their average was weighted as four against the average of the analyses of the hornblende-augite latite and olivine-augite latite which together were weighted as unity. This weighting corresponds approximately to the relative volumetric importance of the different types in the Rossland district.

Table XXII. Comparisons of latites and monzonite.

	2	3	4
	Rossland latites.	Sierra Nevada type latites.	Average of 1 and 2.	Average of world-monzonite.
SiO_2	56 52	58 70	57 85	55 25
TiO_2	1 00	1 06	1 02	60
Al_2O_3 ..	16 96	16 75	16 76	16 53
Fe_2O_3	1 10	3 00	2 44	3 03
FeO	4 51	3 03	3 08	4 37
MnO	11	06	09	15
MgO	4 01	2 50	3 09	4 20
CaO ..	5 03	5 02	5 53	7 19
SrO	13	02	05	
BaO ..	16	19	17	
Na_2O	3 36	3 51	3 61	3 48
K_2O	1 46	4 58	4 02	4 11
H_2O-	11	30	23	66
H_2O	48	82	74	
P_2O_5	31	46	38	43
CO_2	34	......	14	
FeS and Fe_2S_3 ..	23	......	......	......
	99 75	100 00	100 00	100 00

The close correspondence of the Rossland and Sierra Nevada average shows an essential identity of the magmas from which the respective lava crystallized; the justice of correlating the Rossland rocks with the latites i clearly demonstrated. That latite should, as pointed out by Ransome, l considered as the extrusive form of monzonite is indicated in the comparis of Cols. 3 and 4. The two are not strikingly divergent at any point, yet the are differences which together form the exact analogue of the differenc between the world's average syenite and trachyte, or the difference betw the world's average granite and rhyolite, or, in fact, between the world averages of any of the principal plutonic types and its generally recogniz effusive equivalent. In all these cases (as proved by the writer through actu calculation; see chapter XXIV.), the effusive rock is the more salic and som what more alkalic; magnesia, lime, and iron oxides are characteristically low in the surface lava than in the corresponding plutonic. In all the cases it would seem as if magmatic differentiation tends to be mo perfect when magma approaches and reaches the earth's surface, th more salic pole naturally developing at the top of the volcanic vents whe it may be erupted as true surface lava. Without further discussing this theo tical point we may conclude that petrography will gain by accepting full Ransome's highly useful conception of the latites as forming a group important among lavas as the monzonites are important among the pluton types.

Augite Andesite.—At two localities in the volcanic area, lavas belonging to the common species, augite andesite, have been identified. This rock may occur at many other points but its macroscopic similarity to the augite latite makes its discovery very uncertain. As already noted, the writer believes that this type as well as the true basalts are subordinate to the latites in the region covered by the Boundary survey map.

A specimen belonging to what seems to be a massive flow was collected near the Coryell syenite contact on the ridge running northward from Monument 171. It will be observed that this ridge is just west of the body of enstatite-olivine rock which is mapped as harzburgite but may represent a picrite, i.e., an extrusive form of the harzburgite magma. The augite andesite has, notwithstanding its altered character, all the ear-marks of this species of lava. The phenocrysts of augite and labradorite are embedded in a much altered ground-mass in which microlites of those minerals can be detected as the essentials. The alteration products are uralite, chlorite, and quartz and thus differ essentially from those which are so characteristic of the latites. The evidence is quite clear that the ground-mass is not rich in potash.

True augite andesite was found to compose most of the blocks in a very coarse agglomerate capping a ridge lying between Monument 172 and the confluence of Santa Rosa creek and (the western) Sheep creek. The larger blocks are there from three to four feet in diameter. A few fragments in the breccia are exceptional for this volcanic series in being of acid composition, a biotite-quartz porphyry.

Basalts.—A typical olivine basalt was discovered on Mt. Catmorac, the broad divide between Malde creek and Little Sheep creek. It seems to form there a very thick and massive flow interbedded in specially voluminous basic breccias. The phenocrysts are labradorite, augite, and olivine. The ground-mass is the usual holocrystalline aggregate of augite and feldspar. The feldspar is here much more altered than the femic minerals; this is just the contrary of the rule with the latites, in which the feldspars are almost always not so badly altered as the augite, hornblende, or olivine.

In the col on the trail southwest of Lake mountain an equally typical olivine-free basalt forms at least two thick flows separated by a two-foot layer of basic tuff. The contact-planes show here a strike of N. 10° E. and a dip of 75° to the westward; the series has evidently been greatly deformed at this point. The distinction of this basalt from the augite latite is easily made, as the fairly fresh ground-mass is the typical diabasic. A very similar rock though distinctly vesicular, was collected at the edge of the volcanic area on the west side of Twelve-mile creek; it may, however, easily belong to the series of lavas included in the Beaver Mountain group.

Flow of Liparitic Obsidian?—Throughout the whole area covered by the Boundary belt in the Rossland mountains, acid lavas are extremely rare. Fragments of biotite-quartz porphyry, probably a liparite of extrusive origin, are, as we have seen, enclosed in the coarse agglomerate at one point. The only other

2 GEORGE V, A. 1912

possible occurrence of acid lava observed by the writer was noted as a 30-foot intercalation in the Cretaceous (?) argillite at the crossing of the Boundary line and Little Sheep creek. This is a white, aphanitic, massive rock showing a fairly distinct banding which in the field was taken for bedding. The rock weathers yellow to brown. Under the microscope the one thin section made from the rock showed no certain proof of the origin of the rock but the general appearance was that of a devitrified, partially spherulitic obsidian. The very small spherules seem to be poorly developed radial aggregates of quartz and feldspar and their matrix is a very fine-grained granophyric intergrowth of the same minerals. No other minerals have been certainly determined in the thin section. The banding may be a flow-structure.

Since its characters are obscure and largely negative the writer must regard his reference of this rock to the acid obsidians as tentative.

Tuffs and Agglomerates.—Most of the area occupied in the Boundary belt by the Rossland volcanic group is underlain by massive flows of the latites, andesites, and basalts. A considerable tract, estimated as covering at least fifteen square miles, is, however, underlain by a thick, more or less continuous mass of coarse volcanic agglomerate. This pyroclastic composes the majority of the outcrops between Lake mountain on the east and the top of Sophie mountain on the west, besides extending for several miles along Record mountain ridge, northward, from the Boundary line.

The constituent fragments are angular to subangular, ranging in size from dust-particles to blocks four feet in diameter. The deposit is usually without stratification but consists of a tumultuous, massive aggregation of fragments which were evidently never sorted by water-action. Most of them are composed of augite latite, biotite-augite latite, or, to a less extent, of basalt and augite andesite. Besides these, abundant angular blocks of fossiliferous white crystalline limestone occur in the breccia throughout the whole eastern slope of Sophie mountain, and are likewise conspicuous in the breccia on the eastern side of Little Sheep creek valley.

A basic agglomerate macroscopically similar to the Sophie mountain type but lacking the limestone blocks, crops out three miles to the westward, between the Boundary line and Santa Rosa creek. As already noted, microscopic examination of the fragments showed them to be chiefly augite andesite, with a notable proportion of blocks of dark coloured biotite-quartz porphyry.

These breccias bear numerous intercalations of the massive lava flows, thin basic ash-beds and a few, thin beds of black, carbonaceous shale. In a few localities the dip could be taken; in general it is high and ranges from 70° to 90°, showing that the whole group has been heavily mountain-built.

DUNITES CUTTING THE ROSSLAND VOLCANICS.

At various points the andesites encircling the Coryell batholith within the ten-mile belt are cut by dikes and irregular masses of dunite, now partly serpentinized. The largest body occurs on Record mountain ridge, one mile north of the Dewdney trail. It extends over the ridge downward into Little Sheep creek valley. The fresher specimens show the presence of much olivine and some undoubted chromite, but the rock has largely gone to serpentine, talc, and magnetite.

A similar irregular mass occurs on the Red mountain railway west of Rossland. A large dike of rather thoroughly serpentinized dunite cuts the andesitic greenstones south of Castle mountain summit, and a five-foot dike of the same rock cuts the small stock of crushed granite immediately to the southward.

Since this rock is very apt to escape detection among the old volcanics, it is fair to suppose that only a portion of the whole number of occurrences has been discovered. The region has evidently been the scene of fairly numerous intrusions of this very basic type. From the various local relations the dunite has, in part at least, been injected at a relatively late date, possibly as late as the Cretaceous or Tertiary, when it cut the breccias and traps of Record mountain ridge.

DUNITE ON McRAE CREEK.

On McRae creek about three miles above its mouth, the section along the railway crosses 350 yards of a massive, dark, greenish-gray homogeneous intrusive which proved, on microscopic examination, to be a dunite. It cuts biotite schist and a tough, old-looking andesitic breccia. The body probably has the pod form. The olivine occurs in a fairly fresh anhedra varying from 0·4 mm. to 2 mm. in greatest diameter. The alteration products are talc, tremolite, magnetite, and a little carbonate, probably dolomite. No chromite could be recognized in thin section. An analysis of a relatively fresh specimen (No. 528) gave Mr. Connor the following result:—

Analysis of McRae Creek dunite.

SiO_2	41·36
TiO_2	none
Al_2O_3	1·21
Fe_2O_3	9·18
FeO	not det.
MnO	·10
MgO	42·90
CaO	1·34
SrO	none
BaO	none
Na_2O	·04
K_2O	·04
H_2O at 110°C	·16
H_2O above 110°C	1·94
P_2O_5	·04
CO_2	1·40
Cr_2O_3	·15
NiO	·15
S	·50
	100·51
Sp. gr.	3·160

2 GEORGE V, A. 1912

The presence of sulphur interferes with the determination of the relative proportions of the iron oxides in this rock. The analysis clearly corroborates the microscopic evidence that we here have a common type of dunite.

PORPHYRITIC HARZBURGITE (PICRITE?).

At the Dewdney trail south of the head-waters of Santa Rosa creek, the older andesitic traps of the Rossland volcanic group enclose a mass coloured on the map as harzburgite. It is a massive, deep green rock, bearing on its surface abundant cleavage-faces of idiomorphic enstatite, which is embedded in a compact base of olivine and its derivative, serpentine. Many outcrops are characterized by spheroidal weathering, and the rock has assumed a strong brown colour. Here and there it is sheared and thus locally converted into nearly pure serpentine.

The enstatite phenocrysts measure 1 cm. or more in length by 1 to 2 mm. in diameter. Besides olivine the only other constituents are chromite and magnetite; the latter may be entirely secondary from the altered olivine. The enstatite is generally fresh but has yielded some secondary talcose material. The olivine occurs in unusually small grains, which vary from 0.02 mm. to 0·6 mm. in greatest diameter, with an average diameter of probably not more than 0·1 mm. This fine texture of the olivine ground-mass suggests that the mass did not crystallize under a heavy cover. In the field the mass was taken for a thick flow and it is quite possible that it does represent the lava corresponding to a peridotite. A second visit to the locality might solve this interesting problem of relations; meanwhile the rock may be called a harzburgite, and is described among the intrusives.

Mr. Connor has analyzed the fresh specimen (No. 392) collected, with result as follows:—

Analysis of porphyritic harzburgite (effusive?).

		Mol.
SiO_2	42·99	·716
TiO_2	tr.	....
Al_2O_3	1·11	·011
Fe_2O_3	1·87	·012
FeO	5·91	·082
MnO	·05	....
MgO	43·14	1·079
CaO	·10	·002
SrO	none	....
BaO	none	....
Na_2O	·29	·005
K_2O	·13	·001
H_2O at 110°C.	·51	....
H_2O above 110°C.	4·00	·222
P_2O_5	·04	....
NiO	·15	....
	100·29	
Sp. gr	3·075	

The calculated norm is:

Orthoclase..	·56
Albite..	2·62
Anorthite..	·55
Corundum..	·31
Olivine ..	68·08
Hypersthene..	20·68
Magnetite..	2·78
Water..	4·51
	100·09

According to the Norm classification the rock enters the unnamed permagnesic subrang of the unnamed perniric section of the unnamed permirlic rang, in the unnamed section of the perfemone order, maorare.

According to the older classification the rock is a porphyritic harzburgite, if of intrusive origin—an enstatite harzburgite; or an enstatite picrite, if it be a true lava which has crystallized on the surface of the earth.

GABBROS AND PERIDOTITES NEAR CHRISTINA LAKE.

The Sutherland schist-complex is cut by two large masses, (stocks or chonoliths) of gabbroid rocks, each of which has a peridotitic facies. The position and extent of these bodies is roughly indicated on the map.

The one mass, covering about 1·5 square miles, occurs at Fife railway station and is mapped under the name, Fife gabbro. It is a deep green to greenish-black, medium grained hornblende-augite gabbro, composed essentially of bytownite and the two bisilicates. The structure is the hypidiomorphic-granular. On one of its margins the gabbro passes into a typical greenish-black peridotite composed of dominant diopsidic to diallagic augite and the green hornblende. Apatite and ilmenite are constant accessories. This mass is often gneissic through crushing.

Two miles farther north on the railway the section crosses the southern edge of the second of the two gabbro masses, which seems to have about the same area; it is mapped under the name, Baker gabbro. It is composed of biotite, diallage, and basic labradorite (Ab, An,) with usual relations; titaniferous magnetite and apatite are accessory. This rock is quite fresh though locally crushed and gneissic.

The Baker gabbro is cut by an irregular intrusion of very coarse biotite-diallage-olivine peridotite, with accessory magnetite, apatite, and a very little basic plagioclase—a combination of minerals which shows a close genetic connection with the gabbro. The peridotite is fresh, with a specific gravity measured at 3·133, and shows little evidence of crushing. The abundant biotites measure 1 cm. or more in diameter; the diallage and olivine, from 0·5 mm. to 2 cm.

ROSSLAND MONZONITE.

For the reason already noted the complex intrusive mass on which the city of Rossland is situated was not studied in detail during the progress of the

2 GEORGE V, A. 1912

Boundary survey. The following description of the monzonite is taken from a manuscript written by Dr. G. A. Young, one of the joint authors in the forthcoming report on the mining geology of the Rossland district. The writer is very greatly indebted to Dr. Young for the favour of using this material in advance of publication, as well as to Mr. R. W. Brock, who has generously supplied the hitherto unpublished results of Mr. Connor's analysis of the monzonite. Quotation marks indicate the part of the present text supplied by Dr. Young.

'The monzonite body underlies about one-half of the total area of the map sheet (accompanying the special report) and as already stated, represents only the western portion of a roughly oval mass about five miles long in an east and west direction and having a maximum width of about one and three-quarter miles. That part of the monzonite mass lying inside of the area of the map has a very irregular boundary which, commencing on the summit of Deer Park Ridge, first trends northeasterly and then north, passing along the western side of Center Star gulch. The boundary swings across this valley a short distance beyond the northern boundary of the area and pursuing a very irregular course, follows along the top of Monte Cristo mountain and thence diagonally down the southern face of C. and K. mountain, sending a tongue across the summit of the latter. Beyond the eastern limits of the map sheet, the boundary of the monzonite curves around the east face of C. and K. mountain towards the great body of Nelson granodiorite on the north, then turning back on itself, extends eastward across the valley of Trail creek to the slopes of Lookout mountain. The southern boundary of the monzonite from the greatest eastern extension of the body, takes a general westerly course, entering the area under discussion, along the side of Cherry ridge near the southeastern corner of the map sheet and with a bow to the north, strikes westward to the top of Deer Park mountain near the southwestern corner of the area.

'Within the mass thus outlined are several intrusive bodies of porphyritic monzonite and pulaskite as well as a few areas of the bedded series and of the augite porphyrite. The greater part of the monzonitic body is surrounded by the Carboniferous sediments and associated augite porphyrite, the igneous mass cutting sharply across the general strike of these formations. Towards its western end the monzonite is limited by the considerable area of Nelson granodiorite found in the valley of Sheep creek.

'The large area of monzonite with its very irregular boundary, is not occupied by a simple body but by a number of varieties of rock having certain characteristics in common but still presenting much diversity in general appearance and composition. In colour they vary from nearly black to light gray; in grain from very fine to coarse; and in structure from granular to semi-porphyritic. Different types at times cut one another and along the contacts, the younger varieties not infrequently are crowded with inclusions of the older ones, yet in other instances, types of quite diverse appearance seem to pass gradually into one another. The different varieties in some cases occupy large areas to the exclusion of other types, while in other places they appear as dike-like or quite irregular bodies within one another.

' It was not thought profitable to attempt to map separately the different varieties of monzonite, especially as they are all believed to be closely related in origin and composition and to have been nearly contemporaneous. As regards the relative ages of the different varieties it would seem that, in general, the coarser types are younger than the finer and the more feldspathic and lighter coloured varieties are younger than the darker.

' The coarsest type of monzonite and the one most readily separated in the field from the other varieties, occupies a large area stretching from the shaft of the Great Western mine to near the head-works of the LeRoi. Smaller areas of a similar type are common on the south face of Monte Cristo mountain and also along the southwestern border of the monzonite body. This coarse type is usually of a dark colour and consists largely of dark, nearly black prisms of pyroxene or secondary hornblende, flakes of biotite, and a light coloured feldspar, that gives the appearance of lying between the other constituents. In many instances the augite and hornblende form the bulk of the rock, occurring in both large and small, often ragged, prismatic forms frequently varying between one quarter and one half an inch in length. The dark brown biotite, though never as plentiful as the other dark silicates, is abundant and forms large irregular flakes. The feldspars are usually white or slightly greenish in colour and appear to lie between the prisms of augite and hornblende, though when seen in thin sections they often have sharply rectangular outlines; they are chiefly labradorite, with interstitial orthoclase in more subordinate amount.

' This type of monzonite frequently shows local variations along bands where the feldspars sometimes almost disappear, the rock then assumes a greenish black colour and is composed nearly altogether of coarsely crystalline hornblende and pyroxene with much biotite. Sometimes this type seems to end abruptly against the surrounding varieties of more normal monzonite, while at other times it presents transitional forms in which the feldspars increase in amount while the dark coloured constituents decrease in both size and quantity; the remaining larger individuals of pyroxene or hornblende may then give a porphyritic aspect to the rock. Along the southern border of the monzonite body this type or a related one, holds large poikilitic biotite flakes measuring a quarter of an inch or more in diameter and there cuts and holds inclusions of a finer grained variety of monzonite.

' The remaining varieties of monzonite present characters that often remain fairly constant over considerable areas and while examples from different localities may appear quite dissimilar, yet they possess certain features in common and it would be quite possible to select a series of

2 GEORGE V, A. 1912

specimens showing a gradation from any one type to any other. The different kinds are, on the whole, fine and even grained aggregates of white feldspars and dark, nearly black pyroxene, hornblendes, and biotite flakes. The various components usually are distributed uniformly so that on moderately fresh surfaces, the rocks present the appearance of being composed of a finely granular, white ground peppered with tiny dark grains and larger but still small, prismatic individuals of the dark coloured constituents. In both the finer and coarser grained varieties, the relative amounts of the dark and light coloured components vary from place to place, and where the augite or hornblende is exceedingly abundant, the rock assumes a very dark grayish, almost black colour, especially noticeable in the case of the finer grained varieties. On the other hand, with increasing proportions of feldspars, the general colour becomes a lighter gray, a colour more often shown by the coarser than the finer grained kinds.

'Though the rocks are predominantly of a fine and even grained type, yet it often happens that the dark pyroxene or hornblende occur partly in larger prismatic individuals scattered through the finer, uniform material of the bulk of the rock. Very small scales of dark mica are usually present but as a rule in small proportions. Sometimes the minute, shining flakes of this mineral become quite abundant and in some instances larger, ragged individuals with diameters up to one half of an inch are present and enclose the other constituents as in a meshwork.

'When thin sections of the monzonite are viewed under the microscope, the pyroxene is seen to be a pale green augite often forming prismatic individuals seldom measuring more than an eighth of an inch in length. The augite with secondary hornblende is always the chief, and in some cases, virtually the only coloured constituent. At times it forms a large proportion of the whole rock while in other cases, it is completely overshadowed by the feldspars. Brown biotite is usually present in the form of small scales or larger, irregular poikilitic flakes. The feldspars are predominantly, sometimes altogether, of plagioclase varieties. The individuals are generally lath-shaped and in many instances appear to be of the composition of acid labradorite. An alkali feldspar is often present and sometimes is quite abundant, either in irregular grains or in larger, plate-like bodies enclosing the plagioclase laths. Some of the varieties of monzonite contain much magnetite, others scarcely any, while small apatite crystals are almost universal.

'The monzonite is older than, and is cut by, the porphyritic monzonite, the Nelson granodiorite, the pulaskite, and by a large series of dykes. It apparently also is invaded by the diorite porphyrite. The monzonite body, though having a sinuous outline, seldom seems to send offshoots of any size into the older Carboniferous sediments and associated porphyrites which so largely surround it. At three localities only, possible exceptions to this general rule were observed. Within the augite porphyrite near the southern boundary of the area and just to the east of the westerly band of the sediments, there is a small and apparently isolated outcrop of rather coarse monzonite like that of the neighbouring main body. Two small seemingly isolated masses at least partly surrounded by augite porphyrite, occur within the city limits and along the line of the Great Northern railroad near the border of the large monzonite area. Also, a tongue-like extension of the monzonite is shown on the map as extending across the summit of C. and K. mountain; this body is probably directly connected with the main area.

'The border of the central monzonite mass is concealed by drift along the slopes of C. and K. mountain but towards the eastern margin of the map, it may be seen to lie close to the contact of the augite porphyrite and the area of porphyritic monzonite there exposed. From this position, proceeding eastward beyond the limits of the sheet, the line of contact of the monzonite with the older formations, swings around to the north on the slopes of C. and K. mountain, which drop rapidly to the east. On this eastern face above the contact of the monzonite with the augite porphyrite occupying the summit of the hill, are a number of tunnels commencing in the porphyrite but whose dumps are composed largely of monzonite. It would seem that the porphyritic volcanic of C. and K. mountain is a comparatively shallow body occupying the upper portion of the hill but underlain by monzonite which, proceeding westwards, gradually outcrops at successively higher levels along the south face of the ridge and finally occurs in what appears as a dike-like extension across the top of the hill. That is, the top of the body and a portion of the covering of the monzonitic mass seems still to be preserved at this point. This idea furnishes a reasonable explanation for the occurrence of the comparatively large area of the bedded series exposed in the northern part of the city of Rossland within the monzonite and which probably represents a roof-pendant. The same mode of origin may be true of the neighbouring smaller, detached area of similar rocks and also of the two small outcrops of augite porphyrite on the lower slopes of Monte Cristo, or they may represent fragments torn from the formations once overlying or surrounding the monzonite.

'The larger part of the monzonite mass lies in the valley of Trail creek while its greatest extension in a north-east direction are respectively up the Center Star gulch and over the low country east of the slopes of C. and K. mountain. This possible connection between the distribution of the monzonite and the lower lying portions of the country, may be purely fortuitous but when considered in relation with the apparent capping of the body on C. and K. mountain and the possible occurrence of roof-pendants, it points to the conclusion that, within at least the area mapped, the exposures of monzonite belong to a section near the upward limits of the body. It is, nevertheless, possible that at some point or points, the monzonite extended on upwards through the overlying Carboniferous and probably later rocks and may have appeared at the surface as a volcano.'

2 GEORGE V, A. 1912

'The area of the monzonite thus appears to represent the upper portion of an igneous body in places still capped by its old rock roof or holding detached portions of it. The mass is not homogeneous but is composed of many varieties of what seem to be closely related types, the earliest of which are generally the finest in grain and darkest of colour, while the later are coarser, as if they had cooled more slowly and are more feldspathic, perhaps as the result of differentiation processes. In places the intruding varieties have cut portions that apparently already had solidified, since the boundaries are distinct and well defined; in other cases they seem to have invaded masses still partly fluid, since no abrupt change then separates the different kinds. Perhaps some of the finer masses represent portions that had early solidified along the upper bounding surfaces of the igneous mass and afterwards sank into the lower, more central, still fluid portions.

'No direct evidence seemed to be offered in the field as to the methods by which the older sediments and augite porphyrite were removed to make place for the monzonite mass; neither did there appear to be any indications of the absorption of material by the monzonite. Possibly the somewhat abrupt change in the strike of the strata respectively north and south of the axis of the igneous body may indicate some more profound structural break pursuing a general east and west direction and which guided the upward penetrating magma and gave rise to its elongated cross section.

'The monzonite is undoubtedly younger than the Carboniferous sediments and associated augite porphyrite. The structural relations as shown on the accompanying geological map, indicate that the igneous rock was intruded after the major epoch of disturbances whereby the surrounding rocks were tilted and folded. The date of these prominent earth movements has already been discussed and the conclusion reached that they probably took place in Jurassic times. As a result of the line of reasoning adopted, it follows that the monzonite was intruded in the Jurassic or a later period. That the intrusion took place not later than Jurassic times is indicated by the fact that the monzonite is cut by the Nelson granodiorite, itself of Jurassic or early Cretaceous age. The deduction that the monzonite body was formed in Jurassic times is strengthened somewhat by the fact that within the great granite area to the north, the Nelson granodiorite at times presents a monzonitic facies. Possibly the Rossland monzonite was closely connected in origin with the granodiorite and appeared as a forerunner of it.'

Mr. Connor's analysis of a specimen of the granular monzonite, taken at the LeRoi mine, resulted as follows:—

Analysis of Rossland monzonite.

		Mol.
SiO_2..	54.49	.908
TiO_2..	.70	.009
Al_2O_3..	16.51	.162
Fe_2O_3..	2.79	.018
FeO..	5.20	.072
MnO..	.10	.001
MgO..	3.55	.089
CaO..	7.06	.126
Na_2O..	3.50	.056
K_2O..	4.36	.047
H_2O at 110°C..	.97	....
H_2O above 110°C..	1.18	....
P_2O_5..	.20	.001
CO_2..	.10	....
S..	.23	.007
CuO..	none	....
	100.04	

The calculated norm is:—

Orthoclase..	26.13
Albite..	29.34
Anorthite..	16.40
Diopside..	14.58
Hypersthene..	2.76
Olivine..	3.04
Magnetite..	4.18
Ilmenite..	1.36
Pyrite..	.84
Apatite..	.31
Water and CO_2..	1.35
	100.27

According to the Norm classification the rock enters the sodipotassic subrang, monzobose, of the domalkalic rang. monzonase, in the dosalane order, germanare. According to the older, Mode, classification it is a typical monzonite.

The chemical relations of this rock to the Rossland latites and to the calculated world-average for monzonite (reduced to 100 per cent) are shown in Table XXIII.

DEPARTMENT OF THE INTERIOR

2 GEORGE V, A. 1912

TABLE XXIII.

	Rossland Monzonite.	Average of four Rossland Latites.	Average of 12 types of Monzonite elsewhere
SiO_2	54·49	56·52	55·25
TiO_2	·70	1·00	·60
Al_2O_3	16·51	16·96	16·53
Fe_2O_3	2·79	1·10	3·03
FeO	5·20	4·51	4·37
MnO	·10	·14	·15
MgO	3·55	4·01	4·20
CaO	7·08	5·93	7·19
SrO	....	·13	....
BaO	....	·16	....
Na_2O	3·50	3·86	3·48
K_2O	4·36	4·46	4·11
H_2O-	·07	·11	·66
H_2O+	1·18	·48	
P_2O_5	·20	·31	·43
CO_2	·10	·34	....
S	·23		....
FeS_2 and Fe_2S	....	·23	....
	100·04	99·75	100·00

The table shows how faithful is the chemical resemblance of the stock rock to the average monzonite and to the lavas. As usual with lavas and corresponding plutonic species, the average latite is slightly higher in silica than the monzonite.

Dr. Young has suggested a possible Jurassic age for this monzonite. His chief ground for the reference is found in the fact that the body is cut by the Trail (Nelson) granodiorite, which is considered by him as either Jurassic or early Cretaceous in date of intrusion; he also points out that the Nelson granodiorite has monzonitic phases in other parts of the West Kootenay district. Since the monzonite is chemically almost identical with the surrounding latites, we may fairly regard them as contemporaneous in date of eruption; in fact, McConnell stated the view that the monzonite is occupying the actual site of the volcanic vent through which the latites were poured out.

BASIC MONZONITE AND HORNBLENDITE ON BEAR CREEK.

At the confluence of Bear-creek with the Columbia river a small patch of probably Paleozoic schists is exposed. It is cut by a small irregular basic mass, which has been itself tremendously shattered by the granite magma of the Trail batholith. Part of the basic mass is monzonitic and mineralogically and chemically allied to the Bitter creek intrusive, next to be described. Green hornblende, biotite, orthoclase, and andesine, near Ab_1An_1, are the essential constituents, with quartz, apatite, magnetite, and titanite as accessory. The specific gravity is 2·809.

The monzonite merges insensibly into a coarser peridotite, made up of dominant green hornblende, subordinate deep green biotite, and the same accessories as in the feldspathic phase. The specific gravity of this biotitic hornblendite varies (in two specimens) from 3·144 to 3·260.

SHONKINITE TYPE AT BITTER CREEK.

At the crossing of the Dewdney trail and Bitter creek a small, boss-like intrusion of a peculiar, very basic rock cuts the Rossland volcanics. It is coarse-grained, blackish green in colour, and of peridotitic habit. Though friable under the hammer, it is very fresh and apparently quite uncrushed. This rock was collected by an untrained assistant in the camp, who was not capable of mapping the body or of determining its relations to the surrounding lavas of the Rossland group. Unfortunately no opportunity presented itself whereby the writer could visit the locality, so that no statement can be made concerning the essential facts of the field. Petrographically the rock has interest as affording a type transitional between the monzonites and the peridotites. A note concerning its composition will advertise the occurrence and it is hoped that some other geologist will visit the locality and study this rock-body more fully.

Under the microscope the essential constituents are seen to be green hornblende, diallagic augite, and biotite, all more or less perfectly idiomorphic, along with subordinate amounts of sodiferous orthoclase, microperthite, and basic andesine, $Ab_1 An_1$. Apatite, titanite, magnetite, and a very little interstitial quartz are the accessories. The specific gravity is 2·951. The structure is the hypidiomorphic-granular.

By the Rosiwal method the mineral composition was found to be, by weight, approximately:—

Hornblende	56·7
Diallage	12·3
Biotite	8·7
Orthoclase	13·4
Andesine	7·2
Apatite	·8
Magnetite	·4
Quartz	·5
	100·0

The presence of alkaline feldspar in so femic a type is unusual. Chemically this rock must be rather similar to shonkinite.

GRANITE STOCK EAST OF CASCADE.

A greatly metamorphosed mass of granite covers a small area on the heights just east of the Kettle river at Cascade. The body is not well exposed but it appears to form an elongated stock over a mile long and about 800 yards wide. It cuts the older (probably Palæozoic) traps mapped under the colour of the Rossland volcanic group, but the granite must be older than the Coryell syenite or the younger lavas of the Rossland group. The granite is so

2 GEORGE V, A. 1912

thoroughly crushed and altered that its exact original nature cannot be discovered from an examination either of the ledges studied in the field or of the the specimens collected to represent the stock. The microscope shows that the rock was probably a common, medium-grained biotite granite. In its present granulated and altered condition it offers little of petrographic novelty or interest.

The rock is coarse-grained and now gneissic. The original essential minerals seem to have been orthoclase (now microcline), plagioclase (probably oligoclase), and biotite. The abundant secondary minerals are red garnet, muscovite, epidote, and kaolin.

It is possible that this stock is a satellite of the gneissic batholith at and west of Cascade.

TRAIL BATHOLITH.

Definition.—There are but few regions of the world where post-Archean granites are exposed on such a grand scale as in the West Kootenay district of British Columbia. Part of the district is included in the West Kootenay reconnaissance sheet of the Canadian Geological Survey, a map covering about 6,500 square miles. More than two-thirds of this area is underlain by extensive granites probably all of post-Carboniferous age. The whole group form a composite batholith, including various types and bodies called by Mr. Brock, Nelson granite, Valhalla granite, Rossland alkali-granite, alkali-syenite, etc.

The delimitation of these different bodies has been accomplished in part, but, from the nature of the surveys so far carried on, much work still remains to be done before the composite batholith is fully mapped and its anatomy understood. A leading difficulty in drawing boundary lines about the constituent intrusive bodies is found in the occurrence of many included masses of crystalline schists and gneisses which may be in part of pre-Cambrian age but to some extent are certainly metamorphosed post-Cambrian sediments or else sheared phases of the intrusive granites themselves. Many large areas of schistose rocks are thus not easy to classify. Among them is a group of gneisses and schists occurring along the Columbia river from Sullivan creek northward. They have been coloured as Archean on the West Kootenay sheet, although McConnell, who carried on the reconnaissance of this part of the district, states that these foliated rocks are largely 'contemporary in age with the main granite area of the district' *i.e.* the Nelson granite.[**] The probability is, therefore, that the great granite body surrounding the town of Trail is a direct offshoot of the vast batholith forming the central part of the West Kootenay district and that, in the area bordering the Columbia between Robson and Sullivan creek, it has been crushed to the gneissic condition. To that portion of the composite batholith which forms a continuous mass with the granite about Trail and belongs to the one date of intrusion, the name 'Trail batholith' may be given.

[*] See West Kootenay sheet.
[**] Marginal note on map of part of Trail Creek Mining Division, Geol. Surv. of Canada, Preliminary edition, 1897.

Of the hundreds of square miles which may represent the total area of the Trail batholith, only thirty-five are included in the ten-mile belt along the International line; the following description of the Trail granite is found on studies made in this portion of the mass.

Petrograph.—The dominant phase is a medium-grained to somewhat coarse-grained rock of pure gray colour. It is rich in the femic constituents, dark green hornblende, and biotite as well as in plagioclase, which, with the likewise macroscopically visible quartz and orthoclase, completes the list of essentials found in a typical tonalite or granodiorite. The rock has, in general, the unmistakable habit of a granodiorite. It generally shows evidences of strain and, in places, is distinctly gneissic through pressure.

Under the microscope, the necessary minerals are seen to be the usual magnetite, apatite, and titanite, with rare zircons. The orthoclase is often replaced by microcline; the plagioclase is often zoned and averages basic andesine, near $Ab_1 An_1$. The characters of the minerals and the rock-structure are those common in granodiorite and do not need special description. Epidote is a common metamorphic product in the crushed phase of the batholith.

A typical fresh specimen (No. 509), collected in a railway cutting two miles west of Trail, was analyzed by Mr. Connor with the following result:

Analysis of granodiorite, Trail batholith.

		Mol.
SiO_2	62.08	1.035
TiO_2	.73	.009
Al_2O_3	16.61	.163
Fe_2O_3	1.53	.009
FeO	3.72	.051
MnO	.11	.001
MgO	2.44	.061
CaO	5.20	.093
SrO	.08	....
BaO	.09	.001
Na_2O	3.18	.052
K_2O	3.29	.035
H_2O at 110°	.16	....
H_2O above 110°	1.00	....
P_2O_5	.30	.002
	100.47	
Sp. gr.	2.754	

Calculated norm:—

Quartz	15.48
Orthoclase	19.46
Albite	27.25
Anorthite	21.13
Diopside	11.53
Magnetite	2.08
Ilmenite	1.36
Apatite	.62
Water	1.18
	100.06

2 GEORGE V, A. 1912

The mode (Rosiwal method) is approximately:

Quartz	25·9
Andesine	28·1
Orthoclase and microcline	19·2
Biotite	13·4
Hornblende	12·3
Magnetite	·6
Apatite	·3
Titanite	·2
Zircon	trace
	100·0

In the Norm classification the rock enters the sodipotassic subrang. harzose, of the alkalicalcic rang. tonalase, in the dosalane order. austrare; although the ratio of potash molecules to soda molecules is very close to the limit separating harzose from tonalose. According to the older classification the rock is a basic granodiorite.

The average specific gravity of four fresh specimens of the rock is 2·749.

The dominant granodioritic type often passes gradually into a more acid biotite granite or hornblende granite in which the feldspar is chiefly orthoclase, sometimes microperthitic. These types sometimes form streaks in the main body but are chiefly developed in the numerous apophyses. They furnish a transition to the very abundant aplitic dikes, also apophysal from the batholith. Many of the larger apophyses illustrate the differentiation of two quite different rocks in the same fissure. The middle part of each of these dikes is composed of granodiorite or hornblende-biotite granite, while along each wall, a zone of aplite, gradually passing into the more basic rock of the middle zone, is developed. The feldspars of the aplite are orthoclase and microperthite, with accessory oligoclase-albite. Quartz and a little biotite are the remaining essentials. The aplite zones make up one-quarter to one-half of these apophyses. Other dikes are composed entirely of the aplite. Its composition and specific gravity (2·592 for one fresh specimen) closely resemble those of the younger Sheppard granite (spec. grav. 2·600—2·617).

Differentiation in Place.—At the batholithic contact where it crosses the railway branch between Trail and Rossland, there is a large body of relatively acid granite which is regarded as genetically connected with the granodiorite. The body measures about 400 yards in width. The contacts are hidden and it is uncertain whether this more acid rock represents a contact-phase of the batholith or a slightly later intrusion. The structure of the smaller body is, so far as known, throughout porphyritic and the balance of probability is in favour of its being a late differentiate of the batholith and intruded into the zone of contact of the granodiorite and the volcanics. The microscope shows that the porphyritic rock is an alkaline biotite-hornblende granite. The pheno crysts are orthoclase (sometimes microperthitic), oligoclase, biotite, and horn blende. The ground-mass is typical granophyre. A little magnetite and apatite are accessory.

PLATE 33.

Two views of shatter-belt about the Trail batholith, Columbia River.

The chief interest of this body lies in the fact that it bears very numerous basic segregations which appear to be themselves differentiated from the magma from which the granite porphyry crystallized. The segregations are round, about one foot in maximum diameter, and seemingly quite uniform in composition. They are of a dark greenish-gray colour, fine-grained or compact and in the field have the appearance of fragments torn off the volcanic formation close by. This was, indeed, the tentative field interpretation, although it was there recognized that these small masses had also all the characteristics of basic segregations. Microscopic study showed that the latter view is probably the correct one. The rock is somewhat porphyritic, with much altered phenocrysts of orthoclase and some of an oligoclase. No femic phenocryst was to be seen. The ground-mass is a mass of minute, idiomorphic green hornblende prisms embedded in small, interstitial crystals of orthoclase, with which a few soda-lime feldspars may be mixed. A little titanite, less magnetite, and abundant apatite in very minute prisms are the accessories. Chlorite, epidote, and kaolin are the chief secondary minerals. These segregations have, thus, the composition and most of the structural features of typical vogesite. Their rounded and embayed outlines suggest magmatic resorption. They seem to represent a lamprophyric derivative of the granodiorite and the almost aplitic granite porphyry in which they lie would, on that view, correspond to the other pole of the differentiation. More study needs to be given to this case but it is worth while to point out this locality as an easily accessible and perhaps fruitful one where magmatic differentiation in place may be discussed.

Shatter-belt -- The Trail batholith illustrates on a great scale the mechanical disturbances which are so characteristically produced in country-rocks by the intrusion of stocks and batholiths. The shatter-belt is not only unusually broad but it is finely displayed along the eastern side of the Columbia river. (Plate 33 and map sheet No. 8.) This belt has been briefly described in the American Journal of Science (Vol. 16, 1903, p. 123), and will be again referred to in the following theoretical chapter (XXVI.) on the mechanics of igneous intrusion.

So far as observed within the ten-mile belt, the exomorphic influence of the Trail batholith is more strikingly evident in the form of mechanical disruption than in the way of recrystallizing the invaded rocks. One reason for this is that it is practically impossible to distinguish the thermal effects of the intrusion from those produced by the heavy regional metamorphism which had previously affected the traps. Yet it is probable that the strong schistosity and the present composition of the hornblende-biotite gneisses and schists (amphibolites) adjoining the granodiorite to the west of Sayward are, in largest part, inheritances of the contact-metamorphism.

Near the Columbia river contact north of Sayward, the batholith is cut by dikes of monzonite porphyry and of camptonite. On the railway west of Trail it is cut by dikes of hornblende-biotite gabbro.

2 GEORGE V, A. 1912

Conglomerate Formations.

Because of the extremely rare occurrence of water-laid deposits in the Rossland mountains, the discovery of fossiliferous horizon-markers is, throughout the mountain group, a field problem of special difficulty. For that reason certain small patches of conglomerate with sandy and shaly interbeds, which may yield useful fossils, have the particular interest of the geologist who attempts to understand the structural tangle of this region. Four small and quite detached areas of this conglomerate appear within the Boundary belt; these will be described in order from east to west.

Conglomerate at Lake Mountain.—One mile southwest of Lake Mountain summit, a patch of the conglomerate covering about a third of a square mile has been mapped by McConnell and re-traversed by the writer. The rock is there chiefly a coarse, massive conglomerate, dipping at an average angle of 20° to the northeast and showing an apparent thickness of about 300 feet. The mass is truncated by an erosion-surface, so that 300 feet is a minimum thickness at the locality. At no point was the conglomerate found in actual contact with the Rossland volcanics which surround it. There are two possibilities as to the relation between the two formations: the conglomerate may overlie the volcanics, as postulated by McConnell, or, secondly, it may represent a pre-volcanic conglomerate forming a knob which was first buried under the lavas and since uncovered by their denudation. The choice between these alternatives is not ensured by any known fact. The comparatively low dips suggest that the first view is the correct one. At the same time, there are no lava-fragments among the pebbles of the conglomerate, which are composed of gray and greenish-gray quartzite, silicious grit, vein quartz, phyllite, and slate. A few badly altered pebbles of a rock like granite are also present.

Practically all of the material observed in the pebbles could have been derived from the Palæozoic and pre-Cambrian terranes now exposed in the Selkirk range, twenty-five miles to the eastward; in the absence of any other known source, that place of origin appears probable. The pebbles are of all sizes, up to the diameter of one foot. They are of rounded, subangular, and angular shapes. In places the deposit approximates a true breccia in appearance. The imperfect rounding, and, in addition, the generally tumultuous aggregation of the pebbles suggest rapid deposition, as if by a rapid mountain stream. Small irregular lenses of quartz-sandstone and grit form the only breaks in the pebbly mass. Similar arenaceous material composes the cement of the conglomerate, which is also quite highly ferruginous. One dike of basic andesite or latite (character not determined) and a tree (mapped) apophysis of the Sheppard granite cut the conglomerate.

Conglomerate at Sophie Mountain.—A second body of coarse conglomerate covering a square mile or more, crowns the summit of Sophie mountain at the International line. In structure, size of pebbles, and composition this rock resembles the conglomerate at Lake mountain very closely, but here there are a few pebbles of the neighbouring trap-rock as well as some of blackish chert and others of fine-grained granite, while the pebbles are more generally rounded than at Lake mountain. The cement is arenaceous. The sandy lenses range from six inches to two feet in thickness and are never continuous for any great distance on the outcrop. One hundred yards northeast of the Boundary monument a bed of sandy shale, containing poorly preserved dicotyledonous leaves, was found. These obscure fossils were examined by Professor Penhallow who reported as follows:—

> 'The impression of a leaf is certainly a very poor one to found an opinion upon, and the difficulty is complicated by the crossing impressions of superimposed leaves. All I can do is to make a very wide guess. After very careful examination and consideration, I am inclined to think the leaves are those of *Ulmus speciosa*, Newb. If this determination is at all correct, then the age is Tertiary and possibly Miocene; I do not think it can be Cretaceous. Assuming this guess to be correct, I find the specimen to be quite in harmony with specimens in Mr. Lambe's collection from Coal gully, since in both cases the species is the same and the matrix has been similarly metamorphosed.'

At the Boundary monument the conglomerate dips northwest at an average angle of 31°. Seven hundred yards to the northwest of the monument the dip was again determined on sandy intercalations as 80° to the southeast. Along the Velvet mine wagon-road the average dip is about 75° S.E. The attitude of the bedding is, on account of the massiveness of the conglomerate, very difficult to determine, but these readings suffice to show that the conglomerate has been greatly disturbed. The exposures are not sufficiently continuous to warrant a statement as to the thickness of the conglomerate; it is certainly a heavy deposit, possibly a thousand or more feet thick. Just south of the monument it is seen, at one point, to be apparently resting on the older Rossland volcanics and in spite of the general lack of satisfactory contacts, this relation can scarcely be doubted. At one horizon a 20-foot amygdaloidal sill (?) or flow of augite-biotite latite is interbedded with the conglomerate.

At monument 174 the conglomerate is cut by several dikes of augite-biotite monzonite porphyry in composition similar to the flow just mentioned and to latite occurring on Record mountain ridge to the northward.

Conglomerate Area at Monument 172.—The third occurrence of conglomerate was found on the Boundary line at monument 172, a distance of five miles west of the Sophie mountain monument. The stratified deposit forms part of the roof of an irregular intrusion of syenite porphyry which will be described on a later page. Erosion has greatly broken that roof so that the conglomerate crops out now in the form of a number of detached blocks which are apparently immersed in the porphyry. The largest block measures 250 feet by 750 feet in ground-plan. About 200 feet of thickness is represented in this heaviest mass of the conglomerate. The strike of the bedding is N. 30° W.; the dip, 28° N.E. The conglomerate is much brecciated in an east-west zone 100

2 GEORGE V, A. 1912

feet wide, and the zone is impregnated with small quartz veins. This fracturing of the conglomerate may have been contemporaneous with the intrusion of the porphyry.

The conglomerate is not so coarse as that at Sophie mountain or Lake mountain and carries more sandy layers. A second difference was seen in the occurrence of a higher proportion of pebbles derived from the adjacent volcanics. Here, too, there are pebbles of an equigranular biotite granite. Quartz, quartzite, slate, and chert are the other staple materials of the pebbles.

Conglomerate Area at Monument 169.—Another five miles farther west, at monument 169, the Boundary slash crosses a patch of coarse conglomerate, covering about one-quarter of a square mile. This mass has been upturned, with strike N. 80° W., and an average northerly dip of 75°. The well-rounded pebbles, ranging from two inches or less to ten inches in diameter, are chiefly composed of altered porphyritic latite (or andesite?), most probably derived from the Rossland lavas in the immediate vicinity. Compared to them the quartzitic and slaty pebbles are quite subordinate, but lenses of dark-gray, quartzitic sandstone, like many such lenses in the eastern areas of conglomerate, are occasionally intercalated. At this locality contemporaneous latite flows seem to be interbedded with the conglomerate.

Correlation and Origin.—These conglomerate areas have all been mapped under the same colour, though it may well be that they are of different ages. Proceeding from east to west the pebbles of the different occurrences are composed more and more often of material which in the field is indistinguishable from the adjacent Rossland lavas. At the same time the pebbles become more rounded. The local character of the four areas, their alignment and the similarity in the composition of the quartzitic, phyllitic, and slaty pebbles to the rocks forming the Summit series and Priest River terrane as well as the Pend D'Oreille group of the Selkirks—these facts suggest the hypothesis that the conglomerate everywhere represents a heavy mass of river gravels, and that one or more streams flowing westward from the site of the present axis of the Selkirk range were responsible for the accumulations. The deposit of dicotyledonous leaves in the coarse Sophie mountain conglomerate strongly indicates the freshwater origin of that mass at least. It is clear, however, that we have nothing clear or decisive regarding the correlation of the conglomerate bodies with one another or with the recognized systems of rocks. The high probability is that they are all pre-Miocene and post-Jurassic.

BEAVER MOUNTAIN GROUP.

General Description.—In 1898 Mr. R. W. Brock made a brief reconnaissance of the moratains situated between Beaver creek, and Salmon river and south of the Nelson and Fort Sheppard railway. As a result of his work he has mapped a portion of the volcanic rocks of the district as belonging to a special division, the 'Beaver Mountain Volcanic Group.' A very brief description of the group appears in the marginal Explanatory Notes on the West Kootenay

SESSIONAL PAPER No. 25a

Sheet of the Canadian Geological Survey (1904). It reads as follows:—
'These rocks consist of beds of andesites, tuffs and ash-rock, which overlie the
surrounding Rossland volcanics. Their age is not definitely known, they
appear to be comparatively recent. . . . The andesites of Record and Old
tilory mountains may be of the same age, though these have not been differentiated from the Rossland volcanics on the map. The Beaver Mountain volcanic
rocks are occasionally mineralized to some extent.' Mr. Brock makes no mention of associated sedimentary rocks.

In 1902, the present writer, without the knowledge that Mr. Brock had traversed these mountains, made an independent examination. Like Mr. Brock
he was struck with the relatively recent appearance of the lavas and pyroclastic
rocks about Beaver mountain and became convinced that they are considerably
younger than many of those composing the typical Rossland group. This
repeated recognition of a possible subdivision of the volcanic complex is believed
to be quite justified, and Mr. Brock's nomenclature is adopted in the present
report. To future workers in the district it may be proposed that the name
'Beaver Mountain volcanic group' be extended to all the lavas and pyroclastics
of the complex which are contemporaneous with those shown typically on and
in the vicinity of Beaver mountain. The area ascribed to the rocks of the
Rossland volcanic group, within the ten-mile belt, would thus in the future
be diminished as the various patches of the Beaver Mountain rocks are
separated. It will take several seasons of special work to bring about a satisfactory delimitation of the younger and older members of the complex, even if
the dense forest cover and the almost infinitely involved structural difficulties
do not forever prevent this desired mapping.

The Beaver Mountain group is shown on the map as covering about
twenty square miles in the northern part of the ten-mile belt. It is to be
understood that the boundaries are very roughly drawn, for it is often impossible to tell when one passes from the younger rocks either to the latitic masses
or to the older porphyrites and other volcanics of greenstone-like facies.

Sediments.—In this area two patches of water-laid clastics contemporaneous
with the volcanics, are mapped. A small outcrop of them also occurs on the
railway near the water-tank at Beaver. These strata may be called the Beaver
Mountain sediments. They consist of black to dark gray and brown thin-
bedded shales, and gray and greenish, thin-bedded to quite massive sandstones.
A massive conglomerate (granite, quartz, and slaty pebbles) crops out just
west of Champion station. The sandstones often graduate into typical, thick
masses of ash-beds and coarse agglomerates, alternating with vesicular flows
of basalt and augite andesite. The shales and sandstones bear fragments of
plant stems and leaves but no fossil of diagnostic value has been found. More
than 1,000 feet of the sediments are exposed in a section running from Champion station eastward into Beaver mountain. There the dips are always to the
south and vary from 12° to 32°, steepening as the mountain is ascended.
Toward the top of the stratified series heavy flows of porous andesite and much

2 GEORGE V, A. 1912

thicker masses of agglomerates are interbedded with the shales and sandstones. The relations are similar to those of the Mesozoic shales and sandstones which dip under the Sophie mountain-Malde mountain breccia at Little Sheep creek. The characters of sediments, lavas, and pyroclastics are also suggestively like those of the formations at that fossiliferous locality. The agglomerates interbedded with sandstones and shale in the more southern of the two sedimentary areas, i.e, at the south end of the long Beaver Mountain ridge, carry a few small fragments of white marble which is like that in the Sophie mountain agglomerate. There is thus some ground for referring the Beaver mountain sediments and volcanics to the Mesozoic.

In none of the traverses made, either by Mr. Brock or by the writer, has it proved possible to construct a trustworthy columnar section of these rocks. The group is greatly disordered by faulting and by the intrusion of dikes and sills. The greatest difficulty was, however, due to the lack of sufficiently continuous exposures. It is known only that the clastic rocks are of the notable thickness of over 1,000 feet and that they conformably underlie a great thickness of lava and associated pyroclastic material. All these rocks have been upturned and dip at all angles up to that of 90°. The strike is highly variable.

Volcanics.—So far as known, the lavas of the group belong only to the two related species, augit andesite or olivine-free basalt, and in largest part to the former species. The agglomerates and ash-beds, which are exposed on a great scale, are chiefly accumulations of the same basic lavas in pyroclastic condition, but along with those fragments there occur variable amounts of black shale, slate, and gray sandstone, with a little vitreous quartzite and white marble. Petrographically, the andesite and basalt are indistinguishable from the same types where these were observed in the area mapped as underlain by the Rossland volcanic group. Notwithstanding the profound disturbance which the Beaver Mountain rocks have suffered, they are seldom or never schistose over any considerable area. The metasomatic changes are rarely so great as to obscure the true nature of the lavas, though the extreme freshness of the Miocene lavas west of Midway was not observed in any thin section cut from these rocks.

SHEPPARD GRANITE.

One of the youngest intrusives of the Selkirk and Columbia ranges is alkaline biotite granite, which forms a small stock at the head of Sheppard creek, there cutting the older traps of the Rossland volcanic group. This acid type may, for convenient reference, be named the Sheppard granite. The same granite composes a larger stock south of Lake mountain, a small lenticular mass near the summit of that mountain; also a stock and some large, dike-like masses on the lower Pend D'Oreille river, as shown on the maps. In all the bodies the granite is generally quite uniform in character and the following description of the rock, where outcropping at the head of Sheppard creek, will suffice for all of the occurrences.

The granite is a pinkish, medium-to fine-grained, aplitic aggregate of quartz, microperthite, orthoclase, and oligoclase, near Ab_4An_1, and a very little, generally chloritized biotite; in the granite of the stock on the Pend D'Oreille, biotite is replaced by diopsidic augite which is also hardly more than an accessory. A little magnetite and well crystallized titanite, with a few minute zircons, are always present. Apatite seems to fail. The structure is the eugranitic, tending to the panidiomorphic.

The specific gravities of three fresh specimens vary from 2·600 to 2·617. Chemical analysis of the granite by Mr. Connor (specimen No. 500) gave the following proportions:—

Analysis of Sheppard granite.

		Mol.
SiO_2	77·09	1·285
TiO_2	·05	....
Al_2O_3	13·04	·127
Fe_2O_3	·82	·005
FeO	·26	·004
MnO	tr.	....
MgO	·12	·003
CaO	·63	·012
SrO	none	....
BaO	none	....
Na_2O	3·11	·050
K_2O	4·50	·048
H_2O at 110°C	·03	....
H_2O above 110°C	·07	....
P_2O_5	·10	·001
	99·82	
Sp. gr.	2·600	

The calculated norm is:—

Quartz	40·50
Orthoclase	26·69
Albite	26·20
Anorthite	2·50
Corundum	2·04
Hypersthene	·30
Magnetite	·93
Hematite	·16
Apatite	·31
Water	·10
	99·73

In the Norm classification the rock enters the sodipotassic subrang, alaskose, of the peralkalic rang, alaskase, in the persalane order, columbare. According to the older classification it is an aplitic, alkaline biotite granite. For this rock the norm cannot differ much from the mode, which has therefore not been estimated.

The Sheppard granite clearly cuts the Pend D'Oreille schists, the Rossland volcanics, the Trail granodiorite, and a conglomerate of Lake mountain. It is quite uncrushed and probably belongs to a date of intrusion well up in the

Tertiary. The great resemblance of this granite to the aplitic type apophysal from the Trail batholith suggests a genetic connection between that batholith and the stocks, as if the latter are satellites from the former, in the same fashion as the Summit stocks, Lost Creek body, and Bunker Hill stock are satellitic to the great Bayonne batholith. Yet it is possible that the Sheppard granite is not closely associated in age with the Trail granodiorite. The two are genetically connected perhaps only in the sense that the same underground conditions under which the aplite of the batholithic apophyses was developed, prevailed also at the later date when a new magmatic invasion affected the region. If this be true, the Sheppard granite may have been a differentiate from a more basic magma like the Trail granodiorite but younger than that batholith and not exposed in the Boundary belt.

PORPHYRITIC AUGITE-OLIVINE SYENITE.

Just south of the point where the Crowsnest line of the Canadian Pacific railway turns out of McRae creek valley and enters that of Christina lake, the railway cuttings for some 600 feet cross a peculiar basic rock which deserves special note. It is a stock-like body, intrusive into crystalline limestone and schists of the Sutherland complex. To each side of the railway track the exposures are poor and the exact ground-plan of the body could not be discovered in the time available for its study. A larger mass of the same rock occurs as an intrusive mass, about a half mile in diameter, at the head of Fifteen-mile creek north of the Pend D'Oreille river (see map). Erratic boulders of the rock are to be found on the flat west of the Alice mine, north of Creston, and clearly come from a third locality. The repeated discovery of this unusual rock at widely separated points shows that its peculiar structure is not merely a local accident but the persistent product in the crystallization of a definite magmatic type.

The rock at the Christina lake locality is a fresh, dark gray to greenish or bluish-gray, medium-grained to rather coarse-grained aggregate of augite, olivine, biotite, plagioclase, and orthoclase; in this aggregate relatively enormous phenocrystic foils of dark green biotite lie embedded at all angles. The irregular surfaces of the mica-foils range in diameter from 1 cm. to 3 cm. or more, while their thickness is seldom over 1 mm. As the rock is fractured under the hammer the broken surfaces are so generally bounded by the cleaved phenocrysts that the rock is decidedly facetted in a striking way.

Though the rock is almost perfectly fresh, the lustre of the large biotites is rarely higher than the metallic; the lustre is impoverished by a very marked magmatic corrosion of the biotites which, in thin section, have an apparent poikilitic structure in consequence. The more minute biotites of the groundmass have the usual high lustre of the micas. The optical angle of the phenocrystic mica is (2 E) about 15°; that of the ground-mass mica is over twice as large (2 E = about 35°). Both are highly pleochroic in tones from dead-leaf yellow to deep reddish-brown.

SESSIONAL PAPER No. 25a

The olivine of the ground-mass (up to 3 mm. in diameter) is abundant and very fresh; it has a bottle-green colour in the hand-specimen and is colourless in thin section. It is sometimes slightly serpentinized along cracks. The augite is a diopside, occurring in stout prisms up to 2 mm. in length. The plagioclase is labradorite between Ab_1An_1 and Ab_1An_2. Orthoclase is plentiful and, like the labradorite, is extremely fresh. The feldspars are much smaller than the olivine or augite, the plagioclase averaging about 0·2 mm. in diameter, the orthoclase about 0·4 mm. Titaniferous magnetite and apatite are fairly abundant accessories. The order of crystallization appears to be: 1. The accessories. 2. Olivine. 3. Augite. 4. Biotite of ground-mass and biotite phenocrysts. 5. Labradorite. 6. Orthoclase.

The rock of the Fifteen-mile creek mass is, in all essentials, like the one just described, but its facetting is even more pronounced. The mica phenocrysts range from 2 cm. to 5 cm. in diameter by 0·5 mm. to 1 mm. only, in thickness. Curiously enough, the grain of the ground-mass is, in the specimens collected, inversely proportional to the size of the mica phenocrysts. In the Fifteen-mile creek body the phenocrysts are about double the size of those in the Christina lake rock, while the diameters of the respective ground-mass feldspars average about as one to two. That is, these feldspars in the rock with the larger mica phenocrysts have an average volume only about one-eighth of the average volume of the feldspars in the rock with the smaller phenocrysts.

The average specific gravity of two specimens from Christina lake is 2·841; the average for two specimens collected at the eastern locality is 2·815.

Mr. Connor has analyzed a specimen (No. 354) from the Christina lake body, with the following result (Col. 1, Table XXIV.):—

Table XXIV.—Comparison of basic syenite and average monolite of region

	1.	1a. Mol.	2.
SiO_2	52·95	·882	51·78
TiO_2	·70	·009	·87
Al_2O_3	14·00	·137	14·54
Fe_2O_3	2·57	·016	1·77
FeO	5·55	·077	5·29
MnO	·13	·002	·11
MgO	7·29	·182	6·89
CaO	6·93	·123	7·35
SrO	·11	·001	·07
BaO	·32	·002	·25
Na_2O	2·73	·044	2·74
K_2O	5·09	·054	4·60
H_2O at 110 C.	·16	....	·44
H_2O above 110 C.	·50	....	1·91
CO_2	....	....	·88
P_2O_5	·47	·004	·83
	99·50		100·32
Sp. gr.	2·872		

MICROCOPY RESOLUTION TEST CHART

(ANSI and ISO TEST CHART No 2)

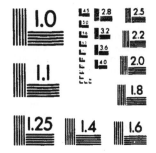

APPLIED IMAGE Inc

1653 East Main Street
Rochester, New York   14609   USA
(716) 482 - 0300 - Phone
(716) 288 - 5989 - Fax

2 GEORGE V, A. 1912

Calculated norm:—

Orthoclase	30·02
Albite	20·96
Anorthite	10·84
Nephelite	1·13
Diopside	16·74
Olivine	12·45
Magnetite	3·71
Ilmenite	1·36
Apatite	1·24
Water	·66
	99·11

The mode (Rosiwal method) is approximately:—

Sodiferous orthoclase	45·1
Labradorite	10·6
Augite	24·5
Olivine	10·7
Biotite of ground-mass	1·7
Biotite of phenocrysts	3·5
Magnetite	3·5
Apatite	·4
	100·0

According to the Norm classification the rock enters the sodipotassic sub-rang, monzonose, of the domalkalic rang, monzonase, in the dosalane order, germanare. According to the older classification the rock is a porphyritic biotite-bearing augite-olivine syenite of an unusual variety. Chemically it is closely allied to olivine monzonite and, in many respects, to shonkinite. It is almost identical with the average minette of this region. The average of the three minette analyses noted in the last chapter is given in Col. 2 of Table XXIV. From the comparison of Cols. 1 and 2 one is led to suspect that this abnormal syenite really represents a minettic magma which, because of the large size of each of the bodies in which it occurs, crystallized with its peculiar structure. It may be true, however, that the remarkably fresh and quite uncrushed olivine-syenite is geologically much younger than most of the minette dikes of the Selkirks.

CORYELL SYENITE BATHOLITH.

Between Record mountain ridge and Christina lake the greater part of the ten-mile belt is covered by a batholith of syenitic rock which is generally quite uncrushed and, like the Sheppard granite, is among the most recent of the intrusives in the Columbia mountain system. The Coryell railway station is situated near the northern contact; the intrusive mass may be called the Coryell batholith.

Though this batholith is marked in the West Kootenay reconnaissance sheet with the same colour as the various stocks of the Sheppard granite, there

SESSIONAL PAPER No. 25a

are constant and important differences in composition between the stocks and the batholith, so that their sharp distinction is necessary in a detailed survey of the region.

Dominant Phase.—Macroscopically, the dominant phase of the Coryell batholith is a medium-to coarse-grained, occasionally somewhat porphyritic, light reddish to brownish-pink rock of typical syenitic habit. It is usually fresh and is generous to the collector of fine specimens. Greenish-black, lustrous hornblende, and brilliant biotite are the visible femic constituents; they occur scattered through the feldspars which are both striated and unstriated. In thin section the principal feldspar is seen to be microperthite, associated with much sodiferous orthoclase and subordinate plagioclase, averaging andesine, $Ab_3 An_2$. A few small idiomorphic crystals of diopsidic augite, a little interstitial quartz, rare grains of allanite, and the usual apatite, titanite, and (probably titaniferous) magnetite are accessory.

The structure is eugranitic; the order of crystallization is apparently: – Apatite, magnetite, titanite, augite, plagioclase, hornblende, biotite, alkaline feldspars, and quartz; with some overlapping in the generation-periods of the hornblende and plagioclase. The specific gravity of six fresh specimens ranges from 2·648 to 2·729, with an average of 2·675.

This chief phase of the batholith has not been specially analyzed for the present report, but the following analysis of a type specimen (collected at a point north of Record mountain) has been made by Professor Dittrich for Mr. Brock:—

Analysis of Coryell syenite.

SiO_2	62·59
TiO_2	0·54
Al_2O_3	17·23
Fe_2O_3	1·51
FeO	2·02
MnO	tr.
MgO	1·30
CaO	1·99
Na_2O	5·59
K_2O	6·74
P_2O_5	0·11
H_2O (direct)	·30
CO_2	trace
Cl	trace
SO_3	trace
	99·83

Another typical specimen collected by the present writer on the 6,820-foot summit about four miles north of the Boundary and thus well toward the center of the batholith has been studied quantitatively according to the Rosiwal method. The weight percentages of the constituents were found to be approximately as follows:—

2 GEORGE V, A. 1912

Quartz	5·1
Sodiferous orthoclase and microperthite	51·2
Andesine, Ab_3An_1	17·9
Hornblende	20·2
Augite	1·5
Magnetite	1·7
Titanite	1·6
Apatite and zircon	·8
	100·0

From these proportions the chemical composition of this specimen has been roughly calculated. It is assumed that the hornblende has the same composition as that of the hornblende in the 'quartz-monzonite' of Mt. Hoffmann, Cal., and that the alkaline feldspars are present in the ratio of two of orthoclase to one of albite. The result is as follows:—

SiO_2	59·2
TiO_2	·7
Al_2O_3	15·9
Fe_2O_3	2·2
FeO	2·7
MgO	2·8
CaO	4·9
Na_2O	3·6
K_2O	5·8
H_2O	·4
P_2O_5	·4
Remainder	1·4
	100·0

The soda is probably too low, yet the calculation seems to show that the analysis made for Mr. Brock would correspond well with that of the typical specimens collected during the Boundary survey.

The rock is evidently a typical hornblende-biotite pulaskite. Sometimes the biotite is almost or entirely absent, though the composition suffers no other essential change—giving an alkaline hornblende syenite. More rarely, the interstitial quartz increases notably and the syenite has the composition and habit of the more acid hornblende-biotite nordmarkite.

Basic Phase at Contact.—The most notable change in composition is found in a strong basification along the main contact. This was observed at the southern contact along the Dewdney trail, on Record mountain and on the western contact north of Sutherland creek, but the most signal illustration or on the northwestern contact near Coryell. In the last locali y the railway tings and the high bluffs to the south of the railway track display the basified zone very finely. It is there at least a mile wide and much wider than on the other sides of the batholith. The basic phase appears to merge gradually into the normal pulaskite on the south. The rock is rather dark gray, coarse-grained, and strikingly rich in the femic minerals, hornblende, diopside, and biotite, named in order of decreasing abundance. Andesine, near Ab_5An_3, is the prevailing feldspar, and accompanies orthoclase and microperthite. The accessories here include apatite, magnetite, and abundant titanite. Quartz seems to fail entirely. The order of crystallization is the same as in the dominant pulaskite.

A large fresh specimen (No. 517) has been analyzed by Mr. Connor with the following result:—

Analysis of basic contact phase (monzonite), Coryell batholith.

		Mol.
SiO_2	52·38	·873
TiO_2	1·10	·014
Al_2O_3	15·29	·150
Fe_2O_3	2·99	·019
FeO	5·53	·076
MnO	·10	·001
MgO	5·84	·146
CaO	7·30	·130
SrO	·15	·001
BaO	·25	·002
Na_2O	3·68	·060
K_2O	3·84	·040
H_2O at 110°C.	·21	....
H_2O above 110°C.	·63	....
P_2O_5	·75	·005
	100·04	
Sp. gr.	2·847	

The norm was calculated to be:—

Orthoclase	22·24
Albite	28·30
Anorthite	13·90
Nephelite	1·70
Diopside	15·20
Olivine	9·44
Magnetite	4·41
Ilmenite	2·13
Apatite	1·55
Water	·84
	99·71

The mode (Rosiwal method) is approximately:—

Sodiferous orthoclase and microperthite	21·5
Andesine	26·3
Hornblende	26·4
Augite	10·7
Biotite	9·5
Magnetite	3·0
Titanite	2·2
Apatite	·7
	100·0

In the Norm classification the rock enters the sodipotassic subrang. monzonose, in the domalkalic rang. monzonase, of the dosalane order. germanare. According to the older classification the rock is a typical hornblende-augite-biotite monzonite.

2 GEORGE V, A. 1912

Apophyses.—Finally, a more acid phase of the batholith is represented in the numerous apophyses cutting the Rossland volcanics all about the batholith. These offshoots have the composition of an alkaline syenite porphyry, which is very poor in the femic (phenocrystic) minerals, hornblende and biotite. The plagioclase (basic oligoclase, near $Ab_x An_x$) is here quite subordinate to the alkaline feldspars. The accessories are those of the monzonite but are in very small amounts. The specific gravity of a type specimen is only 2·601.

The apophyses are clearly not direct injections of the basified magma now found in crystallized form just inside the main contact-line. They are differentiates of the main mass of magma and are analogous to the familiar aplites given off by granitic intrusive bodies. The differentiation was doubtless aided by the special abundance of magmatic fluids, which lowered the viscosity, so that these feldspathic dikes have run out many thousands of feet from the main contact. The strong mineralization often observed in the traps cut by these apophyses may be connected with the act of expelling the fluids during the crystallization of the porphyry.

Contact Metamorphism.—The batholith has exerted notable mechanical and thermal effects on its country-rocks. The traps of the Rossland volcanic group have been converted into hornblende-quartz schist and hornblende-biotite-epidote schist. These rocks, with their schistosity-planes directed peripherally about the batholith, can be traced through a distinct exomorphic zone from three to six hundred yards or more in width. In the old sediments of the Sutherland complex, andalusite has been liberally developed, but it is not easy to say how far the metamorphism of those rocks was due to regional processes, and is thus of older date than the batholithic intrusion.

SYENITE AND GRANITE PORPHYRIES SATELLITIC TO THE CORYELL BATHOLITH.

Along the southern border of the Coryell batholith a long area some two square miles in extent has been mapped as underlain by syenite porphyry. This body has every appearance of being simply a late differentiate of the Coryell syenite, a mass which has been injected along the contact of the main batholith with the volcanic rocks. The porphyry is younger than the coarse syenite, since it encloses blocks of the latter and sends tongues into it. Yet the mineral components of the two bodies are similar. The porphyry has phenocrysts of augite, sometimes of brown hornblende, as well as of biotite, alkaline feldspar, and a few soda-lime feldspars. The ground-mass is essentially like that in the chonolithic rock about to be described; in fact, these two rocks are so alike in mineralogical and chemical constitution that the account of the chonolithic rock, which has been chemically analyzed, will serve for both. The principal difference between them consists in the fact that the larger body has a coarser grain.

South and southeast of the batholith there are very numerous dikes and one irregular intrusion (chonolithic in its relations) which respectively cut the Rossland volcanics and the Sophie Mountain conglomerate. These have rather constant characters and for them also the description of the one, the chonolithic rock, will suffice.

Chonolith.—Just north of the Boundary slash at Monument 169 a relatively large body of the porphyry cuts the Rossland volcanics and the overlying conglomerate which has been described on a previous page. The form of the body

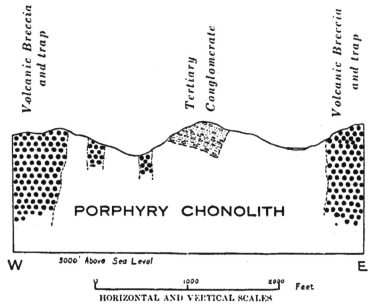

is highly irregular, though there is little doubt that it is an injected mass, thrust into the invaded formations, much as a dike is injected. The body is, thus, probably, a chonolith, rather than a stock or other subjacent body which enlarges downward indefinitely. Figure 24 gives a diagrammatic profile-section of the chonolith, the roof of which is still partly preserved. The rock of the

2 GEORGE V, A. 1912

chonolith will be specially described, as it also represents the type of the porphyry where exposed in the many dikes between Sophie mountain and the water-divide eight miles to the westward.

At distances of fifty feet or more from a contact the porphyry of the chonolith is a rather light gray rock having abundant phenocrysts of soda-orthoclase, andesine (near Ab₃ An₁), biotite, and augite, embedded in a fine-grained, feldspathic base. The feldspar phenocrysts are in thick-tabular Carlsbad twins and are characteristically glassy, like sanidine; they have lengths varying from 5 mm. to 15 mm. The lustrous, highly idiomorphic, black foils of biotite measure from 1 mm. to 2 mm. in diameter; the likewise idiomorphic, diopsidic augite stands out in stout prisms 2 mm. to 3 mm. in length. The andesine phenocrysts are often surrounded by a thick shell of orthoclase, the two feldspars then having common basal cleavage. The ground-mass is a fine-grained hypidiomorphic-granular aggregate of orthoclase individuals, associated with a little oligoclase and considerable interstitial quartz. The accessories, titanite, titaniferous magnetite, and apatite, also form part of the ground-mass. A fresh specimen (No. 409) of this phase has been chemically analyzed by Mr. Connor with result as here noted:—

Analysis of syenite porphyry.

		Mol.
SiO₂	60·51	1·008
TiO₂	·60	·008
Al₂O₃	16·71	·164
Fe₂O₃	1·72	·011
FeO	3·34	·046
MnO	·10	·001
MgO	2·53	·063
CaO	3·62	·064
SrO	·12	·001
BaO	·10	·001
Na₂O	4·64	·075
K₂O	5·20	·055
H₂O at 110° C	·03	
H₂O above 110°C	·27	
P₂O₅	·16	·001
	99·65	
Sp. gr.	2·667	

In the Norm classification the rock enters the sodipotassic subrang, monzonose, of the domalkalic rang, monzonase, in the dosalane order, germanare. The norm is as follows:

Quartz	2·40
Orthoclase	36·08
Albite	39·30
Anorthite	9·45
Hypersthene	6·81
Diopside	6·55
Magnetite	2·55
Ilmenite	1·21
Apatite	·31
H₂O	·30
	99·46

According to the older classification the rock is a typical augite-biotite syenite porphyry.

Nearer the contact of the chonolith the porphyry assumes a much finer grain and a deeper colour, namely, dark greenish-gray. The phenocrysts are soda-orthoclase, plagioclase, and biotite. Augite is absent, both among the phenocrysts and in the ground-mass. The phenocrystic plagioclase regularly affords the extinction-angles of labradorite ($Ab_4 An_6$ to $Ab_3 An_5$) and seems, therefore, to be persistently more basic than the plagioclase of the augite-bearing phase. The ground-mass, in everything but size of grain, seems to be like the ground-mass of the analyzed rock. This augite-free phase is an alkaline biotite syenite porphyry. Its specific gravity was measured and found to be precisely the same as that of the augite-biotite syenite porphyry, namely, 2.667.

Dikes.—Most of the porphyry dikes of the region carry augite among the phenocrysts and, in mineralogical and chemical characters, are practically identical with the analyzed phase of the chonolith. A few dikes have hornblende in place of augite and a few others carry only biotite as femic phenocrysts. The dikes are exposed in great size and number on the west slope of Sophie mountain and some of them are clearly connected with the mineralization of the rocks whence the Velvet and Portland mines have drawn their ore-supplies. (See R. W. Brock, Summary Report, Geol. Survey of Canada for 1900, page 75A). All these injected porphyries are at least as recent as the Coryell batholith intrusion and may be contemporaneous with it.

A dozen or more syenite porphyry dikes, ranging from 8 to 20 feet in width, cut the Sheppard granite stock south of Lake mountain. The microscopic examination of one specimen has shown a strong resemblance to the porphyry of the dikes and chonolith west of Sophie mountain. So far as such a fact may be used for correlation, it affords evidence that the syenite porphyry displayed in the fringe of dikes south of the Coryell batholith and perhaps the Coryell syenite itself are younger than the Sheppard granite. This view is corroborated by the fact that the Sheppard granite stock on the north side of the Pend D'Oreille river is cut by numerous dikes of a rock which appears to be greatly altered porphyritic olivine-augite-biotite monzonite of gabbroid habit, i.e., dikes which may possibly be correlated with a younger member of the Rossland volcanic group. So far as known, the syenite porphyry dikes, though abundant and often well exposed, are never cut by gabbroid or monzonitic rocks nor is the Coryell batholith cut by them.

The evidence is thus fairly good that these syenitic rocks are, next to the granite porphyry, now to be noted, the youngest intrusives of the region.

A few dikes of gray biotite-granite porphyry cut the Rossland volcanics and at two points, clearly cut dikes of the syenite porphyry. One of these two localities is immediately northeast of the section through the chonolith just described, and it is probable that the syenite porphyry traversed by the granite porphyry forms part of the chonolith itself. This granite porphyry dike is only five feet wide but bears orthoclase phenocrysts up to 2.5 cm. in length,

2 GEORGE V, A. 1912

along with smaller ones of quartz, biotite, and acid oligoclase, near Ab, An,. The ground-mass is a microcrystalline granophyre of quartz and feldspar, carrying a little accessory apatite and magnetite.

A dike of about the same size and essentially of the same composition (though with minute biotite in the ground-mass) cuts a thick sill-like dike of syenite porphyry out-cropping just north of the Boundary line on the Velvet Mine wagon-road. This dike follows a master-joint plane in the older porphyry. Two hundred yards farther north on the wagon-road a thicker intrusion of the same granite porphyry follows the bedding of the Sophie mountain conglomerate.

MISSOURITE DIKE.

In the col between Record mountain and Granite mountain, west-northwest of Rossland, the Coryell syenite is cut by a five-foot dike of rock, which in composition is unique among all the specimens collected during the Boundary survey. It is a dark brownish-green, fine-grained, somewhat porphyritic trap, apparently corresponding mineralogically and chemically to an olivine-free missourite, bearing phenocrysts of pseudoleucite. In the hand-specimen a few, small black crystals and innumerable glints of light from minute foils of mica may be discerned. The pseudoleucite phenocrysts are conspicuous but do not constitute more than five per cent of the rock by weight.

In thin section the pyroxene is seen to occur in highly idiomorphic prisms from 1.5 mm. to 0.1 mm. or less in length. The pale greenish colour of the mineral, a lack of pleochroism, and a big angle of extinction indicate that it is a common augite. The mica is a strongly pleochroic, brown biotite and is also thoroughly idiomorphic. Abundant cubes of magnetite (probably titaniferous) and many, relatively large prismatic crystals of apatite are accessory. All of these minerals are embedded in a pale greenish to brownish matrix, largely composed of the same material as that forming the phenocryst-like areas referred to pseudoleucite.

The diagnosis of the phenocrysts and of the related ground-mass of the rock has offered considerable difficulty. The phenocry. ranging from 1 mm. to 3 mm. in diameter, have roundish, polygonal outlines of the order expected from idiomorphic leucite. They are habitually wrapped about with foils of fresh, primary biotite, arranged tangentially about the round phenocrysts. Notwithstanding the perfect freshness of augite and biotite, none of the original substance of the large phenocrysts seems to remain. Each phenocrystic mass is chiefly made up of pale greenish-gray, spherulitic aggregates of fibrous material showing aggregate polarization, with the black cross in parallel polarized light. The spherules are not clean-cut but fade into each other most irregularly. The long, hair-like elements of the spherulitic substance are so thin that it is difficult to be sure of their proper colour or of their full reaction to polarized light. The single element is probably quite colourless. It always shows negative optical character with respect to its length, and the extinction angle of the crystallite is never more than about 5°, corresponding to the range of extinctions in an orthoclase crystal elongated parallel to the *a* axis. The birefringence is low and is like that of orthoclase in very thin sections. In fact, it seems highly probable that most of the material which has replaced the original phenocrysts, is orthoclase. This conclusion is upheld by the study of the chemical analysis of the rock.

The spherulitic substance is regularly mixed with a small amount of obscurely granular material, showing characters like those of hydronephelite, and with other, similarly obscure, pale greenish-gray substance in minute leaf aggregates which have the optical properties of serpentine. A zeolite, like stilbite or desmine, may also be present. All of these materials form a matrix in which very small microlites of augite, biotite, magnetite, and apatite - inclusions in the original minerals are embedded.

To the writer the best interpretation of these round bodies is that they represent pseudomorphs after phenocrystic leucite; their optical resemblance to the described pseudoleucites is certainly great. The alteration of the leucite seems to have taken place as a kind of magmatic after-action, rather than as the result of ordinary weathering, for the ferromagnesian minerals are ideally fresh.

The ground-mass in which the large pseudoleucites and the other idiomorphic minerals lie, is generally quite like that of the pseudomorphs except that there are no outlines even remotely suggesting the crystal form of leucite. Neither here nor in the phenocrystic bodies is there any certainly isotropic material, nor any structure which could have been inherited from the twinning bands of leucite. Nevertheless, the similarity of ground-mass and phenocryst indicates that they were originally composed of the same material, chemically if not mineralogically. The simplest assumption is that the ground-mass of the rock was chiefly allotriomorphic leucite, which, like the leucite of the phenocrysts, was unstable during the cooling period following crystallization.

This view, cannot, with the material in hand, be proved, but it is strongly upheld by the close chemical parallel between this rock and the typical missourite described and named by Pirsson.§ In that species the constituent minerals are apatite, iron ore, olivine, augite, biotite, leucite, and some zeolitic products. The leucite is there unquestionably present and is interstitial. The ob ious differences between the Record mountain dike (as originally crystallized) and the type missourite consist in the presence of about five per cent of phenocrystic leucite and the absence of olivine in the British Columbia dike. The presence of olivine in a rock of this kind is not a matter of principal importance, for, as n has pointed out, biotite may be considered as the chemical equivalent of a mixture of leucite and olivine.

In Table XXV the result of Mr. Connor's analysis of the dike specimen No. 541, Col. 1; molecular proportions in Col. 1a) and the analysis of the type missourite from the Highwood mountains, Montana, (Col. 2), are given.

§ L. V. Pirsson, Bull. 237 U.S. Geol. Survey, 1905, p. 115.

2 GEORGE V, A. 1912

Table XXV.—Analyses of missourite.

	1.	1a. Mol.	2.
SiO₂	42·31	·705	46·08
TiO₂	2·00	·025	·73
Al₂O₃	11·40	·112	10·01
Fe₂O₃	4·07	·026	3·17
FeO	6·11	·085	5·61
MnO	·11	·001	tr.
MgO	11·31	·283	14·74
CaO	11·02	·196	10·55
SrO	·16	·002	·20
BaO	·84	·004	·33
Na₂O	·82	·013	1·31
K₂O	3·69	·039	5·14
H₂O (at 110°C)	2·28	···· }	1·44
H₂O (above 110°C)	2·72	····	
P₂O₅	1·44	·010	·21
SO₃	··	····	·05
Cr₂O₃	·05	····	····
Cl	····	····	·03
CO₂	tr.	····	····
	100·13		99·57

Sp. gr. .. .. 2·817

The calculated norm is:—

Orthoclase	21·68
Nephelite	3·39
Anorthite	16·68
Diopside	24·57
Olivine	15.16
Magnetite	6·03
Ilmenite	3·80
Apatite	3·10
Water	5·00
	99·41

In the Norm classification the rock enters the dopotassic subrang, absarokose, of the alkalicalcic rang, camptonase, in the salfemane order, gallare.

The actual mineral compos. on of the dike (determined by the Rosiwal method) and that of the type missourite agree with the chemical comparison in indicating the place of the Record mountain dike in the missourites of the prevailing classification.

	Shonkin Stock, Montana.		Record Mt. Dike.
	(by weight.) %		(by weight.) %
Augite	50·0	Augite	37·5
Biotite	6·0	Biotite	20·0
Leuc.....	16·0	Pseudoleucite, etc.	33·8
Olivine	15·0	Magnetite	5·7
Analcite	4·0	Apatite	3·0
Zeolites	4·0		
Iron Ore	5·0		100·0
	100·0		

This dike may, thus, be described as a somewhat porphyritic, olivine-free missourite. No other intrusion of this rock has as yet been found in the Rossland region, but closer search may lead to the discovery of other bodies.

VARIOUS OTHER DIKES.

There are undoubtedly many thousands of dikes in the area covered by the Boundary belt within the Rossland mountains. Most of them are more or less clearly apophyses of the stocks and batholiths of the region or else their more aplitic derivatives. As such the types have already been briefly described. The intrusion of certain of the dikes has been accompanied, or closely followed, by the formation of large mineral deposits. Many others have been observed which are of importance in showing the relative ages of the formations. From a purely petrographic point of view, however, most of the ... s have few features which make them worthy of special description.

The relation of the Beaver Mountain group to the Rossland latites may possibly be indicated by the occurrence of a dike of monzonitic porphyry cutting the Beaver Mountain sediments at the 2,800-foot contour on the spur running up eastward from Champion railway station. The dike (or sill?) is about 100 feet wide and seems to strike due east and west. It is a fresh, medium-grained, dark greenish-gray rock, porphyritic through the prominence of large, lustrous biotites. The mass of the rock is a hypidiomorphic-granular aggregate of augite, hornblende, biotite, labradorite, and orthoclase, with accessory ilmenite, apatite, and titanite. The specific gravity is 2·867. The habit of this rock is much like that of the chemically analyzed porphyritic olivine syenite, though the plagioclase seems here to be relatively much more abundant. The rock has been classified as a porphyritic hornblende-augite-biotite monzonite with phenocrysts of biotite. It may be contemporaneous with the Rossland monzonite, which in that case, would be younger than the Beaver Mountain sediments. The tuffs associated with those sediments are cut by diabasic dikes and by labradorite porphyrite dikes which are doubtless the intrusive equivalents of some of the bas ... and andesitic flows in this volcanic group. One of the labradorite porphyrite dikes, bearing phenocrysts of labradorite in a diabasic ground-mass of ... zite and basic plagioclase, forms one of the walls of an auriferous quartz-vein at the Princess mining claim.

The fern-bearing argillite at the Little Sheep creek locality is cut by a thirty-foot, north-south, vertical dike of typical augite-biotite monzonite porphyry, which may fairly be regarded as apophysal from the large Rossland stock of monzonite. Quite similar dikes cut the older members of the Rossland volcanic group on the railway between Trail and Rossland. Dikes of hornblende-bearing augite-biotite monzonite porphyry cut the great intrusion-breccia about the Trail batholith at the Columbia river.

Close beside the last-mentioned dike but without evident age relation to it is a somewhat unusual rock occurring in the form of a two-foot dike cutting the Trail granodiorite near its contact, and striking N. 15° W., with a dip of 80° to the eastward. It is a dark gray-green rock, compact and diabasic

2 GEORGE V, A. 1912

or trappean in look. Under the microscope it is seen to be the exact analogue to a common fine-grained diabase except that a brownish-green hornblende replaces the usual augite. The hornblende has been partly altered to a fibrous condition, and has a pale greenish tint; but there is no evidence that augite has ever been present in the rock. It was, of course, necessary to determine whether the uralitic secondary product had been derived from pyroxene. A careful study of the thin section has led to the conclusion that it rather represents this clearly uncommon form of alteration from hornblende. In any case, there can be no question that much original, green hornblende has in this rock an intersertal relation to the other essential, labradorite-bytownite. Texturally the rock bears the same relation to camptonite that diabase bears to certain dike-gabbros in which the pyroxene crystallized before the feldspar.

South of Trail the granodiorite is cut by many trap dikes which, unfortunately, have not been studied microscopically. Perhaps such a study would declare with certainty the age relations of the batholith and the various members of the Rossland volcanic group.

The Sutherland schist complex is cut by a number of dikes of hornblende-biotite monzonite porphyry, which may be contemporaneous with the Rossland latites and monzonite.

True lamprophyre dikes are rare along the traverses made by the writer. The minettes cutting the Pend D'Oreille phyllites and slates near the Columbia and the Boundary line have already been noted. In the immediate vicinity of the Rossland mines mica-lamprophyres are very common and have been studied by Young and Brock.

About one-quarter mile west of the forty-fourth mile-post on the railway between Coryell and Cascade, a four-foot, porphyritic dike of camptonitic habit cuts a second dike of gabbro which itself cuts a volcanic breccia belonging to an old member of the Rossland volcanic group as mapped (though probably pre-Cretaceous in age). The younger dike is composed of beautifully crystallized, idiomorphic crystals of green hornblende, augite, and plagioclase, embedded in a microcrystalline, trachytic ground-mass of plagioclase and the same femic minerals. Like the Coryell syenite the rock is very fresh and quite uncrushed, and it may represent a rather acid camptonite which has been derived from that batholith.

SUMMARY OF STRUCTURAL RELATIONS IN THE ROSSLAND MOUNTAINS.

According to their degree of deformation the stratified rocks of these mountains may be classified in three divisions. The first, characterized by highly complex crumpling and by mashing, includes the formations of pre-Mesozoic age; all of them seem to be Palæozoic, as there is no suggestion anywhere of the occurrence of a pre-Cambrian terrane in the Rossland mountains. The second division includes the flows, pyroclastics, and interbedded sediments of the Rossland volcanic group, which have usually high dips but lack the chaotic structure due to orogenic mashing. The third division covers only the various patches of conglomerate and sandstone, mapped on Sophie mountain, Lake mountain, etc.; the dips of these beds may be locally high but on the average much lower than the dips of either of the other two rock-divisions. It will be recalled that considerable areas, coloured on the map as 'Rossland Volcanic Group,' are really underlain by the traps and greenstones of Paleozoic age. The separation of these from the Mesozoic portion of the Rossland group has so far proved impossible.

The unconformities demonstrated within the Boundary belt are two in number. The one occurs between the Paleozoic complex and the Mesozoic members of the Rossland volcanic group and associated sediments; the other, between the coarse conglomerates of Sophie mountain, etc., and the older members of the Rossland volcanic group.

Excepting the minute crumples, folds are seldom decipherable in any part of these mountains. The much broken anticline (?) at Little Sheep creek is, in fact, the only element in the belt which shows the semblance of the arch-trough structure characteristic of simpler ranges. Faults are certainly very numerous but their mapping was out of the question in the time allowed for this part of the Boundary section; an obvious difficulty in the way of making a useful map of the faults is the general absence of horizon-markers. The primary importance of the breaks and slips in the igneous rocks particularly to the economic geology of the district is emphasized by Messrs. Brock, Young, and others. The Velvet mine on the western slope of Sophie mountain is located on a zone of master faulting, the dislocations occurring along a number of nearly vertical, meridional faults. This zone of faults has determined the location of (the western) Sheep Creek valley. Another master fault or zone of faulting is strongly suspected along the axis of the deep valley of Christina lake, whereby the traps on the east have been brought into contact with the Cascade gneissic batholith; this hypothesis cannot as yet be proved.

The structural relations of the igneous bodies have already been discussed in the respective descriptions of the formations. It will suffice here to note the salient facts. The Trail and Coryell batholiths are typical cross-cutting bodies, with the usual appearance of having replaced their country-rocks for many thousands of feet of depth, in each case. The contact shatter-zone of the Trail batholith is perhaps the finest, because the widest and also best exposed, in the whole Boundary belt. The shatter-zone of the Coryell batholith is not so conspicuous at any point, though this body likewise encloses blocks of the invaded traps and schists. Igneous bodies which have been injected without replacing country-rocks in the sense of assimilating them in some fashion, are extremely numerous; they include the thousands of dikes, the volcanic neck (?) at Rossland, as well as chonolithic masses, such as the one of syenite porphyry south of the Coryell batholith, the body of abnormal olivine syenite near Christina lake, and the various bodies of dunite and other peridotites.

The whole region has evidently been under the powerful control of igneous action, both volcanic and batholithic. The batholithic masses, including stocks, are clearly related in their genesis to periods of intense mountain-building which cannot be well dated from facts derived from this local study and must be dated from analogies with outside regions. Using such additional informa-

25a—vol. ii—24½

tion the writer is inclined to recognize three periods of strong mountain-building for these mountains; one, late Jurassic; a second, post-Laramie and pre-Miocene; the third, late Miocene. Each period seems to have been immediately followed by batholithic intrusion. Apart from the Cretaceous and earlier vulcanism, as well the important erosion-periods registered in the unconformities, the structures of the Rossland mountains are largely explained by the grand events just enumerated. So far as recorded in the exposed rocks the region appears to have been above sea since Carboniferous times, though at any time discoveries may show the presence of Mesozoic or Tertiary marine strata in the volcanic complex.* Meantime, it can be stated that this mountain group owes its principal structures to the repeated orogenic crushing of a very heavy volcanic pile and of its metamorphosed Paleozoic foundation.

TIME RELATIONS.

With present knowledge, the chronicle of geological events in the Rossland mountains can be only partially deciphered. The difficulties in the way of completing a systematic survey of their history are the usual ones encountered in the Boundary survey as, indeed, in most areas of complex mountains. Imperfect exposures, the rarity of sedimentary formations and the even more notable scarcity of fossils form part of the difficulties, but the great variety and obscure relations of the igneous rock-bodies, both extrusive and intrusive, are responsible in special degree for the uncertainties still affecting the geology of these mountains. A tentative scheme of the geological events will be offered in the present section. The grounds on which the scheme is based belong to three classes: first, those which may be rated as observed facts; secondly, those which are regarded as more or less strong probabilities; and, thirdly, those which are to a large extent theoretical, embodying principles derived from other fields.

Observed Facts.—It is convenient to survey the known facts of relation in outline, as follows. Certain of the associated probabilities will be noted in direct connection with these statements.

1. The crystalline limestone of Little Sheep creek valley and a bed of calcareous quartzite at the O.K. mine west of Rossland are obscurely fossiliferous. The limestone is crinoidal and is lithologically similar to that occurring as fragments in the volcanic breccia of Sophie mountain. In those fragments McConnell found fossil remains which have been regarded as probably of Carboniferous age. All these limestones are lithologically similar to the Pend D'Oreille limestone across the Columbia river. The fossils discovered by Brock in the O.K. mine quartzite are tentatively referred by Ami to the Carboniferous.

2. The much less metamorphosed Sophie mountain conglomerate, and the argillite-sandstone series occurring at the crossing of Little Sheep creek and the Boundary line are also fossiliferous; in each case the remains are those of

* The more massive phases of the Rossland volcanic group resemble the Nicola Triassic lavas on the South Thompson river.

land-plants. Those found at the creek have been tentatively referred by Penhallow to the Lower Cretaceous. For present purposes it seems safer to refer these beds, more broadly, to the Mesozoic. Those occurring in the conglomerate are too poorly preserved to be of stratigraphic service. The relatively unmetamorphosed Beaver Mountain sediments carry abundant, carbonaceous plant-stems but no useful fossil has been discovered.

3. The volcanic breccias of Malde and Sophie mountains overlie the Carboniferous (?) sediments unconformably.

4. The Trail granodiorite batholith cuts schists which are almost certainly the equivalent of the Pend D'Oreille schists. It cuts the older lavas (andesites and basalts) of the Rossland volcanic group and the ultra-basic monzonite and hornblendite at the Columbia river.

5. The Sheppard granite cuts the Trail granodiorite, the Pend D'Oreille schists and the conglomerate on Lake mountain.

6. The Coryell syenite batholith cuts the youngest recognized members of the Rossland volcanic group as well as the Sutherland schistose complex.

7. Numerous dikes of biotite-augite syenite porphyry cut the Rossland volcanics, the Sophie mountain conglomerate and the conglomerate at Monument 169. A few dikes of this porphyry cut the Coryell syenite.

8. Dikes of biotite-granite porphyry cut the syenite porphyry just mentioned.

9. The Coryell syenite is cut by at least one dike of missourite and by narrow dikes of syenite-aplite.

10. The Sutherland schists are cut by at least one dike of unsheared camptonite, which may be a lamprophyric derivative of the Coryell syenite.

11. The Sophie mountain conglomerate is cut by dikes of monzonite porphyry; others of the same kind of dikes cut the Beaver Mountain lavas.

12. The small stock of crushed biotite granite east of Cascade cuts the greatly metamorphosed andesitic rocks or greenstones mapped as part of the Rossland volcanic group but probably of Paleozoic or, at least, pre-Cretaceous date.

13. This biotite-granite stock is cut by dikes of dunite. Masses of dunite cut the older andesitic lavas of the Rossland group.

14. The Fife and Baker gabbros and peridotites cut the old greenstones just mentioned.

15. A small mass of biotite-olivine-diallage peridotite cuts the Baker gabbro.

16. Dikes of the syenite porphyry mentioned under "7" cut the small body of "harzburgite" northwest of Monument 172.

17. Large dikes of biotite-granite porphyry (probably apophysal from the Trail batholith) cut the Pend D'Oreille schists near the crossing of the Columbia river and Boundary line; dikes of augite minette cut both schists and granite porphyry. A dike of hornblende-augite minette cuts the dikes of augite minette.

18. Many dikes of minette and some of kersantite cut the Pend D'Oreille schists east of the Columbia river. According to Brock and Young, dikes of

2 GEORGE V, A. 1912

minette, kersantite, odinite. spessartite, and vogesite cut the monzonite and 'augite porphyrite' at Rossland, where the mica-lamprophyres are of two ages, separated by a period of ore-formation.

19. The Rossland monzonite cuts the older (andesitic) members of the Rossland volcanic group. The monzonite is cut by intrusives lithologically identical with the Coryell syenite and by dikes of alkaline syenite porphyry (probably equivalent to the biotite-augite syenite porphyry mentioned under '7'). McConnell, Brock, and Young consider that the granite cutting the Rossland monzonite is equivalent to the Trail ("Nelson" or "Older") granodiorite.

20. The peculiar porphyritic (facetted) olivine syenite forms small irregular masses cutting the older members of the Rossland volcanic group and the still older Sutherland schists.

21. Some evidence on age relations may be derived from the amount of crushing and dynamic metamorphism suffered by each of the different formations. The observed facts may be here summarized.

The sedimentaries of the Rossland mountains are all strongly deformed; in nearly all of the bodies high to vertical dips have been measured. The Pend D'Oreille group of rocks, the Sutherland complex, and the fossiliferous limestone-chert-quartzite series in Little Sheep creek valley are mashed and intensely metamorphosed. The plant-bearing (Cretaceous?) argillites and sandstones in Little Sheep creek valley are crumpled and faulted greatly but are not much metamorphosed. The Sophie mountain conglomerate has been energetically upturned but is not much metamorphosed at any observed point. The other bodies of conglomerate show lower dips and an induration of about the same order as those seen at Sophie mountain. The Beaver Mountain sediments show dips rarely surpassing 50° and are not metamorphosed beyond the point of decided induration.

The eruptives may be divided into three classes according to the amount of crushing and metamorphism effected in each body, the evidences being controlled by microscopic examinations.

The greenstones of the Pend D'Oreille and Rossland volcanic groups and the biotite-granite stock east of Cascade have been intimately crushed and largely recrystallized.

The second class, representing bodies which have been sheared only locally and are little metamorphosed, includes many of the andesites and basalts of the Rossland group; the Fife and Baker gabbros, the Rossland monzonite and latites (rarely sheared); the Beaver Mountain volcanics (very rarely sheared); the Trail batholith (often strained); the Sheppard granite (very rarely sheared or strained); some of the minettes and other lamprophyres.

The third class includes bodies which are not noticeably (in field or laboratory) crushed, strained, or metamorphosed dynamically. These are: the Coryell syenite and the dikes cutting it (biotite-augite syenite porphyry, biotite granite porphyry, aplite, missourite); many minettes and other lamprophyres; the porphyritic olivine syenite.

Probable Relations.—The more detailed descriptions of the formations contain statements of certain conclusions which can only be regarded as a fairly strong balance of probability in each case. A short summary of these views will be of use in approaching the final correlation. The evidences in their favour will, for the most part, be found on earlier pages.

1. The fern-bearing Mesozoic argillites and sandstones of Little Sheep valley overlie the Paleozoic (Carboniferous?) limestone-chert-quartzite series unconformably and are overlain, with apparent conformity, by the Mahle mountain-Sophie mountain breccia (Rossland group).

2. The conglomerates shown in the four mapped areas are equivalent in age and are younger than some at least of the andesites and basalts of the Rossland volcanic group, while perhaps slightly older than the latites.

3. The monzonite porphyry dikes cutting the Sophie mountain conglomerate, the Beaver Mountain sediments, and the Little Sheep creek (Cretaceous?) sediments are of the same age as the Rossland monzonite.

4. The latites are genetically connected with the Rossland monzonite and both are younger than the great mass of the Rossland andesite and basalt.

5. The gabbros, the dunites, and other peridotites between Rossland and the Christina lake-Kettle river valley are genetically connected with the andesite-basalt (not greenstone) phase of the Rossland volcanic group.

6. The syenite-porphyry dikes cutting the Sheppard granite south of Lake mountain are the equivalents of the apophyses from the Coryell syenite batholith.

7. It is assumed that the last great orogenic revolution which has affected this region was that at the close of the Laramie. Vertical to very steep dips are taken, therefore, to mean that the rocks so deformed are of pre-Eocene age. All the sedimentary formations and, so far as known, all the lavas and pyroclastics of the ten-mile belt often show dips which are much higher than those characterizing, for example, the Oligocene beds west of Midway. Moderate folding and faulting probably affected the Rossland mountains during or at the close of the Miocene (as in the region west of Midway), but it has not proved possible to distinguish the results of that deformation from those due to the post-Laramie revolution. The Pend D'Oreille group and the other (probably) Paleozoic sedimentaries of the region were doubtless more or less deformed near the close of the Jurassic, when these rocks may have been crumpled and mashed to a degree rivalling their present condition.

8. The tentative correlation is partly based on the law that granitic (batholithic) intrusion follows periods of more or less intense mountain-building and seldom or never affects undeformed strata. The uncrushed   ell syenite batholith and its satellites, lamprophyres, and aplites are refer.   , the post-Eocene orogenic period. The older, partially sheared but not gre.. tly crushed Trail batholith, the majority of the lamprophyric intrusions of the region, and the greatly crushed biotite-granite stock east of Cascade are referred to the late Jurassic period of deformation and intrusion.

2 GEORGE V, A. 1912

Correlation.—Combining facts and probabilities, the following table of the formations occurring in the Rossland mountains has been prepared. The formations are named in groups which are not to be considered as strictly contemporaneous but are to be interpreted in the light of the foregoing statements. The table carries a heavy burden of hypothesis and every chronological table for this district must carry the burden until the sedimentary formations are more closely dated. The general sequence of the formations is more certain than their correlation with the recognized geological systems. The table is to be read and used only in the light of the many doubts expressed or implied in the foregoing pages. On that basis the table is offered as embodying the stronger probabilities; so understood it may perhaps be of service in suggesting future observations on these difficult terranes.

Uncrushed....	Biotite-granite porphyry dikes..... .... Biotite-augite syenite porphyry dikes and chonolith.. ... Syenite-aplite dikes............. ... . Camptonite dikes........ .... Missourite dike........................... Coryell syenite batholith .... Porphyritic olivine syenite masses..... .. .. Some of the mica-lamprophyres (?)..... ... ... Sheppard granite stocks and dikes.... . ... Beaver Mountain group..................... Conglomerate of Sophie mountain, Lake mountain, etc.	*Tertiary (Miocene ? to Eocene),*
Deformed and locally crushed	Trail granodiorite batholith........................... Granite stock east of Cascade..... .... Rossland monzonite; shonkinitic rocks; hornblendite at Columbia river..... .... Latites of Rossland volcanic group. .... Dunites; harzburgite (effusive?); Fife and Baker gabbros and peridotite.......................... Much andesite and basalt of Rossland volcanic group, with argillitic interbeds............................ Plant-bearing argillite at Little Sheep creek....:.......	*Mesozoic.*

UNCONFORMITY.

Mashed and metamorphosed.	Greenstones and older andesitic rocks of area mapped as underlain by Rossland volcanic group; fossiliferous limestone, chert and quartzite of Little Sheep creek valley; phyllite, quartzite, limestone, etc., of Pend D'Oreille group; Sutherland schist complex........	*Paleozoic (Carboniferous at least in part).*

CHAPTER XIV.

FORMATIONS IN THE MOUNTAINS BETWEEN CHRISTINA LAKE
AND MIDWAY (*Middle part of Columbia Mountain System*).

GENERAL DESCRIPTION.

As one crosses the Kettle river-Christina lake valley he immediately
encounters, in the Boundary belt, a new formation which does not appear in
the Rossland mountains. It consists in a thoroughly metamorphosed, highly
gneissic granite batholith, here named for convenience, the Cascade gneissic
batholith. The eastern limit of this body, marked as it is by the strong valley
occupied by lake and river, is also a natural dividing line between the rock
formations. The batholith belongs, in fact, to a complex of formations which
centre about the 'Boundary Creek mining district,' just as the formations
east of Christina lake centre about the Rossland mining camp. Within
the five-mile Boundary belt the formations occurring between the lake and the
mountain slopes just east of Midway are believed to be all of pre-Tertiary age.
The Midway (volcanic) formation, described in the next chapter, seems to be
clearly referable to the Tertiary. It covers a relatively large area at and west
of the town. We may therefore appropriately place the western limit of the
area discussed in the present chapter, at the eastern limit of the Midway
formation. (See Maps No. 9 and 10.)

For the information here published regarding this area the writer is very
largely indebted to the printed preliminary reports on the geology of the
Boundary Creek mining district, by R. W. Brock (Summary reports of the
Director of the Geological Survey of Canada for 1901 and 1902). Mr. Brock
spent nearly all of two arduous seasons in a detailed geological study of the
Boundary belt (here about 13 miles broad) between Grand Forks and Midway.
It seemed therefore inadvisable for the present writer to attempt a thorough
survey of this stretch. He has, accordingly, simply made two rapid traverses
across the mountains between the towns mentioned, so as to attain a general
acquaintance with the rocks as described by Mr. Brock. In addition, the writer
has made closer studies of the rocks between Christina lake and Grand Forks
as well as of the Midway volcanics. It is quite possible that most of the rocks
occurring between Grand Forks and Christina lake are much older than any
of the formations which are volumetrically important in the area described
by Mr. Brock. Partly for this reason as well as to preserve in some measure
the east to west order of treatment which is being followed in this report, the
formations of the Christina range will be first described. There will follow
an abstract of Mr. Brock's results which are here recorded as seems best to

2 GEORGE V, A. 1912

suit the purposes of the present report. For convenience in discussion and in correlation a few special names are given to the formational groups as defined by Mr. Brock; this is done with the co-operation of Mr. Brock himself.

The rocks in this part of the Boundary belt are, so far, entirely unfossiliferous and it is impossible to date the different formations with assurance. The writer's experience during the Boundary survey, especially during the mapping of the Rossland mountains and of those lying between Midway and the Skagit river, has suggested certain correlations with the recognized geological periods which are somewhat different from those made by Mr. Brock. His chronological table will be reproduced in order to show the differences of conception; the table will at once be useful in illustrating the character of the rocks encountered in the 'Boundary Creek District.'

Geological Formations of the 'Boundary Creek District' (Brock).§

Pleistocene)...............	Glacial and recent deposits.
Tertiary	Injections of intrusive sheets, dikes and plutonic masses. Ore deposits, volcanic flows. Tuffs, ash beds, volcanic conglomerates, sandstone and shales, with a little lignite.
Jurassic?	Granodiorite.
Paleozoic?...............	Serpentine. Green porphyrite.
Paleozoic?...............	Green porphyrite. Volcanic conglomerates, tuffs, ash beds, with arenaceous limestone. Serpentine. Limestones, argillites, quartzite.
Crystalline schists?.......	Gneisses and schists.

Within the five-mile Boundary belt the only rocks corresponding to the Tertiary group listed by Mr. Brock belong to the Midway formation, which is not considered in the present chapter. In the sheet accompanying this report the 'Paleozoic?' volcanics are mapped under the name 'Phoenix volcanic group'; the 'Paleozoic?' argillites and quartzites are mapped under the name 'Attwood group'; the 'Crystalline schists?' are recognized as in part made up of a schistose complex, mapped under the name 'Grand Forks group' and for the rest (within the five-mile belt), made up of a highly gneissic granite intrusive into the Grand Forks complex and mapped under the name 'Cascade gneissic batholith.' At Grand Forks these crystalline rocks are cut by a small intrusive stock, mapped under the name 'Smelter granite.' The 'Green porphyrites' are here mapped with the same colour as the Phoenix group, with which the porphyrites are genetically connected and from which they are very hard to differentiate in the field.

GRAND FORKS SCHISTS.

The dominant country-rocks of the Cascade granite—the Grand Forks schists—include a series of schistose types, which have been completely or

§ Summary Report, Director of Geological Survey of Canada for 1902, p. 95.

SESSIONAL PAPER No. 25a

almost completely recrystallized, so that their primary nature is often in doubt. For the most part they seem to have been originally basic extrusives of andesitic and basaltic character; in less degree, intrusive and dioritic or gabbroid, or sedimentary, argillaceous rocks. These have been metamorphosed to ever-varying phases of amphibolite, fine-grained orthoclase-bearing hornblende schist, hornblende-ep' 'ote-plagioclase schist, actinolite schist, and biotite-diorite gneiss. Along with these, thick lenses or pods of white crystalline limestone are interbedded. The limestone is, as yet, unfossiliferous but resembles the Carboniferous limestone occurring about Rossland. It crops out on each side of the Kettle river east of Grand Forks and is tentatively mapped as forming there one large body. Mr. Brock also reports small lenses of limestone in the basic schists of Observation mountain.

Concerning the complex Mr. Brock writes (in his report for 1902, page 96): 'These rocks have a strong lithological resemblance to the Archean rocks of the Shuswap series, and are the oldest rocks found in the area covered by the present map-sheet, but they may possibly be more highly metamorphosed argillites and limestones such as are found elsewhere in this district.' The present writer has found no new facts with which to raise the doubt expressed in this sentence, but provisionally and in the interests of greater simplicity in the geological interpretation of the region, takes the second of the alternative views.

CASCADE GNEISSIC BATHOLITH.

General Description.—From Cascade to Grand Forks the five-mile belt is chiefly underlain by a relatively old intrusive body of gneissic granite which extends an unknown though but short distance to the northward and an unknown distance to the southward of the belt. Within the belt itself this mass—the Cascade gneissic batholith—covers about forty square miles.

Its eastern contacts with the Rossland volcanics and with the Sutherland schists are hidden, so that it is impossible to state, with full confidence, the relation of the granite to these other two formational groups. However, as already noted, the older traps of the Rossland group are cut by a small stock of crushed gneissic granite on the southern flank of Castle mountain, two miles east of Cascade. While the stock granite is greatly altered it seems originally to have resembled the rock of the Cascade batholith in essential respects and the correlation of the two, in a tentative way, seems permissible. The Sutherland schists, as exposed along the railway east of Christina lake, are traversed by dikes of crushed granite porphyry which may also be regarded as possibly apophysal from the Cascade batholith. At the same time, there is no means of determining whether the main contacts of the batholith with these eastern schists or traps are now intrusive contacts; the present contacts may have been established as a result of meridional faulting along the valley of Christina lake, whereby the granite has been faulted up against the trap-schist complex.

2 GEORGE V, A. 1912

The western contact of the batholith is well exposed at several points in the five-mile belt and it clearly illustrates many of the familiar phenomena of batholithic intrusion with the basic schists, gneisses, and greenstones about Grand Forks. One of the best and most accessible localities for observing this relation occurs on the railway, four miles east of Grand Forks station.

The batholith has been so intensely crushed that it is generally gneissic in high degree. So prominent is this structural feature that the whole mass has been mapped, in the reconnaissance West Kootenay sheet of the Canadian Geological Survey, as belonging to a group of Archean crystalline schists. The writer believes, however, that the batholith was intruded possibly, if not probably, long after the Cambrian period and that the gneissic structure was developed in post-Paleozoic time.

So thorough has been the shearing and mashing of the granite that not a single ledge of undeformed rock was recognized in the whole five-mile belt. In places the granite appears fairly massive, but a careful examination of the outcrop and especially the microscopic study of typical hand-specimens of this phase, show that the constituent minerals have been strained, warped, granulated, or recrystallized. The specimens which seem most nearly to approximate the original granite are light gray, medium-grained, gneissic, though not banded, aggregates of quartz, feldspar, and biotite, with a small, variable amount of accessory apatite, magnetite, titanite, and rare zircon. The original rock was thus most probably a biotite granite. It was a type differing from the most common mica granite only in carrying rather more plagioclase (andesine-labradorite, near $Ab_4 An_1$) than orthoclase. Quartz was present in large amount. If hornblende or pyroxene were essential, such a rock would form a typical granodiorite. The specific gravity of four specimens of the fairly massive phase varied from 2 674 to 2·718 and averaged 2·680.

Nature and Origin of Banding.—Much of the batholitic mass has, however, been metamorphosed into a well banded gneiss (Plate 34). The bands differ, in mineralogical and chemical composition, not only from each other but also from the rarer, more massive and less metamorphosed phase. Representative samples of the banded gneiss were taken at several localities and subjected to microscopic examination. The bands are found everywhere to belong to either one of two kinds, respectively light-coloured and dark-coloured. Except for a few isolated grains of epidote and yet rarer garnets the bands are composed of the same minerals that form a massive gneissic granite. The banding is here simply produced by the varying concentration f the mineral.

In the light-coloured and more acid bands the constituents are chiefly quartz, orthoclase, and andesine-labradorite, with quite subordinate biotite and only the barest traces of the accessories, apatite, magnetite, and titanite. The last two almost entirely or quite fail to appear in the thin sections. By the Rosiwal method a rough estimate of the weight percentages was made for a typical specimen collected on the railway track about two miles west of Cascade. The proportions are as follows:—

Sheared Cascade granodiorite, showing banded structure.   Two-thirds natural size.

SESSIONAL PAPER No. 25a

Quartz	43·3
Orthoclase	37·8
Andesine-labradorite	14·7
Biotite	3·8
Apatite, etc	·5
	———
	100·0

The specific gravity of this band is 2·636.

A similar estimate of the weight percentages of the minerals in a typical dark band gave a strongly contrasted result:—

Quartz	17·7
Orthoclase	9·5
Acid labradorite, Ab,An,	44·1
Biotite	22·5
Garnet	3·3
Magnetite	1·9
Titanite	·6
Apatite	·4
	———
	100·0

The specific gravity of this dark band is 2·980.

In the development of the banded structure there has evidently been an advanced segregation of the basic minerals, including the accessories in the dark bands, with a corresponding concentration of the quartz and orthoclase in the light bands. The specific gravities are directly related to these concentrations and differ, respectively, from the specific gravity of the more massive, less altered phase of the batholith (2·689).

The dark bands are, on the average, much narrower than the light ones, thicknesses of more than one or two inches being quite uncommon. Often they are separated by rock which, although it is gneissic, has nearly the composition of the original unsheared granite. The distribution of the dark bands is that which would characterize zones of shearing in such a batholithic mass, and it is probable that this highly micaceous phase of the gneiss has been produced through a leaching of the more basic material from the original granite, followed by the recrystallization of that material in the zones or planes of shearing. This hypothesis will be more fully presented in connection with the precisely similar phenomenon of banding in the sheared batholiths of the Cascade range. The hypothesis merits attention since it implies the idea of the efficiency of lateral secretion on a colossal scale.

Over large areas the batholith is free from intrusive dikes. A few narrow basic dikes were observed on the Canadian Pacific railway track. A thin section of a specimen from one dike occurring near the western contact of the batholith showed evidence of profound alteration. The dike-rock is now a mass of chlorite, pyrite, and plagioclase, and originally was probably of diabasic composition.

SMELTER GRANITE STOCK.

Immediately northwest of Grand Forks the Cascade batholith and the Grand Forks schists are cut by a boss or small roundish stock of a quite different

2 GEORGE V, A. 1912

granite. The stock covers about 1.5 square miles; its eastern contact runs close to the smelter buildings, and for the purpose of distinction, the body may be called the Smelter stock.

This granite is a light flesh-pink to pinkish-gray, fine-grained rock, generally massive but sometimes showing a parallelism among the constituents. Quartz, feldspar, lustrous black amphibole, and dark-green pyroxene are discernible in the hand-specimen. Each of the last two is quite subordinate both in size and number of individuals. Under the microscope the feldspar is seen to be very largely orthoclase, with a very small, accessory amount of plagioclase which is probably a highly acid oligoclase. Considerable titanite in euhedra and anhedra, and a little magnetite are the other accessories. The amphibole is a pleochroic green hornblende; the pyroxene is a strongly coloured, green monoclinic augite, apparently related in its optical properties to ægerite. Of three microscopic preparations of the rock not one gave sections favourable for discovering the absorption scheme or angle of extinction of either mineral. Both the femic minerals, where not granulated by pressure, are greatly corroded by the magma and always occur in a more or less scrappy condition.

The structure is the typical panidiomorphic-granular, but a secondary gneissic structure due to crushing has often been superinduced.

The schistosity never approaches the perfection of that in the neighbouring Cascade granite and banding was never observed. This aplitic hornblende-pyroxene granite is, in age, almost certainly pre-Miocene and is probably post-Triassic, but the evidence for a close dating of the intrusion is lacking.

The specific gravities of three fresh specimens vary from 2·626 to 2·652, averaging 2·645.

ATTWOOD SERIES.

As already noted, this name is proposed for the assemblage of metamorphosed sedimentary rocks which together compose the oldest series exposed within the 'Boundary Creek district' proper, unless we except the Grand Forks schists and the Cascade batholith. This new name is taken from that of Attwood mountain which is situated within the five-mile belt and is largely composed of the rocks in question. These have been described by Mr. Brock as follows (Report for 1902, p. 96) :—

'The limestones, argillites and quartzites, cut by serpentines, form a series which closely resemble the Cache Creek series (Carboniferous) of the Kamloops district. They occur in areas of greater or less extent in almost all parts of the district. They are always more or less metamorphosed; the limestone is generally white and crystalline, although occasionally a core of black or drab limestone is to be seen; the argillites are or were somewhat carbonaceous but are frequently altered. A hornblende or mica schist found in the Long Lake region seems to be an alteration form. Frequently both the limestone and argillites are altered by silicification which, when complete, produces a quartzite-like rock. In the argillites, quartz films and bands are often found parallel to the fissility. Some apparently true quartzites occur. The rocks also show the effects of mechanical deformation. The limestone is in places brecciated. These sedimentary rocks are among the oldest in the district. They are cut and greatly disturbed by the later intrusions of eruptive rocks so that little can now be determined regarding their thickness and original stratigraphical relationships. They seldom form large continuous bands but generally appear as islands of greater or less extent in the intrusive rocks. They probably form parts of a once extensive series of sediments which covered southern British Columbia.'

In the sheet accompanying the present report the limestone of the Attwood series is mapped with the same colour as that showing the position of the limestone in the Grand Forks complex. The Attwood limestone is believed to compose all the masses so coloured in the area west of the North Fork of the Kettle river. It is possible that some of the more completely altered greenstones of the belt represent contemporaneous basic lava flows or ash-beds in the true sediments of this group, but on account of the initial difficulty in the field, such greenstones have not been differentiated from the younger volcanics (Phœnix group) on the map.

As to the age of this unquestionably very thick group of sediments nothing is known with absolute certainty. The present writer is, however, strongly of opinion that Mr. Brock is correct in regarding them as equivalent to the fossiliferous (probably Carboniferous) series of argillitic, quartzitic, and limestone rocks occurring in the Rossland mountains. Lithologically the two groups are extremely alike; their distance apart geographically is but slight; and their relations to the respectively associated volcanics and intrusive rocks are strikingly similar. Hence the writer follows Mr. Brock in his tentative correlation of the Attwood series with the Carboniferous system, though of course, recognizing the possibility that some Triassic or even Jurassic sediments may be included in the group as actually mapped.

CHLORITE AND HORNBLENDE SCHISTS.

Mr. Brock has mapped small patches of chlorite and hornblende schists and remarks in the legend that their origin is uncertain. Two of these patches occur in the five-mile belt along the eastern contact of the Midway volcanic formation. Mr. Brock's brief reports do not contain any additional information concerning these metamorphic rocks. They may be the equivalents of certain phases of the Grand Forks schists and, like the latter, may be highly altered masses of the older greenstones of the district.

PHŒNIX VOLCANIC GROUP.

The town of Phœnix is situated in the midst of a large though interrupted area of basic volcanics, a series for which the name 'Phœnix group' has been proposed for the purposes of the present report. At the town itself there is a

2 GEORGE V, A. 1912

small mass of volcanic rock which has been interpreted by Mr. Brock as a true neck, which is much later in date than the main body of massive lava and pyroclastic material round about. To the latter only is the proposed name to be applied. Mr. Brock's description of the group is best given in his own words (Report for 1902, p. 97):—

'The older pyroclastic rocks and porphyrites are widespread; in fact they are the commonest rocks in the district.

'This series of rocks consists of green tuffs and volcanic conglomerates and breccias, fine ash and mud beds, flows of green porphyrite, and probably some interbedded limestones and argillites. The tuffs, conglomerates and breccias consist of a mixture of pebbles and boulders of porphyrite material with a great many fragments (probably a large proportion) of the rocks through which the volcanics burst. Pebbles and boulders of limestone, argillites, jasper and chert are common. Such of serpentine and old granite and old conglomerates are much rarer. In form the pebbles and boulders are rounded, subangular, angular and of irregular and fantastic outli . Sometimes they are somewhat sorted but often they are tumultuously arranged (agglomeratic). Beds of mud, ash and tuff alte⁻ 'te rapidly with coarse volcanic conglomerates and agglomerates. Sor   .es the matrix seems to be formed of porphyrite injected between the boulders. Limestone, now crystalline, seems occasionally to have been interbanded with them. It is often arenaceous, bands containing rounded sand grains and pebbles alternating with pure limestone. The sand and pebbles are well sorted and these arenaceous bands are sharply defined from the pure limestone. The matrix of these bands is white crystalline limestone. Argillites are also interbanded to a limited extent, although it is not always possible to distinguish the volcanic muds from such sedimentary material.

'The porphyrite seems to be a little later than most of the pyroclastic rocks although some of it may-be interbanded. Owing to the alteration in these rocks through mountain-building processes and contact metamorphism, it is not possible to separate the porphyrites from the pyroclastic rocks, on the map. The porphyrite is usually too highly altered to make out its original character, but it seems to have been an augite-porphyrite similar to that of the West Kootenay district. In places it is agglomeratic.

'The great changes produced by mountain-building processes and later igneous intrusions, make it difficult or impossible to discover the history of these rocks. The first part of this period of volcanism seemed to have been one of heavy explosions with periods of sedimentation, and to have been followed by a period of more quiet lava flows. The amount of material extruded must have been very great.

'A very striking feature in these rocks is the way ' which islands or irregular masses of the older sedimentary rocks appear in them. In part, these are included fragments, in part they may represent infolded masses in truncated anticlines, or inequalities in the surface on which this old volcanic series was deposited. Appressed anticlines and faults can be seen in them, but the grand features of their structural relationship are lost through the effects of the later igneous intrusions. Some of the limestone inclusions are to be explained as squeezed intercalated beds. Under pressure, the limestone flows and from a thin bed a line of inclusion-like lenses may be formed. This series of pyroclastic and volcanic rocks seems to have been formed immediately after the sedimentary series, and is therefore probably Palæozoic. In the Palæozoic formations of the Kamloops district, also, green effusive rocks occur.

'As already remarked some of the serpentine appears to be of later age than this series.'

The problem of correlation of the Phœnix volcanics with the standard systems of rocks is practically the same as that found in the Rossland mountains, where the Rossland volcanic group was placed by McConnell and Brock in the Carboniferous. The view expressed in the last chapter, that the great bulk of the Rossland volcanics is of Mesozoic age, is founded on arguments which are in part of the same nature as those deducible from the facts recorded by Mr. Brock for the Phœnix area. The present writer's very limited knowledge of the belt between Grand Forks and Midway forbids his taking any definite position different from that taken by Mr. Brock. Yet it seems possible that the Phœnix group also is largely of Mesozoic age and contemporaneous with the similar andesitic members of the Rossland group. Mr. Brock has not reported any latitic phases here, such as are so abundant in the eastern district, but their presence in limited quantities may be declared when chemical analysis has been applied to the 'Boundary Creek district' formation. In any case, however, the resemblance of the Phœnix and Rossland groups lithologically and the parallelism of their dynamic histories are sufficiently patent to make their direct correlation highly probable. The abundance and nature of limestone and argillitic fragments in the agglomerates of the Phœnix group suggest that those sediments were thoroughly consolidated, if not metamorphosed before the major eruptions took place. This leads one to suspect that the Phœnix volcanics are really in distinct unconformity to the Attwood series, as the Rossland volcanics are unconformable upon the Carboniferous rocks of Little Sheep creek and upon the formations of the Pend D'Oreille group.

As a suggestion, rather than as a conclusion, the writer has therefore correlated the Phœnix group with the Rossland group in its middle part and thus holds the hypothesis that both belong to the Mesozoic. The question is, however, in the writer's opinion, wide open.

SERPENTINE.

Many bodies of serpentine have been mapped in the Boundary Creek district Mr. Brock has given a short account of them in immediate connection with the Attwood series of argillites, etc. He writes (Report for 1902, p. 96) :--

'The serpentine occurs as bands and masses cutting these sedimentary rocks. The intrusive nature of the serpentine is shown in the way in

25a—vol. ii—25

2 GEORGE V., A. 1912

which it cuts across the bedding of the older rocks and in the contact metamorphism it produced. In places traces of the structure of the original eruptive rock can be made out in the serpentine. In Central camp the serpentine is occasionally somewhat fibrous, approaching asbestos. Near the Koomoos-McCarren Creek divide it seems to pass into a soapstone or talc. Often it is altered to a rusty aggregate of dolomite (and perhaps other carbonates) and white quartz veins. It is doubtful if all the serpentine in the district is of one age. Boulders of serpentine are found in the green volcanic conglomerates which would indicate that some of it was older than these pyroclastic rocks. On the other hand, some of it seems to be intrusive in the green porphyrite which is of a little later age than these volcanic conglomerates.'

No detailed petrography of the serpentine has been published. The fact that the majority of the perfectly analogous serpentine masses of the Rossland district, and, as we shall see, of the Rock Creek district west of Midway, have been derived from typical dunite, it appears that the 'Boundary Creek district' serpentines have also originated from intrusions of that rock type. The serpentines of all three districts have been correlated and seem to be best regarded as genetically associated with the great andesitic (porphyrite) extrusive masses of the respective districts.

GRANODIORITE.

Finally, it remains to note the many small stocks and other intrusive bodies of granodiorite, which is the youngest rock in the five-mile belt between Grand Forks and Midway except the Midway volcanic formation itself. Mr. Brock's summary account of the rock may be given in full (Report for 1902, pp. 98-99):—

'At various points throughout the whole district, bosses, irregular masses and dykes of a light gray granitoid rock make their appearance. It is a quartz-bearing biotite-hornblende rock, in places apparently granitic, in others rather dioritic. It is probable that it will prove to be, generally, a granodiorite. It sends out numerous dykes throughout the country, especially in the southern portion of the district. These have usually a porphyritic structure with a micro-granitic groundmass. Some are granite porphyries, but a great number are quartz-diorite-porphyrites, as are also some of the smaller bosses. On McCarren creek, north side, are some basic hornblende gabbro-porphyritic dykes which may belong to the same intrusion. In places these shade off into pure hornblende rocks.

'This granodiorite is evidently intrusive, cutting all the rocks above mentioned. The mechanism of its intrusion is extremely interesting, for it unquestionably forced its way up through the overlying rocks by digesting them and rifting off fragments. This is proved by its contacts, both along the sides and roofs of the masses. These are, except in the case of the dykes, rarely sharply defined, but are irregular and suture-like. The

SESSIONAL PAPER No. 25a

intrusive holds inclusions of the surrounding rocks, and the surrounding rocks are often filled with granitic material. The composition of the intrusion seems to be affected by the digested material of the rock into which it has forced itself. It is also shown by the way in which the granodiorite is exposed in small, more or less circular but irregularly bounded masses, in different parts of the district, such as in Wellington camp and on Hardy mountain. In many cases no definite boundary can be assigned to the granitic mass. From the way in which the rock makes its appearance in all parts of the district, it is evident that the whole of it, at no great depth, is underlain by this rock. This rock has some strong resemblance to the Nelson granite of the Kootenay district, both in composition and in its relationship to the surrounding rocks. The Nelson granite, which has been carefully studied, is a sort of granite representative of the monzonite group of rocks, intermediate between the alkali and the lime-soda series of rocks, and about on the boundary line between granite and diorite. Its composition* is as follows:—

Analysis of granodiorite.

	Per cent.
SiO$_2$	66·46
TiO$_2$	·27
Al$_2$O$_3$	
Fe$_2$O$_3$	
FeO	1·83
MgO	1·11
CaO	3·43
Na$_2$O	4·56
K$_2$O	4·58
H$_2$O	·29
P$_2$O$_5$	·06
	99·93

—Analysis by Dr. M. Dittrich, Heidelberg.

'The Boundary Creek rock will be found on analysis to contain a greater percentage of alkaline earths, but this may be due to the material it has acquired from the rocks into which it has been intruded, and may represent only a local peculiarity. As the Nelson granite occurs to the north and east of this district and probably also to the west, the Boundary Creek rock in all probal." belongs to the same great intrusion. If so, its age will be about Jurassic. This agrees with its stratigraphical position in this district.'

CORRELATION.

There are a few certainties and many uncertainties regarding the relative geological ages of the formations in the region considered in the present chapter. It is known (1) that the Cascade batholith cuts the Grand Forks

---

* As represented in a specimen from the Kokanee mountains, West Kootenay (R. W. Brock).

2 GEORGE V., A. 1912

schists; (2) that the Smelter granite stock cuts both those formations; (3) that all three formations mentioned have been crushed and that the two older bodies have been intensely metamorphosed during orogenic movements; (4) that the heavy masses of agglomerates and tuffs of the Phœnix group contain many fragments of the Attwood limestone and other sediments, apparently indicating that the latter rocks were already well consolidated, if not somewhat metamorphosed, before the Phœnix agglomerates were formed; (5) that the Attwood sedimentaries and, in apparently less degree, the Phœnix volcanics are intensely crumpled and sheared; and (6) that the granodiorite of the region cuts both those groups of rock and has been itself not seriously deformed since its consolidation from the magmatic condition.

It should be also noted that, in the northern part of the Boundary Creek district and thus outside the five-mile Boundary belt, Mr. Brock has mapped and described many small bosses, dikes, and sheets of porphyritic alkaline syenite of the pulaskite type. This intrusive cuts all the other formations of the district, including conglomerate, shale, and lignite-bearing sandstone with associated volcanics, all of which Mr. Brock regards as almost certainly of Tertiary, and perhaps of Oligocene, age. He points out the great similarity of the pulaskite to that composing the Coryell batholith and other large bodies of West Kootenay.

On the whole, therefore, it appears clear that the succession of rock-formations in the Rossland mountains and in the 'Boundary Creek district' is strikingly similar. This is, of course, not surprising, in view of the fact that the two districts lie side by side. The point is specially stated again, since it is chiefly this parallelism in the histories which has emboldened the writer to draw up the following tentative scheme of correlation for the rocks between Christina lake and Midway.

Correlation, Christina Lake to Midway.

Glacial and Recent deposits	*Pleistocene.*
Syenite and syenite porphyry (north of five-mile belt)	*Miocene ?*
Conglomerate, sandstone, shale, and lignite (north of five-mile belt)	*Oligocene.*

UNCONFORMITY.

Serpentine, intrusive bodies	
Phœnix group:	
Pyroclastics, flows?, and	*Chiefly Mesozoic ?*
Contemporaneous intrusions of porphyrite	
Smelter granite stock (aplitic satellite from Cascade batholith?)	
Granodiorite stocks, dikes, etc	*Jurassic.*
Cascade gneissic batholith	
Attwood series (argillite, quartzite and limestone)	*Palæozoic, probably Carboniferous.*
Chlorite and hornblende schists	?
Grand Forks schists	?

PLATE 35

Park land on Anarchist plateau east of Osoyoos Lake.

CHAPTER XV.

FORMATIONS OF THE FIVE-MILE BELT BETWEEN MIDWAY AND OSOYOOS LAKE (MIDWAY MOUNTAINS AND ANARCHIST MOUNTAIN-PLATEAU).

INTRODUCTION.

A large body of the Midway volcanic formation covering the area about Midway town forms a natural geological province by itself. These Tertiary lavas and pyroclastics are piled upon an unconformable base of presumably Paleozoic sediments, for which the name 'Anarchist series.' is proposed. The older series is exposed on a large scale in the belt between the main Kettle river and Osoyoos lake and outcrops at a few points within the Midway volcanic area. The whole of this stretch, where the Boundary belt crosses the volcanic area and the extensive exposure of its foundation rocks, forms a convenient geographical unit, the geology of which will be described in this chapter.—(See Maps, Nos. 10, 11 and 12.)

Besides the two rock groups just mentioned a fossiliferous, Tertiary series of rocks will be described under the name Kettle River formation. After the sediments the igneous rocks, extrusive and intrusive, will, as usual, be described in their respective order of age. Some intrusive phases of the lavas will be considered before the corresponding extrusives. The largest exposed intrusive body within this part of the belt (excepting the Osoyoos batholith, which will be described in the next chapter) is named the 'Rock Creek chonolith.' It is intrinsically of special importance and its petrography throws much light on certain members of the Midway extrusive rocks.

ANARCHIST SERIES.

General Description.—From a point near the confluence of Rock creek and Johnston creek westward to the Osoyoos granite batholith—a distance of about twenty miles—the Boundary belt is almost entirely underlain by a highly metamorphosed, chiefly sedimentary group of rocks. These compose the Anarchist-mountain plateau (Plate 35) and may be called the 'Anarchist series.' The name is literally not inappropriate, for these rocks cannot as yet be reduced to stratigraphic order or structural system. The series is also represented in detached areas between Johnston creek and Midway and unquestionably underlies the Kettle River formation and the Midway lavas. It is separated from those Tertiary formations by a profound unconformity. West of Osoyoos lake the Anarchist series is probably represented by a yet more extensively meta-

2 GEORGE V., A. 1912

morphosed group of rocks bounding the Osoyoos batholith on the west. In all, some eighty square miles of the five-mile belt is known to be underlain by this series.

Notwithstanding the considerable area covered by the series, a rather prolonged field and laboratory study of its constituent rocks has failed to produce satisfactory details as to their succession or position in the geological time-scale. One of the major difficulties in the field work was found in the exceptional continuity of the Glacial-drift cover which mantles the bed-rock to a degree unequalled in the rest of the Boundary belt. An example of the rarity of outcrops may be cited from the field notes of the season of 1904, during which a traverse of ten miles east and north of Sidney post-office led to the discovery of only three small outcrops. For this reason, although the country is very accessible, the facts to be cited concerning the Anarchist series are merely those of a geological reconnaissance.

The rocks of the series belong to four classes,—quartzite, phyllitic slates, limestones, and greenstones. Where more metamorphosed these become, respectively, micaceous quartzite, mica schist, marble, and amphibolite. The dominant species are quartzite and phyllite, apparently in about equal proportion. Greenstone is next in importance, while the limestone is represented only in a few local, pod-like masses generally from 200 to 100 feet or less in thickness.

The quartzite is a gray to green, very hard rock, commonly sheared so as to simulate the associated, more argillaceous rock in its fissility. Under the microscope the quartzite is quite normal, presenting the usual appearance of recrystallized, interlocking quartz-grains with variable ar ounts of biotite, sericite, and chlorite in minute foils. Pyrite is a common metamorphic accessory. The rock is sometimes slightly calcareous and in a few thin beds passes over into silicious limestone. Much more often it is greatly enriched in micaceous elements which have evidently been derived from relatively abundant argillaceous matter; recrystallization is so thorough that in none of the thin sections examined was original argillaceous substance to be found.

The same is true of the old argillites proper. They are now holocrystalline, though with the fine grain of true phyllite or of metargillite. Where the metargillite crops out it is possible to determine the attitude of the true bedding, but such fortune is exceptional and the only structural plane usually to be discerned is the schistosity. The colour of the phyllite or metargillite varies from dark gray to bluish-gray and greenish, with dark slate-gray as the dominant tint. The essential minerals are the same as in the quartzite, simply occurring here in different proportions. The dark colour of the rock is doubtless chiefly due to the inclusion of carbonaceous matter in small amount. Like the quartzites the phyllitic rocks are traversed by multitudes of quartz veinlets, and at many points by huge veins of white quartz.

The limestone pods can never be followed far across-country; they represent beds that have been sheared into great fragments as the enclosing argillaceous rock underwent its heavy mashing and dynamic metamorphism. The limestone is generally massive, rather pure, and of a light bluish-gray colour; sometimes

It is a white marble. From its instant and violent effervescence with acid the rock must be considered as low in magnesia. At a couple of localities it is tinted a dark gray, as if by included carbonaceous matter. In one thin section the polygonal structure of a coral-fragment was found, but, in spite of long search, no useful fossil was detected at any point. The lack of organic remains is amply accounted for by the wholesale recrystallization of the limestone, which has generally lost all trace of bedding.

West of the Kettle river none of the limestone pods is known to be over 200 feet in thickness. On Deer Hill, three miles west of Midway, a more considerable marble-like mass occurs, but it is cut off on all sides by igneous rocks. Still larger bodies of what is probably the same limestone have been mapped by Mr. Brock in the belt between Grand Forks and Midway.

The greenstone occurs in broken, massive to schistose bands throughout the whole length of the Boundary belt from the Kettle river to the Osoyoos batholith. Their structural relations are even more elusive than those of the limestones. It is probable that both injected and effusive basic rocks are represented but the latter are believed to be the more important in volume. They form beds in phyllites and quartzites. Like the limestones the lavas can seldom be traced far along the strike; they have evidently undergone the profound faulting, stretching, and mashing which has affected the other members of the series. The accompanying alteration has been so great that original tuffaceous or vesicular phases have been almost entirely obliterated. So far it has proved impossible to locate the top or bottom of any of the lava flows, to correlate the different bands or to find the aggregate thickness of the lavas. It appears certain only that the aggregate thickness represents many hundreds of feet and possibly several thousand feet.

The greenstones are notably uniform in their present composition and were probably as nearly uniform originally. The lavas must have been basic, either basalt or basic andesite. Fourteen thin sections have been examined under the microscope. In no case was original material present in any large quantity. Both massive and schistose phases are composed in ever-varying proportions of secondary, actinolitic amphibole; quartz; plagioclase of medium basicity; epidote; calcite; chlorite; and magnetite. The rock-types thus include massive greenstone, chloritic schist, epidotic schist, hornblende schist, and true amphibolite. The constant recurrence of these banal characters and varieties seems to show pretty clearly the common origin of the greenstone members throughout the Midway mountains and Interior Plateaus as sampled along the Boundary line—a derivation from the same basic magmatic types whence have come so largely the greenstones of the world.

Nature of the Metamorphism.—A prominent feature of all the members of the Anarchist series is their notable metamorphism. The cause of such recrystallization is by most geologists found in 'dynamic metamorphism.' Of late years Termier and others have raised the question whether this principle has had an essential part in the production of the crystalline schists. In fact,

2 GEORGE V, A. 1912

Termier denies the efficiency of dynamic metamorphism, in the accepted meaning of that term, in the development of these rocks. After forcibly presenting his arguments for the case of the Alps, Termier states his conviction that the mineralogical changes suffered by the Alpine sediments are due only to thermal metamorphism aided by magmatic emanations. One must believe, however, that he goes too far in holding that dynamic metamorphism 'does not exist.'§

The petrographic study of the Anarchist series seems to show facts that do not substantiate his view. It will scarcely explain the striking uniformity of the phyllites throughout the greater part of their area in the Boundary belt. Their only serious variation from the normal occurs in a narrow contact zone about the Osoyoos batholith. The batholithic magma has there produced relatively large-foiled muscovite along with tourmaline and other familiar contact products—all minerals which are regularly absent in the twenty-mile belt from Osoyoos lake eastward. There is, indeed, a decided difference of quality between the obvious contact metamorphism and that change wh'   has affected the main body of old argillite so drastically. The yet more ancient argillites of the Rocky Mountain Geosynclinal were completely recrystallized by the action of their own fluids acting under conditions of dead weight and deep burial. In that case igneous intrusions are extremely rare at the present surface, and there is no indication that they have ever affected the Cambrian rocks or in many cases come within several miles of them. If, then, static metamorphism can cause the more or less perfect recrystallization of argillaceous rocks, it seems most reasonable to believe that similar rocks, also deeply buried, charged with fluids and certainly heated during orogenic movements of great intensity, would, in the process of time, crystallize so as to form phyllites or mica schists.

.A complete discussion of the problem would be out of place in this chapter, but with this note the writer wishes to record his belief in the soundness of the time-honoured conception of dynamic metamorphism.

ROCK CREEK PLUTONIC ROCKS.

Near the forks of Rock creek a small area of plutonic rocks occurs between the Paleozoic sediments of the Anarchist series and the Tertiary lavas north of the creek. The plutonics include granodiorite, basic diorite, and serpentinized dunite. All of these cut the Paleozoics. The diorite is cut by the granodiorite which has furnished arkose material to the Kettle River Oligocene formation. Both granodiorite and diorite are therefore of pre-Oligocene age and are probably post-Carboniferous. The dunite has been vigorously crushed and sheared, indicating that it also dates from pre-Oligocene time.

Diorite.—The diorite is a dark green, medium-grained rock. It is greatly altered, but the original essential constituents seem to be biotite, green hornblende, and plagioclase of medium acidity. Magnetite, apatite, titanite,

§ Cf. P. Termier. Congrès géologique internationale, Compte Rendu, Ninth session, Vienna. 1903. pp. 571-586.

and interstitial quartz are the primary accessories. The rock may be classified as a biotite-hornblende diorite.

Granodiorite.—The granodiorite sends apophyses into the diorite and at a few points encloses blocks of it, so that their relative age is determined. The granodiorite is rather coarse-grained and of a reddish-gray colour much lighter than that of the diorite. The constituent feldspars include basic andesine near $Ab_1 An_1$ (dominant), with orthoclase and microperthite. The other essentials are quartz, biotite, and hornblende, each of which has optical characters like those of the respective minerals in the older diorite. This rock too is notably crushed and altered. Epidote, calcite, and kaolin are very abundant secondary constituents.

Diorite and granodiorite disappear on the east under the Tertiary arkose and lavas, so that the whole original extent of these plutonic bodies is not known.

A larger area of somewhat crushed granodiorite is exposed on the slope south of the confluence of Rock creek and Kettle river. In that area some three square miles of the Boundary belt are underlain by the granodiorite which seems to have the field relations of a typical stock or batholith. The body extends for an unknown distance south of the Boundary line. It sends a conspicuous, wide apophysis northward to the Kettle river at the 'Riverside Hotel.' This great dike cuts the Anarchist phyllites and seems to be continued beyond the river by a strong dike of the same rock, cutting limestone and quartzite.

Petrographically this granodiorite is practically identical with that cropping out at the forks of Rock creek but has a basified contact-phase recalling the quartz diorites.

Dunite.—Dunite, generally altered to serpentine, occurs at two different localities in the valleys of Rock creek and Kettle river. One mile up the river from their confluence several large outcrops of heavily slickensided serpentine and talc, shown microscopically to have been derived from a pure olivine-chromite rock, were found. The structural relations and total area of this body could not be accurately determined. On the west and southwest it is either covered by Tertiary conglomerate or cut off by the chonolith of rhomb-porphyry. On the east the serpentine disappears beneath the river gravels and it was not found in place beyond the river.

In the deep gorge of Rock creek, immediately below Baker creek, a somewhat less obscure occurrence of the dunite was discovered. In spite of the profound crushing to which the dunite and its country-rocks have been subjected, it seems clear that the dunite forms an intrusive dike at least 100 feet wide, striking in a general northeast-southwest direction. A yet larger dike, measuring 300 feet in width, crops out on the slope a half-mile west of the forks of the creek. Both dikes cut the Anarchist phyllite and associated rocks. The dunite is like that on the Kettle river, which is, therefore, probably also a great dike and contemporaneous with the dikes on Rock creek.

The dike at the canyon has been studied with more care than the others and has been subjected to chemical analysis. In general, the rock is massive,

2 GEORGE V., A. 1912

compact, and of a greenish-black colour, mottled with abundant areas of
rather pale green talc. Olivine is the only visible primary essential and occurs
with its usual granular habit. Neither chromite nor picotite could be certainly
detected in any of three sections cut from the hand-specimens from this
locality. The analysis shows, however, the presence of a small amount of
chromic oxide. The secondary products are the usual ones, serpentine, talc,
tremolite, magnetite, and a carbonate which is probably dolomite. Small grains
of pyrite, doubtless introduced during the alteration of the rock, were also seen
in the hand-specimen.

Professor Dittrich's analysis of this typical, though partially hydrated and
otherwise altered, dunite (specimen No. 282) gave the following result:—

Analysis of Rock Creek dunite.

SiO_2	40.25
TiO_2	tr.
Al_2O_3	1.10
Fe_2O_3	4.61
Cr_2O_3	.15
FeO	3.04
MnO	.11
MgO	37.91
CaO	1.16
Na_2O	.18
K_2O	.16
H_2O at 110°C.	.32
H_2O above 110°C.	9.08
CO_2	1.95
	100.32
Sp. gr.	2.868

The hydration of this rock is evidently so pronounced that a calculation of
the proper norm is not directly possible. The place of the rock in the Norm
classification cannot, therefore, be stated.

KETTLE RIVER FORMATION.

General Description.—The map illustrates the fact that the Kettle River
formation in its present distribution within the Boundary belt occurs
only in shreds and patches. It has been cut to pieces by faults and by dikes,
sills, and chonoliths of the various porphyries; it has been deeply buried
beneath the Midway lavas. Extensive erosion has in many places uncovered
the sediments but has also largely destroyed their continuity by penetrating
the entire formation and laying bare much of its Paleozoic floor. In consequence of all these events the Kettle River beds now form detached, slab-like
masses seldom more than a few hundred yards in width. At least seventeen
isolated patches of the formation have been found between Midway and the
forks of Rock creek, a distance of fifteen miles in an air line. The formation
does not crop out farther west, but small patches of it are probably represented
in the northern part of the area mapped by Mr. Brock as the 'Boundary Creek
District.'

No single section gives the whole thickness of the formation and all of the different sections together do not afford a strict idea of the total thickness nor of the exact strength of each member. Moreover, from the nature of the formation it is highly probable that none of the members originally held a given thickness for many miles across country. A complete columnar section cannot, therefore, as yet be constructed. All that is now possible is to state the general succession of the beds and the minimum thickness of each member in its thickest section, and thus to indicate a minimum thickness for the whole formation where most completely developed.

The columnar section worked out on this basis may be described as follows:

Columnar section of Kettle River formation.

Top, conformable contact with overlying Midway lavas.

1,000+feet— Fossiliferous, gray, feldspathic sandstones with thin interbeds of shale.
900+ "   Coarse conglomerate.
200+ "   Coarse arkose-breccia (a local deposit).
—————
2,100+ "

Base, unconformable contact with underlying Anarchist series and with pre-Tertiary plutonic intrusives.

The basal arkose-breccia crops out only in one place, in the form of an elongated area cresting the north wall of Rock creek canyon between the forks and Johnston creek. It is composed of angular to subangular blocks of the diorite and granodiorite on which the breccia lies, so that it is likely that few or none of the blocks have travelled far from their parent Tertiary ledges. The blocks are of variable size, many of them being four feet or more in diameter. Very few are rounded; it is not certain that any at all are water-worn. The cement is simply the disintegrated, highly feldspathic material of the granitic rocks. From its evidently local nature one would not expect the breccia to form the base of the formation generally. In fact, that member is wanting at the lower contacts where the formation rests on the Anarchist quartzites and phyllites; such contacts were discovered at several points to the south of the Kettle river and of Rock creek.

The conglomerate is well exposed in Rock creek canyon from one to two miles upstream from its mouth, and again, on the summit immediately south of the Riverside hotel. In the former locality it forms part of the roof of the Rock creek chonolith of rhomb-porphyry. At the latter locality the conglomerate lies, with evident unconformity, on the Anarchist quartzites and green-stones. In neither case nor at any other locality was the top of the conglomerate recognized; its uppermost beds have been eroded away or faulted out of sight.

The rock is a well consolidated, gray to brownish mass of rounded pebbles and boulders of all diameters up to three feet. These are almost always well water-worn. In composition they reflect the formations on which the conglomerate rests; gray quartzite, white and gray limestone, vein quartz, slate, phyllite, greenstone, amphibolite, altered porphyrites, and granodiorite are all more or

2 GEORGE V, A. 1912

less abundantly represented among the pebbles. Most of them were manifestly derived from the Anarchist series and have probably not been carried far by the moving water. The cement is sandy and often somewhat calcareous. On the hill south of the Riverside hotel the conglomerate is notably uniform, with only a very few, small lenses of sandstone. In the .. nyon sections, beds of both grit and sandstone, reaching three or four feet i.. thickness, interrupt the coarser sediment. These finer-grained beds are characterized by sliver-like, subangular fragments of black argillite, from one-half to one inch in length. Such fragments are identical in appearance with smaller ones occurring in the thick sandstone member and serve as a kind of fossil in suggesting that sandstone and conglomerate belong to one conformable series of beds.

N.W.   S.E.

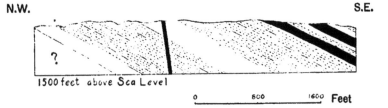

1500 feet above Sea Level

0   500   1600 Feet

FIGURE 25.—Section northeast of bridge over Kettle River, six miles above Midway.
Legend :—Dot and-line, Kettle River sandstone. Blank, pulaskite porphyry sills. Solid black, rhomb-porphyry.
At southeast end a composite sill of pulaskite porphyry and rhomb-porphyry ; at northwest end data lacking, owing to land-slide.

Strata which seem to represent the base of the sandstone member are seen to overlie conformably the conglomerate in the canyon section. About 100 feet of these beds are there exposed and in all respects are like the sandstones exposed in the much thicker sections along the Kettle river and on Myer's creek. In all cases the sandstones carry plant-stems and the thin interbeds of shale are carbonaceous.

The best sections in the uppermost member include one on the Kettle river wagon-road at its abrupt turn four miles west of Midway, and a second, occurring just above the river alluvium two miles to the northward. The former locality was long ago noted by Bauerman.§ It is illustrated in Figure 25.

The sandstone is generally of medium to fine grain and of colour ranging from whitish, through the dominant light gray, to light brown. Even to the naked eye it normally appears highly feldspathic. A thin section of a typical specimen was found to consist of angular to subangular fragments of quartz,

§H. Bauerman, Report of Progress, Geol. Surv., Canada, for 1882-3-4, part B. p. 39.

PLATE 36.

Fossil plants in the Kettle River sandstone. In upper view, cast of a pine cone of Tertiary age. Natural size.

25a—vol. ii—p. 396.

orthoclase, microcline, and abundant plagioclase (averaging labradorite)—all enclosed in a light-coloured, argillaceous base. The sandstone has thus the composition of a bedded arkose.

The argillite interbeds are thin-layered, often papery. The colours are light to dark gray and blackish, the variation depending on the relative content of carbonaceous matter. The lighter coloured shales are rather highly silicious and weather almost white.

A bed of coal is reported to have been found near the base of the sandstone in the canyon section but it was not accessible at the time of the writer's visit to the section. No reliable statement as to its thickness could be obtained from the settlers, but it is doubtless thin; on account of the fact that the area of the enclosing sandstone is very small, this coal could have little practical importance unless it were of relatively great total thickness.

Reviewing the characteristics of the different members of the formation, we are prepared to find that the fossils contained indicate a fresh-water origin for this series of beds. Such is, in fact, the most probable view of the Kettle River formation. Some horizons (the paper-shales particularly) suggest lacustrine sedimentation; others suggest fluviatile sedimentation. As the average lake is an enlarged river-channel, so the larger rivers in flood are temporary lakes. It is here very difficult, if not impossible, to distinguish in most of the beds those which were laid down during river-floods from those laid down on the floor of a permanent lake.

Geological Age.—The Kettle River beds are seldom entirely free from traces of fossil plants and at a few horizons useful material was collected during the seasons of 1902 and 1905. (Plate 36.) The different collections have been grouped under the locality numbers 250 (where specimens were taken in both years), 271, 1001 and 1007. These localities are marked on the MS map. No animal remains were found at any point, though special search was made for them at the many ledges exposed. The plant remains were sent to Professor D. P. Penhallow of McGill University, who reported at length on the collection. His paper was published in the Transactions of the Royal Society of Canada, 1908, and is reprinted as Appendix B of the present report. The reader is recommended to read this important study of the British Columbia fossil plants recently collected; it will be found that the treatment of the Kettle River fossils is specially full. Four new species belonging to the genera Picea, Pinus, and Ulmus are named and described. The present writer is under deep obligation to Professor Penhallow for the special pains which he took with this set of collections. His report is of immediate value in the present connection since it contains a full discussion of the Kettle River flora as compared with the floras of the Similkameen and other formations of the west. As a result of his work Professor Penhallow concluded that the Kettle River beds should be referred to the Oligocene and so they have been mapped in the accompanying sheet. Their general correlation with the Tertiary formations of the United States and Canada is also discussed in the paper by Professor Penhallow.

2 GEORGE V., A. 1912

MIDWAY VOLCANIC GROUP (IN PART).

General Description.—The town of Midway lies well within a large area of basic volcanic rocks which extend along the Forty-ninth Parallel continuously from a point about three miles due east of the town to a point eight miles west of it. On east and west alike the volcanics are bounded by the much older Paleozoic sedimentary complex, so that the lavas may be said to lie in a great syncline-like depression between the Paleozoic rocks of the Phœnix mining district and the Paleozoics of Anarchist-mountain plateau. Mr. Brock has shown that the Midway volcanics extend for at least fifteen miles to the northward of the Boundary line in the longitude of Midway. It is not known how far they extend to the southward.

The entire group of volcanics is believed to be of post-Cretaceous age. At several points they rest with apparent conformity on the Oligocene Kettle River sandstones and show no evidence of having undergone the intense crushing and profound metamorphism which have affected the Paleozoic lavas in their immediate vicinity. Though of such relatively recent date—late Oligocene or post-Oligocene—the Midway lava formation is considerably faulted and tilted, the older lavas having shared the disturbances that have affected the underlying sandstones.

The upturning has in a measure facilitated the discovery of the stratigraphic relations but it has been found that the exposures within the Boundary belt are too imperfect to declare the whole stratigraphy. Nine different types of lava and several horizons of agglomerate and tuff are represented. These rocks are cut by basic dikes and sheets which have solidified into porphyrites not to be easily distinguished from the compact phases of the extrusives. The structural complexity has been heightened by the injection of many dikes, sheets, and more irregular bodies of syenite porphyry. As a result of these various processes the succession and relative volumes of the different lavas are only partly determined.

The nine types of lava include olivine basalt, augite andesite, hornblende-augite andesite, biotite-augite andesite, hornblende-augite-biotite andesite, biotite andesite, trachyte, extrusive rhomb-porphyry, and an analcitic lava. The first six species have normal characters and their description need not be detailed; the last two species named are quite unusual types and will be described at greater length.

Tuffaceous beds are not rare, though they are subordinate to the massive flows. The pyroclastic phases seem to be most commonly associated with, and composed of, the trachyte.

Petrography of Subalkaline Lavas.—The *olivine basalt* occurs on the slopes northwest of Midway and at various other points on the north side of the Kettle river. A notable area of this rock is also found on the rolling plateau between the river and Myer's creek canyon. The basalt has the usual deep gray-green to blackish colour, with phenocrysts of augite, olivine, and, generally, basic labradorite. The ground-mass shows the usual variation from the diabasic aggregate of basic plagioclase and augite to the glassy paste cementing microlites of those minerals. The specific gravities of four compact, comparatively fresh specimens vary from 2·663 to 2·759, thus roughly indicating the variable proportion of glass in the rock. At times this lava is highly scoriaceous; calcite and chlorite usually fill the gas-pores.

Augite andesite is yet more abundant than the basalt. In many essential respects excepting in the absence of phenocrystic olivine and in the lower percentage of the pyroxene and iron oxide, this andesite seems to be lithologically similiar to the basalt. One fresh compact specimen has the specific gravity, 2·733. Like the basalt and, in fact, like all the other Midway lavas, the andesite is not crushed or sheared but it is considerably altered as if by ordinary weathering.

The *hornblende-bearing andesites* were found at only one locality. The top of the low mountain immediately to the northwest of the confluence of Myer's creek and the river, is composed of a vesicular to compact andesite which bears phenocrysts of labradorite, hornblende, less abundant augite, and sometimes biotite. The biotite-free phases seem to be the commonest in the sections actually traversed. The sporadic appearance of the mica among the phenocrysts is a phenomenon not well understood but it was observed that, in one thick flow finely exposed in the cliff overlooking the river, the compact interior of the mass carried all three femic phenocrysts, while the vesicular surface shell of the flow carried only hornblende and augite.

Macroscopically these two andesites are very similar, with a fairly light, greenish-gray to brownish-gray colour except where the lustrous-black phenocrysts interrupt the general surface of the rock. In each case the ground-mass is habitually hypocrystalline or else a microcrystalline aggregate of andesine, hornblende, rare augite granules, and considerable primary quartz.

In the precipitous cliff just mentioned the hornblende-bearing andesites are seen to overlie vesicular flows of *biotite-augite andesite*. The vesicles of these conformable flows are here and there arranged in rude layers which indicate a probable local strike of the series of N. 30° E. and a dip of 10°-20° to the northwestward.

The augite-biotite andesite is much darker coloured and clearly more basic than the overlying flows. Labradorite is a never-failing phenocryst. The biotite is sometimes more abundant than the augite but the reverse relation often holds. The ground-mass is generally glassy, with microlites of feldspar and augite. Primary quartz seems to be quite absent. The specific gravity of a non-vesicular, fresh specimen is 2·633, showing the presence of considerable glass. As usual with these lavas both phenocrysts and ground-mass are often much altered. The gas-pores of the vesicular parts are filled with quartz, calcite, or a yellow zeolite, probably delessite. In two thin sections of this andesite the base was found to carry orthoclase in considerable amount. It occurs in minute, allotriomorphic individuals which are intersertally arranged with respect to the ground-mass microlites of andesine. The abundance of orthoclase and the presence of biotite suggest a content of potash like that of some latites.

2 GEORGE V., A. 1912

The augite-biotite andesite occurs at various other points both north and south of the river—on the slope southwest of Midway, on the hills south of Myer's creek, and on the slopes of the Kettle river valley opposite Rock creek.

Biotite andesite was found at only two points, cropping out in the wall of the box-canyon on Myer's creek three miles from the main river, and again in the blocks of a volcanic agglomerate found on the left bank of the river three miles north of the first locality. In both cases the rock is highly vesicular and varies in colour from a dark greenish-gray to a more ashy hue. Dark brown biotite and basic andesine form the phenocrysts. The ground-mass is chiefly glass in which minute plagioclase microlites are embedded. Chemically this rock may be equivalent to the augite-biotite andesite.

The *biotite-trachyte* is extensively developed on the heights north of the Kettle river and east of Rock Creek post-office. It also crops out in the box-canyon of Myer's creek five miles above the confluence with the river. A somewhat doubtful and certainly local body of it was noted a few hundred yards northeast of the bold cliff at the elbow of the Kettle river four miles west of Midway. At all the localities the trachyte is notably uniform in field-habit and in composition. Chemically it is unquestionably very similar to the analyzed syenite porphyry to be noted as forming the great sills and dikes northeast of the Kettle river bridge. One can scarcely doubt that the trachyte and syenite porphyry belong to the same eruptive period.

The trachyte is a brownish-gray, commonly vesicular rock with conspicuous phenocrysts of feldspar and biotite, though the latter mineral is not always macroscopically visible. In one thin section augite was seen to occur among the phenocrysts. The feldspars range from one to three millimetres in diameter; the biotites, from 0·5 mm. to 1·5 mm. in diameter of foils. In each case the average and maximum sizes are much smaller than in the intrusive, syenite-porphyry phase of the same magma. The phenocrystic feldspars were found to be quite variable in composition. They may be made up of labradorite wholly surrounded by thick shells of orthoclase or soda-orthoclase; or of true microperthite and soda-orthoclase without associated plagioclase; or of orthoclase and andesine occurring together but not intergrown. The biotite is deep brown and intensely pleochroic.

The ground-mass is generally microcrystalline and is then composed of minute feldspars of tabular form. These are almost invariably murky with alteration-products, so that their determination is not easy. Most of them are untwinned and are probably sodiferous orthoclase. A little interstitial quartz is often visible. Magnetite and apatite are the well individualized accessories. Some brownish glass is occasionally present; it is never abundant.

The rock is to be classed among the alkaline biotite-trachytes, the effusive form of a typical pulaskite.

ROCK CREEK CHONOLITH.

STRUCTURAL RELATIONS.

Just above its confluence with Kettle river, Rock creek flows through a narrow steep-walled gorge in which excellent sections, not only of the Oligocene sandstones and conglomerates but also of an intrusive porphyry cutting the sediments, may be studied. For nearly two miles the creek flows between walls composed of the porphyry, or of the porphyry overlain by the sandstone-conglomerate formation. A somewhat prolonged field study of this mass of porphyry showed that it covers about five square miles within the five-mile Boundary belt. It is the largest known body of this rock in the region.

At several points the porphyry splits the Tertiary clastics after the manner of a thick sill or laccolith. At other points the intruded mass followed the contact-surface of the unconformable Tertiary and Paleozoic rocks. At still other points the intrusive contact cuts across Tertiary conglomerate and Paleozoic quartzite or schist alike. For long distances the contact-line of the porphyry is, as shown on the map, remarkably straight but is characterized by several rectangular bends. These elements of form in the contact-surface are such as might be produced if the magma has entered an opening bounded by two systems of faults or master joints cutting each other at right angles. The intrusive is never equigranular and nowhere displays the characteristics of a plutonic rock of the batholithic order. The steady persistence of the porphyritic structure even in ledges farthest removed from contacts is, on the other hand, a good indication that the whole mass has been injected and does not enlarge downwardly, like a stock or batholith. The exceedingly irregular form of the body and its relation to the invaded formations forbids our classifying it among the laccoliths. It has, indeed, been taken as a type of the intrusive bodies called 'chonoliths' by the writer.[*]

The porphyry has two strongly contrasted phases. One of these characterizes the principal part of the chonolith; the other is regularly found along its walls and roof and is the product of rapid chilling on the contacts.

DOMINANT ROCK-TYPE.

General Description.—The principal phase is a greenish-gray, nearly or quite holocrystalline, fine-grained rock which is often so abundantly charged with phenocrysts as to appear equigranular. The microscope shows, however, that the porphyritic structure is always present. The phenocrysts always include rhomb-shaped feldspar, and augite, which are generally accompanied by biotite and a few small olivine crystals. In some cases biotite is a more important phenocryst than augite. The feldspar rhombs vary from 2 mm. to 5 mm. in length; the augite prisms, from 1 mm. to 3 mm. in length; the biotite foils, from 1 mm. to 4 mm. in diameter. The olivines are seldom clearly visible to the naked eye, partly because of their being serpentinized and partly because

---

[*] Chap. **XXV.**; also Jour. Geology, Vol. 13, 1905, p. 499.

2 GEORGE V., A. 1912

of their small diameters of 1·5 mm. or less. The ground-mass is composed of a fine-grained aggregate of alkaline feldspars and numerous microlites of augite and biotite, along with some titanite, abundant accessory titaniferous magnetite, and large prisms of apatite. In several of the thin sections nephelite, occurring in small, stout, hexagonal prisms, enters the list of accessories, but it has generally been altered to hydronephelite. Calcite and serpentine are the principal accessory products; the former apparently resulting from the alteration of the rhomb-feldspars, the latter from the alteration of olivine. A variable amount of isotropic matter, almost certainly glass, forms a base within the ground-mass.

Notwithstanding the development of the secondary products noted, the rock must be regarded as fresh. It is strong and breaks with a sonorous, phonolitic ring. The changes suffered by the olivine and rhomb-feldspars are doubtless due to the action of imprisoned magmatic water, rather than to ordinary weathering. The augite and biotite and the feldspars of the ground-mass are, in the specimens collected, generally quite unaltered—a testimony to the freshness of the rock.

The specific gravity of this phase varies from 2·647 to 2·751. The higher value applies to a holocrystalline specimen rich in phenocrystic augite. The lower values correspond to specimens with some glass in the ground-mass.

Rhomb-feldspar.—The phenocrystic feldspar of this phase is more or less opaque and generally of a brownish colour, through the inclusion of small augite, magnetite, and apatite crystals and granules. Twinning is not visible macroscopically. The bounding surfaces of the crystals are never smooth but are affected by irregular shallow bays filled with the material of the ground-mass. Yet the surfaces approximate closely in their positions and relations to the crystal planes characteristic of the phenocrystic feldspar in the Norwegian rhomb-porphyry, viz. (110), ($\bar{1}$10), ($\bar{2}$01). Both the basal and clinopinacoidal cleavages are well developed. Individual phenocrysts cleaved parallel to the base are roundish or rectangular; those cleaved parallel to (010) have acute-rhombic outlines similar to those of the well-known anorthoclase in the Norwegian porphyry and in the lavas of Kilimandjaro.

Under the microscope these feldspars are generally seen to be zoned; others are unzoned and then have the same optical properties and chemical composition as the cores of the zoned individuals. The core composes from 50 to 90 per cent or more of each zoned crystal, averaging about 80 per cent of it.

In its general properties the core is somewhat allied to anorthoclase. The double.refraction is relatively low; the single refraction is somewhat greater than that of orthoclase or that of the outer shell of the zoned individuals. Very often the sections transverse to the zone of cleavages display the fine mesh so characteristic of anorthoclase and due to the simultaneous development of albite and pericline twinning. On cleavage plates parallel to the base the extinction is nearly parallel to the trace of the pinacoidal cleavage but it may be as much as 1° from strict parallelism. Two sections nearly perpendicular to the acute bisectrix showed that the optical angle is relatively small (2V estimated to be between 40° and 50°).

The first notable difference from true anorthoclase was discovered in sections nearly or quite parallel to (010). Such sections are abundant in the slides and are readily recognized by their rhombic outlines. The obtuse positive bisectrix emerges centrally in these sections. The extinction as determined in sixteen zoned individuals, all cut nearly parallel to (010), varies from 2° to 11°. In these cases equal illumination of core and outer shell occurred at angles varying from 30° to 40° with respect to the trace of the basal cleavage. Using this principle of equal illumination after the method invented by Michel-Lévy, it seems possible to orientate the directions of extinction on (010); they are characteristically negative and read, on the average, about -10°. The core is generally inclosed in a single thin shell with extinction on (010) varying from +4° 30' to +12°, but there is often a yet thinner intermediate shell with an extinction close to 0°. The outer shell is never twinned, is glass-clear, and has the single and double refraction of orthoclase, or, in many cases, soda-orthoclase. The intermediate shell is optically and, doubtless chemically, a feldspar transitional between the core feldspar and the outermost orthoclase.

Though in other respects resembling anorthoclase the feldspar of the cores shows the anomalous average extinction of -10° (maximum at -14°) on (010), thus contrasting with the angles of +6° 30' to +10° for anorthoclase. This behaviour suggested that the mineral might contain a notable amount of the barium feldspar (celsian) molecule which, in hyalophane, has the property of developing negative angles of extinction; these angles increase in size as the celsian constituent of hyalophane increases in amount. A quantity of cleavage fragments of the rhomb-feldspars were accordingly broken out of the rock, cleansed from adhering material of the ground-mass and submitted to Mr. Connor for analysis. The analysis resulted as follows:

SiO_2	54·60
TiO_2	·60
Al_2O_3	22·17
Fe_2O_3	2·00
MgO	1·30
CaO	4·62
SrO	·80
BaO	1·09
Na_2O	·4·46
K_2O	5·58
H_2O	2·50
	99·72

The material was evidently impure, the microscope showing that augite and titaniferous magnetite are the more important primary inclusions in the feldspar. On account of the presence of the minute but heavy inclusions, separation by heavy solutions is, in this case, a highly objectionable method. It seemed better to use the hand-picked material and then calculate the feldspar s composition after eliminating the oxides introduced into the analysis by the inclusions. For this purpose the augite is regarded as having the composition

25a—vol. ii—26½

of that in analcite basalt from Colorado,[*] the analysis of which gives essential oxides in the following percentages: SiO_2, 49·26; Al_2O_3, 6·01; Fe_2O_3, 3·31; FeO, 4·23; MgO, 12·40; CaO, 21·79. The iron oxide not entering into the constitution of the augite is regarded as contained in the titaniferous magnetite. The calcium of calcite is considered as derived from the feldspar; the corresponding CO_2 is subtracted from the total.

Recalculating the analysis after these eliminations we have:—

	Per cent.	Molecules.
SiO_2	57·98	·966
Al_2O_3	23·28	·218
CaO	2·74	·049
SrO	·93	·009
BaO	1·28	·008
Na_2O	5·24	·084
K_2O	8·55	·070
	100·00	

The analysis shows that the feldspar is relatively rich in barium and strontium oxides and seems to confirm the view that the peculiar optical properties are to be correlated with the presence of those oxides.

The molecular proportions in the 'purified' feldspar correspond to the following proportions among the four recognized feldspar molecules and a hypothetical 'strontium feldspar' molecule which is analogous to that of celsian.

Albite	42
Orthoclase	36
Anorthite	12
Celsian	5
" Strontium feldspar "	5
	100

It should be noted that, since a certain though small proportion of the outer shells (composed of apparently pure orthoclase) of some of the phenocryst fragments are included in the analyzed sample, the proportion of the potash molecule is somewhat higher than it would be found in the anomalous cores of the phenocrysts. The core feldspar (making up, as estimated, about 80 per cent of the average phenocryst) is clearly an unusual species, carrying, as it does, essential amounts of five different bases. It is perhaps best described as an abnormal anorthoclase, rich in the barium-strontium feldspar molecules.

Other Constituents.—The phenocrystic pyroxene is greenish-black in the hand-specimen and very pale green in thin section. It is not sensibly pleochroic and has the optical characters of common augite. It incloses magnetite and apatite but is itself of earlier generation than the phenocrystic feldspar. The biotite is apparently a common, highly ferruginous variety of mica, with intense pleochroism in colour-tints from pale to very deep leaf-brown. The rather rare olivine phenocrysts are everywhere entirely altered to brownish-yellow serpentine and dolomite but the shape and general habit of the pseudomorphs leave little doubt as to the nature of the original mineral.

[*] Bull. 228, U.S. Geol. Survey, p. 165 and Jour. Geology, Vol. 5, 1897, p. 684.

SESSIGNAL PAPER No. 25a

The ground-mass feldspars never display rhombic outline and form either thin tabular crystals or stouter individuals of rectangular sections. Many of them are twinned on the albite law but the twinning lines are seldom clean-cut, straight, and continuous through the crystal-section as in the case of the soda-lime feldspars. Both anorthoclase (in these twinned individuals) and sodiferous orthoclase seem to be represented. The difficulty of diagnosing the often very minute microlites is great and there is no certainty the soda-lime feldspar may not also occur. The microlites of augite and biotite react optically like the corresponding phenocrystic minerals.

The glass in which all the crystalline constituents are embedded is colourless to pale brownish. It may be quite isotropic but, in most cases, there are faint changes of colour as the slide is rotated between crossed nicols and under the gypsum plate. This anisotropic behaviour is apparently attributable to incipient zeolitization of the glass. It may be noted that, in the thirty or more thin sections of this intrusive porphyry, not one shows outlines of analcite in crystallized individuals. It will be seen that in a closely associated extrusive lava, primary analcite forms one-third of the rock. In the porphyry of the chonolith, however, the isotropic base is, seemingly, an entirely interstitial and amorphous substance, a glass. If analcite is present in the base it must be secondary and without crystal form.

Chemical Composition and Classification of the Rock.—Professor Dittrich has chemically analyzed a specimen nearly representing this central phase of the chonolith. The specimen was taken from a ledge on the northern brink of Rock creek canyon, about 2,500 yards upstream from the confluence with Kettle river. The chonolith here contacts with two heavy masses of conglomerate which formed part of its roof. The specimen was collected at a point 200 feet, measured perpendicularly, from the conglomerate. Even at that distance the ground-mass of the rock carries a considerable amount (perhaps 20 per cent by volume) of glass. The specific gravity of the analyzed specimen is only 2·621. From the careful study of many thin sections, the writer has concluded that the completely holocrystalline phase must give an almost identical analysis. Professor Dittrich's work shows the following proportions among the oxides (specimen No. 1054):—

Analysis of principal phase, Rock Creek chonolith.

		Mol.
SiO₂	51·83	·863
TiO₂	·86	·011
Al₂O₃	18·25	·178
Fe₂O₃	4·26	·027
FeO	1·46	·020
MnO	tr.	....
MgO	3·28	·082
CaO	4·08	·073
SrO	·42	·004
BaO	·43	·003
Na₂O	4·68	·076
K₂O	5·75	·061

2 GEORGE V., A. 1912

Analysis of principal phase. Rock Creek chonolith.—Continued.

		Mol.
H₂O at 110°C	·27	....
H₂O above 110°C	3·15	·175
P₂O₅	·55	·004
CO₂	·43	·010
	99·70	
Sp. gr	2·621	

In many essential respects this rock is mineralogically and chemically like the classic types described by Brögger, Rosenbusch, and Bäckström.* For convenience the average analysis of seven of those types is given in Col. 4, Table XXVI. Column 5 shows the range of variation in the oxide proportions. The chemical difference between the Norwegian and British Columbia rocks are largely explained by the higher proportion of femic minerals in the latter. The large amount of water liberated above 110° C. is almost certain to have come principally from the glassy base. Since the rock is so fresh it does not seem possible that this water was introduced during weathering; it must, seemingly, be regarded as of primary origin, though it has been responsible for, or, at least co-operated in, the partial zeolitization of the glassy base. The British Columbia rock rock is further notable for relatively high percentage of BaO and SrO, two oxides which were not determined in the Norwegian specimens.

Table XXVI.—Analyses of rhomb-porphyries and related rocks.

	1	2	3	4	5	6
SiO₂	52·43	51·83	52·13	57·19	54·0—69·72	51·02
TiO₂	·86	·86	·86			·56
Al₂O₃	19·18	18·25	18·72	19·44	16·48 - 22·15	16·82
Fe₂O₃	3·51	4·26	3·90	6·44	2·08—10·79	3·78
FeO	2·08	1·46	1·77			3·66
MgO	2·61	3·28	2·95	1·28	·70 - 1·93	3·38
CaO	3·71	4·08	3·90	3·10	2·42 - 4·01	6·06
SrO	·42	·42	·42			·30
BaO	·35	·43	·39			·49
Na₂O	4·85	4·68	4·77	6·32	3·04— 8·39	3·49
K₂O	5·95	5·75	5·85	4·44	3·24— 6·30	7·44
H₂O -	·27	·27	·27	1·34	·60 - 3·25	2·75
H₂O	3·19	3·15	3·17			
P₂O₅	·42	·55	·49			·49
CO₂	tr.	·43	·22			
	99·83	99·70	99·81	99·55		100·04

1. Chilled phase, Rock creek chonolith.
2. Central phase, Rock creek chonolith.
3. Average of 1 and 2.
4. Average of seven types of Norwegian rhomb-porphyry.
5. Range of oxides in seven types averaged in 4.
6. Average of four types of basic syenite (borolanose) from the Highwood mountains; L. V. Pirsson, Bull. U. S. Geol. Surv., No. 237, 1905, pp. 172-3.

* Analyses taken from Osann's Beitraege zur Chemischen Petrographie, 1905, pp. 31 and 138, and from Rosenbusch's Elemente der Gesteinslehre, 1901, p. 286.

Chemically the rock of the chonolith is most nearly matched by the 'basic syenite' and 'syenite-porphyry' from the Highwood mountains, as described by Pirsson.[*] The average of four analyses of the latter rocks is noted in Col. 6 of Table XXVI. Pirsson's list of the component minerals is: iron ore, apatite, biotite, olivine, augite, orthoclase often surrounded with rims of soda-orthoclase, and sometimes demonstrable nephelite. In the Highwood porphyry the feldspar phenocrysts have tabular habit and do not form rhombs. Anorthoclase is not described as occurring in any of the four analyzed types. The relatively high barium and strontium oxides suggest that the alkali feldspars, as in the chonolith porphyry, are charged with the celsian molecule at least, but the orthoclase is described as free from an appreciable admixture of the anorthite molecule.

In spite of the differences noted, the rock of the chonolith must be regarded as remarkably similar to the uncommon basic syenite of the Montana occurrence. In the Norm classification, the two enter the same subrang, namely, the sodipotassic borolanose, of the domalkalic rang, essexase, in the dosalane order, norgare. The norm of the chonolithic rock has been calculated as follows:—

Orthoclase..	33.92
Albite..	24.43
Anorthite..	11.40
Nephelite..	10.51
Diopside..	6.32
Hypersthene..	5.60
Magnetite..	2.00
Hematite..	2.88
Ilmenite..	1.67
Apatite..	1.24
Water..	3.42
	99.48

Mineralogically and structurally, however, the nearest relative to the chonolithic rock is the rhomb-porphyry described by Brögger and others, and the rock may be referred to henceforth under that name. In the Norm classification the typical Norwegian rhomb-porphyry is referred by Washington to laurvikose, the dosodic subrang of the perfelic, persalane order, canadare. Brögger's 'nephelite rhomb-porphyry' enters the dosodic subraug, viezzenose, of the peralkalic rang, miaskase, in the lendofelic, persalane order, russare. The three types thus represent three different subrangs in as many different rangs, orders, and classes. The norm-chemical system of classification evidently fails to bring out those similarities among the three types which must impress the field-geologist in the highest degree. The older, more purely mineralogical classification, laying emphasis on the dominance of the rare feldspar, anorthoclase, in the at least equally rare rhombic form, obscures the obvious differences in magmatic relationships. Neither as borolanose nor as rhomb-porphyry is the British Columbia rock ideally classified. That can only be done when what may be called a mode-chemical system of classification is

---

* L. V. Pirsson, Bull. 237, U.S. Geol. Surv., pp. 89 ff, 1905.

DEPARTMENT OF THE INTERIOR

perfected. Meantime, in naming the rock rhomb-porphyry the need of the field geologist is the better satisfied. It may thus be defined as a basic rhomb-porphyry rich in augite and biotite; where vitrophyric, it is charged with much original water resident in the more or less zeolitized glass.

CONTACT PHASE OF THE CHONOLITH.

For distances under fifty feet, measured perpendicularly from the contact, the chonolithic rock has a field-habit notably unlike that of the phase just described. This difference is due simply to chilling on roof or walls. Within this chilled zone the rock is a dark bluish-gray or slaty-gray, very fine-grained porphyry. The phenocrysts are again chiefly rhomb-feldspars which are here quite glassy and less charged with inclusions and secondary calcite than the feldspar of the central phase. They measure from 2 mm. or less, to 6 mm. in length. With them, a few augite prisms up to 2 mm. or 3 mm. long, and hexagonal biotite foils from 1 mm. to 2 mm. in diameter, may be seen in the hand-specimen. At other localities biotite is not phenocrystic but is abundant in the ground-mass.

A typical specimen of this phase will serve as a basis for its further description. It was collected at the same group of ledges from which the analyzed specimen of the central phase was taken but at a point only ten feet from the invaded conglomerate. Under the hammer the chilled phase breaks with a sonorous, almost metallic ring which, of itself, indicates the signal freshness of the rock at the average outcrop.

Under the microscope this type specimen of the chilled phase shows a few small serpentinized olivines among the more conspicuous phenocrysts. The accessories are the same as in the central phase with the exception that nephelite was nowhere recognized in the thin section. The ground-mass is hyalopilitic with many minute, acicular augites, thin biotite foils, and feldspar microlites, embedded in a base which is almost certainly a true glass somewhat zeolitized. This base, composing at least 40 per cent of the rock by volume, generally polarizes with exceeding faintness, the anisotropic property being only appreciated with the gypsum plate. Just how far the anisotropy is due to devitrification and how far, possibly, to the straining of the glass, cannot be declared. In spite of close study, with high magnification, there is no certain clue as to the nature of the devitrification product or products. That a large amount of glass still remains is indicated by the low specific gravity of the specimen, viz., 2·608. Other fresh hand-specimens have specific gravities of 2·564 and 2·645. The average specific gravity of this phase is about 2·600, and contrasts with that of the central phase, which is about 2·740. Excepting the small amount of serpentine derived from the olivines, here and there a minute point of calcite, and the undetermined zeolite of the base, there are no noteworthy secondary products visible in thin section. The rock is exceptionally fresh.

Professor Dittrich's analysis of this specimen (No. 1053) gave the following result:—

Analysis of chilled contact-phase, Rock Creek chonolith.

		Mol.
SiO_2	52·43	·874
TiO_2	·86	·011
Al_2O_3	19·18	·188
Fe_2O_3	3·51	·022
FeO	2·06	·029
MnO	tr.	....
MgO	2·61	·065
CaO	3·71	·066
SrO	·42	·004
BaO	·35	·002
Na_2O	4·85	·078
K_2O	5·95	·064
H_2O at 110°C	·27	....
H_2O above 110°C	3·19	·178
P_2O_5	·42	·003
CO_2	tr.	....
	99·85	
Sp. gr.	2·608	

The oxides are present in proportions essentially similar to those in the central phase. The almost perfect freshness of this rock shows even more clearly that the water, great in amour' as it is, is an original constituent and is derived chiefly from the base of the rock, a small part of it emanating from the mica.

The norm of the chilled phase of the chonolith has been calculated as follows:—

Orthoclase	35·58
Albite	20·96
Anorthite	12·79
Nephelite	10·79
Hypersthene	4·90
Diopside	3·46
Magnetite	4·18
Ilmenite	1·67
Hematite	·64
Apatite	·93
Water	3·46
	99·36

The rock is borolanose in the Norm classification; a basic rhomb-porphyry in the older classification.

OTHER INTRUSIONS OF RHOMB-PORPHYRY.

A smaller intrusive mass of the porphyry occurs on the slope south of the Kettle river about two miles below Rock Creek post office. The exposures are not extensive but the body is elongated and seems to have a dike-like form, at least 1·5 miles long by 400 to 800 yards in width. It cuts the Paleozoic phyllites and quartzites as well as a small patch of Tertiary sandstone at the southern end of the body. The northern end cannot be seen, as it is covered by the Kettle river gravels. On account of the imperfect exposure, the exact structural relations were not determinable, but the mass appears to form a cross-cutting, injected body and may fall in the chonolith class rather than in that of dikes or sills.

A third large area of the porphyry crops out on the slope north of Rock creek and west of the large chonolith. It is quite possible that this more westerly mass really forms part of the chonolith and that the two are connected underground. They are separated by Tertiary conglomerate overlying pre-Tertiary granodiorite.

Petrographically, these porphyries are practically indistinguishable from the porphyry of the large Rock Creek chonolith. Both the 'central' and 'contact-chilled' phases are represented with typical characters.

The Tertiary sandstones of the wagon-road section four miles west of Midway and those exposed east of the bridge two miles to the northward, are cut by dikes and sills of the porphyry. In all these thinner bodies the porphyry has the habit of the 'chilled' phase of the chonolith above described. A small amount of nephelite is almost always present; it occurs in the ground-mass and often forms small idiomorphic, phenocryst-like crystals. In no case, however, does the nephelite rival the rhomb-feldspar, the augite, or the biotite in abundance. As a rule the ground-mass is not vitrophyric and in this respect the dike and sill-rocks are more like the central phase of the Rock Creek chonolith.

Other dikes or sheets appear to cut the basalts on the heights north of the Kettle river above the bridge. Still others cut the Anarchist schists at various points between the Rock Creek chonolith and Osoyoos lake. The most westerly occurrence discovered is that on the summit four miles east of Osoyoos lake and 1·5 miles north of the Boundary line. At that point the Anarchist phyllite-amphibolite complex is cut by several north-south, nearly vertical dikes of the porphyry, several of which approximate a hundred feet in width. These dikes are twelve miles or more west of the Rock Creek chonolith and twenty miles west of the most easterly dikes of the rhomb-porphyry within the Boundary belt. The relatively wide distribution of this peculiar rock-type is paralleled in the classic Norwegian district. Here as there, similar conditions of magmatic differentiation and of crystallization seem to have prevailed over a large area. In neither case, however, does it appear necessary to believe that the region was underlain by a continuous, deep *couche* of the magma respectively represented by the porphyry actually intruded. It is at least as probable that each of these porphyry types is the product of splitting from one or more deep-seated masses of more usual composition.

EXTRUSIVE PHASE OF THE RHOMB-PORPHYRY.

Along its northern edge the normal porphyry of the Rock Creek chonolith lies in contact with a highly vesicular and doubtless extrusive phase of the same rock. The relation between the two is very obscure. In the field no sharp line of demarcation, separating the phases, can be drawn. The same is true of the contact between the lava and the intrusive body of porphyry west of the large chonolith. It is possible that the phases merge into each other, so that the same magma which has intrusive relations on the south, found free vent to the surface at the northern extremity of the chonolith. This lava, often highly vesicular, composes the heights along the northern limit of the map on the west side of Kettle river and forms a continuous mass for nearly four miles measured east and west. It is not known how far it extends northward, outside the five-mile Boundary belt. No safe clue could be found as to the dip or as to the total thickness of the lava; the latter must, however, measure many hundreds of feet. It is always massive and without such partings as would be expected if the mass were the result of many successive flows. In any case the flows are thick and, apparently, are unaccompanied by pyroclastic materials.

Petrographically this lava closely resembles the chilled phase of the chonolith. The colour of the rock varies from slate-gray to the more common brownish-gray; in the more vesicular lava the tint becomes lighter and more distinctly brown, changes doubtless incidental to simple weathering in a porous rock. The vesicles are of all sizes up to those 2 cm. or more in length. They are regularly filled with calcite, more rarely by isotropic silica, probably opal.

The lava is always porphyritic; the phenocrysts are rhomb-feldspar ('anorthoclase' like that in the phenocrysts of the chonoliths), augite, and biotite, though, in some slides, the biotite is quite rare among the phenocrysts. The ground-mass is mostly composed of the same minerals, along with orthoclase. All are in microlitic development and are embedded in an abundant colourless to pale greenish or brownish glass. As in the intrusive phase the feldspar of the ground-mass never shows rhombic sections. The accessories are, here also, apatite, titaniferous magnetite, and probably titanite in minute grains. The apatite occurs in unusually large prisms and is fairly abundant. The glassy base is sometimes zeolitized and is then brownish and polarizes faintly, as in the analyzed specimens of the chonolith. The glass is roughly estimated to compose from 20 to 35 per cent or more of the whole volume. These estimates agree with those which can be roughly made from the density of the rock. For this purpose three fresh specimens of the non-vesicular lava were specially chosen and their specific gravities determined; the values are, respectively, 2·597, 2·602 and 2·624. In this respect as in the mineralogical composition and general habit, the similarity of the lava to the chilled phase of the chonolith is very manifest. So patent is this resemblance that no chemical analysis of the lava seemed necessary.

ANALCITIC RHOMB-PORPHYRY ('SHACKANITE').

North of Rock Creek.—As one climbs northward from Rock creek canyon up the mountain lying immediately east of the north fork of the creek and about five miles north of the Boundary line, he first passes over Tertiary conglomerate, then over the most westerly of the three large bodies of intrusive rhomb-porphyry, and, near the 4,000-foot contour, reaches the edge of the great lava mass which has just been described. The first summit of the mountain

2 GEORGE V., A. 1912

is composed of that lava. At the col just north of that summit and for a half mile farther north, the ridge is capped by a great mass of lava of a type related to the one described but distinguished from it both chemically and mineralogically.

This second kind of lava is often highly vesicular, very massive, and of the same range of dark colours and general habit as the first type. Though very compact, it always bears small glassy phenocrysts of rhomb-feldspar and a few augites visible to the unaided eye. In the field it was not suspected that the second lava was to have any special interest not shared by the type above described. As a consequence of this view, only two hand-specimens of this lava were collected. The desired facts of field occurrence were likewise not obtained in the measure in which they might have been if the writer had been conscious of the unusual character of the rock. It is known only that it occurs in one or more very thick flows and that the lava appears, from its topographic position, to overlie the first, the more normal rhomb-porphyry lava, though the two almost certainly belong to the same epoch of extrusion. Whether the two types are separated by a sharply defined surface of contact is not known. This rarer type seems to cover about a third of a square mile within the five-mile Boundary belt and at least as much more beyond its northern limit. How much additional area is covered by it is also unknown. It is thus clear that a second visit to the mountain and a careful field-study are required before a satisfactory account of this occurrence of lava can be given. All that is now possible is to furnish a brief note on the character of the lava and thus suggest to some future geologist one more point for study in this complicated and interesting region.

The real character of this lava, unique among all the formations occurring in the whole transmontane section, was revealed only after its optical and chemical analysis had been performed. Macroscopically, as already noted, it is much like the rhomb-bearing lava to the south. The rock is usually very fresh and breaks sonorously under the hammer. The colour is usually a very dark slate-gray, tending to dark brown on weathered surfaces. The rhomb-feldspar phenocrysts vary from 1 mm. to 3 mm. in length while the few augite prisms are even shorter. Under the microscope a few small olivines, altered to brownish-yellow serpentine or to carbonate, and exceedingly rare foils of deep brown biotite are seen to form phenocrysts, but both of them are only accessory constituents. The leading peculiarity of the rock is found in the ground-mass, which is largely composed of minute but perfectly formed analcite crystals in dodecahedral development. With these are associated many feldspar microlites and the accessories, apatite, magnetite, a few grains of pyrite, and probably titanite. All of these minerals are embedded in an abundant, transparent, pale brownish glass which contains a few grains of secondary carbonate, and may be somewhat zeolitized. With the exception of the alteration of the olivine and glass, the rock seems to be practically as fresh as the day it first solidified from fusion.

SESSIONAL PAPER No. 25a

To describe the phenocrysts and most of the constituents of the ground-mass would be merely to repeat the description of the corresponding minerals in the normal rhomb-porphyry. A principal difference between the two rocks is, however, found in the fact that here biotite is only a very rare accessory and does not occur in the ground-mass. The phenocrystic 'anorthoclase' retains all its peculiarities; the same mineral is the more abundant feldspar of the ground-mass, the other being glass-clear (probably highly sodiferous) orthoclase, giving the familiar rectangular sections as in many alkaline porphyries. Plagioclase feldspar seems to be entirely absent from the rock.

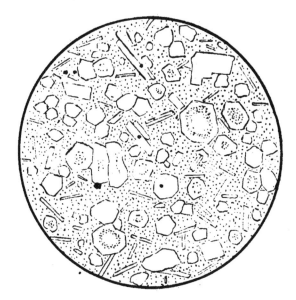

FIGURE 26.—Partly diagrammatic drawing from thin section of ground-mass of "sbackanite" (neither phenocrysts nor the rare pyroxene granules of ground-mass shown). The roundish, often polygonal crystals are analcite, with and without minute inclusions. The rectangular, trapezohedral, and lath-shaped crystals are alkali feldspar. Magnetite grains represented in actual proportion. All these crystals lie in a matrix of brown glass and serpentine (dotted). Diameter of circle, 1 mm.

The quite isotropic analcite forms sharply marked polygonal or round, colourless crystals varying from 0·02 mm. to 0·2 mm. in diameter (Fig. 26). Some of them seem to be quite devoid of inclusions but the majority are charged

2 GEORGE V., A. 1912

with minute prisms or grains of apatite, magnetite, pyroxene, and probably glass. These inclusions have a tendency to concentric arrangement within their host. The extremely sharp definition and idiomorphism of the crystals and the all but perfect freshness of the rock are, of themselves, sufficient proof of the primary nature of the mineral. The well-known demonstrations of Lindgren, Pirsson and others, as to the occurrence of analcite as a primary, magmatic product in igneous rocks seem amply sufficient to show the possibility that the mineral with these optical properties may be truly analcite. The convincing fact as to the truth of this conclusion is furnished by the chemical analysis of one of the specimens. The analysis was made by Professor Dittrich with all his accustomed accuracy and thoroughness. The high percentage of combined water indicates that the isotropic mineral is analcite, rather than a member of the sodalite family. Moreover, Mr. Connor found, by test, that chlorine shows no more than a mere trace.

A rough determination of weight percentages by the Rosiwal method gave the following result:—

Feldspar phenocrysts ('anorthoclase')	10·5
Augite phenocrysts	5·5
Olivine phenocrysts	2·6
Analcite	29·2
Feldspar of ground-mass ('anorthoclase' and sodiferous orthoclase)	23·0
Glass	25·0
Magnetite	2·0
Apatite and titanite	2·0
Biotite	·2
	100·0

A principal error in this estimate may consist in the figures for glass and analcite. These were distinguished in the thin section only where the crystal form of the analcite was clearly seen; where the crystal form was not seen, the colourless, isotropic, or nearly isotropic material was referred to glass. It is probable, therefore, that the percentage for glass is too high and that for analcite is too low in the foregoing table.

Professor Dittrich's analysis of a non-vesicular specimen (No. 1064) resulted as follows:—

Analysis of 'shackanite.'

		M.J.
SiO_2	52·24	·871
TiO_2	·73	·009
Al_2O_3	19·28	·189
Fe_2O_3	4·34	·027
FeO	1·13	·015
MnO	tr.	....
MgO	1·85	·046
CaO	4·43	·079
SrO	·42	·004
BaO	·36	·002
Na_2O	6·34	·102
K_2O	2·40	·026

Analysis of ' shackanite '—Continued.

		Mol.
H_2O at 110°C	·80	....
H_2O above 110°C	4·63	·258
P_2O_5	·59	·004
CO_2	·35	....
	99·59	

Sp. gr.	2·528

A second, non-vesicular specimen with more abundant augite had a specific gravity of 2·637.

Two different norms were calculated. For the one the combined water was, as usual, neglected and all the soda was referred to albite and nephelite molecules. The result is:—

Orthoclase	14·46
Albite	43·49
Anorthite	16·96
Nephelite	5·40
Hypersthene	3·50
Diopside	2·37
Hematite	3·36
Magnetite	1·39
Ilmenite	1·37
Apatite	1·24
Water and CO_2	5·78
	99·32

A second norm was calculated on the assumption that the combined water should not be neglected; in this case analcite is a standard mineral, taking the place of nephelite as the lenad. This norm is essentially like the first except that the albite is here 33·54 and the analcite replacing the nephelite is 16·72.

According to the first calculation this rock enters the dosodic subrang, akerose, of the domalkalic rang, monzonase, in the dosalane order, germanare—as defined in the Norm classification.

According to the second calculation the rock enters the dosodic subrang, essexose, of the domalkalic ... .se, in the dosalane order, norgare.

In the older classification the rock may be called an analcitic rhomb-porphyry; if any systematist wishes a more compact, single-word name for this new petrographic type, he could refer to it as ' shackanite.' This word is coined from the name of the railway station (Shackan) at the southern foot of the ridge which is crowned by the analcitic rock. Such naming would have the advantage of avoiding the use of ' porphyry ' as a systematic designation for another species of effusive lava. The writer would prefer to see the term 'porphyry' restricted, in technical petrography, to the porphyritic intrusive rocks.

Other Occurrences.—The analcitic rhomb-porphyry lava, generally vesicular, also crops out liberally on the strong ridges north and northwest of the Kettle River bridge, six miles above Midway. The upper part of each ridge, through

2 GEORGE V., A. 1912

an east-west distance of three miles, has excellent bed-rock exposures which afford some indication of the mode of occurrence of the lava. On the mountain due north of the bridge it is seen to form massive flows, each a hundred feet or more in thickness. These are conformably interbedded with several equally heavy flows (100-200 feet thick) of a vesicular trachyte of quite different habit and composition. The two lavas vary in their power of resistance to weathering; the great beds strike regularly N. 45° E. and dip at an average angle of 30° to the southeast. As a result the mountain is strongly ribbed with alternating scarps and back-slopes with crests of the corresponding ridges trending N.E.—S.W. (See Plate 73, B.)

On a specially sharp, meridional, 3,100-foot ridge, situated 2.5 miles farther west, the analcitic lava with typical character was observed resting on the Tertiary sandstone. At this point the strike is nearly due north and south; the dip, 35° E. It is most probable that, between the two localities, the lavas and sandstone are repeated in outcrop by a number of strike-faults. The sandstones are cut by large dikes of the rhomb-porphyry with the habit of the chilled phase of the chonolith. These dikes were probably among the feeders of the lava flows.

The analcitic porphyry is practically identical in character with that described on the west side of the Kettle river. It is here somewhat more vesicular, the flattened pores reaching an inch in greatest diameter. They are filled chiefly with calcite but a very few carry a yellowish zeolite.

INTRUSIVE ROCKS CUTTING KETTLE RIVER STRATA.

All the nine types of the Tertiary lavas were necessarily erupted through fissures or other vents in which the chemically corresponding 'hypabyssal' rocks have crystallized. Most of the lavas are in fact, paralleled in the dike and sill rocks which at many points within the boundary belt cut the Tertiary sediments and the Paleozoic formations.

Porphyrites—It has been noted that most of the andesitic species are closely similar in their chemical composition; their distinction as species is chiefly based on mineralogical characters which are doubtless due in largest part to differences of physical conditions during crystallization. Where the same magmas were intrusive, temperature and some other conditions must have been more uniform than were those prevailing during the solidification of the surface lavas. This is the probable reason why few distinct types have been found among the dikes or sills corresponding to the Midway andesitic flows.

These types are two in number—hornblende porphyrite and augite-biotite porphyrite. The former composes a poorly exposed sill cutting the Tertiary shales and sandstones at the gulch which mouths a half mile northwest of the Canadian Pacific railway 'Y' at Midway. This is a normal fine-grained porphyrite bearing phenocrystic labradorite and dark-green hornblende, with a feldspathic base in which a little primary quartz may be discovered under the microscope.

The augite-biotite porphyrite forms a number of great dikes and a large chonolith-like mass cutting the Paleozoic limestone and quartzite on Deer Hill. An apophysal sill from the largest body cuts the tilted Kettle River sandstones at the western base of the hill. The same intrusive appears at several other places in the Boundary belt but the exposures were often not full enough to declare the structural relations of the bodies. In some cases it was not possible to tell if the rock were really not a non-vesicular phase of the surface lava. The Deer Hill dikes and chonolith (?) are mineralogically like the biotite-augite andesite of the flows, with the natural exception that the porphyrite is more thoroughly crystallized. The ground-mass is here holocrystalline and generally carries a notable amount of free quartz along with the feldspar, augite, and biotite microlites. The porphyrite of the apophysal sill mentioned carries a little phenocrystic hornblende as well as the biotite and augite.

No dikes corresponding to the effusive olivine basalt are known to occur in the belt, though it is quite possible that some of the observed basalt is really in dike relation. A 300-foot sill of fine-grained augite gabbro cuts the Tertiary sandstone on the southern face of the conspicuous north-south ridge three miles due east of the mouth of Rock creek. This gabbro is probably an intrusive phase of the same magma which is represented in the vesicular basaltic flows across the river. The rock has the mineralogical composition of a typical, fine-grained diabase but has the hypidiomorphic-granular structure. The pyroxene is common augite without the diallagic parting. A few serpentinized grains of olivine appear in one thin section, suggesting a transition to the olivine gabbros. A few sporadic flakes of biotite are accessory in the same thin section. In the sill-rock generally the only essentials are augite and labradorite.

Pulaskite Porphyry.—In the stretch of eight miles between Ingram creek and the mouth of Rock creek the Tertiary sediments are cut by a large number of thick sills and dikes of a pulaskite porphyry, which is in strong lithological contrast to all the other igneous types of the region except the alkaline trachyte. The porphyry is conspicuous in the field and, if the exposures were sufficient, it could be mapped with relative ease. Most of the intrusions were found on the north side of the Kettle river, but a few dikes of the porphyry cut the andesites on the south side.

Just east of the bridge six miles above Midway the porphyry is specially developed, generally as sills in the fossiliferous Tertiary sandstone. These sills vary from 100 feet or less to about 800 feet in thickness. Toward the top of the broad hill east of the bridge and 1,000 feet above the river, the sandstones are squarely truncated by an apparently continuous mass of the porphyry of which the sills are offshoots. This larger mass has been injected into the sandstones without any definite relation to bedding-planes and is perhaps best described as

2 GEORGE V., A. 1912

a chonolith. The northern limit of the chonolith is so completely covered by glacial drift and soil that there the relations of the body could not be deciphered. On the ... the relatively coarse-grained porphyry seems to pass under a thick cov ... wn effusive phase, the vesicular biotite trachyte. Yet it is quite ... the two rocks pass into each other gradually, or, thirdly, that they ... faulted into contact. The difficulty of deciding between these ... curiously paralleled in the problem above noted in connection w... th-porphyry chonolith on Rock creek, where, again on the north sid. it ... contact with thick vesicular lavas of the same chemical composition ...

... is very uniform in habit and composition. It is a ... rock, carrying abundant phenocrysts of pale, flesh-pink, the ... which reach 1 cm. or more in diameter. A few biotites, gen... : diameter form the only other phenocrysts. Under the microsco ... .coloured ground-mass is seen to be essentially a typical trachytic ma... lar feldspars, with which a few small biotites and rare prisms of green hornblende are associated. A little interstitial quartz and small amounts of titanite, apatite, and magnetite are accessory. The ground mass is highly miarolitic with actual cavities between the feldspars. The resulting porosity goes far to explain the low specific gravity obtained for the fresh analyzed specimen, namely, 2·497.

The feldspars are all more or less cloudy with decomposition-products, calcite and kaolin. The phenocrysts are chiefly soda-orthoclase with extinction on (010) of 12°+. This feldspar has a tendency to a perthitic structure. Carlsbad twins are common. A few andesines (Ab, An.), generally surrounded with thick shells of soda-orthoclase, occur in some thick sections. The feldspar of the ground-mass is generally twinned on the same Carlsbad law and seems also to be chiefly soda-orthoclase. A little plagioclase may also be there present, though none was certainly identified under the microscope.

Professor Dittrich has analyzed a typical specimen (No. 1010) from a sill cutting the sandstones 400 yards northeast of the Kettle river bridge. The result is as follows. (Table XXVII, Col. 1.):—

SESSIONAL PAPER No. 25a

TABLE **XXVII.**

Analysis of pulaskite porphyry and related rock.

	1	2	
		Mol.	
SiO₂	62 04	1 034	62 20
TiO₂	72	009	54
Al₂O₃	17 63	173	17 23
Fe₂O₃	1 98	013	1 51
FeO	1 57	022	2 02
MnO	tr.	...	tr.
MgO	99	025	1 50
CaO	1 75	031	1 30
Na₂O	4 73	076	5 50
K₂O	6 74	071	6 71
H₂O, at 110°C	12		30
H₂O, above 110°C	1 18	...	30
P₂O₅	17	001	11
CO₂	30		tr.
	99 82		99 53
Sp. gr.	2 497		

The norm was calculated to be:—

Quartz	4 38
Orthoclase	39 48
Albite	39 82
Anorthite	7 23
Hypersthene	2 30
Diopside	43
Magnetite	3 02
Ilmenite	1 37
Apatite	31
Water	1 30
	99 64

In the Norm classification the rock enters the sodipotassic subrang, pulaskose, of the domalkalic rang, pulaskase, of the persalane order, canadare. In the older classification it is a typical pulaskite porphyry. Col. 2 of Table XXVII shows the result of Professor Dittrich's analysis of the Coryell batholith syenite (made for Mr. Brock: see page 359). The striking chemical resemblance of the two rocks and their analogous positions as the youngest or nearly the youngest intrusives of their respective districts, suggest that they are of approximately contemporaneous origin. For what it is worth this argument tends to substantiate the view stated on page 376 that the Coryell batholith is of Tertiary and post-Oligocene date.

2 GEORGE V., A. 1912

ORDER OF ERUPTION OF THE MIDWAY LAVAS.

A summary of the facts actually determined for the relations of the various lavas of the Midway group may be given in a few words. All are clearly younger than the older Kettle River Oligocene beds but it is quite possible that the basalt and augite andesite, if not much of the biotite-bearing andesite, were contemporaneous with the younger sandstones. All of the andesites and basalts, apparently without exception, are diked by the trachyte or its equivalent, the pulaskite porphyry. The relation of the trachyte to the extrusive rhomb-porphyry and the analcitic lava is not quite clear. In the section east of the Kettle river bridge a dike of the rhomb-porphyry, as shown in Figure 25, cuts a great sill of the pulaskite porphyry. At two other points the relation seems to be reversed, though the field evidences are not there so compelling as in the first mentioned case. The flows of trachyte, rhomb-porphyry, and analcitic lava (shackanite) are closely associated and are in apparent alternation. All three types seem to belong to one eruptive period which probably opened with the extrusion of trachyte and closed with the analcitic rhomb-porphyry (shackanite).

The probable succession of the lavas is, then, as follows:—

Lavas.	Corresponding intrusives of region.
Youngest group. { Analcitic lava (shackanite).	Rhomb-porphyry.
Extrusive rhomb-porphyry .	
Alkaline trachyte .	Pulaskite porphyry.
Middle group { Biotite andesite	
Biotite-augite andesite.	Augite-biotite porphyrite
Hornblende-augite-biotite andesite	
Hornblende-augite-andesite	
Oldest group.... { Augite andesite.	Augite porphyrite.
Olivine basalt	Augite gabbro.

The oldest and middle groups of the lavas are believed to be of Oligocene age. The youngest group may conceivably belong to the late Oligocene but it is more probable that these peculiar lavas together with the chonoliths, dikes, and sills of rhomb-porphyry and pulaskite porphyry, were erupted during or just after the deformation of the Oligocene Kettle River beds. Similar deformation of older Tertiary strata elsewhere in British Columbia and in Washington State has been credited to the close of the Miocene period. The rhomb-porphyries and the trachyte are provisionally referred to that stage of geological history. On this view the Midway volcanic group is a compound formation involving products of two distinct volcanic epochs in this region.

STRUCTURAL RELATIONS OF THE COLUMBIA MOUNTAIN SYSTEM WEST OF CHRISTINA LAKE.

In its complexity the group of mountains discussed in this and the preceding chapter is much like the Rossland mountain-group. The western group has, however, an extra series of volcanic and sedimentary rocks of Tertiary age, which are almost certainly not represented at any point in the Rossland mountains.

The most heavily crumpled and metamorphosed sediments are those of the Attwood and Anarchist series, which appear to correspond directly with the Pend D'Oreille and Sutherland schist series of the Rossland mountains. Here again the Paleozoics are so thoroughly mashed, kneaded, and welded that very little can be said as to the detailed structure of these rocks all the way from Christina lake to Osoyoos lake.

Most of the volcanic rocks occurring in the 'Boundary Creek district' are much less deformed than the Paleozoics, though generally showing high dips. The relations are so like those of the Rossland volcanics that the writer is provisionally assigning the Phoenix volcanic group also to the Mesozoic. The crushing of the Paleozoics is assigned to the late Jurassic orogenic revolution; the sharp upturning of the Phoenix volcanics to the post-Laramie revolution.

As already pointed out, the Midway volcanic formation and the associated Kettle River (Oligocene) sediments form a mass of rock which either fills a broad syncline prepared on an earlier Tertiary surface, or else represents a down-faulted block of the Oligocene rocks which have thus subsided relatively to the Paleozoic terranes of Attwood mountain and Anarchist-mountain plateau. The former relation seems the more probable. Post-Oligocene, probably late Miocene faulting and moderate uptilting have affected the Kettle River and Midway formations which, in contrast to the other two divisions of the rocks in the western part of the Columbia mountain system, show low dips.

At least two unconformities are registered in the relations shown in the Boundary belt. The Kettle River beds are in striking unconformity to the underlying Paleozoics and the (probably Jurassic) plutonic bodies. The relatively little sheared and altered basic volcanics of the 'Boundary Creek district' mapped as the Phoenix group, are believed to be unconformable upon the crumpled Attwood series. A third unconformity may exist between the tilted Oligocene Midway volcanics and the alkaline flows and breccias composed of rhomb-porphyry, trachyte, and shackanite.

The region where traversed by the Boundary belt, does not seem to show a single unbroken fold of any importance. The various strata are either mashed into an undecipherable complex or are faulted, with displacements which are registered best in the bedded rocks of the Tertiary formations. The many westward-facing and northwestward-facing scarps on the lava beds forming the ridges north of the Kettle river between Ingram creek and Rock creek, are in part to be explained by a number of faults. These separate long narrow blocks which appear to be successively downthrown on the northwest. (See Plate 73, Figure B). The gravel and sand of the river bottom between the Kettle river bridge and Rock creek covers the trace of a strong east-west fault separating the uptilted sandstones on the north from the more flat-lying basaltic lavas on the south of the river. Much of the dislocation so manifest to north and south of Rock creek itself has been accomplished by sharp faulting which is thought to be contemporaneous with the intrusion of the Rock Creek chonolith.

Though far less important here, batholithic intrusion has affected the Paleozoic rocks in a way quite similar to that in which the oldest terrane of the Rossland mountains has been affected. In the Boundary Creek district the

2 GEORGE V., A. 1912

number of small stocks is so great that Mr. Brock believes that further erosion would disclose a large, continuous batholith of granodiorite, of which the stocks and associated dikes are roof features. (See page 387.)

The injected igneous bodies, i.e., those which have come into place by displacing rather than by replacing the country-rocks, include a vast number of dikes, as well as a few bodies which have been described as chonoliths. It is also suspected that the poorly exposed pulaskite porphyry east of Kettle river bridge is chonolithic in its relations and that the dunites of this region may be fairly classed with the chonoliths.

A fuller statement as to the structural relations among the many formations of the western Columbia system is to be found in the foregoing descriptions of the different rocks, in the section on correlation which closes this chapter, and in Mr. Brock's reports. These relations are of intrinsic interest but they are also important in throwing light on the later events of the neighbouring mountain systems where strata of Oligocene age have not been discovered.

CORRELATION.

In a general way the succession of geological events which are registered in the rocks of the five-mile belt between Midway and Osoyoos lake has been discovered. A partial correlation of the formations may be made, though much remains to be accomplished, especially in the analysis and proper dating of the thick members which have been assembled under the name 'Anarchist series.' This oldest group is almost certainly the same as that which crops out at intervals between the Columbia river and Midway, and, in the Rossland district, bears obscure fossils referred to Carboniferous species. Though the lithological similarity of the Anarchist series to these Rossland rocks and, as we shall see, to the very thick, fossiliferous, undoubtedly Carboniferous rocks found in the Skagit range, may be an accidental and illusory resemblance, it seems best to correlate the Anarchist series, or much of it at least, with the Carboniferous rocks of western British Columbia.

The dunite (serpentine) bodies of this region are intensely sheared and metamorphosed and hence seem to be much older than the dunites and other peridotites of the Rossland mountains. Dawson has described many masses of serpentine as being nearly or quite contemporaneous with the basic effusive rocks of the Carboniferous Cache Creek series of western British Columbia. In the present instance it is known that the dunite cuts the phyllites of the Anarchist series and is never seen to cut the Rock Creek diorite or granodiorite; further, it appears practically certain that the dunite is much older than the Kettle River beds. For the present the dunite may be tentatively correlated with the probably Carboniferous greenstones of the Anarchist series. The Rock Creek gabbro and diorite may have direct genetic connection with the dunite, though in the field their associations are with the granodiorite.

The crushed and gneissic Osoyoos granodiorite is described in the next chapter, where evidence is given for referring it to the late Jurassic period. The Rock Creek granodiorite is not so much sheared but it is distinctly strained and

SESSIONAL PAPER No. 25a

altered as if by strong orogenic pressures. Somewhat naturally, then, it may be correlated with the Osoyoos batholith. It will be recalled that Mr. Brock has provisionally referred the intrusion of the many granodiorite bodies of the Boundary Creek district to the Jurassic.

The Kettle River beds are clearly unconformable upon the Rock Creek granodiorite, diorite, and gabbro, and upon the phyllites, limestones, greenstones, and quartzites of the Anarchist series. The discovery of relatively abundant fossils in the Kettle River formation facilitates its final correlation as well as that of the Midway volcanics with their corresponding intrusive phases.

These various fixed and tentative conclusions are stated in the following table, which has its quota of essential doubts:—

Pleistocene.... ....Glacial and Recent.

Miocene. ....
{ Midway volcanic group (in part):
    Rhomb-porphyries; chonoliths, sills, dikes, and effusive forms
    (including "shackanite").
    Trachyte and pulaskite porphyry; flows, sills, and dikes.

Oligocene...............
{ Midway volcanic group (in part):
    Various mica-andesites and hornblendic andesites; porphyrite dikes.
    Basalt and augite andesite.
  Kettle River formation; sandstone, conglomerate, shale, traces of lignite

UNCONFORMITY.

Jurassic (intrusive)....
{ Rock Creek granodiorite; Osoyoos granodiorite.
  Rock Creek gabbro and diorite.

Carboniferous?.... ....
{ Dunite (serpentine).
  Anarchist series; phyllite, quartzite, limestone, and greenstone.

CHAPTER XVI.

FORMATIONS OF THE OKANAGAN RANGE AND OF KRUGER MOUNTAIN PLATEAU.

From the eastern slope of the wide valley occupied by Osoyoos lake to the Pasayten river, a distance of just sixty miles along the Boundary, the mountains are composed of almost continuous plutonic rocks. (Maps No. 12 and 13). This strip of generally rather rugged mountains forms part of a huge batholithic area of heterogeneous rocks which will be adequately mapped only after many more seasons of arduous field-work. The geological findings within such a belt as now to be described would be much increased in value if they could be systematically compared with field studies throughout the whole batholithic province. For many reasons such a complete survey is now impracticable. The present chapter is thus a sort of a report of progress on the geology of these crystalline rocks of the northern Cascades. Nevertheless discoveries of prime importance to the geology of the entire range have been made within even the limited area of the five-mile belt. Certain of the broader conclusions there deduced may, it is believed, be relied on, and will not need serious emendation as the exploration of the mountains continues. In the following pages there is offered another class of considerations which are theoretical and need the facts of the field, especially of the whole Cascade field, for their full discussion. In these matters particularly, a five-mile belt can not speak for the whole Okanagan range, except as geological experience in that belt accords with verified geological experience the world over.

General Description of the Batholithic Area.

To simplify the following discussion it will be well review the general geographical relations among the different geological units. To the same end it is convenient to adopt a special name for each unit. The cross-section, Figure 27, shows the units in their relative positions.

The most easterly component body occupies both slopes of Osoyoos Lake valley; it is the southern part of a great batholithic mass of granodiorite and may be called the Osoyoos batholith. The most westerly unit extends from Pasayten river to within a mile or so of Cathedral Peak. It is also a batholith of granodiorite and seems to compose the cliffs of the conspicuous Mount Remmel, five miles south of the Boundary. This mass may be called the Remmel batholith. Immediately to the eastward of the Remmel a third large batholith, this time composed of a quite different rock, true biotite granite,

2 GEORGE V., A. 1912

FIGURE 27.—*Diagrammatic east-west Section through the Okanagan Composite Batholith.*

7.—Schists and associated Palæozoic rocks.
Cr.—Pasayten Lower Cretaceous arkose
  sandstones.
1a.—Chopaka peridotite.
1b.—Basic complex.

2.—Ashnola gabbro.
3a.—Remmel batholith, Western phase.
3b.—Remmel batholith, Eastern phase.
3c.—Osoyoos batholith.
4.—Kruger alkaline body.

5.—Similkameen batholith.
6a.—Cathedral batholith, Older phase.
6b.—Park granite stock.
7.—Cathedral batholith, Younger phase.

The components of the batholith are numbered in order of intrusion. Horizontal scale, one inch to ten miles;
vertical scale, one inch to two miles. The vertical exaggeration makes contact lines generally too steep. On a
natural scale the basic bodies, 1a, 1b, and 2, would be shown as extending deeper into the granites; the actual
distortion is intended to illustrate the 'pendant' nature of each body.

SESSIONAL PAPER No. 25a

underlies all of the belt as far as Horseshoe mountain, on the divide between the Ashnola and main Similkameen rivers. This may be called the Cathedral batholith—named after the fine monolithic mountain occurring within the limits of the granite. The fourth principal unit lies between the Cathedral and Osoyoos batholiths; it is composed of a basic hornblende-biotite granite which is trenched by the deep valley of the Similkameen river, and an appropriate name for it is Similkameen batholith. These four principal units make up five-sixths of the whole area here described.

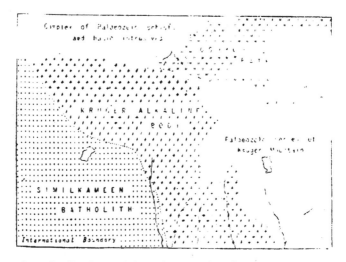

FIGURE 28.—Map showing relations of the Osoyoos, Similkameen, and Kruger igneous bodies and the invaded Paleozoic formations. Scale 1 : 110,000.

The subordinate geological members (excluding dikes) within the batholithic area are eight in number.

The largest of these consists of apparently-Paleozoic schists, quartzites, greenstones, and other rocks forming the ends of two tongues that enter the belt respectively from north and south (see Figure 28). These rocks occur on the roughly tabular Kruger mountain. The two schist tongues adjoin the Osoyoos batholith and nearly cut it off completely from direct contact with other plutonic units in the belt.

Between the schists and the Similkameen batholith is a comparatively small area of highly composite intrusives belonging to the malignite and nephelite-

2 GEORGE V., A. 1912

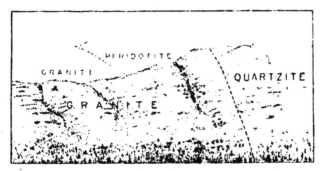

FIGURE 29.—Plunging contact surface between the Similkameen batholith and the Chopaka roof-pendant. Contact shown in broken lines. The vertical distance between the two ends of the contact line seen on the nearer ridge is 1,000 feet. Drawn from a photograph; looking east.

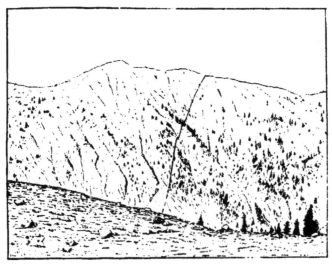

FIGURE 30.—Outcrop of the same intrusive contact surface shown in Figure 29. The vertical distance between the two ends of the contact line as drawn is 00 feet. Granite on the right; quartzite and schist on the left. Drawn from a photograph; looking west.

syenite families (see Figure 28). These crop out on the western summits of the Kruger-mountain plateau and may be referred to as the Kruger alkaline body.

The Similkameen granite preserves what seem to be remnants of its once complete roof (see Figure 31). Chopaka mountain is crowned with a large patch of schist. This Chopaka schist is cut by a strong body of gabbro apparently transitional into pure olivine rock—the Chopaka basic intrusives. The whole forms a huge irregular block of roof rock surrounded by the Similkameen granite. Excellent exposures show that the contact surface between the granite and the schist-gabbro mass dips beneath the invaded formations (Figures 29 and 30). The writer has little doubt that the relations indicated in the figures are typical of the whole boundary of the older terrane and that the granite underlies the visible block in every part. In a similar section more than a half mile in length the granite can be seen actually underlying the schist occurring on Snowy mountain.

ROOF-PENDANTS.

Each of these schist-blocks, once a downwardly projecting part of a roof in stock or batholith, may be named a 'roof-pendant' or simply 'pendant.' It is analogous to the pendant of Gothic architecture.

A brief digression on this conception may be permitted. Unusually fine examples of roof-pendants are illustrated in the great slabs of bedded rocks interrupting the areas occupied by the batholiths of the Sierra Nevada. One of the most recent descriptions is published by Messrs. Knopf and Thelen, following the lead of Lawson in a study of Mineral King, California.[*] Other examples, so well treated by Barrois, were found during the detailed geological survey of Brittany.[†] In all these and many other cases, and yet more clearly than on the Forty-ninth Parallel, the masses of country-rock (invaded formation) form respectively parts of a once continuous roof. The often perfect preservation of the regional strike in each of many examples very strongly suggests that these slabs have not sunk independently in their respective magmas. Such partial foundering would have almost inevitably caused some twisting of the block out of its original orientation. Granite and block have come into present relations because the magma, and not the block, was active. The point is of importance, as it bears on the mechanism of intrusion in these instances. It is further worthy of note that determination of roof-pendants and their distribution may sometimes lead to the discovery of the approximate constructional form of batholiths.

A small pendant, composed of amphibolitic and micaceous schists and of quartzite, occurs on the north slope of Horseshoe mountain; another of similar constitution flanks the summit of Snowy mountain.

In all three cases the pendants appear in the highest portions of the batholith as now exposed in the belt; yet each block projects downward, deep into the heart of the granite mass.

---

[*] Bulletin, Department of Geology, University of California. Vol. 3, No. 15, 1904, and Vol. 4, No. 12, 1905.

[†] C. Barrois: Annales, Société Géologique du Nord, many volumes, especially Vol. 22, 1894, p. 181

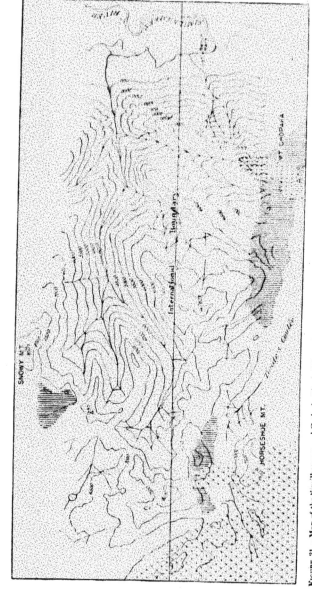

FIGURE 31.—Map of the Similkameen and Cathedral batholiths and the Chopaka intrusive body, as shown on the Boundary belt. The Similkameen batholith (left blank) bears three roof-pendants of schist (vertical lining), and is cut by the Cathedral granite (inverted carets). The Chopaka intrusive body (gabbro and dunite), cutting schists but older than the Similkameen granodiorite, is marked by crosses. Contour interval, 500 feet. Scale :— 1 : 118,000.

A long slab of gabbro, ranging with the Cathedral fork of Ashnola river, is similarly a roof-pendant of the Remmel batholith; it may be called the Ashnola gabbro (see Figure 32). A still larger pendant, composed of gabbros and peridotites, lies in the Remmel batholith just west of the main valley of the Ashnola. On account of the extraordinary diversity of rocks and of rock structures in this pendant, it may be called the 'Basic Complex' (see Figure 33).

Northeast of the complex is an elliptical stock of biotite granite, intrusive into both the Remmel granodiorite and the Basic Complex. The white, massive outcrops of the granite are very conspicuous on the northern spurs of Park mountain; the rock may be referred to as the Park granite (see Figure 33).

Within the five-mile belt these various rock bodies occupy areas shown in the following table. The bodies are noted in order from east to west, beginning on the east:—

	Square miles.
Osoyoos batholith..	40
Anarchist series in Kruger mountain..	15
Kruger alkaline body..	9
Similkameen batholith..	75
Chopaka schist (Anarchist series)..	2
Chopaka basic intrusives..	1½
Horseshoe schist (pendant; Anarchist series)..	1
Snowy schist (pendant; Anarchist series)..	¼
Cathedral batholith..	61
Remmel batholith..	64
Ashnola gabbro (pendant)..	1½
Basic Complex (pendant)..	6½
Park granite stock..	9
	———
Total..	296

The batholiths and the rocks of the Anarchist series extend far to the north and to the south of the belt, so that the total area of each is much greater than is shown in the table. The figures given for all the other bodies represent nearly their respective total areas. Less than 7 per cent of the belt is underlaid by rocks not clearly plutonic in origin, and of that 7 per cent perhaps half is greenstone or other igneous rock. The 3 or 4 per cent of non-igneous rock is chiefly quartzite and phyllite of the Anarchist series. The sedimentaries have been completely cut asunder by the plutonics; it is now possible to walk from one end of the belt to the other, the whole distance of 60 miles, and not once set foot on bed-rock which is other than of deep-seated, igneous origin (see Figure 27).

UNITY OF THE COMPOSITE BATHOLITH.

Barring a few patches, the enormously thick pre-Paleozoic, Paleozoic, Mesozoic, and Tertiary sediments and schists represented in the Cordillera elsewhere are wanting in this part of the Cascade system. With thicknesses running into tens of thousands of feet, they once unquestionably composed the Okanagan range, and of them the ancestors of these Boundary mountains were built. Erosion has removed some of the formations, attacking the earth's sedimentary crust from above. There is every reason to believe that perhaps even more of

2 GEORGE V., A. 1912

the old mountain substance was removed during the successive batholithic intrusions. Thus the sedimentary crust has also been attacked from beneath; its integrity has been destroyed through the displacing or replacing of sediments by igneous magma. In bringing about this gigantic result all the batholiths have acted together. Though they are of very different ages, their energies have been devoted to a common work. Their effects are so integrated that in causing the nearly complete disappearance of the ancient strata they have imitated on a larger scale what occurs with any homogeneous batholith. From this point of view the Boundary belt, stretching from the eastern contact of the Osoyoos batholith to the western contact of the Remmel batholith, forms a small segment of one composite batholith somewhat broader than the Okanagan range. To emphasize this primary fact, the whole plutonic mass has been called 'The Okanagan Composite Batholith.'

SEDIMENTARY ROCKS AND ASSOCIATED BASIC VOLCANICS.

Within the five-mile belt the only rocks of sedimentary origin are those which, with much probability, may be regarded as part of the Anarchist series already described. The largest area is found in Kruger-mountain plateau, where the dominant types in the country-rock of the plutonic masses are cleaved, micaceous quartzite and still more abundant sheared greenstone or amphibolite. The description of these rocks would be largely a repetition of that given for the Anarchist series as developed to the eastward of Osoyoos lake. The chief differences consist in the lower proportion of true phyllite in Kruger mountain and in the somewhat higher degree of crystallinity (metamorphism) shown in the western mass. Furthermore, no limestone has been found in Kruger mountain. These differences are, however, not of the kind to forbid direct correlation of the two terranes. The proximity of the two and the very positive resemblances of the rocks and associations on the two sides of the lake make the correlation probable in high measure. The greenstones and amphibolitic rocks carry thin interbeds of a once-argillaceous type, now phyllite, as well as thicker bands of the dominant quartzite. One thin section seems to prove that part of the greenstones are pyroclastic and basaltic or andesitic in original composition. These igneous rocks are almost certainly contemporaneous with the silicious sediments. The quartzite was occasionally seen to be thinly banded and cherty, recalling some of the normal types in Dawson's Cache Creek (Carboniferous) series.

The quartzites of the Chopaka roof-pendant (inclosed in the Similkameen batholith) have been examined microscopically. They show the usual characters of a metamorphosed quartzite, being rich in shreds and minute foils of a green biotitic mica; feldspar was not discoverable in either of two thin sections. The amphibolites of the Horseshoe and Snowy pendants, like those of the Kruger-mountain mass, are of quite usual microscopic characters, indicating the derivation of these rocks from basic volcanics. Their full description would be tedious and unnecessary.

Tertiary (?) Rocks at Osoyoos Lake.

In passing, it may be noted that a coarse conglomerate unconformably overlying the Anarchist series of Kruger mountain was found on the western side of Osoyoos lake at a point about two miles south of the Boundary line. The rounded, angular or subangular pebbles are often large and bouldery. They consist chiefly of granite, gneissic granite, and quartz or quartzite. The cement is feldspathic and arkose-like. This formation has been briefly described by Messrs. Smith and Calkins, who regard it as probably of Tertiary age.[*] The relations are much like those of the Kettle River beds (Oligocene). This deposit does not continue as far as the Boundary line and accordingly is not mapped in the sheet. The bedding of the conglomerate is obscure but the probable dip is about 70° in a northerly direction; the mass has been notably deformed.

PETROGRAPHY OF THE COMPOSITE BATHOLITH

Before proceeding to a detailed statement of the structure and history of the composite batholith, a brief petrographical description of its components will be necessary. Much of the usual petrographical detail has been omitted as not bearing directly on the main problems.

The eruptive rocks will be described as nearly as possible in the order of their respective dates of intrusion.

Richter Mountain Hornblendite.

The Anarchist series of greenstones is cut by at least one large body of an ultra-basic, greenish-black (when fresh), coarse-grained rock which weathers freely on the slopes of Richter mountain northwest of the Boundary line. The microscope shews that this rock is composed of dark green hornblende (apparently primary) and a diopsidic pyroxene which is nearly colourless in thin section; the former mineral is generally in excess. Apatite, pyrite, magnetite, and some titanite are the accessories. No feldspar was noted. Uralite or uralitic, pale-coloured amphibole, quartz, zoisite, epidote and calcite are secondary products. The pyroxene does not show the diallage structure; the rock may be classified as a pyroxene-rich hornblendite. The specific gravity of the freshest of three specimens is 3.302. This rock body has been so altered that its minerals are generally considerably altered. Its boundaries have not been so clearly determined that it could be advisedly mapped. The best exposures have, in fact, been found just north of the five-mile belt limit, on the top of the mountain. The body seems to cover more than the half of a square mile at least.

Chopaka Basic Intrusives.

The basic and ultra-basic intrusives of the Chopaka roof-pendant have been described by Smith and Calkins as uralitic gabbro, serpentines, and pyroxenites.

---

[*] G. O. Smith and F. C. Calkins, Bull. 235, U.S. Geol. Surv., 1904, p. 33.

2 GEORGE V., A. 1912

Within the area covered by the Commission map (Figure 31), the present writer has found no pyroxenite, but has referred all the massive intrusives of the Chopaka pendant (excluding dikes) to two rock types and their metamorphic derivatives.

Most of the rock within the area is feldspathic and seems to belong to a fairly steady type—normal gabbro transitional to metagabbro. It is a dark gray-green, medium-grained, hypidiomorphic-granular rock, originally composed of essential labradorite (Ab, An,) and diallage and accessory apatite, with a little magnetite. Crush metamorphism, supplemented by ordinary weathering, has largely changed the diallage into actinolitic amphibole, both compact and smaragditic. The specific gravity of the least altered rock is 2·959.

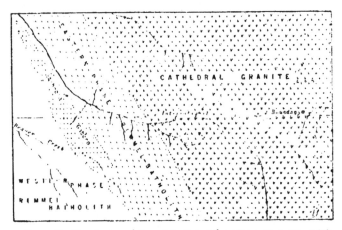

Figure 32.—Map showing relations of the Cathedral and Remmel batholiths and the Ashnola gabbro. The Younger Phase of the Cathedral granite is shown by stippling. The remarkably straight contact line of the Cathedral granite lies sensibly parallel to the gneissic banding in the Remmel batholith, Eastern Phase. Scale :—1 : 118,000.

That common rock type is associated with a large body of a dark greenish-gray, fine-grained rock of which several specimens show the composition very clearly. It was originally made up entirely of granular olivine without any certain accessory constituent. No trace of chromite has been found. Serpentine, talc, tremolite, and magnetite are present in most of the thin sections, but apparently in all cases as decomposition products of the olivine. The specific gravity of the rock varies from 3·100 to 3·173. It is a dunite without chromite.

The field relation of the gabbro and olivine rock has not been determined. They may belong to distinct intrusions or they may be due to differentiation within a single body. Though transitions seemed to appear in the actual out-crops, the search for final evidence in these rocks, crushed and obscured as they are, has so far proved unavailing. Analogous occurrences in other parts of the Boundary belt suggest that the gabbro and olivine rock were intruded at different dates.

ASHNOLA GABBRO.

Throughout its five miles of length (Figure 32) the Ashnola gabbro body is homogeneous in composition, but often varies abruptly in grain from median to quite coarse. The colour is uniformly a peculiar deep fawn, which is the dominating tint of the feldspar. This colour is rather remarkable, as the rock proves under the microscope to be quite fresh, with feldspars of glassy clearness. The essential constituents are a green augite, often colourless in thin section, brownish-green hornblende, brown biotite, and labradorite, Ab_5 An_4. Abundant apatite, some magnetite (probably titaniferous), and a very little interstitial quartz are the accessories. The structure in the original rock is the hypidiomorphic-granular, though the augite is often, especially in the coarser grained phases, poikilitic. Regular intergrowths of the augite and hornblende are common.

A fresh type specimen (No. 1403), taken near the contact with the Remmel granite about three hundred yards north of the Boundary line, has been analyzed by Mr. Connor, with the following result:—

Analysis of Ashnola gabbro.

		Mol.
SiO_2	47·76	·796
TiO_2	2·20	·028
Al_2O_3	18·58	·182
Fe_2O_3	2·19	·014
FeO	9·39	·131
MnO	·29	·004
MgO	4·15	·104
CaO	9·39	·168
SrO	·03	....
BaO	·02	....
Na_2O	3·61	·058
K_2O	·47	·005
H_2O at 105°C	·12	....
H_2O above 105°C	·53	....
P_2O_5	·78	·006
	99·51	
Sp. gr.	2·957	

2 GEORGE V., A. 1912

The calculated norm is: -

Orthoclase	2·78
Albite	30·39
Anorthite	33·08
Hypersthene	10·71
Diopside	6·54
Olivine	6·70
Ilmenite	4·26
Magnetite	3·25
Apatite	1·86
Water	·63
	100·24

The mode (Rosiwal method) is approximately:

Labradorite	57·5
Hornblende	21·8
Augite	12·0
Biotite	3·0
Magnetite	3·6
Apatite	1·6
Quartz	·5
	100·0

In the Norm classification the rock enters the presodic subrang, hessose, of the docalcic rang, hessase, in the dosalane order, germanare. In the older classification it is an augite-hornblende-biotite gabbro. The specific gravities of two fresh specimens average 2·946.

Although the gabbro is older than the Remmel granodiorite and has shared in the great dynamic metamorphism which, as we shall see, has profoundly affected the more acid rock, there is far less crushing action manifest in the gabbro than in the granodiorite. Gneissic structures were indeed sometimes seen in the ledges, but banding was never discovered, and the granulation is seldom comparable to that of the Remmel. It is, moreover, suspected that some of the gneissic arrangement of minerals in the gabbro is due to fluidal alignment of its tabular feldspars in the original magmatic period. For some unknown reason the gabbro has resisted crushing and shearing better than the granodiorite.

BASIC COMPLEX.

Petrographically and structurally, the Basic Complex is perhaps the most steadily variable plutonic mass in the entire Boundary section from the Great Plains to the Pacific. It covers an area stretching from Ashnola river westward over Park Mountain ridge, a distance of five miles. The extreme north-and-south diameter is about three miles, and the total area is nearly seven square miles. The Remmel granodiorite once completely surrounded the Complex, which, as above noted, is in pendant relation to the batholith. The pre-Remmel extent of the Complex was greater than the area now exposed; how much of it was destroyed during the Remmel intrusion it is impossible to say. The part thus remnant was still further diminished during the intrusion of the Park granite, which now, as illustrated on full three miles of contact line, projects strongly into the body of the Complex. A large block of the latter formation, measuring about 400 yards in length by 200 yards in width, was found within the Park granite mass itself; it may represent a roof-pendant in the stock, and thus a small analogue to the larger basic body in its relation to the Remmel batholith.

The Basic Complex is made up of a remarkable assemblage of basic plutonic rocks of at least three different periods of intrusion. The oldest types are coarse-grained. They include highly irregular bodies of hornblendite, which in the field is often seen to be transitional into a labradorite-bearing hornblende-augite peridotite; this in its turn merges into hornblende-augite gabbro. All of these rocks are believed to be of contemporary origin. Their occurrence is so sporadic that it is difficult to say how much of the whole basic area they really cover—possibly one-quarter of it by rough estimate. These rocks are cut by many large dikes and more irregular masses of hornblende-gabbro, augite-hornblende gabbro, and hornblende-biotite-quartz gabbro. Such types are of medium to coarse grain. Their specific gravity varies from 2.873 to 2.986.

As there is no discoverable system in the differentiation of the earliest intrusive members, varying as they do most capriciously from ledge to ledge, so there is no discoverable system in the trends or occurrence of the countless later injections of the gabbros. The complication has been still further increased by the intrusion of thousands of narrow and broader dikes of granite. Much of the granite is apophysal or aplitic from the Remmel batholith; some of it is apophysal from the magma supplying the Park granite stock, while many dikes of acid pegmatite locally traverse the whole mass. The complication was finally made perfect through the enormous crushing which the Basic Complex underwent, both during the intrusion of the granites and during the orogenic revolution when the Remmel granodiorite itself was sheared into banded gneisses.

In the shearing of the Basic Complex its material was metamorphosed and, in part, it migrated. The mode of migration is believed to be that which will be briefly discussed in connection with the petrographic descriptions of the crushed Osoyoos and Remmel batholiths. The metamorphism has developed many schistose phases, among which hornblende-biotite-diorite gneiss (specific gravity, 2.766 to 2.863) and well foliated hornblendite are common.

As a result of this long and varied history, scarcely any two ledges within the area of the Complex accord in composition. The constitution of what appears to be the commonest phase of the complex, the augite-hornblende gabbro, and the peculiar fawn colour of its feldspar, furnish a probable correlation of part of the whole mass with the Ashnola gabbro. There is no certainty of similar correlation with the basic rocks of Mount Chopaka.

Nodule-bearing Peridotite Dike.—The Complex is cut by a remarkable forty-foot dike which is excellently well exposed on the north slope of Park mountain. The exact locality is found on the divide west of the Ashnola river, about 1,500 yards southeast of the Line monument and on the 7,250-foot contour. The dike is sensibly vertical and strikes east-southeast. Its wall-rocks are typical

2 GEORGE V., A. 1912

members of the Complex, coarse hornblende-peridotite and hornblende-gabbro, cut by medium-grained gabbro. All of these rocks are more or less gneissic but the dike itself is neither schistose nor otherwise visibly affected by dynamic action.

The dike is highly conspicuous through the whole length of its outcrop. The salient feature is the studding of its surface with hundreds of nodules which form about 40 per cent of the rock. (Plate 37.) These nodules are ellipsoidal or potato-shaped, measuring from 3 cm. to 6 cm. in diameter. They resist destruction by weathering more effectively than their matrix, so that they stand out prominently on the ledges.

The nodules are light-green, granular aggregates of interlocked olivine crystals from 1 mm. to 10 mm. in diameter. Rarely, a small anhedron of pyroxene, probably diallage, appears as an accessory, interstitial constituent of the nodule. No other primary mineral except abundant minute microlites of chromite or picotite, is present. The olivine is often surprisingly fresh but along cleavage cracks it has gone to serpentine, talc, magnetite, and tremolitic amphibole. In other cases these secondary minerals compose most of the nodule. The deep brown inclusions of spinel or chromite have the usual sharp crystal-form and parallel arrangement in the individual grains of olivine.

The matrix is much darker-coloured than the nodules and is considerably more altered. It was originally composed chiefly of a granular aggregate of hypersthene, with which a green hornblende was associated. Now, however, the matrix is mostly a felted mass of colourless amphibole, often assuming a greenish tint like that of actinolite. Much magnetite, bastite, talc, and some serpentine also occur in the felt. Limonite often stains the thin section. The hypersthene has the usual colour, pleochroism, and other properties and has a great abundance of interpositions which, under the microscope, have the same optical properties as the spinel-like microlites of the olivine. Minute granules of what appears to be magnetite also occur in normal parallel arrangement in the residual cores of the hypersthene. The deep green hornblende seems without question to be of primary origin and thus of origin and nature quite different from those of the tremolitic amphibole.

Between the nodules and their matrix there is almost invariably a kelyphitic shell of a colour yet lighter than that of the nodule. The shells vary in thickness from a couple of millimetres or less to 15 mm. They are composed chiefly of tremolite and magnetite, the amphibole prisms often radiating outward from the nodule. Some talc and serpentine also appear in these 'reaction rims.'

Plate 37 shows the relations of nodule, reaction-shell, and matrix. The kelyphitic phenomenon is well known to petrographers and needs no further description. The peculiarity of the dike consists in the fact that it is a peridotite crowded with large olivine nodules. So far as known to the writer no similar dike has been described in petrographic literature. It may be noted that the nodules preserve their average size throughout the cross-section of the dike; the average is not affected by the proximity of the walls. In this respect also the conditions of crystallization were unusual.

PLATE 37

Specimens of nodule-bearing peridotite from forty-foot dike cutting schistose rocks of Basic Complex. The nodules are granular aggregates of olivine, regularly surrounded with kelyphitic shells, as shown. Two-thirds natural size.

PLATE 38.

Western slope of Anarchist mountain plateau, viewed from west side of Osoyoos lake. Glaciated surface of the sheared Osoyoos granodiorite. Sloping terrace left after retreat of the lake waters.

SESSIONAL PAPER No. 25a

The dike-rock may be classed as a hornblende-bearing harzburgite. The specific gravity of a large, relatively fresh specimen is 3·099.

Vesicular Andesite Dikes.—On the 7,718-foot summit north of Peeve Pass a half-dozen andesite dikes, varying from a foot to six feet or more in width, cut the Basic Complex. These dikes are nearly vertical and strike between N. 45° E. and N 90° E. All of them are more or less vesicular. The material of the dikes seems to be uniform—a rather light gray, amygdaloidal lava, either aphanitic or porphyritic, with phenocrysts of altered plagioclase, (probably labradorite) and of augite. The ground-mass is a felt of minute plagioclase microlites, largely chloritized augite granules, and abundant glass. The rock is almost certainly an augite andesite and in any case must vary but little from that common type. The amygdules are composed of calcite; like the phenocrysts they are generally arranged parallel to the dike-walls. This arrangement is probably a flow-structure and is not due to crush-metamorphism. In fact, the lava-like rock does not seem to have been appreciably squeezed at all.

The field evidence thus went to show that the andesite was intruded after the wholesale shearing of the Complex had taken place. A vesicular dike of olivine basalt cuts the Cathedral granite. It is probable that the andesite dikes were injected very late in the history of the composite batholith. In both cases the vesicularity of the dike-rocks suggests that they were intruded near the surface; if so, they belong to the Recent period or to the latest Tertiary.

Osoyoos Batholith.

That part of the Okanagan valley in which Osoyoos lake lies has been largely excavated in a body of intrusive, granitic rock to which the name ' Osoyoos batholith ' has been given. The northern and southern limits of the body were not determined but they are known to occur well outside the Boundary belt. (Plate 38).

Original Granodioritic Type.—The batholith has undergone such drastic alteration through dynamic metamorphism that it is difficult to find ledges or even hand-specimens of the original rock. Considerable sampling of the mass within the five-mile belt has led the writer to conclude that, while the body was of distinctly variable composition at the time of its crystallization from the magma, yet that the staple or dominant rock was originally a rather typical, medium- to coarse-grained granodiorite.

The colour is the familiar light gray characteristic of monzonites, granodiorites, and some other granular rocks rich in plagioclase. In the likewise fresh though somewhat metamorphosed phases the rock assumes a light greenish-gray tint due to the dissemination of metamorphic biotite or to the abundant development of epidote. All phases weather light brownish-gray. The essential constituents are deep green hornblende, brownish-green biotite, orthoclase, quartz, and unzoned andesine. Ab_3An_1. The accessory minerals are apatite, zircon, magnetite, and titanite; none of these may be called abundant. Allanite in rather large amount is accessory in the basified contact zone. Colourless epidote is invariably present, but is regarded as of metamorphic origin. Where

2 GEORGE V., A. 1912

it becomes abundant the iron ore has partially or wholly disappeared; then probably entering into the composition of the epidote. Biotite is generally dominant over hornblende and plagioclase over orthoclase.

A fresh specimen nearly representing the original granodiorite was collected at a point two miles north of the Boundary line and about two miles from the eastern contact of the batholith. This specimen is rather coarse-grained and is gneissic, though not so schistose as the average rock of the batholith in the observed exposures. The essential and accessory minerals are those named in the foregoing list; biotite is more abundant than hornblende and plagioclase than orthoclase. This specimen (No. 295) was analyzed by Mr. Connor, with the following result :—

Analysis of Osoyoos granodiorite.

		Mol.
SiO₂	68.43	1.110
TiO₂	.20	.003
Al₂O₃	15.80	.155
Fe₂O₃	1.06	.007
FeO	1.85	.025
MnO	.10	.001
MgO	1.46	.036
CaO	4.08	.073
SrO	.02	....
BaO	.09	.001
Na₂O	3.47	.056
K₂O	2.51	.027
H₂O at 105°C	.05	....
H₂O above 105°C	.53	....
P₂O₅	.07	.001
	99.72	
Sp. gr	2.68	

The calculated norm is :—

Quartz	26.82
Orthoclase	15.01
Albite	29.34
Anorthite	19.74
Hypersthene	5.61
Diopside	.23
Magnetite	1.62
Ilmenite	.46
Apatite	.31
Water	.58
	99.72

The mode (Rosiwal method) is approximately :—

Quartz	37.0
Orthoclase	7.5
Andesine	33.1
Biotite	10.8
Hornblende	3.1
Epidote	8.0
Titanite	.3
Apatite	.1
Zircon	.1
	100.0

In the Norm classification the rock enters the dosodic subrang, yellowstonose of the alkalicalcic rang, colorodose, in the persalane order, britannare

During the metamorphism which even this type specimen has suffered, some of its basic matter has probably been removed. The silica is thus believed to be slightly higher than it would be in an analysis representing the original average rock - perhaps by as much as four or five per cent higher. In the older classification the rock is a granodiorite verging on quartz diorite.

Along the eastern contact of the batholith the average plagioclase is labradorite, Ab_3An_2, and it so far replaces the orthoclase that the rock becomes a true quartz diorite. In the hand-specimen this somewhat basified contact phase is indistinguishable from the true granodiorite. The limits of the orthoclase-poor zone were therefore not closely fixed in the field. It is probable that the zone is not more than a few hundred yards in width, and that the original rock of the batholith was, in the large, homogeneous. A second exceptional phasal variation is founded on the disappearance of hornblende in rock that shows decided cataclastic structure, other constituents remaining the same as in the normal granodiorite. This phase — gneissic biotite granite rich in andesine occurs sporadically in the heart of the batholith. Possibly it is not of original composition, the hornblende having been removed through metamorphic action.

Dynamic and Hydrothermal Metamorphism of the Granodiorite. — Superimposed upon the original variations in the batholith are the much more striking effects of intense orogenic strains. Even the most massive phases show, under the microscope, the varied phenomena of crushing stress – granulation, bending of crystals, undulatory extinctions, and recrystallization. Because of the creasing, the average rock is no longer the original rock. The granodiorite has been changed into several metamorphic types, of which three may be noted.

The commonest transformation is that into a *biotite-epidote-hornblende gneiss*, with essential and accessory constituents like those in the original granodiorite, but in somewhat different proportions. The colour is light gray, with a green cast on surfaces transverse to the schistosity; parallel to the schistosity a dominant and darker green colour is given by abundant fine-textured leaf aggregates of biotite. These aggregates are not simply crushed and rotated original mica foils, but, like the epidote, represent true recrystallization and the incipient migration of material within the granulated plutonic rock. At the same time much of the original hornblende, apatite, and magnetite have been removed.

A second metamorphic type is a yet more highly schistose *biotite-epidote gneiss* often transitional into biotite schist. The essential constituents are biotite, epidote, orthoclase, andesine, and quartz. The accessories include very rarely apatite and magnetite, while titanite seems to have entirely disappeared along with the hornblende. Orthoclase seems here to be more abundant than plagioclase. The quartz and feldspars are intensely granulated and, with polarized light, are full of strain shadows. The rock is more richly charged with biotite than the hornblende-bearing gneiss.

The third metamorphic type occurs in immediate association with the gneiss just described, being interbanded with it. It is a fine-grained, strongly schistose

MICROCOPY RESOLUTION TEST CHART

(ANSI and ISO TEST CHART No. 2)

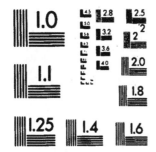

APPLIED IMAGE Inc

1653 East Main Street
Rochester, New York 14609 USA
(716) 482 - 0300 - Phone
(716) 288 - 5989 - Fax

2 GEORGE V., A. 1912

dark greenish gray *hornblende gneiss of basic character.* The essential minerals are idiomorphic green hornblende and allotriomorphic feldspars in mosaic with considerable interstitial quartz; the last is hardly more than accessory. The feldspar is mostly unstriated and not easy of determination. Orthoclase seems to be dominant, but, as shown by extinctions on (010), approaches soda-orthoclase in composition. The plagioclase is possibly andesine. Titanite, apatite, and well crystallized magnetite are accessory in large amounts. The hornblende prisms are often twinned parallel to (100). That crystallographic plane now lies characteristically parallel to the plane of schistosity. Except for the soda content of the orthoclase, the minerals all appear to have the same characters as in the granodiorite.

This third phase occurs in zones of maximum shearing in the batholithic mass. It is believed to represent a new secondary rock formed by the recrystallization of the materials leached out of the other two metamorphic phases just noted and out of the granodiorite as it was crushed. The recrystallization either accompanied or followed the very closing stage of the orogenic crushing. This fact is demonstrated by the entire absence of granulation or even undulatory extinctions in the mineral components.

The probable history of the metamorphism may now be summarized. After the complete solidification of the original granodiorite, very intense crushing stresses affected the whole body. The straining and granulation of the minerals exposed them to wholesale solution, whether in water and other fluids inclosed in the rock or in fluids of exotic origin. This process of solution was hastened by the rise of temperature incident to violent crushing. All the minerals must have been affected, but it appears that the hornblende, biotite, magnetite, apatite and titanite were most likely to be dissolved and so migrate with the fluids that slowly work their way through the rock in its mechanical readjustments.[*] Escape for the mineral-laden fluids (perhaps chiefly water freed from combination in biotite or from solid solution in hornblende) was most ready in the zones of maximum shear. Thither the fluids were drawn, and there some of the dissolved material recrystallized so as to develop the darker coloured bands of biotite-epidote gneiss, biotite schist, and hornblende gneiss.

Where the granulation was least the granodiorite retains nearly its original composition, though epidote may be formed; the specific gravity averages 2·730. Where the granulation was more pronounced, as in the first metamorphic type described, much of the hornblende, titanite, magnetite, and apatite have been leached out and abundant metamorphic biotite and epidote have formed; the result is a biotite-hornblende-epidote gneiss with a density less than that of the original

---

[*] This conclusion has in this instance been deduced from the study of thin-sections. In general it accords with the results of experiment. Müller has found that in carbonated water hornblende and apatite are much more soluble than either orthoclase or oligoclase. Magnetite is less soluble than any of those minerals, but the relatively minute size of its crystals in granodiorite would allow of its complete solution and migration before the essential minerals had lost more than a fraction of their substance. It is also possible that magnetite would suffer especially rapid corrosive attack from fluid in which the chlorine-bearing apatite has gone into solution. Cf. R. Müller in Tschermak's Miner. und Petrog. Mittheilungen, 1877, p. 25.

SESSIONAL PAPER No. 25a

granodiorite because of the loss in heavy constituents (specific gravity, 2.692). A further stage of granulation and energetic shearing led to the formation of perfect schistosity in rock made up of the quartz-feldspar ruins of the original rock, cemented by very abundant biotite and epidote—the biotite-epidote gneiss (specific gravity, 2.783). The fissures and fluid-filled cavities developed in the zones of maximum shear are now occupied by the strongly schistose hornblende gneiss (specific gravity, 2.939) and similar products of complete solution, migration, and subsequent complete recrystallization.

The granodiorite has thus become not only mechanically crushed, but to a large extent rendered heterogeneous. It is now not only gneissic, but banded in zones of new rock markedly varied in composition. The schistosity and banding everywhere agree in attitude; the strike varies from N. 70° W. to N. 75° W., but over large areas, as indeed over the whole batholith east and west of Osoyoos lake, averages N. 45° W. almost exactly. Neglecting minor crumplings, the dip varied from 70° N.E. to 90°, averaging about 82° N.E. This average attitude is close to that observed in the schists cut by the granodiorite, but represents an exceptional strike among the main structural axes of the Cordillera. It may be noted that shearing is much more manifest on the east side of Osoyoos lake than on the west side.

REMMEL BATHOLITH.

From the Pasayten river to the western base of Cathedral Peak the larger part of the Boundary belt is underlain by the Remmel batholith. This granitic body is like the Osoyoos batholith in exhibiting a well-developed gneissic and banded structure, along with a great heterogeneity in chemical and mineralogical composition. The causes of this variable constitution are here again two in number. The one is original or magmatic; the other is secondary and due to metamorphism. The metamorphic action has been most marked in a band immediately adjoining the Cathedral batholith. This part, comprising one-seventh of the total area in the Boundary belt, is called the Eastern phase. The rest of the body as exposed in the belt is called the Western phase. Each phase is variable in itself but the two are contrasted by general characteristics which persist throughout most of each area. At the Pasayten river the batholith is unconformably overlain by the Lower Cretaceous Pasayten series of strata.

Western Phase.—The least metamorphosed part of the batholith is to be found in the Western phase. None of the collected specimens can, however, be confidently regarded as illustrating the precise average of this phase or of the batholith as a whole. The writer has, however, selected for analysis one fresh specimen which approximates the probable average rock of the Western phase as originally constituted. The specimen was taken from a ledge two miles south of the Boundary line and 2,000 yards from the contact with the Ashnola gabbro.

This rock has the look of a medium- to coarse-grained, slightly porphyritic, gray granite. Lustrous black biotites in conspicuous, often quite idiomorphic foils reaching 1 cm. or more in diameter, are the only phenocrysts. Otherwise the structure of the rock, though somewhat obscured by crushing, seems originally to have been the normal eugranitic. A deep brownish-green hornblende, quartz, andesine (averaging Ab, An), and orthoclase are the other essential constituents. Titanite, magnetite, and apatite are the accessories. Except for the often unusually perfect idiomorphism of the biotite the rock has, thus, the general habit of a common granodiorite.

Mr. Connor's analysis (specimen No. 1405) afforded the following result:—

Analysis of Remmel batholith, Western phase.

		Mol.
SiO₂	63·30	1·055
TiO₂	·50	·006
Al₂O₃	17·64	·173
Fe₂O₃	1·58	·010
FeO	3·08	·043
MnO	·47	·007
MgO	1·23	·031
CaO	5·03	·089
SrO	none.	....
BaO	·05	....
Na₂O	4·56	·074
K₂O	1·16	·013
H₂O at 105°C	·14	....
H₂O above 105°C	·51	....
P₂O₅	·27	·002
	99·52	
Sp. gr.	2·721	

The norm was calculated to be:—

Quartz	18·24
Orthoclase	7·23
Albite	38·78
Anorthite	22·80
Corundum	·40
Hypersthene	7·59
Magnetite	2·32
Ilmenite	·79
Apatite	·62
Water	·65
	99·12

The mode (Rosiwal method) is approximately:—

Quartz	27·0
Orthoclase	7·2
Andesine	50·7
Biotite	5·7
Hornblende	4·3
Magnetite	3·8
Titanite	·6
Apatite	·5
Epidote and zircon	·2
	100·0

In the Norm classification the rock is the dosodic yellowstonose of the alkalicalcic rang, coloradase, in the persalane order, britannare. According to the older classification the rock enters the class of quartz mica diorites but verges on typical granodiorite.

Seven other specimens of the batholith as exposed to the westward of the Ashnola gabbro were studied microscopically. They were found to include yet more basic diorites and also types which belong to the biotite granites rich in plagioclase. The specific gravities of the seven specimens range from 2.641 to 2.775, averaging 2.706.

Where strong shear-zones occur in the Western phase they are occupied by dark greenish-gray, fine-grained, fissile hornblende gneiss very rich in hornblende and similar to the metamorphic filling of shear zones in the Osoyoos granodiorite. Between these narrow shear zones the more normal rock usually shows mechanical granulation and fracture rather than extensive recrystallization.

Roughly estimating the relative volume of each type, the writer has concluded that the Western phase is, on the average, a granodiorite which is very close to a quartz diorite. At the western side of the exposed batholith where it disappears beneath Cretaceous sediments, the granitic rock is relatively uncrushed, poor in orthoclase and rather abundantly charged with phenocrystic biotite and with hornblende. Toward Park mountain the zones of intense shearing become more and more numerous. The rock then loses its porphyritic appearance and tends to be a gneissic biotite granite, in which hornblende is wanting and orthoclase has increased at the expense of the soda-lime feldspar. Near the long hand of Ashnola gabbro the Western phase carries bands of crushed rock which is indistinguishable from the staple rock of the Eastern phase.

Eastern Phase.—East of the roof-pendant of Ashnola gabbro the batholith shows evidence of having undergone its maximum shearing and metamorphism. It there consists of narrow bands of highly micaceous gneiss alternating with parallel, much broader bands of less micaceous gneiss. These bands are generally more acid than the typical rock of the Western phase.

A specimen fairly representing the average of the Eastern phase was collected at a ledge 1.8 miles south of the Boundary line and in the middle of the zone of the batholith composed of this phase (Figure 32). The rock is in macroscopic appearance a light gray, medium-grained, somewhat gneissic granite, weathering light brown. Quartz, biotite, orthoclase, and plagioclase (probably andesine, near Ab_2An_1) are the essential components. Rare apatite, zircon, and magnetite grains are the accessories. A few reddish garnets are occasionally developed. There is seldom any indication of straining or crushing of the minerals constituting the band whence the specimen was taken. Microscopic study leaves the impression that the material of this and similar bands has been wholly recrystallized. The structure is now the hypidiomorphic-granular.

This specimen (No. 1398) was analyzed by Mr. Connor with result as follows:

2 GEORGE V., A. 1912

Analysis of Remmel batholith, Eastern phase.

		Mol.
SiO₂	70·91	1·182
TiO₂	·20	·003
Al₂O₃	16·18	·159
Fe₂O₃	·51	·003
FeO	1·09	·015
MnO	·04	....
MgO	·37	·009
CaO	2·92	·052
BaO	·10	·001
Na₂O	1·33	·021
K₂O	5·53	·059
H₂O at 105°C	·03	....
H₂O above 105°C	·12	....
P₂O₅	·11	·001
	99·44	
Sp. gr	2·654	

The norm was calculated to be:—

Quartz	34·68
Orthoclase	32·80
Albite	11·00
Anorthite	14·73
Corundum	2·75
Hypersthene	2·09
Magnetite	·70
Ilmenite	·46
Apatite	·31
Water	·15
	99·67

The mode (Rosiwal method) is roughly:—

Quartz	34·3
Orthoclase	37·1
Andesine	25·5
Biotite	2·3
Magnetite	·5
Apatite, zircon, and epidote	·3
	100·0

In the Norm classification this rock is transitional between the, as yet, unnamed dopotassic subrang of the alkalicalcic rang, coloradase, in the persalane order, britannare, and the corresponding, likewise unnamed subrang of the order, columbare. In the older classification the rock has the chemical and mineralogical composition of a common biotite granite. It is, however, improbable that this type is an original product of crystallization from the batholithic magma.

The specific gravities of four fresh specimens of the less micaceous bands of the Eastern phase vary from 2·644 to 2·654, averaging 2·651. These narrow limits of variation agree with the microscopic study of the same specimens in showing that the lighter bands are relatively uniform in composition.

The darker bands have not been systematically scrutinized with the microscope but their field habit is that of common mica gneiss, often passing over into feldspathic mica schist; they never seem to carry any hornblende. They occupy probably no more than five per cent of the area covered by the Eastern phase.

These zones were regarded in the field as located along planes of maximum shearing. They accord very faithfully in attitude with a strike of N 2° to 25° W. and a dip nearly vertical, but sometimes 75° or more to the east-northeast — structural elements induced by regional orogenic movements in the Cordillera. It is improbable that the banding represents peripheral schistosity about the Cathedral batholith. The chief reason for excluding this view is that peripheral schistosity is lacking in the great Similkameen batholith, which is also cut by the Cathedral granite. It appears, on the other hand, that the Remmel batholith was already crushed and its banding produced before either the Similkameen or Cathedral granite was intruded.

Interpretations of the Two Phases.—Three interpretations of the two phases are conceivable. They may be supposed to be distinct intrusions of two different magmas; or, secondly, original local differentiation products in the one batholith; or, thirdly, distinguished in their present compositions because of the unequal dynamic metamorphism of a once homogeneous magma. Against the first view is the fact that the two phases, where in contact, seem to pass insensibly into each other. In favour of the third view are several facts which do not square with the second hypothesis, and the writer has tentatively come to the conclusion that the third hypothesis is the correct one. Among these facts are the following:

1. The Eastern phase covers that part of the Remmel body which has suffered the greatest amount of dynamic stresses exhibited either in the Remmel or in any other of the larger components of the Okanagan composite batholith. It has been seen that the less intense though still notable dynamic metamorphism of the Osoyoos granodiorite led to the special excretion of most or all of the hornblende, apatite, magnetite, and titanite from that rock and the secretion of those leached-out compounds in the free spaces of the shear zones. The biotite was similarly segregated, but its mobility was found to be considerably less than that of the hornblende. If the metamorphism had been yet more energetic in the Osoyoos body, the more soluble compounds would have been carried away completely and the whole would have crystallized in the form of acid biotitic gneiss banded with especially micaceous schists in the zones of maximum shear. Such appears to the writer to be the best explanation of the Eastern phase of the Remmel batholith.

2. The composition of the rock and the fact that, as above mentioned, it seems to have been thoroughly recrystallized into a strong, well knit, banded gneiss without cataclastic structure agree with this view.

3. The conclusion is substantiated in the study of more moderate shearing in the Western phase itself. There the strongly granulated and not recrystal-

2 GEORGE V., A. 1912

lissi granodiorite shows impoverishment in the more mobile hornblende and accessories, which are segregated into intercalated recrystallized bands. Thus rubble-less-tree, crushed rock indistinguishable in composition from the rock the Eastern phase occurs sporadically in many local areas within the normal crushed granodiorite of the Western phase.

In summary, then, the Remmel granodiorite, gneissic biotite granite, biotite gneiss, biotite-quartz diorite, and hornblende gneiss appear to belong to a single batholithic intrusion. The mean of the two chemical analyses corresponds to the analysis of a fairly typical granodiorite. In view of the greater volume of the Western phase it appears that the average original rock of the whole batholith was a granodiorite quite close in its composition to a quartz-hornblende-biotite diorite.

This mass has been dynamically and hydrothermally metamorphosed with intense shearing in zones trending N. 20 to 25° W. Over most of the batholith so far investigated these zones of physical and chemical alteration are not so well developed as to obscure the essential nature of the primary magma (Western phase). The shearing and transformation are much more profound in a wide belt elongated in the general structural direction N. 25° W. Here the rocks are well banded biotite gneisses, the material of which is residual after the deep-seated, wholesale leaching of the more basic mineral matter from the crushed granodiorite (Eastern phase).

KRUGER ALKALINE BODY.

General Description. All the way from the Great plains to the Pacific waters nepheline rocks are extremely rare on the Forty-ninth Parallel. The Boundary section is now so far completed that it can be stated that in the entire section the Kruger body is the only plutonic mass bearing essential nepheline; it is likewise the most alkaline plutonic mass.

One of its principal characteristics is great lithological variability. It varies signally in grain, in structure, and above all in composition. (Plate 39). All the varietal rock types carry essential feldspars of high alkalinity—microperthite, microcline, soda-orthoclase, and orthoclase. Nepheline, biotite, olive-green hornblende, a pyroxene of the aegerite-augite series, and melanite complete the general list of essentials. Titanite, titaniferous magnetite or ilmenite, rutile, apatite, and acid andesine, Ab An (the last entirely absent in most of the rock phases), form the staple accessories, though any one or more of the coloured silicates may be only accessory in certain phases. Muscovite, hydronephelite, kaolin, calcite, epidote, and chlorite are secondary, but on account of the notable freshness of the rocks are believed to be due to crush-metamorphism more than to weathering.

According to the relative proportions of the essential minerals, at least ten different varieties of alkaline rock have been found in the body. These are:—

Augite-nephelite malignite,
Augite-biotite-nephelite malignite,
Augite-biotite-melanite malignite,
Hornblende-augite malignite,
Augite-nephelite syenite,

Hornblende-nephelite syenite,
Biotite-melanite-nephelite syenite,
Augite-biotite-nephelite syenite,
Porphyritic augite syenite,
Porphyritic alkaline biotite syenite.

TYPES FROM THE KRUGER ALKALINE BODY.

A.—Porphyritic alkaline syenite ; one-half natural size.
B. Nephelite syenite (salic variety) ; two-thirds natural size.
C. Malignite ; two-thirds natural size.

There is a question as to how far this list of varieties actually represents the original magmatic variation within the body. The evidence is good that the augite and hornblende and a part of the biotite, along with the feldspars and nephelite, crystallized from the magma. It is not certain in the case of melanite which, in the Ontario malignite, as described by Lawson, appears to be a primary essential.* Microscopic study shows that much of the melanite in the Kruger rocks is of magmatic origin, but that perhaps much more of it has replaced the pyroxene during dynamic metamorphism. In such cases the pyroxene, where still in part remaining, is very ragged, with granular aggregates of the garnet occupying the irregular embayments in the attacked mineral. A further stage consists in the complete replacement of the augite by the melanite aggregates which are shot through with metamorphic biotite. These peculiar reactions between the pyroxene and the other components of the rock are widespread in both syenite and malignite.

All the phases so far studied in this natural museum of alkaline types can be grouped in three classes—granular malignites, granular nephelite syenites, and coarsely porphyritic alkaline syenites. The malignitic varieties are always basic in look, dark greenish-gray in colour, and medium to coarse in grain (specific gravity, 2·757 to 2·967). The nephelite syenites are rather light bluish-grey in tint, medium to fine-grained, and break with the sonorous ring characteristic of phonolite (specific gravity, 2·606 to 2·719). The third class of rocks is much less important as to volume; they are always coarse in grain, of gray colour, and charged with abundant tabular phenocrysts of microperthite which range from 2 to 5 centimetres in length. These phenocrysts as well as the alkaline feldspars of the coarse groundmass are usually twinned, following the Carlsbad law—a characteristic very seldom observed in the malignites, or nephelite syenites. (Plate 39, A)

The nephelite syenites often send strong apophysal offshoots into the malignites, but such tongues are highly irregular and intimately welded with the adjacent basic rock as if the latter were still hot when the nephelite syenites were intruded. Moreover, there are all stages of transition in a single broad outcrop between typical malignite and more leucocratic rock indistinguishable from the nephelite syenite of the apophyses. Similarly, even with tolerably good exposures, no sharp contacts could be discovered between the coarse, porphyritic syenites and the other phases. The porphyritic rocks almost invariably showed strong and unmistakable flow structure, evidenced in the parallel arrangement of undeformed phenocrysts; these generally lie parallel to the contact walls of the body as a whole. The phasal variety of the Kruger body and the field relations of the different types seem best explained on the hypothesis that the phases are all nearly or quite contemporaneous—the product of rapid magmatic differentiation accompanied by strong movements of the magma. These movements continued into the viscous stage immediately preceding crystallization. (Plate 39, A).

Three specimens representing as many principal types were submitted to Professor Dittrich for analysis.

*A. C. Lawson, Bulletin, Dept. of Geology, University of California, vol. 1, 190.
25a vol. i—29

2 GEORGE V., A. 1912

Augite-biotite Malignite.—The first specimen was collected at a ledge about 50 yards west of the contact with the older Kruger-mountain-schists and 1,200 yards west of the small lake-on the top of the mountain-plateau. This rock is dark-coloured, medium to fairly coarse-grained, and of gabbroid habit. (See Plate 39, Figure C). The essential minerals are augite, (with rare outer shells of olive-green hornblende), biotite, micro-perthite, microcline, nephelite, and probably soda orthoclase. Apatite, a little titaniferous magnetite or ilmenite, and titanite are original accessories. Melanite is also 'an abundant accessory but in this case all of the garnet may have been derived from the pyroxene through crush-metamorphism. A little hydronephelite and more abundant muscovite, which seems to replace nephelite, are present as secondary products, but on the whole the rock is to be described as fresh.

The order of crystallization among the original minerals is: apatite; iron ore; titanite; augite; feldspars; nephelite. The order is unusual in that the nephelite follows the feldspars.

The chemical analysis of this specimen (No. 1100), by Professor Dittrich, resulted as follows:—

Analysis of malignite, Kruger alkaline body.

		Mol.
SiO_2	50.49	.842
TiO_2	.92	.011
Al_2O_3	15.83	.155
Fe_2O_3	6.11	.038
FeO	3.04	.042
MnO	.11	.001
MgO	3.38	.084
CaO	7.99	.148
Na_2O	3.12	.050
K_2O	6.86	.073
H_2O at $110°C$	.29	....
H_2O above $110°C$	1.20	....
P_2O_5	.42	.003
CO_2	.07	....
	99.83	
Sp. gr.	2.849	

The calculated norm is:—

Orthoclase	40.59
Albite	7.34
Nephelite	10.22
Anorthite	8.90
Diopside	18.11
Wollastonite	1.97
Magnetite	7.42
Ilmenite	1.67
Hematite	.96
Apatite	.93
Water	1.49
	99.63

The mode (Rosiwal method) gives nearly:-

Microperthite	}	
Microcline	}	36.3
Soda orthoclase	}	
Augite		36.5
Biotite		11.0
Melanite		9.5
Nephelite		5.4
Apatite		1.0
Magnetite and titanite		.3
		100·0

In the Norm classification the rock enters the sodipotassic subrang, boro-lanose, of the domalkalic rang, essexase, in the dosalane order, norgare.

According to the principles of the older classification the nearest relatives to this rock are found in the malignites of Ontario, as described by **Lawson** (see Table XXVIII, Cols. 5, 6 and 7). This Kruger mountain rock differs from the Ontario types chiefly in the fact that here potash greatly preponderates over the soda. Though in this respect the rock is an extreme member of the group named by Lawson; it is, apparently, best classified as an augite-biotite malignite.

Femic Nephelite Syenite.—The second analyzed specimen was collected near the contact with the Kruger Mountain schists at a point about 1,000 yards northwest of the locality where the first specimen was found. This second rock is a bluish-gray, medium-grained, somewhat porphyritic type. The phenocrysts are tabular crystals of microperthite, reaching 1 cm. or more in length. Qualita-tively the mineralogical composition is like that of the specimen just described. Here, however, the femic constituents are decidedly less abundant, while the feldspars and nephelite have notably increased. The order of crystallization and the decomposition-products are, respectively, the same as in the first speci-men. In the thin section of the second specimen it was observed that the garnet and biotite interpenetrate so intimately as to suggest a primary origin for the former, though decisive proof of that has not been found. Like the first speci-men, this one has been somewhat crushed, so that a metamorphic origin of the garnet is quite possible. The rock powder gelatinized strongly on heating with acid showing that nephelite is abundant. Optical tests seemed to show that some free albite here accompanies the other feldspars.

Professor Dittrich's analysis (specimen No. 1116) gave:-

Analysis of femic nephelite syenite, Kruger alkaline body.

		Mol.
SiO₂	52.53	·875
TiO₂	·07	·001
Al₂O₃	19·05	·186
Fe₂O₃	4·77	·030
FeO	2·10	·029
MnO	·13	·001
MgO	1·99	·050
BaO	·09	·001
CaO	5·75	·103
SrO	·19	·002

2 GEORGE V., A. 1912

Analysis of femic nephelite syenite, Kruger alkaline body--Continued.

		Mol.
Na₂O..	4·03	·065
K₂O..	7·36	·078
H₂O at 110°C..	·13	....
H₂O above 110°C..	1·49	....
P₂O₅..	·2	·002
CO₂..	·27	....
	100·17	
Sp. gr..	2·719	

From the microscopic study it is very probable that the titanic oxide is notably higher than is shown in the foregoing table. Otherwise the chemical analysis corresponds well with the optical analysis. Rough calculation has shown that the garnet must be low in alumina and high in lime and iron, and is thus, as already suggested by its colour, a true melanite. The appreciable amounts of barium and strontium oxides suggest that some of the feldspar mixtures may in complexity rival the phenocrysts of the Rock Creek rhomb-porphyry. We have here one more illustration of the rule that these two oxides tend to occur in relatively high proportion in the highly alkaline rocks.

The norm calculated from the analysis is :-

Orthoclase..	43·37
Albite..	11·00
Anorthite..	11·95
Nephelite..	12·50
Diopside..	10·80
Wollastonit..	·70
Magnetite..	6·73
Ilmenite..	·15
Hematite..	·16
Apatite..	·62
Water..	1·62
	99·60

The mode (Rosiwal method) is approximately :-

Feldspar..	63·9
Nephelite..	15·1
Biotite..	11·1
Melanite..	8·8
Apatite..	·6
Titanite..	·5
	100·0

In the Norm classification this rock must be classified with the first specimen as borolanose. In the older classification it may be best named a biotite-melanite-nephelite syenite, transitional to malignite.

Nephelite Syenite.--The third specimen was taken from a ledge 2,300 yards due west of the southern end of the lake on the plateau and 1·5 miles north of the Boundary line. It represents a specially large, relatively homogeneous mass a mile long and 400 yards wide, which crowns the 4,200-foot summit west of the lakelet. This mass is made up of the leucocratic phase of the alkaline body. (Plate 39, B).

The rock is a light bluish-gray, rather fine-grained syenite, breaking with a sonorous ring. In the hand-specimen it shows a weak parallel structure, probably due to flow in the late magmatic period. A few small hornblendes and many small feldspars twinned on the Carlsbad law, are arranged parallel to the planes of the flow-structure. Minute biotites can also be detected macroscopically.

Under the microscope the fairly abundant hornblende is seen to be a strongly pleochroic, olive-green variety of great absorptive power. The biotite is scarcely more than accessory. Nephelite, orthoclase, microperthite, microcline, and probably soda-orthoclase [extinction of 8° on (010)] are the light coloured essentials. The list of accessories includes melanite, apatite, and titanite. Iron oxides are absent or are present in but the barest traces.

The rock is very fresh, even the nephelite showing little alteration. In this case the relations of the melanite point to its being a primary mineral. The rock has been little, if at all, crushed since it crystallized. The garnet is generally poikilitic, enclosing feldspar granules, and seems to have been the last product of crystallization. A little anatase, probably derived from the titanite, was observed.

The chemical analysis of this specimen (No.1109) gave Professor Dittrich the following result:—

Analysis of nephelite syenite, Kruger alkaline body.

		Mol.
SiO_2	55·11	·918
TiO_2	·48	·006
Al_2O_3	21·28	·209
Fe_2O_3	2·64	·016
FeO	1·29	·018
MnO	·08	·001
MgO	·59	·015
CaO	2·82	·050
Na_2O	6·24	·101
K_2O	8·36	·089
H_2O at 110°C.	·14	....
H_2O above 110°C.	·58	....
P_2O_5	·27	·002
CO_2	·08	....
	99·96	
Sp. gr.	2·606	

The calculated norm is:—

Orthoclase	49·48
Albite	13·62
Anorthite	5·28
Nephelite	21·30
Diopside	3·23
Wollastonite	1·04
Magnetite	3·02
Ilmenite	·91
Hematite	·48
Apatite	·62
Water	·72
	99·70

2 GEORGE V., A. 1912

In the Norm classification the rock enters the sodipotassic subrang, beemerose, of the peralkalic rang, miaskase, in the persalane order, russare. In the older classification it is a hornblende (biotite) nephelite syenite. Partly on account of the fine grain of this rock its actual mineralogical composition or mode could not be determined by the Rosiwal method.

Summary.—Table XXVIII facilitates a rapid review of the chemical variety of the Kruger alkaline body. Col. 4 shows the average of all three analyses and is doubtless not far from the average for the whole body. This average recalls the analysis of a typical leucite syenite and also that of the borolanite described by Horne and Teall.§ In mineralogical composition, however, the average rock of the body would more closely approximate the malignites of Ontario. It seems best, therefore, to consider the average rock of the body as a malignite passing into nephelite syenite. Differentiation of the corresponding magma has yielded true nephelite syenite of granitic structure; coarse augite and biotite syenites of porphyritic structure; and various types of malignite, in which, however, the potash is in distinct excess over the soda. In the latter respect the Kruger body is in contrast with the average malignitic type of Ontario.

TABLE XXVIII.

Analyses of malignites and nephelite syenites.

	1	2	3	4	5	6	7
SiO₂	50·43	52·53	55·11	52·71	47·85	51·88	51·38
TiO₂	·92	·07	·48	·49	......	·33	·12
Al₂O₃	15·83	19·05	21·28	18·72	13·24	14·13	15·88
Fe₂O₃	6·11	4·77	2·64	4·31	2·74	6·45	1·48
FeO	3·04	2·10	1·29	2·14	2·65	·94	4·37
MnO	·11	·13	·08	·11			
MgO	3·38	1·99	·59	1·99	5·68	3·14	4·43
CaO	7·99	5·75	2·82	5·52	14·36	10·81	8·62
SrO	......	·19					
BaO		·09					
Na₂O	3·12	4·03	6·24	1·46	3·72	6·72	7·57
K₂O	6·86	7·30	8·36	7·42	5·25	4·57	4·20
H₂O	·29	·13	·14	·18	2·74	·18	·42
H₂O	1·20	1·49	·58	1·09		·96	·98
P₂O₅	·42	·28	·27	·32	2·42		
CO₂	·07	·27	·08	·14			
	99·83	100·17	99·96	99·60	100·65	100·41	99·45
Sp. gr.	2·849	2·719	2·668		2·873	2·88	

1. Augite-biotite malignite, Kruger mountain.
2. Biotite-melanite-nephelite syenite, Kruger mountain.
3. Hornblende-nephelite syenite, Kruger mountain.
4. Average of 1, 2, and 3.
5. Nephelite-pyroxene malignite, Poohbah lake, Ontario.
6. Garnet-pyroxene malignite, Poohbah lake, Ontario.
7. Amphibole malignite (garnet-free), Poohbah lake, Ontario.

§Trans. Roy. Academy, Edinburgh, Vol. 37, 1892, p. 163.

PLATE 40.

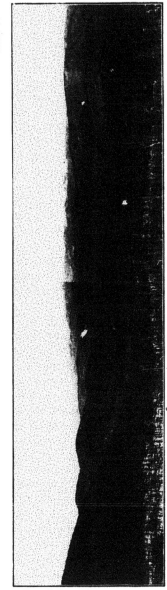

View looking southwest from slope of Mount Chopaka over plateau-like surface of Similkameen batholith, six thousand feet above sea-level.

SESSIONAL PAPER No. 25a

The average specific gravity of thirteen fresh specimens of the Kruger body is 2·750.

Metamorphism.—Few of the specimens collected are free from signs of crushing. This has sometimes induced a decided gneissic structure, and almost always the microscope shows fracture and granulation. The abundant development of metamorphic melanite and biotite and perhaps also the occasional production of large poikilitic scapolites indicate some recrystallization through dynamic metamorphism. The abundance of microcline and the corresponding subordinate character of the orthoclase is another, yet more familiar, relation brought about through the crushing. The mechanical alteration of these rocks is far from being as thorough as in the case of the Osoyoos batholith. This is a principal reason for believing that the alkaline mass was intruded after the Osoyoos granodiorite had been itself well crushed. No other definite field evidence for or against that view has been discovered. However, the magmatic relationships between the uncrushed Cathedral and Similkameen batholiths and the Kruger body also suggest that all three belong to one eruptive epoch of several stages—an epoch long subsequent to the intrusion of the Osoyoos and Remmel batholiths. The Similkameen granite is clearly intrusive into the Kruger alkalines, which may owe their strained and often granulated condition to the forceful entrance of that immense and immediately adjoining body of granite (see Figure 28).

SIMILKAMEEN BATHOLITH.

General Character.—The staple rock of the Similkameen batholith (Plate 40) is a medium- to coarse-grained, light pinkish-gray soda granite. Its essential constituents are hornblende, biotite, quartz, basic oligoclase (averaging $Ab_7 An_3$), and the alkaline feldspars, microperthite, microcline, microcline-microperthite, and orthoclase. The last named is characteristically rare; microperthite is the most abundant of the alkaline feldspars. The accessories are magnetite, apatite, and beautifully crystallized titanite. Allanite is a rare accessory; epidote is occasionally present, but apparently is secondary. The structure and order of crystallization are normal for granites, though microperthite is often in phenocrystic development.

A type specimen collected on the wagon-road following the west side of the Similkameen river valley, at a point three miles north of the Boundary slash, was studied microscopically and chemically.

A total analysis of this specimen (No. 1355) was made by Mr. Connor, with result as follows:—

2 GEORGE V., A. 1912

Analysis of dominant phase, Similkameen batholith.

		Mol.
SiO₂	66·55	1·109
TiO₂	·40	·005
Al₂O₃	16·21	·159
Fe₂O₃	1·98	·013
FeO	1·80	·025
MnO	·12	·001
MgO	1·32	·033
CaO	3·86	·069
SrO	·01	....
BaO	·03	....
Na₂O	4·07	·066
K₂O	2·84	·030
H₂O at 105°C	·01	....
H₂O above 105°C	·24	....
P₂O₅	·15	·01
	99·59	
Sp. gr.	2·693	

The calculated norm is:—

Quartz	21·78
Orthoclase	16·68
Albite	34·58
Anorthite	17·51
Hypersthene	4·05
Diopside	·64
Magnetite	3·02
Ilmenite	·76
Apatite	·31
Water	·25
	99·53

The mode (Rosiwal method) is approximately:—

Quartz	22·0
Orthoclase and microcline	6·7
Microperthite	27·0
Oligoclase	29·8
Biotite	5·5
Hornblende	4·2
Magnetite	1·8
Titanite	1·1
Epidote	1·1
Apatite	·8
	100·0

In the Norm classification the rock enteres the dosodic subrang, yellowstonose, of the alkalicalcic rang, coloradase, in the persalane order, britannare.

In the older classification it is a granodiorite, though the dominance of microperthite and the relative acidity of the soda-lime feldspar allies the rock to the alkaline granites.

For many square miles together the great central portion of the batholith is composed of this rock—a soda-rich biotite-hornblende granite or granodiorite of an average specific gravity of 2·706.

At the head of Toude (or Toat) coulee the rock of a large area within the batholith is generally porphyritic and distinctly finer grained than the staple granite, the specific gravity averaging 2 675. The phenocrysts are poikilitic microperthites bearing many inclusions of the other constituents. In the specimens so far examined, orthoclase tends to dominate over microperthite. Near the contacts with the normal equigranular rock, oligoclase replaces the alkaline feldspars to a great extent; yet this phase is always poorer in both hornblende and biotite than the normal phase, which is thus slightly the more basic rock. The finer grained phase was seen at several places only a few feet from the coarser; the contact is there sharp, but the absolute relation between the two phases could not be determined. It is highly probable that both are of nearly contemporaneous origin, the intrusion of the porphyritic phase having followed that of the equigranular rock by a short interval, as if in consequence of massive movements in one slightly heterogeneous, partially cooled magma. The porphyritic phase often shades into the other so imperceptibly that a separation of the two phases on the map is a matter of great difficulty, if not of impossibility.

The material of the batholith is further varied by rather rare basic segregations. These have the composition of hornblende-biotite diorite, being made up of the minerals of earlier generation in the host.

Basic Phase at Contact.—Much more important products of differentiation, as shown by microscopic analysis, are illustrated in a wide zone of contact basification. Here there occur several related types of alkaline or subalkaline syenites. In specimens collected along the contact with the Kruger alkalines, quartz nearly or altogether fails, biotite is absent, and abundant diopsidic augite accompanies the essential hornblende. The feldspars are the same as in the staple rock, with basic oligoclase, $Ab_2 An_9$, yet more abundant than there. Zircon is added to the list of accessories.

A specimen (No. 1107) of the basified shell showing this mineralogical composition was collected at a point two miles north of the Boundary line and about 200 yards from the contact with the Kruger alkaline body. It was analyzed by Professor Dittrich, with the following result:

Analysis of basic contact-phase, Similkameen batholith.

		Mol.
SiO_2	54·06	·901
TiO_2	·86	·010
Al_2O_3	18·75	·183
Fe_2O_3	4·64	·029
FeO	3·10	·043
MnO	tr.	....
MgO	2·75	·069
CaO	7·35	·131
Na_2O	4·60	·074
K_2O	3·00	·032
H_2O at 110°C.	·10	....
H_2O above 110 C.	·41	....
P_2O_5	·55	·004
CO_2	·11	....
	100·22	
Sp. gr.	2·819	

2 GEORGE V., A. 1912

The calculated norm is:—

Orthoclase	17·79
Albite	38·78
Anorthite	21·41
Diopside	8·92
Hypersthene	2·63
Olivine	·45
Magnetite	6·73
Ilmenite	1·52
Apatite	1·24
Water	·51
	99·98

The mode (Rosiwal method) is approximately:—

Orthoclase	22·9
Microperthite	17·2
Oligoclase (Ab$_5$An$_1$)	23·4
Hornblende	22·8
Augite	9·0
Magnetite	1·8
Apatite	1·3
Titanite	1·1
Zircon	·1
Quartz	·4
	100·0

In the Norm classification this rock enters the dosodic subrang, andose, of the alkalicalcic rang, andase, in the dosalane order, germanare.

According to the older classification it is an augite-hornblende soda monzonite. The analysis closely resembles that of the typical rock from Monzoni, except that the soda is strongly dominant over the potash. The specific gravity of the basic shell varies from 2·800 to 2·819.

It is an open question as to how far this basic phase is due to absorption of material from the adjacent malignite-syenite series and how far it is due to magmatic differentiation.

On the contact with the quartzites and schists of mount Chopaka the basification is less pronounced; compared to the staple granite. this phase is poor in quartz and rich in oligoclase-andesine and hornblende. It may be called a hornblende-biotite soda-monzonite of a specific gravity of 2·712–2·748.

For a half mile or more northwest of the contact with a large body of schist forming the Horseshoe pendant (Figure 31) the batholith exhibits a third basic phase. There is an almost complete disappearance of alkaline feldspars, other characters of the rock remaining essentially like those of the granite. This phase is a hornblende-biotite-quartz diorite of a specific gravity of 2·736. Here again there is doubt as to the exact cause of the basification. The Horseshoe pendant is largely amphibolitic in composition, and it is possible that assimilation of material from these schists is partly responsible for the development of the quartz diorite.

Comparison with Kruger Alkaline Body.—The intimate field-association of the Similkameen granodiorite with the Kruger alkaline body naturally suggests the question whether the two masses are consanguineous. The chemical analyses

PLATE 41.

Typical view in higher part of the Okanagan Range. Cathedral Mountain on left.

do not fully or directly answer the question but the mineralogical features of the respective rocks are alike in so many special ways that one must believe in a genetic bond between the masses. On comparing many thin sections from typo-specimens of each mass the writer has found certain significant characters in common, which are briefly noted in the following list:—

a. An unusually beautiful polarization pattern in the essential microcline microperthite feldspars; these minerals are of sensibly identical nature in the two bodies.

b. In each mass the   most recurrence of narrow shells of olive-green hornblende enclosing the pale green augite.

c. The essentially similar nature of the hornblende whether as rims or as independent crystals within the two masses. It varies somewhat in depth of tint but is always of this scheme:

    a=pale grayish green.   b=olive-green.
    c=olive-green.

The extinction on (010) is slightly variable but the measurements always ran between 14° and 22°, indicating in all probability a common variety of hornblende.

d. In each mass the recurrence of essential brown biotite with sensibly constant optical properties.

e. The augite of the basified shell in the batholith is a variety closely similar to if not identical with that characterizing the Kruger body.

We seem justified in concluding that, in spite of the strong chemical contrasts of the two masses, they have family traits suggesting that both belong to one petrogenic cycle.

Dikes Cutting the Similkameen Batholith. —The coarser phase of the Similkameen granite is cut, not only by the younger phase and by apophyses of the Cathedral granite, but also by a few basic dikes. One of these dikes has been examined with the microscope and found to be a medium-grained hornblende-diorite porphyrite, with phenocrysts of hornblende and andesine in a granular ground-mass of plagioclase and hornblende microlites and quartz.

The younger phase of the granite is cut by a few narrow dikes of black, fresh-looking trap which is not porphyritic to the naked eye but, under the microscope, shows phenocrysts of basic labradorite, $Ab_1 An_2$, colourless augite, and dark green hornblende. The rock is an augite-hornblende porphyrite and all of these dikes are probably genetically connected with the porphyrite cutting the coarser phase of the batholith.

CATHEDRAL BATHOLITH.

Older Phase.—The youngest of the batholithic intrusives is petrographically the simplest of all. Its material is singularly homogeneous, both mineralogically and texturally. The rock is a coarse-grained, light pinkish-gray biotite granite of common macroscopic habit. The essential minerals are microperthite, quartz,

2 GEORGE V., A. 1912

oligoclase, Ab, An, orthoclase (often microcline), and biotite; the accessories, apatite and magnetite, with rather rare titanite and zircon. The order of crystallization is that normal for granites. Sometimes, and especially along contact walls, the rock is porphyritic, with the microperthite developed in large idiomorphic and poikilitic phenocrysts, which, as described by Calkins, weather out with a retention of the crystal form.

A type specimen (No. 1388) collected on the Commission trail near the top of Bauerman ridge, has been analyzed by Mr. Connor, with result as follows:—

Analysis of Cathedral granite, Older phase.

		Mol.
SiO₂	71.21	1.187
TiO₂	.16	.002
Al₂O₃	15.38	.151
Fe₂O₃	.25	.001
FeO	1.47	.021
MnO	.06	.001
MgO	.33	.008
CaO	1.37	.024
BaO	.09	.001
Na₂O	4.28	.069
K₂O	4.85	.051
H₂O at 105°C.	.02	....
H₂O above 105°C.	.43	....
P₂O₅	.05	....
	99.95	
Sp. gr.	2.621	

The calculated norm is:—

Quartz	23.46
Orthoclase	28.36
Albite	36.16
Anorthite	6.95
Corundum	.61
Hypersthene	3.18
Magnetite	.46
Ilmenite	.30
Water	.45
	99.93

The mode (Rosiwal method) is approximately:—

Quartz	35.7
Orthoclase	7.0
Microperthite	40.3
Oligoclase	11.0
Biotite	5.0
Magnetite and titanite	.7
Apatite	.3
	100.0

SESSIONAL PAPER No. 25a

In the Norm classification the rock enters the sodipotassic subrang, toscanose, of the domalkalic rang, toscanase, in the persalane order, britannare. In the older classification it is a biotite granite rich in soda. The specific gravities of three fresh specimens vary from 2.621 to 2.637, averaging 2.631.

A local varietal phase, bearing olive-green hornblende as a second essential, was found in the contact zone, 400 yards or more in width, alongside the Similkameen hornblende-biotite granite; here there may also be some slight enrichment in oligoclase at the expense of the microperthite. Neither hornblende nor biotite is abundant. The specific gravity of this phase is 2.644. The cause of the basification must once more be left undecided; it may lie in assimilation, in differentiation, or in both.

The ordinary basic segregation is notably rare in this batholith. A few, with the composition of biotite-quartz diorite, were seen, but they seldom exceeded a few inches in diameter.

Younger Phase.—The coarse granite had been intruded, and apparently so far cooled that joints had developed within its mass, when a second eruptive effort thrust a great wedge of nearly identical magma into the heart of the batholith. This may be called the Younger phase of the Cathedral batholith. It forms a large dike-like mass 3½ miles long and averaging 400 yards in width; its length runs about north 60 degrees west and lies parallel to a system of master joint planes within the Older phase.

The Younger phase has the same general colour as the coarse granite, but is finer-grained, more regularly porphyritic, and more acid. The microperthite of the older granite is here largely replaced by orthoclase and microcline, both sodiferous; at the same time the plagioclase is more acid, being oligoclase near Ab_8An_1. The accessories are the same as in the coarse granite, but are much rarer. Biotite also is here less abundant. The weight percentages are approximately:—

Quartz	38.8
Orthoclase and microcline	33.4
Oligoclase	17.6
Microperthite	5.8
Biotite	3.5
Magnetite	.6
Apatite	.3
	100.0

The Younger phase approaches an aplitic relation to the Older. The contacts between the two were seen at several points; they are sharp, yet the two rocks are closely welded together, and it seems probable that the coarser granite was still hot when the younger granite was injected.

Relation to Similkameen Batholith.—The Cathedral granite is unquestionably consanguineous with the Similkameen granodiorite. Apart from their obviously close association both in the field relations and in the geological chronology, a near magmatic relationship for the two batholiths is indicated by the essential similarities in the optical properties of the respective minerals. These likenesses are observable in the quartz, microperthite, microcline, orthoclase, biotite, and the accessories, as well as in the hornblende which, as we have seen, is very rare in the Cathedral batholith.

It would be a matter of the highest importance if one could demonstrate the cause of this blood-relationship between the two batholiths. To say that they are magmatic differentiates is only to restate the petrogenic problem. The profitable questions are: *What* was differentiated in the two intrusive periods; and, what was the actual differentiating process?

These questions cannot be answered with assurance. All that seems possible now is to indicate the lines on which future investigation is needed. To do even that would anticipate part of chapters XXVI. and XXVII. and the writer will here offer only one conjecture as to the relation between the bodies. The guess is based on the proved efficiency of density differences to explain splitting in a heterogeneous magma, like that which composed the Moyie sills; secondly, on the view that a mediosilicic magma tends to separate into the antagonistic gabbroid (basaltic) and granitic magmas, this separation taking place with special rapidity just before solidification of the original magma could take place.

Let us assume that part of the Similkameen granodiorite long remained molten or was, by whatever means, partly remelted, and then gradually cooled. It is conceivable that during the cooling the basic elements corresponding in total composition to a gabbro, would settle down, leaving a persilicic residue in the upper part of the magma chamber. To develop the hypothesis still further the basic differentiate is assumed to have the composition of the local Ashnola gabbro. Finally, it is assumed that just one-fifth by weight of the remelted granodiorite settles out, this particular proportion being that which would give a residue with silica very nearly equal to that in the Cathedral granite. The residue has thus been calculated and found to be fairly close in composition to the Cathedral granite in all the other essential oxides. The result of the calculation is shown in Col. 3 of Table XXIX. Cols. 1, 2, and 4 respectively state the analyses of the Ashnola gabbro, the Similkameen granodiorite, and the Cathedral granite.

View of cirque head-wall composed of massive Cathedral granite. Scale given by man on the least jointed cliff.

Felsenmeer on Similkameen batholith, about seven thousand feet above sea level. Okanagan Range.

TABLE XXIX.

Showing chemical relation of Similkameen and Cathedral batholiths.

	1	2	3	4
SiO_2	47·76	66·55	71·41	71·21
TiO_2	2·20	·40	·00	·16
Al_2O_3	18·58	16·21	15·65	15·38
Fe_2O_3	2·19	1·98	1·92	·25
FeO	9·39	1·80	·00	1·47
MnO	·20	·12	·07	·06
MgO	4·15	1·32	·61	·33
CaO	9·39	3·86	2·47	1·37
SrO	·03	·01	·00	None.
BaO	·02	·03	·04	·09
Na_2O	3·61	4·07	4·19	4·28
K_2O	·47	2·84	3·41	4·85
$H_2O -$	·12	·01	·00	·02
$H_2O +$	·53	·24	·16	·43
P_2O_5	·78	·15	·00	·05
Remainder			·04	
	99·51	99·59	100·00	99·95

1. Analysis of Ashnola gabbro.
2. Analysis of Similkameen granodiorite.
3. Result of subtracting one-fifth part of each oxide shown in Col. 1 from the amount of the corresponding oxide in Col. 2, and recalculating to 100 per cent.
4. Analysis of Cathedral granite.

The divergence of the oxide proportions in the calculated residue from those in the Cathedral granite is inconsiderable except in the case of potash and lime and even those differences are no greater than those often observed in two analyses from any one batholith in other regions. It may fairly be claimed that the gravitative separation of non-silicic and subsilicic constituents (gabbroid mixture) making up about one-fifth by weight of the Similkameen granodiorite, would leave, in the upper part of the magma-chamber, a more silicious magma quite like that of the Cathedral granite. The composition of the less dense residue would be the same whether the separation took place through fractional crystallization or through true magmatic splitting.

Obviously, little stress can be laid on the actual figures resulting from the calculation just described. It has rather been intended as offering a concrete illustration of the hypothesis. On the other hand, the general principles underlying the hypothesis are, in the writer's belief, worthy of attention, for they seem to be among the most promising among all the principles of modern petrology. The calculation shows that it is not unreasonable to retain the conception that the Cathedral granite is a gravitative differentiate from the Similkameen granodiorite magma, and that a magma allied to gabbro or diabase and thus matching the basaltic and other dikes actually cutting the Cathedral

2 GEORGE V., A. 1912

granite, is the other pole of the differentiation. The chief difficulty of discussing this view, as of all its competitors, lies in the limited nature of the data from the structural geology of the range. Herein lies the importance of a comparison with the magmatic history of the Purcell sills and analogous injections of which the structural relations are well understood. Such comparison will be noted in the theoretical chapter XXVII.

Dikes Cutting the Cathedral Batholith.—Near the highest peak on Baterman ridge the coarse Cathedral granite is cut by a small dike of typical olivine basalt. The dike is exposed for sixty feet, in which distance it varies in width from four feet near the middle of the exposure to less than two feet at each end. The basalt this forms a lenticular mass, standing practically vertical. The strike of the dike is N. 35° E. and in the same quadrant as the average strike of the andesite dikes cutting the Basic Complex. The basalt is even more vesicular than the andesite mentioned. The middle of the dike is abundantly charged with gas-pores one to three millimetres or more in diameter. These are commonly elongated parallel to the walls of the dike. For five or ten centimetres from each wall the pores are very rare and the rock is compact, as if by chilling. The basalt carries xenoliths of the adjacent granite and of large quartz and feldspar crystal fragments also torn from the walls.

The microscope shows that the basalt is exceedingly fresh, not even the olivine being essentially affected by weathering. In view of this freshness it is noteworthy that the vesicles carry no trace of calcite or other filling. It looks as if they had never been filled with mineral matter. These facts together with the vesicular character of the lava, suggest that the basalt was injected near the surface and is therefore of later date than the unroofing of the batholith. In any case it is the youngest eruptive known to occur within the Okanagan composite batholith.

The phenocrysts are greenish augite and colourless olivine, both of which are abundant. The ground-mass consists of bytownite laths and augite granules, with a mesostasis of brown glass.

Two small, parallel, lamprophyric dikes of pod-like form and less than three feet in maximum width, cut the Cathedral granite on the ridge 1,200 yards northeast of Cathedral Peak. These dikes, in contrast with the basalt, are much altered and it is difficult to diagnose them. The original constituents seem to have been plagioclase, green hornblende, diopsidic augite, and possibly some biotite. The grain is fine; the structure, panidiomorphic to eugranitic. The rock may be a greatly altered camptonite or else hornblende diabase.

PARK GRANITE STOCK.

The Park granite stock measures 4 miles in length by 2½ miles in width (Figure 35). This granite is coarse, unsqueezed, and in almost all respects resembles macroscopically the Older phase of the Cathedral batholith, of which the Park granite seems to be a satellite. Under the microscope the rock differs from the coarser Cathedral granite chiefly in the entire replacement of microperthite by orthoclase; so that this granite is a normal biotite granite rather than a soda granite. The greater homogeneity of the dominant feldspar may explain the fact that the Park granite is somewhat more resistant to the weather than the Older phase of the Cathedral batholith. A few prisms of dark green hornblende are accessory in much the same proportion as in the Younger phase

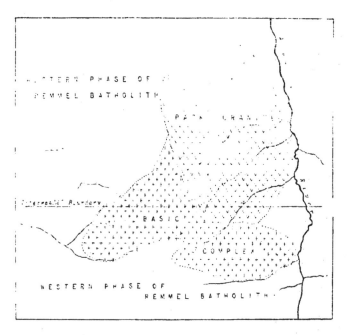

FIGURE 33.—Map showing relations of the Remmel batholith, Park granite, and Basic Complex. Scale :—1 : 110,000.

of the Cathedral. With these exceptions, both essential and accessory constituents are, in individual properties and in relative amounts, practically identical in the type specimens of stock and the Older phase of the batholith. The specific gravity of the Park granite is 2·673.

A second, very small boss of the Park granite occurs within the mass of the Remmel batholith some 5 miles west-southwest of the Park granite stock. This boss is circular in plan and measures not more than 250 yards in diameter.

25a—vol. ii—30

2 GEORGE V., A. 1912

GEOLOGICAL RELATIONS AND GENERAL STRUCTURE.

The Okanagan mountains are among the most accessible in the whole trans-Cordilleran section along the Forty-ninth Parallel. Even without a trail, horses can be taken to almost any point in the 5-mile belt. Owing partly to mere altitude, partly to the general climatic conditions, the summits are often well above the timber line, while the mountain flanks are clad with the woods of beautiful park lands. (Plates 40 and 41). Another special advantage in determining geological relations consists in the freshness of the rocks, which have been heavily glaciated and have not been seriously injured by secular decay. With a little searching, excellent and often remarkably perfect exposures of every formation and of its more important contacts can usually be discovered. Each of the principal field relations now to be noted has been determined not from one contact alone, but through the accordant testimony of several favourable localities.

The oldest rocks within the batholithic area are the quartzites and schists of Kruger mountain, with their associated basic intrusives; and the roof pendants of the Similkameen batholith (Figures 27, 28 and 29). Without doubt, these rocks are of the same age as the similar types found in the Anarchist series. All of these Paleozoic (probably in large part Carboniferous) formations had been crushed and dynamically metamorphosed before the intrusion of the oldest granitic component of the composite batholith (the Remmel or Osoyoos granodiorite).

Analogies drawn from better known parts of the Cordillera suggest that the basic intrusives of Chopaka mountain are of late Paleozoic (Carboniferous) age, though, of course, younger than the schists and quartzites which they cut.

Since the rocks of the Basic Complex are crushed and metamorphosed in as extraordinary degree as any of the above-mentioned formations, the complex is regarded as a Paleozoic parallel to the Chopaka basic intrusives, though perhaps not strictly contemporaneous with the latter. For a reason already noted, the Ashnola gabbro is possibly to be correlated in age with the larger part of the Basic Complex.

The mode of intrusion and therewith the structural relation of each of these basic masses to its original country rock cannot be declared. In the case of two of them—the Basic Complex and the Ashnola gabbro—not a fragment of the invaded formation has been found. It is, however, improbable that any of these bodies ever had batholithic dimensions. Their present isolated positions and the analogy of other similar gabbro-peridotite bodies in the Cordillera suggest that each of them was of relatively small size. The Chopaka body cross-cuts the bedding of the quartzites and schists. It may be in chonolithic relation to these—that is, it may be an irregularly shaped mass magmatically injected into the bedded rocks, but not, as with a true laccolith, following bedding planes. The contacts are insufficiently shown to warrant any decision in the case. The Ashnola gabbro may similarly be the residual part of an injected body. That it was a comparatively small body is suggested by an apparent flow structure still preserved even in the medium grained facies of the gabbro. In a batholithic rock of that texture, fluidal arrangement of the minerals is very rare. The infinitely diverse composition and structure of the Basic Complex much more clearly points to a non-batholithic origin. One imagines rather that the lithological and structural complications are in this case such as might appear at the deep-seated focus of an ancient volcanic area. The geological record has, however, been too largely obscured or destroyed that any of these hypotheses concerning the basic intrusives can be verified.

One fact is certain, that all of the bodies are older than the granites by which they are surrounded. Their contacts with the granites are the sharpest possible; gabbro or peridotite is pierced by many typical apophyses of granite or granodiorite which has often shattered the basic rocks and isolated blocks which now lie within the basic body. Here there is no question of the gabbros being differentiation products from their respective granitic magmas, as so often described in the granodiorite batholiths of California.* There remains, secondly, the conclusion that these basic intrusives were probably not of batholithic size. They show that some time before the real development of the Okanagan composite batholith began, a basic, suberustal magma was erupted on a limited scale—possibly in the form of stocks, possibly in the form of chonoliths.

Undoubted batholithic intrusion began with the irruption of the granodiorites. The familiar phenomena of such intrusion are exhibited along the contacts of the Osoyoos batholith. For several hundred yards from the igneous body the phyllites have been converted into typical, often garnetiferous, mica schists. This collar of thermal or hydrothermal metamorphism would doubtless be yet more conspicuous if at the time of intrusion the Paleozoic series had not already been partly recrystallized in the earlier dynamic metamorphism of the region.

The Remmel batholith is, as we have seen, composed of granodiorite similar in original composition to the rock of the Osoyoos batholith. Fossiliferous Lower Cretaceous arkose sandstones, grits, and conglomerates overlie the Remmel unconformably. The materials for these rocks were in part derived from the secular weathering of the Remmel granodiorite, the weathering being accompanied by rapid deposition of the débris in a local sea of transgression. Arkose sandstones, which alone measure more than 10,000 feet in thickness, were thus deposited in a down-warped marine area just west of the Pasayten river. To furnish such a volume of sediment, there would appear to have been in the region, preferably to the eastward of the Pasayten, a much larger area of granitic rocks than is now represented in the Remmel and Osoyoos batholiths combined. It is possible, indeed, that at that time these two batholiths were part of one huge mass of granodiorite which largely occupied the site of what is now the Okanagan composite batholith. Both Remmel and Osoyoos granodiorites have suffered profound metamorphism, so similar in its effects in the two rock masses that it may most simply be attributed to the same period of orogenic disturbance. The systematic parallelism of the shear zones in each

---

*See many of the Californian folios issued by the U.S. Geological Survey.

25a—vol. ii—30½

2 GEORGE V., A. 1912

batholith and the fair accordance in trends of the zones occurring in both batholiths suggest that there has been but one such revolutionary disturbance since the batholith were irrupted. If this be true, the period is identical with the post-Lower Cretaceous epoch, when the Pasayten Lower Cretaceous was thoroughly folded and crushed into its present greatly deformed condition in the Hozomeen range.

The Osoyoos and Remmel batholiths are thus probably contemporaneous; probably both are post-Carboniferous and certainly pre-Cretaceous. It is best to correlate them with similarly huge bodies of granodiorite determined as Jurassic in California and southern British Columbia.

It should be noted that, since the Remmel granodiorite disappears under the cover of lower Cretaceous at the Pasayten, sixty miles is the minimum width of the Okanagan composite batholith.

In the latter part of the Jurassic the granodiorite batholith was uncovered by erosion, then downwarped to receive a vast load of. quickly accumulated sediments until more than 30,000 feet of the Pasayten Cretaceous beds were deposited in the area between the Pasayten and Skagit rivers. As yet there is no means of knowing how far this filled geosynclinal extended to the eastward, but it doubtless spread over most of the area now occupied by the Okanagan mountain range.

The prolonged sedimentation was followed by an orogenic revolution that must have rivalled the mighty changes of the Jurassic. The Cretaceous formation was flexed into strong folds or broken into fault blocks in which the dips now average more than 45° and frequently approach verticality. It was probably then that the Jurassic granodiorites were sheared and crushed into banded gneisses and gneissic granites essentially the same as the rocks now exposed in the Remmel and Osoyoos batholiths. No sediments known to be of later age than the Lower Cretaceous have been found in this part of the Cascade system; hence it is not easy to date this orogenic movement with certainty. Dawson has already summarized the evidence going to show that many, perhaps all, parts of the Canadian Cordillera were affected by severe orogenic stresses at the close of the Laramie period.[*] It is probable that the stresses were even greater along the Pacific coast than they were in the eastern zone, where the Rocky Mountain system was built. To this post-Laramie. pre-Eocene epoch the shearing of the granodiorites may be best referred.

We have seen that there are good reasons for considering the composite Kruger alkaline body as younger than the granodiorites. It is clearly older than the Similkameen granite, as proved by the discovery of fine apophyses of the granite cutting the nephelite rocks. The Kruger body once extended some distance farther west over an area now occupied by the granite. The former, when first intruded, was an irregularly shaped mass without simple relation to its country rocks, the Paleozoic complex. The mode of intrusion was that of either a stock or a chonolith. In the first case the body was subjacent and enlarged downwardly; in the second case it was injected and its downward

[*] G. M. Dawson, Bull. Geol. Soc. Am., Vol. 12, 1901, p. 87.

cross-section may have diminished. As with so many other instances, the contacts are too meagerly exposed to fix the true alternative. The nephelite syenite was in part injected into the nearly contemporaneous malignite. The common fluidal structure of these rocks also points to a mode of wedge intrusion more like that of dike or laccolith than like that of a stock. The Kruger body may thus represent a composite chonolith, but the problem of its style of intrusion must remain open. The date of the intrusion was post-Laramic. The alkaline magma may have been squeezed into the schists while mountain building progressed or after it had ceased. The crushing and incipient metamorphism of this body are on a scale more appropriate to the thrust resulting from the irruption of the younger Similkameen granite than to the more powerful squeezing effect of the post-Laramic mountain-building.

True batholithic irruption was resumed in the replacement of schists, nephelite rocks, and possibly much of the granodiorite by the Similkameen batholith. This great mass is uncrushed, never shows gneissic structure, and has never been significantly deformed through orogenic movements.

The composite batholith received its last structural component when the Cathedral granite finally cut its way through Remmel granodiorite, Similkameen granite, remnant Paleozoic schists, and possibly through Cretaceous strata, to take its place as one of the most imposing geological units in the Cascade system. The field proofs are very clear that the Similkameen granite was solid and virtually cold before this last granite ate its way through the roots of the mountain range; in the manner shown, for example, in the large intrusive tongues cutting the schist pendant north of Horseshoe mountain (Figure 31). The contacts between the two batholiths are of knife-edge sharpness. The younger granite, persisting in all essential characters even to the main contacts, sends powerful apophyses into the older granite, exactly as if the two batholiths were dated several geological periods apart. Both are of Tertiary age and bear witness to the tremendous plutonic energies set free in a late epoch of Cordilleran history. Quietly, but with steady, incalculable force, this youngest magma worked its way upward and replaced the invaded rocks. During the same time the satellitic Park granite was irrupted with the stock form and relations.

Smith and Mendenhall have described a large batholith of granodiorite, intrusive into Miocene argillites at Snoqualmie pass in the northern Cascades and 100 miles southwest of Osoyoos lake.* This is one of the youngest batholiths yet described in the world. The more basic phases of the Similkameen batholith present similarities to the rock at Snoqualmie pass. It is thus possible that the Similkameen granite was irrupted in late Miocene, or even in Pliocene time.

The Cathedral granite must be of still later date. In this connection the work of Smith and Calkins is of special interest, for they have found that the Snoqualmie granodiorite is intimately connected with a large body of biotite

* Bull. Geol. Soc. Am., Vol. 11, 1900, p. 223; Snoqualmie. Folio, U.S. Geol. Survey, 1906, p. 9.

2 GEORGE V., A. 1912

granite which answers very well in its description to the Cathedral granite. They write:—

'There is included in this formation [Snoqualmie batholith] . . . a mass of more siliceous biotite-granite, which forms Guye Peak, the spur to the west of it, and part of Snoqualmie mountain. Its relation to the granodiorite was not definitely determined, but it is supposed to be derived from the same magma and nearly contemporaneous with it, since granodiorite and granite show the same relation to the adjacent sediments.　.　　.
The biotite-granite, when examined in thin section, is found to contain a large percentage of quartz, about an equal amount of alkali feldspar, somewhat less oligoclase, and a little biotite, largely replaced by chlorite. The alkali feldspar is microperthite, in contrast with the orthoclase of the granodiorite, which is usually not notably perthitic. The accessories are magnetite, titanite, zircon, and apatite.'*

This account of the Snoqualmie batholith suffices to show that there has been a close parallel in the magmatic history of that body and of the compound mass represented by the Similkameen and Cathedral batholiths. The parallel greatly strengthens the belief that these batholiths at the International Boundary are of late Neocene age.

RESUME OF THE GEOLOGICAL HISTORY.

The stages in the development of the formations in the belt between Osoyoos lake and the Pasayten river may now be summarized. We begin with the oldest stage which is of importance in this particular history.

1. Heavy sedimentation during the upper Paleozoic, possibly continued into the Triassic. Contemporaneous vulcanism and injection of dikes, sheets, and larger chonolithic (?) masses of gabbro and peridotitic magmas, the intrusives perhaps dating from the close of this period. The sedimentation and vulcanism produced the rocks of the Anarchist series; tentatively regarded as mostly of Carboniferous age. The intrusive bodies are represented in the Chopaka, Ashnola, Basic Complex, and Richter mountain gabbros and ultra-basic rocks. Some differentiation within the intrusive masses.

2. In Mesozoic time, probably during the Jurassic, intense deformation and metamorphism of most of the rocks so far mentioned. Strong mountain-building.

3. During the somewhat later Jurassic, batholithic irruption of the Osoyoos and Remmel granodiorites. Contact differentiation of quartz diorite in the former, at least.　.

4. Rapid denudation of the granodiorite batholiths in the late Jurassic; local subsidence of their eroded surface beneath the sea, there to be covered with a thick blanket of Cretaceous sediments which are in part composed of débris from the granodiorite itself.

* Snoqualmie Folio, page 9.

SESSIONAL PAPER No. 25a

5. At the close of the Laramie period, revolutionary orogenic disturbance, shearing and crushing the granodiorites and basic intrusives. In the former, development of strong crush-foliation and banding with the formation of new rock types, including biotite-epidote-hornblende gneiss, biotite-epidote gneiss, basic hornblende gneiss, biotite schist, hornblende schist, and recrystallized biotite granite-gneisses; in the basic intrusives, development of metagabbro and various basic (dioritic) gneisses and hornblendites. Simultaneous strong folding of the Cretaceous strata.

6. Either accompanying or following the post-Laramie deformation, the (chonolithic?) intrusion of the Kruger alkaline body, which consists of nearly synchronous masses of nephelite syenite and malignite. In these at least ten different rock types, due in part to the splitting of an alkaline magma and in part to later dynamic metamorphism, have been recognized.

7. In Tertiary time the batholithic irruption and complete crystallization of the soda-rich Similkameen hornblende-biotite granite, its contact basification forming soda-monzonites and quartz diorites.

8. In later Tertiary time the batholithic irruption of the Cathedral biotite granite, Older phase, accompanied by the intrusion of the Park Granite stock, immediately followed by the injection of the Cathedral granite, Younger phase, within the body of the Older phase.

9. Removal by denudation of much of the cover over each intrusive body. Complete destruction of the Cretaceous cover except at the Pasayten River overlap. Certain dikes of olivine basalt injected into the Cathedral and other granites are apparently of Pleistocene age and represent the latest products of eruptive activity in the Okanagan range. The vesicular-andesite dikes cutting the Basic Complex are probably as recent. The porphyrite dikes cutting the Similkameen batholith are possibly contemporaneous with these basaltic and andesitic injections. These dikes are quantitatively of little importance in the development of the composite batholith itself.

SEQUENCE OF THE ERUPTIVE ROCKS.

The reference of the different batholithic members to definite geological periods is tentative and still gives grounds for debate, but the relative order in which the component bodies were intruded is largely, and so far finally, determined. In very few other parts of the world are the conditions so favourable for tracing out the succession in time of an equal number of large-scale intrusive bodies. It is therefore expedient to note the sequence of the eruptive rocks in the Okanagan batholith. The writer believes that a careful study of the sequence here and in similar batholithic provinces can yield valuable results affecting the theory of granitic intrusion.

Of the available chemical analyses those which most typically represent the original composition of the various component bodies have been noted in Table **XXX**. In that table the analyses are arranged from left to right in the order

2 GEORGE V., A. 1912

of decreasing age for the corresponding rocks. It should be noted that analysis No. 3 refers to a crushed and otherwise metamorphosed phase of the Osoyoos batholith; through the leaching out of hornblende, iron oxides, etc., this analysis probably shows higher silica than the original average rock would show. The latter rock would have a silica percentage not far from that shown in the Remmel batholith, Col. 2.

Excepting the Kruger alkaline body there is a pretty definite law governing the series, whereby it shows an increase of silica with decreasing age. This law is yet more clearly appreciated when one considers, first, that the older phase of the Similkameen batholith with 66.55 per cent of silica is immediately succeeded by the younger phase which must have from 67 to 70 per cent of silica; and, secondly, that the Older phase of the Cathedral batholith with 71 21 per cent of silica is succeeded by the Younger phase with about 76 per cent of silica.

Table XXX.—Analyses of members of Okanagan composite batholith.

	1	2	3	4	5	6
SiO_2	47 76	63 30	68 43	52 71	66 55	71 21
TiO_2	2 29	50	29	49	49	16
Al_2O_3	18 58	17 64	15 80	18 72	16 21	15 34
Fe_2O_3	2 19	1 58	1 06	4 31	1 98	25
FeO	9 39	3 08	1 85	2 14	1 80	1 47
MnO	29	47	19	11	12	06
MgO	4 15	1 23	1 46	1 99	1 32	33
CaO	9 39	5 03	4 08	5 52	3 86	1 37
SrO	63	None	62	....	01	....
BaO	02	05	09		03	09
Na_2O	3 61	4 56	3 47	4 46	1 07	4 28
K_2O	17	16	2 51	7 12	2 84	4 85
H_2O	12	14	05	18	01	02
H_2O	53	51	53	1 09	24	43
P_2O_5	78	27	07	32	15	05
CO_2	....	....	....	11	....	
	99 51	99 52	99 72	99 60	99 59	99 95

1. Ashnola gabbro.
2. Remmel batholith, Western phase.
3. Osoyoos batholith, somewhat metamorphosed.
4. Average of three analyses of Kruger alkalines.
5. Similkameen batholith.
6. Cathedral batholith.

We have seen that the Kruger alkaline body is of small dimensions and that it may be an injected, chonolithic mass rather than a true subjacent body. The fact that this body forms an interruption in the regular basic-to-acid series of the plutonics is, therefore, no objection to regarding the law of increasing silica with decreasing age as strictly applying to the recognized batholiths of the region. The succession of undoubted batholithic magmas gave rocks with

SESSIONAL PAPER No. 25a

the following silica percentages, arranged in order, from oldest to youngest:—63·30 and (68.42   .r); 66·55; 67 to 70; 71 21; 76 ±.

On the other hand, if we include the Kruger alkaline body, we have two tandem series, one begun by the Chopaka and other old basic intrusives and the later series begun by the basic-alkaline Kruger intrusive. These two series were separated by millions of years, representing an interval in which the Osoyoos and Remmel batholiths were completely crystallised, crushed, eroded, and then deeply buried beneath the Pasayten geosynclinal -all before the Kruger body was intruded.* It thus seems highly probable that the Remmel-Osoyoos granodiorite and Kruger alkaline body were not in direct genetic connection. Each of the two series of intrusives was inaugurated by a definite revival of plutonic energy and, in each case, the first magma to be injected was a basic magma. In each case the subterranean heat sufficed to prepare large, subjacent masses of granodiorite. The granodiorite closed the first series so far as true batholithic intrusion in this region was concerned. In the second series the conditions favoured the still later generation of the alkaline Cathedral granite in its two phases.

Corresponding to the succession of chemical types, the members of the composite batholith illustrate a law of decreasing density with decreasing age. This relation is shown in the following table (XXXI.), in which the average specific gravities of the respective fresh, holocrystalline rocks are noted. As usual the readings were made at room temperatures. Where possible many large hand-specimens were employed in each determination. Again it is observed that the average specific gravities fall into two regular series, the second being initiated with the value for the Kruger alkaline body. Including only the undoubtedly batholithic members, the sequence from the Osoyoos-Remmel to the younger phase of the Cathedral granite is quite regular. From the known behaviour of holocrystalline rocks as they are melted, it is practically certain that the densities of the successively intruded magmas followed the same law of gradual decrease.

---

* During this long interval the Lower Cretaceous Pasayten agglomerate of augite andesite was erupted from local volcanoes on to the eroded surface of the Remmel batholith. This formation will be described in the next chapter; its occurrence is of interest in the present connection as showing that the two batholithic, granitic series were separated in time by a period in which magma much poorer in silica afforded the staple eruptive of the region. To the probably Tertiary Similkameen batholith the Pasayten andesite eruptive bears a chronological relation which is analogous to that of the late Paleozoic andesites (greenstones) of the Okanagan range and of the Anarchist plateau to the Remmel-Osoyoos batholith. The eruption of the Pasayten agglomerate as well as of the Kruger alkaline body clearly shows the justice of regarding the magmatic history of the Okanagan composite as divisible into two series, each begun by eruptions of basic or relatively basic rock.

2 GEORGE V., A. 1912

Table XXXI.—Correlations and comparisons among members of the Okanagan composite batholith.

Geological Age.	Stage of intrusion.	Name of body.	Observed variation in Specific Gravity.	Average Specific Gravity.	Average Percentage of Silica.
Late Palæozoic (Carboniferous?, possibly Triassic.)	1c	Chopaka basic intrusives	Gabbro, 2.959 Dunite, 3.173.	3.054	47.00*
	2.	Richter hornblendite	3.302	3.302	45.00*
	1c.	Ashnola gabbro	2.935–2.957	2.946	45.75
		Basic Complex (metamorphosed)	2.766 ca. 3.100	2.872	50.00*
Jurassic	3.	Osoyoos granodiorite batholith (metamorphosed)	2.682–2.939	2.750	65.00
	3a and 3b.	Remmel granodiorite batholith; two principal phases due to metamorphism.	2.655–2.680	2.729	64.00*
(Lower Cretaceous		Pasayten andesite			57.00*)
Close of the Laramie, or Tertiary.	4	Kruger alkaline body	Malignites, aver. 2.824 Syenites, average 2.675	2.750	52.71
Tertiary	5a and 5b.	Similkameen granite batholith, Older phase.	2.687–2.729	2.701	66.55
	6a	Cathedral granite batholith, Older phase.	2.660–2.686.	675	68.00*
	6b.	Park granite stock	2.621 2.644	631	71.21
	7.	Cathedral batholith, Younger phase.		655	70.00*
				686	76.00*
(Pleistocene ?		Olivine basalt dikes			48.00*)
(Pleistocene ?		Augite andesite dikes			
		Porphyrite dikes			57.00*)

* Estimated.

The theoretical bearing of this double law underlying the evolution of the Okanagan composite batholith will be discussed more fully in chapter XXVI. At present it may only be pointed out that the proved facts regarding the changes of acidity and density in the batholith are readily correlated with the view that a si-Cambrian granitic magmas are of secondary origin and have been differentiated automatically through density stratification. On his view the basic (gabbro, i.e. basaltic) magma is the original carrier of the heat, and the granites as a class have resulted from the interaction of the superheated basic magma on a different acies, schists, and sediments or on pre-existing granitic terranes. The Osoyoos-Remmel granodiorite is the product of the assimilation of acid Paleozoic and pre-Paleozoic terranes by invading basic magma. The Similkameen batholith is largely the product of the refusion of the Remmel and Osoyoos batholiths. The Cathedral granite is a later differentiate of the magma which had partly crystallized as the Similkameen granite, or was a differentiate from the Similkameen granite when partly remelted.

Among other purposes Table XXXI. will serve to show the correlation of the different formations described in the present chapter, excepting that the oldest of all, the Anarchist series (Carboniferous?), is not entered; nor is the Tertiary (?) conglomerate at Osoyoos lake noted, for its relations are not important to a treatment of the composite batholith.

The principal cause of differentiation has been sought in gravitative adjustment, stratifying the magmatic *couche* according to the law of upwardly decreasing density (meaning, in general, increasing content of silica from below upward in the magmatic strata). Some authors hold that large-scale differentiation may develop basified contact zones by the diffusion of basic materials to the surfaces of cooling. In chapter XXVII., an alternative and preferable explanation of the thicker basic contact-shells is outlined, again with primary emphasis laid on gravitative differentiation.

. It has not proved possible to demonstrate a law of increase of density with depth in the Similkameen granite. A series of fifteen fresh specimens of the rock were collected at altitudes varying from 1,200 to 8,050 feet above sea, and their specific gravities were carefully determined. The difference between the densities of specimens taken near or at the two extremes of vertical distance was found too small to allow of a definite conclusion, though the difference, small as it is, favours the law of density stratification. It must be remembered, however, that the concentration of volatile matter, such as water vapour dissolved in the magma but largely expelled during crystallization, would possibly be greatest at the roof. The specific gravities of the crystallized rocks may therefore not afford direct values for the total density stratification during the fluid state of the magma. Then, too, the observed relative uniformity of the Similkameen granite is a function of the scale of the subcrustal magma *couche*. It was unquestionably very thick; strong density differences are probably not, on any hypothesis, to be expected in a vertical section less than several miles in depth.

The whole petrogenic cycle had already closed and the Cathedral batholith was solidified when the dikes of vesicular basalt and andesite were injected into

2 GEORGE V., A. 1912

the Cathedral granite and the Basic Complex. These dikes represent essentially the same common basic type which forms the Ashnola gabbro and other of the oldest intrusives of the range. As the plutonic energies became exhausted in the formation of the Cathedral granite, the original heat-carrier has alone survived in the molten state and is capable of injection on the small, dike scale. In this feature the history of the composite batholith is similar to that of many other batholithic provinces, where the latest granite is diked by common basalt or by its hypabyssal, chemical equivalent.

Finally, it will be not. 1 that the conditions of crystallization underwent a decided change during the long interval between the intrusion of the Osoyoos-Remmel granodiorite and the Kruger alkaline body. Magmatic stages 1a to 3b inclusive, afforded non-alkaline rocks rich in hornblende and carrying plagioclase, either basic or of medium acidity, as the dominant feldspar. These bodies may be regarded as belonging to one consanguineous series. Magmatic stages 4 to 7 inclusive, afforded alkaline rocks bearing nephelite in the most basic phases and microperthite (orthoclase in 6b and 7) as the dominant feldspar throughout the series except in certain basified contact-zones. This group belongs to a second consanguineous series. The youngest of all the intrusives, the basalt and andesite dikes, belong to a third consanguineous series, closely allied in mineralogical and chemical composition with the earlier, members of the first series. The first and third series each began with a magmatic type which is chemically equivalent to the commonest of extrusive lavas (basalt). The second series began with a basic magma which may have been a peculiar differentiate of the same original basaltic couche or, as seems more probable, of that couche locally modified and controlled in its differentiation by some absorption of sedimentary terranes into which the Kruger body was injected.

METHOD OF INTRUSION.

Year by year the conviction has been growing ever stronger in the minds of many able geologists that such a batholith as any one of those here described has assumed its present size and position by actually replacing an equal or approximately equal mass of the older, solid rock. The Okanagan composite batholith repeatedly illustrates this truth. The writer is unable to conceive that the huge Cathedral batholith, for example, could have been formed by any process of simple injection, without leaving abundant traces of prodigious rending and general disorder in the granites alongside. We have seen, on the contrary, that the Similkameen granite on the east is notably free from such records of orogenic turmoil, while the shear zones of the Remmel batholith on the west most probably antedate the Cathedral granite intrusion. The very scale of these great bodies is suggestive of bodily replacement; it is hard to visualize an earth's crust which would so part as to permit of the laccolithic or chonolithic injection of a mass as great as a batholith.

The problem will be discussed at length in chapter XXVI., in which the many facts won from the study of the Boundary section will be correlated with the essential facts of the field in other parts of the world.

GENERAL SUMMARY.

1. At the Forty-ninth Parallel of latitude the Okanagan mountains and a part of the belt of the Interior Plateaus (the Interior Plateau of Dawson) have been carved by erosion out of an assemblage of plutonic igneous rocks which, in spite of the diverse lithological character of the rocks, should be regarded as an enormous single member of the Cordilleran structure. This plutonic group is named the Okanagan Composite Batholith. The details of its constitution are given in a foregoing résumé of its geological history.

2. This composite batholith was of slow development, beginning with small intrusions in late Paleozoic (or possibly Triassic) time, increased by great batholithic irruptions of granodiorite during the Jurassic, and completed by likewise immense irruptions of alkaline hornblende-biotite granite and biotite granite batholiths of Tertiary age, possibly as late as the Upper Miocene or the Pliocene. The satellitic Tertiary stock of Castle Peak in the Hozomeen range (see next chapter), is composed of normal granodiorite.

3. The local intrusion of a small, composite body of malignites and nephelite syenites; the regular basification along the batholith and stock contacts, giving collars of monzonites and diorites; and the sporadic appearance of certain peridotites (hornblendites and dunites) are probably all incidents of magmatic differentiation and do not directly represent the compositions of general subcrustal magmas.

4. The composite batholith offers striking testimony to the probable truth of the assimilation-differentiation theory of granitic rocks.

5. The composite batholith includes two consanguineous series of intrusions. The older one is non-alkaline; the younger, alkaline. They are separated in time by the whole Cretaceous period, at least.

6. The two consanguineous series nevertheless appear to belong to one comd petrogenic cycle. Throughout the cycle batholithic intrusion has followed sual law of decrease in magmatic density and increase of magmatic acidity with the progress of time.

7. Exposures of contact surfaces in the Similkameen batholith illustrate with remarkable clearness the downward enlargement of such bodies with depth.

8. The Similkameen granite bears three roof-pendants. Their distribution suggests that the present erosion surface of this batholith west of the Similkameen river is not far from coinciding with the constructional, subterranean surface of the batholith.

9. The Osoyoos and Remmel granodiorites have been extensively metamorphosed by orogenic crushing and its attendant processes. The metamorphism was both dynamic and hydrothermal. The granodiorites have been locally, though on a large scale, transformed into banded gneisses and schists. These changes have been brought about through the hydrous solution and migration of the original mineral substance of the granodiorites, especially the more basic minerals. The dissolved material has been leached out from the granulated rock

and has recrystallized in strong shear zones to which the solutions have slowly travelled. The shearing and metamorphism probably began at a time when the Remmel batholith was buried beneath at least 30,000 feet of Cretaceous strata.

10. The intensity of this metamorphism and the development of the great Tertiary batholiths agree with other facts to show that post-Jurassic mountain building at the Forty-ninth Parallel was caused by much more powerful compression than that which is shown in the broader Cordilleran zone passing through California; there the Jurassic batholiths are relatively uncrushed and Tertiary batholiths seem to be lacking.

11. The problems of the Okanagan composite batholith illustrate once again, and on a large scale, the utmost dependence of a sound petrology upon structural geology. A suggested chief problem involves the relation of mountain-building to the repeated development of large bodies of superheated magma only a few miles beneath the surface of the mountain range. The fact of this association is apparent; its explanation is not here attempted. (See Chapters XXIV to XXVIII).

Plate 43

Looking southeast along summit of Skagit Range from ridge north of Depot Creek.

CHAPTER XVII.

FORMATIONS OF THE HOZOMEEN RANGE.

GENERAL DESCRIPTION.

As the section is carried westward across the Pasayten river, we enter a new and more or less distinct geological province. One natural western limit of this province occurs at Lightning creek, but it is convenient to describe in the same connection the formations extending a few miles still farther west, so as to group within this chapter the various facts known about the geology of the Hozomeen range. At the Skagit river there is another abrupt change of formations. The Hozomeen range at the Forty-ninth Parallel is, in fact, an unusually well defined mountain group both in its topographic and its structural relations. (Maps No. 14 and 15).

Within the limits of the Boundary belt the range is composed of a dominant sedimentary group of rocks, here called the 'Pasayten series'; a more subordinate, older group of sediments and greenstones, here named the 'Hozomeen series'; a volcanic member of the Pasayten series, here named the 'Pasayten Volcanic formation'; two small stock-like bodies of 'Lightning Creek diorite,' which cuts the Pasayten series; a larger, typical stock of 'Castle Peak granodiorite,' also cutting the Pasayten series; a chonolith of syenite porphyry, cutting the Pasayten series and probably satellitic from the larger stock; and a few sills and dikes of porphyrite, cutting the Pasayten series and perhaps satellitic from the Lightning Creek diorite. West of the Pasayten river a small area of the Remmel batholith enters the five-mile belt. This plutonic mass is the local, unconformable basement of the great Pasayten series of rocks.

The geographical order of the formations as they are encountered in carrying the section westward, will be roughly followed in the brief descriptions of this chapter. The oldest rocks, those of the Hozomeen series, crop out only in the ridge of Mount Hozomeen itself and will be considered last of all.

PASAYTEN SERIES.

Introduction.—From the Pasayten river to Lightning creek at the eastern foot of Mount Hozomeen—a distance of twenty miles—the Boundary belt is underlain by an extraordinarily thick group of sedimentary rocks, here and there punctured by small bodies of intrusive igneous material. These sediments form a large area which was traversed by Russell and by Smith and Calkins during their respective reconnaissances in the state of Washington. During his journey along the Boundary in the years 1859-61, Bauerman crossed an area of stratified rocks which doubtless represents the northern continuation of the

479

2 GEORGE V., A. 1912

strata now to be described.* G. M. Dawson made a traverse up the northeastern headwater of the Skagit river and described the same body of rocks in greater detail.† The area where crossed by Dawson is about fifteen miles north of the Boundary line. Though he measured one section over 4,400 feet this : and, from paleontological evidence, proved the 'newer Mesozoic' age of the series of beds—later referring to them as Cretaceous—, Dawson did not give a special name to the series. His brief description will be found to correspond quite closely to the following account of the sediments. The present writer adopts the name 'Pasayten series,' thus modifying somewhat the title given to this great group by Smith and Calkins. The change from the original name, 'Pasayten formation,' seems to make one more appropriate to an extremely thick assemblage of strata which range in age from Lower to Upper Cretaceous.

Stratigraphy.—On the whole the Pasayten beds are tolerably well exposed, so that the succession can be made out with fair accuracy. At the Forty-ninth Parallel they compose a gigantic monocline with its base at the Pasayten river and its uppermost beds forming the steep ridges north, south, and west of Castle Peak. Across the strike the monocline measures at least sixteen miles in width. West of Castle Peak the youngest exposed member of the series, a thick mass of argillite, is strongly folded and faulted, giving steep dips. The lack of well marked horizons in this folded belt has rendered it as yet impossible to state its exact structure. Consequently there is much uncertainty as to the precise nature of the general columnar section in its upper part. At Lightning creek the argillite is cut off by a profound fault which brings it into sharp, more or less vertical contact with the Paleozoic rocks of the Hozomeen series. In the Boundary section, therefore, the top of the Pasayten series is not visible and the youngest exposed bed seems to be truncated by an erosion surface. No other area of the series has been examined in detail and the columnar section can be stated only in terms of observations made in the five-mile belt; such observations are necessarily incomplete.

As the writer carried his traverses from the basal unconformity at the Pasayten river westward, he became truly embarrassed by the colossal thickness which characterized the successive members. The cumulative thickness in a plainly conformable and comparatively young formation seemed almost incredible. For this reason special care was exercised in the field to note any possible hints of duplication of strata in the great monocline. It was found, however, that such duplication could have taken place only to a quite limited extent. The upper two-thirds of the series is charged with conspicuous horizon-markers; these would inevitably be repeated visibly among the fine exposures of the rocky ridges, if important duplication through normal faulting or other means had taken place. With a conviction which increased greatly as the field work and then the office study progressed, the writer has concluded that the series must

---

* H. Bauerman, Report of Progress, Geol. Survey of Canada, for 1882-3-4, Part B. p. 14.

† G. M. Dawson, Report of Progress, Geol. Surv. Canada, for 1877-8, Part B, p. 105.

total at least 30,000 feet in thickness. This is a minimum estimate, for the field sections as plotted show a total thickness of over 40,000 feet. The chief uncertainty resides in the determinations for the top and bottom members. As noted in the columnar section their respective strengths, namely, 3,000 and 10,000 feet, are estimated as the lowest possible minima. (Figure 34).

The whole succession is shown in the following table:

Columnar section Pasayten series.

Member.	Thickness in feet.	Lithological Characters.
		Top, erosion surface
L.	3,000	Gray to black argillite, bearing plant-stems and impressions of ammonite shells.
K.	7,100	Gray and green feldspathic sandstones with interbeds of black argillite and thin lenses of conglomerate; fossil plants and animal remains.
J.	1,400	Coarse conglomerate.
I.	300	Black argillite.
H.	3,500	Green feldspathic sandstone with rare argillite interbeds; fossil plants and shells about 200 feet from the top.
G.	200	Fairly coarse conglomerate.
F.	1,500	Gray and green, feldspathic sandstones.
E.	100	Conglomerate.
D.	1,100	Gray and green, feldspathic sandstone.
C.	600	Red argillite and sandstone.
B.	10,000	Very massive, medium-grained, arkose sandstone; fossil plants at about 900 feet from the top and also about 3,500 feet from the base.
A.	1,400	Volcanic agglomerate conformable to sandstone B.
	30,200	Base, unconformable contact with older Remmel batholith.

The volcanic agglomerate was crossed in four different traverses. Sufficient information was obtained to indicate its relations to the neighbouring formations. The agglomerate forms a remarkably straight and clearly continuous band of nearly even width, crossing the whole five-mile belt in a northwesterly direction and thus parallel to the strike of the adjacent sandstone. Though the breccia at every observed outcrop is quite devoid of bedding-planes, there can be little doubt that it is everywhere conformably underlying the sandstone. It is regarded as practically contemporaneous with the lowest beds of *member B.* Within the Boundary belt the agglomerate rests on the eroded surface of the Remmel batholith. The petrographic character of the agglomerate will be described in a special section of this chapter.

For a distance from the agglomerate the granodiorite is thoroughly decolourized and has the look of having undergone secular disintegration before the breccia was deposited. The depth of this shell of ancient weathering was measured near the Pasayten river and found to be about 400 feet. Such a depth means that the pre-volcanic surface was characterized by low slopes on which the rotted rock could lie and slowly increase at the expense of the fresh granodiorite beneath. The straightness of the line showing the contact between

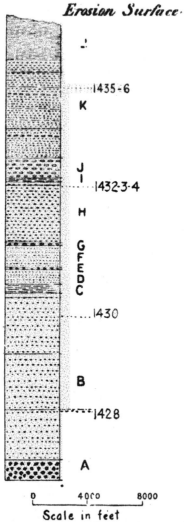

Figure 34.—Columnar section of the Pasayten Series, including the Pasayten Volcanic formation (member A). Approximate horizons of fossils indicated by collection numbers.

agglomerate and granodiorite in a direct indication of it; the best inference was that when the volcanic activity began

With the vulcanism or closely following it there was a strong downwarping of the region. Great changes of slope must have occurred, for the agglomerate is overlain directly by *member B*; a very thick sandstone, seemingly made up of the decomposition products of the granodiorite, which were now swept into the downwarp from its positive areas on the east. The resulting accumulation of arkose and feldspathic sands was to prove amid of long duration.

Striking characteristics of *member B* are its massiveness and uniformity of grain and substance. The massiveness is so great that even in large cuts representing strata fifty feet or more thick, it is often difficult to find the bedding-plane at all. In such cases the writer was at first in doubt as to whether the rock were really detrital, so much did it simulate a decolourized granite. Careful search, however, always showed the presence of true bedding which was best displayed in thin partings of dark shale in the sandstone. These shales and sometimes the sandstone itself were found to carry fossil plants; no further question was possible as to the nature of the whole formation. A few ripple marks were discovered in the upper beds of the member.

On both sides of the Boundary line the numerous readings of strike and dip showed close accordance all across the sandstone through the six miles from the Pasayten river to Chuchuwanteen creek. The strike averaged about N. 39 W.; the dip, about 18 S.W. The apparent thickness of *member B* is at least 15,000 feet. It is possible, however, that the strata have been in part repeated by a northwest-southeast normal fault running along the valley just east of Monument 81, and it has appeared safer to estimate the thickness from the simple monoclinal element between that valley and the band of agglomerate three miles to the eastward. Even this estimate gives 10,000 feet as the minimum.

At the strong, canyon-like valley of Chuchuwanteen creek there is such change of dip (though almost no change in strike) than another normal fault with upthrow on the northeast has been postulated and marked on the map. It is possible that some of the youngest beds of *member B* are represented only on the southwest side of that fault but they are neglected in estimating the minimum thickness of the member as given in the general columnar section.

The sandstone is normally a light-gray, medium to rather fine-grained rock; seldom showing the bedding-planes in the hand-specimen. It weathers gray to brownish-gray, rarely whitish. Excepting for the rare and thin interbeds of argillite already noted there are almost no variations from the monotonous character of the sediment; no conglomerate was found in this member. The sandstone is well consolidated and is often quite tough before the hammer. A typical specimen was sliced and examined microscopically. As expected from the macroscopic appearance the rock was found to be very rich in feldspar fragments. A rough estimate of the weight percentages credits about 30 per cent to quartz, 30 per cent to orthoclase, 35 per cent to plagioclase (andesine to

2 GEORGE V., A. 1912

labradorite) and 5 per cent to biotite, titanite, epidote, and limonite. All but the epidote and limonite are of detrital origin. The feldspars are greatly kaolinized and were doubtless nearly as much altered before the fragments found their places in the bed. The biotite occurs in thin, ragged and crinkled flakes, quite like those which may be seen in micaceous sands of the present day. The specific gravity of the specimen is 2·625.

The mineralogical composition of the sandstone is like that of the secularly weathered shell of the Remmel granodiorite below the agglomerate, *member A*. In both cases hornblende fails to appear, as if it had been leached out completely during the ancient weathering. Otherwise the important constituents of the Remmel batholith are all represented in the sandstone. There can remain no doubt that the sandstone has resulted from the destruction of the batholith. It is probable that the various sandstones overlying *member B* have had a like origin, but, from the lack of microscopic analysis, the proof of this has not yet been completed.

Member C is exposed in but a small area, occurring at the northern limit of the Boundary belt on the eastern slope of the Chuchuwanten valley. Farther south it is faulted out of sight by the Chuchuwanten fault. This member is the most highly variegated portion of the Pasayten series. It consists of a group of rapidly alternating red argillaceous sandstones and grits; gray, feldspathic, often pebbly sandstone and grit; with red, gray and green conglomerate. The beds range from an inch or less to twenty feet in thickness. The pebbles of the conglomerates are composed of hard, gray quartzite, chert, and hard, red and gray slate. Some of the larger, always well rounded boulders are as much as two feet in diameter.

These beds of C have variable attitudes; the rapid changes are probably connected with the adjacent fault. A mile or more east of Chuchuwanten creek the member dips rather steadily about 20 degrees to the north and visibly overlies *member B*. It is itself there overlain by 400 feet of *member D*, which on the north gradually approaches a horizontal position and is terminated above by an erosion-surface. The measurement of the thickness of *member C*, 600 feet, was made at this locality.

Member D is lithologically like the great basal sandstone but tends to assume a dominant green colour.

Member E is well exposed near the Boundary slash in a cliff overlooking Chuchuwanten creek on its west side. The conglomerate is of medium coarseness and seems to contain few pebbles not composed of gray quartzite or vein quartz. It is overlain by *member F*, some 1,500 feet of green and gray feldspathic sandstone of rather dark tints but essentially like the sandstone below the conglomerate E.

Member G is a 200-foot bed of conglomerate recalling E in its general character but abundantly charged with pebbles of an andesitic nature—the only known occurrence of such material in the conglomerates of the series.

Member H is not well exposed; so far as seen, it is a homogeneous, green, feldspathic sandstone. The estimate of the thickness, 3,500 feet, though so great, is believed to be a minimum. This member is best seen on the trail up Castle creek; there abundant, though not well preserved fossils, both shells and plants, were found at a horizon about 200 feet below the top of the member. These fossils will be described on succeeding pages.

The overlying black argillite *member I*, is, so far as known, unfossiliferous. It very clearly overlies *member II* and underlies the conglomerate of *member J*.

On account of its coarseness and great thickness—1,400 feet—member J is a very conspicuous element of the series. It was traced continuously from the summit north of Castle creek to a point well south of the Boundary line. Throughout that distance the conglomerate preserves a strike averaging about N. 22° W., and a dip of from 60° to 65° to the west-southwest. This steady behaviour of so prominent a member tended to make the structural study of the formation west of the Chuchuwanten comparatively easy. Its occurrence only once in the wide monocline has been a principal reason for believing that pronounced duplication of strata by strike-faults has not taken place.

This conglomerate is usually coarse; the pebbles reach eighteen inches or more in diameter. They are highly diverse in character. The list of different materials is long, including; gray, banded quartzite; blackish quartzite; hard, black and gray argillite; gneissic and massive hornblende granite; white aplitic granite; biotite granite; syenite porphyry; amphibolite; fine-grained diorite; coarse hornblende gabbro; greenstone schist; sericite schist; aphanitic quartz porphyry; and a breccia composed of white quartz fragments with jaspery cement. This list shows that most of the staple pre-Cretaceous formations of the region are represented. Among the granite pebbles were a considerable number having the composition of typical, fresh Remmel granodiorite or quartz diorite in its Western phase. Both the hornblende and the large, lustrous-black crystals of biotite in the Remmel are to be seen in these pebbles. The latter must have been derived from fresh, little-weathered ledges.

The cement of the conglomerate is a green feldspathic sand essentially like the green sandstones overlying and underlying this great conglomerate member. The cement yields rather readily to the weather, so that long talc-slopes of the weathered-out pebbles fringe the many cliffs where the conglomerate crops out.

Towards the top the conglomerate grows finer grained and merges into the very thick sandstone *member K*. This sandstone forms a continuous band crossing the Boundary belt. The band is a little over two miles wide and the average dip is about 45° to the west-southwest. The calculated thickness—7,100 feet,—is again enormous but it is a minimum. Three cross-sections of the wide band were traversed. In none of them was there any sign of repetition of beds nor any serious departure from the average strike and dip. The uniformity in the width of the band is another indication of the absence of strong faulting within the area covered by this member.

The sandstone of K is a hard, green to gray, brown-weathering, feldspathic rock much like those in members *D*, *F*, and *H*. It is interrupted by numerous beds of black to rusty argillite and argillaceous sandstone and is itself often more argillaceous than the average sandstone of the older members. Thin lenses

2 GEORGE V., A. 1912

of fine-grained conglomerate also occur at intervals. One of these, about 2,000 feet below the top of the member, carries fossil shells. A few feet away from the locality where the large shells were found, the sandstone encloses plant-remains. From the eastern end of the Castle Peak stock of granodiorite to the Lightning creek fault—a distance of nearly ten miles—the greater part of the Boundary belt is underlain by the black argillite of *member L*. The strata here show dips varying from 25° to 90°. As already noted it has not proved feasible to work out the folds and faults with entire confidence; in consequence, the thickness of the argillite is in doubt. It is known, however, that it must be at least 3,000 feet and may, as estimated in the field, be more than 5,000 feet. At its base it grades rapidly into the conformable sandstone *member K*.

Member L is a rather homogeneous, hard, black or dark-gray shale, in which thin, green and gray sandstone beds are intercalated. The shale weathers gray and brown in varying tints. At three horizons,—one found opposite the mouth of Pass creek, another 700 yards south of the 7,860-foot summit overlooking the creek, and the third on the ridge 1,000 yards east of Frosty Peak,—the shale carries fossil plants and ammonite impressions.

Granitic intrusions have to some extent metamorphosed the argillite. The metamorphic effect is apparently most pronounced about the Castle Peak stock, though the effects are nowhere very striking. The shale inclusions in the stock have been converted into hornfels of common type.

Fossils Collected.—It has been seen that the monotonous chain of failures in the many efforts to discover fossil remains along the Forty-ninth Parallel was seldom broken. The decided novelty of finding them at several horizons within the Pasayten series was specially welcomed, as these discoveries bade fair to clear up many points in the dynamic history of the eastern half of the Cascade mountain system and incidentally to throw light on the history and relations of unfossiliferous formations in the broad Columbia system as well. Many of the correlations noted in preceding chapters have, in fact, been made in the light of the analogies which may be traced between the structure and stratigraphy of the more easterly ranges and the more closely determined structure and stratigraphy of the Hozomeen range.

The conditions of field work during the Boundary survey did not permit of exhaustive collections at any point. As a guide to the future paleontological study of the Pasayten series the exact localities of the different collections of plant and animal remains will be noted. Each locality will be referred to by the corresponding specimen number. In connection with each the stratigraphic and paleontological details will be added.

No. 1428. At the 6,750-foot contour 400 yards southeast of the 6,920-foot peak situated two miles north of the Boundary line and about three miles west of the Pasayten river.

Stratigraphic position: about 3,500 feet above the base of *member B*. Sandstone with shaly interbeds.

Fossils: plants only; determined by Professor Penhallow as:

Gleichenia gilbert-thompsoni Font.
Glyptostrobus sp.
Pinus sp.
Salix sp.

Horizon: Cretaceous of Shasta series; see Appendix B.

No. 1430. At 4,200-foot contour, east side of Chuchuwanten creek canyon, about 400 yards north of Boundary slash.

Stratigraphic position: about 900 feet below top of *member B*. Shale bands in sandstone.

Fossils: plants only; determined by Professor Penhallow as:

Cladophlebis skagitensis, n. sp.
Gleichenia sp.
Aspidium fredericksburgense, Font.
Nilsonia pasaytensis, n. sp.
Cycadites unjiga, Dn.
Populus cyclophylla, Heer.
Myrica serrata, n. sp.
Quercus flexuosa, Newb (?)
Quercus coriacea, Newb.
Sassafras cretaceum, Newb.
Dorstenia (?) sp.

Horizon: Professor Penhallow writes:

'Reviewing this evidence, we observe that there are eleven species of plants from locality 1430. Of these Dorstenia (?), which is of questionable character, and Pinus (sic), which is chiefly represented by seeds and may indicate any one of several horizons, need to be eliminated because not specifically defined. This leaves nine well-defined species, of which three are definitely Lower Cretaceous and six as definitely Upper Cretaceous.'

He concludes that this flora shows two well defined horizons within the Shasta-Chico series. See Appendix B.

Nos. 1432-33-34. 4,700-foot contour, north side of Castle creek valley, four miles down stream from crossing of that stream and the Boundary line; just east of conspicuous band of thick conglomerate, *member J*.

Stratigraphic position: about 380 feet below top of *member H*.

Fossils: plants, determined by Professor Penhallow; animals, determined by Dr. T. W. Stanton.

Plants: 'The only specimen under number 1433 showed on one side, two small fragments of leaves which, from their obviously parallel venation, are to be regarded as belonging to some endogenous plant, the nature of which could not be determined. On the opposite side of 1433 is a single leaf of a pine.' See Appendix B.

2 GEORGE V., A. 1912

Animal remains:
> 1432:
>> *Pecten operculiformis* Gabb.
>> *Trigonia* sp. Fragmentary imprint.
>> *Eriphyla* ? sp. Small casts.
>> *Pleuromya* ? sp. Fragmentary imprint.
>> *Rissoa* ? sp. A small obscure gasteropod with the general form and sculpture of this genus.
>
> 1434:
>> *Serpula* ? sp.
>> *Pecten operculiformis* Gabb.
>> *Trigonia* sp. Related to *T. æquicostata* Gabb, and *T. maudensis* Whiteaves.
>> *Eriphyla* ? sp. Small casts.
>> *Pleuromya papyracea* Gabb.
>> *Ancycloceras remondi* Gabb. ? Fragment.
>> *Ancycloceras* ? sp.
>> *Hamites* ? sp.
>> *Lytoceras batesi* (Trask) ? Fragmentary small specimen.
>> *Belemnites impressus* Gabb. ?. Fragmentary imprint.

Horizon: Regarding the animal remains, Dr. Stanton writes:

'The two lots from Castle creek, numbered 1432 and 1434, evidently belong to the same fauna. The horizon is clearly Cretaceous and apparently within the limits of the Horsetown formation.'

Nos. 1435-36. 7,000-foot contour, 350 yards east of 7,622-foot peak five miles nearly due east of Castle Peak.

Stratigraphic position: 2,300 feet below top of *member K.*

Fossils: plants and (1435) one fossil marine shell.

Professor Penhallow found that the plant remains of 1436 consist, apparently, of fragments of the rachises of ferns which remain indeterminable, although he is inclined to consider them as derived from the one species *Gleichenia gilbert-thompsoni,* thus relating this horizon to that of 1428.

Dr. Stanton writes: 'The specimen numbered 1435, which according to your section comes from a much higher horizon than 1432 and 1434, has not been identified, but it is suggestive of Tertiary rather than Cretaceous. It is a marine shell.' He described the shell thus:

'*Lucina* ? sp. A single large Lucinoid shell whose generic characters have not been determined. Its size and external features suggest some of the Tertiary and living shells that have been referred to Miltha and Pseudomiltha.'

Horizon: probably Cretaceous (Upper Cretaceous), since *member L* includes at least one bed in which impressions or casts of ammonite shells were seen, and there is little doubt that *L* truly overlies *K.*

SESSIONAL PAPER No. 25a

In order to make the relations of these fossiliferous horizons clearer, the diagram of Figure 34 has been prepared.

Evidently much more work needs to be done on this great monocline, but it seems already probable that much if not all of the recognized Shasta-Chico series of California and Oregon is here represented. The southern geosynclinals of this age rival the one of the Hozomeen range in the almost incredible amount of sedimentation which is manifested.*

PASAYTEN VOLCANIC FORMATION.

This formation has already been referred to as member *A* of the Pasayten series, occurring at the very base of the Cretaceous series. It occurs in only one part of the Boundary belt, on the densely thicketed slopes of the Pasayten river valley. The exposures are not numerous but those observed were found near the bottom and top as well as in the middle of the formation. At nearly all of the outcrops the mass is composed of typical andesitic breccia. One large outcrop near the Boundary slash showed a compact phase which may represent a thick flow of somewhat vesicular andesite; this phase could not be followed any notable distance through the brush. Elsewhere the breccia is clearly dominant, so that it seems safe to describe the formation as essentially a breccia of rather uniform composition.

The breccia is extremely massive; at none of the outcrops was it possible to find undoubted evidences of stratification. The estimated thickness—1,400 feet—has been deduced on the assumption that the breccia conformably underlies the sandstone of member *B* of the Pasayten series; this assumption seems quite justified by the fact of parallelism between the out-ropping bands of the two members.

The volcanic mass consists very simply of angular blocks of porphyritic andesite, cemented by a well consolidated ash. The blocks are of all sizes up to those 12 to 15 inches in diameter. At one point the smaller fragments were seen to be rounded as if water-worn. In general, however, evidences of sorting or rounding by water-action were entirely absent. No other material than andesite composes the fragments. The breccia was nowhere seen to be rendered schistose through pressure.

The blocks of the agglomeratic mass are dark greenish or brownish-gray. The phenocrysts of altered pyroxene and plagioclase and occasionally of fresher hornblende were usually conspicuous in the blocks, especially the larger ones. Under the microscope the feldspar phenocrysts proved to average labradorite, near $Ab_1 An_1$. Some are zoned, with more basic labradorite in the cores and basic oligoclase in the outer rims. The pyroxene seems to be a common green augite; it is generally pretty thoroughly altered to uralite and chlorite. The hornblende is a common green variety; it is not so abundant as the augite.

*See J. S. Diller and T. W. Stanton on 'The Shasta-Chico Series'; Bull. Geol. Soc. America, Vol. 5, 1891, pp. 435-464.

The ground-mass of the rock varies in structure from the microcrystalline to the devitrified-glassy. Microlites of basic andesine and of augite, with grains of magnetite and apatite, are the determinable original minerals, but the ground-mass is generally altered to the usual obscure mass of secondary chlorite, calcite, etc.

In spite of the alteration it seems possible to recognize two original, closely related types. One of these, the commoner, is normal augite andesite (specific gravity, 2·673); the other is a hornblende-augite andesite with dominant augite.

About six miles east of the Pasayten river and 2,500 yards south of the Boundary line the Remmel batholith is interrupted by a circular, pipe-like mass of volcanic agglomerate about 350 yards in diameter. This rock is quite similar to that at the river, excepting that the breccia here is somewhat coarser, blocks two feet in diameter being common, and, secondly, that, besides the lava-blocks, it carries a large proportion of angular fragments of the Remmel granodiorite.

The lava of the blocks is often vesicular. It consists of altered andesites apparently belonging to the same two species as those above noted at the river. There is much probability that the two masses of agglomerate are contemporaneous and genetically connected.

Two possibilities are open. The small, eastern body may be a part of the once continuous volcanic cover locally down-faulted and thus preserved against erosion at the higher level; or the smaller body may occupy one of the actual vents through which the Pasayten andesites were ejected. The rounded ground-plan of the eastern body, its greater coarseness of texture and the abundance of granitic blocks of evidently local derivation, all suggest that the second interpretation is the correct one. If so, we have here a volcanic neck, a type of igneous-rock body which is by no means common in the northern half of the Cordillera.

LIGHTNING CREEK DIORITE.

On the divide between the south and main forks of Lightning creek and thus from two to three miles west of the Castle Peak stock, the upturned argillites of the Pasayten formation are cut by two intrusive masses of diorite. The map (No. 14) shows the ground-plan of these two bodies. Each is elongated and both of them cross-cut the sedimentary rocks after the manner of true stocks. The more easterly body is dike-like, being about seven times as long as it is broad. The other body is nearly of the same length, 1·5 miles, but is broader, with a maximum width of 0·6 mile. From the evident similarity of their lithology and of their geological relations, it is reasonable to suppose that the two bodies are connected underground. If so, they represent the partially denuded top of a considerable stock with a roof of irregular form. The total area of diorite as exposed is a little more than a square mile.

Both bodies have been somewhat, though not greatly, squeezed and sheared, so that the diorite generally has a gneissic structure. The secondary planes strike on the average about N. 40° W. and their dip is about vertical. In consequence of this dynamic action the diorite is notably more altered than is the uncrushed Castle Peak granodiorite. These facts suggested in the field that the diorite intrusions are of older date than the Castle Peak stock. Microscopic study has, however, shown much similarity between the diorite and the basified contact shell of the larger stock, so that a nearly contemporaneous origin of all three bodies seems possible. On this second hypothesis the orogenic stress responsible for the shearing of the diorite might be regarded as having been local and connected with the profound Lightning creek fault or with the intrusion of the slightly younger Castle Peak granite; or the stress might be considered as having been more wide-ranging and strong enough to shear the smaller bodies but too feeble visibly to affect the much larger stock. No facts were observed which would enforce a decision between these two hypotheses, though the former seems the more probable. It is known that all three bodies are of post-Pasayten (post-Laramie) age and almost certainly of pre-Pliocene age. If, as seems probable, the Castle Peak stock dates from the mid-Miocene, the diorite is perhaps also best referred to the Miocene.

Petrographically the two diorite bodies are alike. Each shows a conspicuous, basified contact-shell. Each shell is rich in femic constituents and averages, for each body, about 100 feet in thickness. The great bulk of each body is a hornblende diorite bearing accessory orthoclase. The rock is of a light gray colour and of medium grain. The essential minerals are: zoned plagioclase, averaging acid labradorite near $Ab_1 An_1$, and a highly idiomorphic, green hornblende. The accessories include a little apatite with titaniferous magnetite, abundant titanite, and some inter-stitial quartz and orthoclase. Pyrite is present but may be secondary, like the abundant calcite, chlorite, and kaolin. Biotite seems to be entirely wanting in this principal phase.

The structure is the normal eugranitic; under the microscope the minerals are seen to be strained but, in all the thin sections examined, are surprisingly free from granulation. The specific gravity of a type-specimen of this phase is 2.763.

The basic contact-shell is composed of a darker gray diorite with much more hornblende than in the principal phase and with a moderate amount of brown biotite among the essentials. The feld-spar averages labradorite, $Ab_1 An_1$; orthoclase seems to be entirely absent. The accessory minerals are the same as in the principal phase but are somewhat less abundant. Epidote and zoisite are here added to the list of secondary materials. The structure is again the eugranitic. The specific gravity of a type-specimen is 2.832. This phase is a basic hornblende-biotite diorite.

OTHER BASIC INTRUSIONS CUTTING THE PASAYTEN FORMATION.

At the Boundary slash near the divide, the younger Pasayten sandstones are cut, at one horizon, by a prophyritic mass which follows the strike of the beds and measures thirty-five feet or more in width. It may be a sill or a dike, the outcrops not sufficing to fix the alternative. The rock is dark greenish-gray and highly porphyritic, with abundant large, white phenocrysts of plagioclase

2 GEORGE V., A. 1912

and smaller ones of green hornblende. The ground-mass is feldspathic and microcrystalline, bearing many microlites of the green hornblende. The feldspar is murky with alteration-products but seems to be an acid labradorite. The rock may be classed as a basic hornblende porphyrite.

Three miles down-stream from the Boundary slash Castle creek crosses a 40-foot sill of basic rock clearly intrusive into the green Pasayten sandstone. This sill is lithologically related to the dike (?) just described but is more basic (spec. gravity of a type-specimen, 2·950). The rock is dark-green, coarse-grained and almost peridotitic in appearance. Macroscopically only one constituent, hornblende, is clearly visible. Under the microscope this mineral is seen to be extremely abundant, composing more than half of the rock by weight. It occurs in thoroughly idiomorphic prismatic crystals measuring as much as 10 mm. or more in length. These phenocrysts are embedded in a fine-grained matrix of soda-lime feldspar (probably labradorite) with which a little magnetite, apatite, and interstitial quartz are associated. The largest feldspar laths are about 0·5 mm. long, and the average lath is much smaller.

The rock may be classified as a hornblende porphyrite but it is an anomalous member of that species, carrying an extremely high percentage of hornblende. Chemically this porphyrite must be much like the peculiar gabbro of the Moyie and other sills of the Purcell range.

Concerning the date or dates of these small porphyrite intrusions no more can now be said than that they are both post-Lower Cretaceous. They may well be specially basic derivatives of the magma represented in the Lightning Creek diorite stocks.

CASTLE PEAK STOCK.

Its Special Importance.—A brief description of the Castle Peak stock was given in the preliminary paper on the Okanagan composite batholith. The outcrop of the body covers about ten square miles; it is the largest intrusive mass exposed in the Hozomeen range where crossed by the Boundary belt. The stock deserves special study and will be rather fully illustrated, for its contacts are more perfectly displayed than are those of any other stock or of any batholithic mass in the entire Forty-ninth Parallel section. Erosion has bitten deeply into the formations composing this part of the range. The upper slopes of Castle Peak and of the neighbouring mountains are above tree-line. For a double reason, therefore, the geologist can see relatively far down into the depths where this granitic body was intruded. At various points around the periphery of the stock the contact-surfaces can be followed downwards with the eye for a thousand or more feet, measured vertically. It happens also that the folded Cretaceous strata forming the country rock of the intrusive are so arranged that the relations to the stock in plan are as plainly evident as the relations in vertical sections. Since all these structural relations are of primary importance to the theory of the intrusion and since the contact-relations of stocks and batholiths are very seldom seen with equal clearness, the

SESSIONAL PAPER No. 25a

figures and text illustrating this stock in the preliminary paper will be fully reproduced in the present chapter.

Finally, the stock merits attention as it throws light on the problem of the age of the petrographically similar Similkameen batholith and therewith aids in discovering the difficult geological chronology of the whole Okanagan composite batholith.

Dominant Phase.—The staple rock of the Castle Peak body is a fresh, light gray, granitic type of medium grain. In the ledge the mass of dominant quartz and feldspar is speckled with fairly abundant, lustrous-black hornblende and biotite. Under the microscope the principal feldspar is seen to be plagioclase, often zoned and averaging andesine, $Ab_4 An_3$. Orthoclase, probably sodiferous, is a less abundant essential. The hornblende is deep green and is sensibly identical with that of the Similkameen granite. A few of the hornblende crystals contain small cores of augite or of felted uralite apparently derived from augite. No free pyroxene is present. Magnetite or ilmenite, apatite and titanite are the accessories.

A typical fresh specimen (No. 1441), collected on the southwest spur of Mt. Frosty, has been analyzed by Mr. Connor, with result shown in Table XXXII, Col. 1.

TABLE XXXII.

Analyses of the Castle Peak and Similkameen granodiorites.

	1.	Mol.	2.
SiO_2	66·55	1·109	66·55
TiO_2	·60	·008	·40
Al_2O_3	15·79	·155	16·21
Fe_2O_3	·15	·001	1·98
FeO	3·08	·043	1·50
MnO	·06	·001	·12
MgO	2·14	·053	1·32
CaO	3·47	·062	3·86
SrO	·01	....	·01
BaO	·03	....	·03
Na_2O	4·39	·071	4·07
K_2O	2·80	·030	2·84
H_2O at 105°C	·05	....	·01
H_2O above 105°C	·40	....	·24
P_2O_5	·04	....	·15
	99·56		99·59
Sp. gr.	2·678		2·693

The calculated norm is:—

Quartz	18·24
Orthoclase	16·68
Albite	37·20
Anorthite	15·01
Hypersthene	9·02
Diopside	1·71
Ilmenite	1·21
Magnetite	·23
Water	·45
	99·75

2 GEORGE V., A. 1912

The mode (Rosiwal method) is approximately :—

Quartz	18.2
Orthoclase	17.5
Andesine	41.7
Biotite	11.5
Hornblende	9.5
Magnetite	.9
Titanite	.3
Apatite	.4
	100.0

In the Norm classification the rock enters the dosodic subrang lassenose, of the domalkalic rang, toscanase, in the persalane order, britannare. For convenience the analysis of the specimen representing the principal phase of the Similkameen batholith is entered in Col. 2 of Table XXXII. It will be noted that, although these two analyses are exceedingly alike and differ by not so much as would two random specimens from either of the two bodies, yet, according to the system of the Norm classification, the batholith rock must be classified as yellowstonose in the alkalicalcic rang, coloradase, and thus in a quite different pigeon-hole from that assigned to the dominant phase of the Castle Peak stock. One may seriously question the value of a classification which obscures the fact that the two bodies are chemically almost identical. The writer believes this fact to be of primary importance in the discussion of their geological relations.

In the older classification this principal phase of the stock is a typical granodiorite.

Basic Contact Phase.—The stock has a distinctly basified contact-shell, from 200 to 500 yards wide along the existing outcrops. In this shell, quartz is not visible or at least conspicuous, to the naked eye; orthoclase is only a rather rare accessory. The plagioclase averages the basic andesine, $Ab_1 An_1$. Hornblende and biotite are present in higher proportion than in the principal phase. In grain and structure the two phases are similar. The specific gravity of the basic phase was found to be 2.811. This value agrees very closely with the specific gravity of the contact phase of the Similkameen batholith (2.819), showing another indication of the direct genetic connection between the two bodies. This basic phase of the stock has not been analyzed, but it is clearly a hornblende-biotite diorite rich in accessory quartz.

Structural Relations.—The area and ground plan of the stock are shown in Figure 35.

The country rocks are the Cretaceous argillites and sandstones, so folded and faulted as to present dips varying from 40° to 90°. Lines of strike and characteristic dips are illustrated in the diagrammatic map.

It can be seen from the map that the stock is not in laccolithic relations; but only in the field, as one follows the wonderfully exposed contact line, does one appreciate the fullness of the evidence that the plutonic mass is a cross-cutting body in every sense. Even where the contact line locally coincides in direction with the strike of the sediments, as at the eastern end of the stock, the dipping strata are sharply truncated by the granodiorite (Figure 35). Moreover, the granodiorite was not introduced by any system of cross faults or peripheral faults dislocating the sedimentary rocks. Owing to the special attitudes of the latter, the strike and dip of the beds would be peculiarly

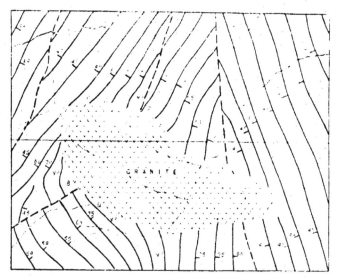

FIGURE 35.—Map showing relation ... the Peak stock to the deformed Pasayten formation. Strike and ... ... degrees shown; faults in broken lines. Scale:—1 : 115,000.

sensitive to such dislocation. The faulting actually displayed in the Cretaceous beds is strike faulting and was completed before the granodiorite was intruded. (Figure 35.) The igneous body is thus neither a bysmalith nor a chonolith. The magma entered the tilted sediments, quietly replacing cubic mile after cubic mile until its energies failed and it froze *in situ*.

Not only so; the superb exposures seen at many points in the deep canyons trenching the granodiorite illustrate with quite spectacular effect the downward enlargement of the intrusive body. At both ends and on both sides of the granodiorite body the steep mountain cliffs exhibit the intrusive contact

2 GEORGE V., A. 1912

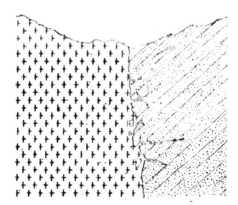

FIGURE 36.—Contact surface between the Castle Peak granodiorite and tilted Cretaceous sandstones and argillites. Section
in wall of glacial cirque at eastern end of the stock, the point
marked "A" in Figure 35. Scale:—one inch to 185 feet.

FIGURE 37.—Plunging contact surface between intrusive granodiorite of Castle Peak
stock and Cretaceous argillites and sandstones of Passayten series. Drawn from photograph looking south. Contact shown by heavy line in middle of view. Granodiorite
on left; sediments on right. The vertical distance between the two ends of the contact
line as drawn is 1,500 feet. Castle Peak on the left.

SESSIONAL PAPER No. 25a

surface through vertical depths of from 300 to 2,200 feet. In every case the contact surface dips away from the granodiorite, plunging under sandstone or argillite and truncating the beds. The angle of this dip varies from less than 20° to 80° or 85° (Figures 36 to 40). On the north side of the granodiorite a

Figure 38.—Plunging contact surface between intrusive granodiorite (on the right) and Pasayten formation (on the left). Drawn from a photograph taken on the north side of the Castle Peak stock, near the point "U", Figure 35. View looking east. Contact shown by heavy line in middle of view. Granodiorite on right of the line, which represents 1,700 feet of depth at nearer ridge. Contact also located in the background, with broken line.

section of the domed roof of the magma chamber still remains (Figure 40). It is noteworthy that a well developed system of rifts or master joints in the granodiorite seems, with its low dip, to be arranged parallel to the north sloping roof, as if due to the contraction of the igneous rock on losing heat upward by conduction.

25a—vol. ii—32

2 GEORGE V., A. 1912

This *form* of downward enlargement makes it still more surely impossible to conceive *that* the granodiorite was injected into the sediments by filling a cavity *with* orogenic energy. A visible section even 2,200 feet deep does no *provide continuance* of downward enlargement with depth; yet there is no log. *no* doubt that at least the steeper observed dips of the igneous cont. *are* but samples of its dips for several miles beneath the present

FIGURE 39. Plunging contact surface, Castle Peak stock, south side, near point "D", Figure 35. View looking east. Granodiorite on left of line showing contact. The vertical distance between the two ends of this line as drawn is 880 feet. The highest summit is Castle Peak.

land surface. Moreover, if the granodiorite made its own way through the stratified rocks and was not an injected body, passively yielding to ordinary orogenic pressures, there must have been free communication between the now visible upper part of the magma chamber and the hot interior of the earth. Downward enlargement is not only proved in visible cliff sections; it is demanded as a necessary condition of heat supply during spontaneous intrusion.

SESSIONAL PAPER No. 25a

The Castle Peak plutonic body thus appears to be a typical stock, an intrusive mass (a) without a true floor, (b) downwardly broadening in cross-section, and (c) intruded in the form of fluid magma, actively, though gradually, replacing the sedimentary rocks with its own substance. It is the most ideally exposed stock of which the writer has any record.

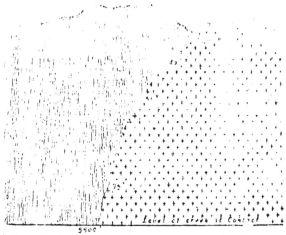

FIGURE 40. Intrusive contact between granodiorite and nearly vertical Pasayten argillite. Sketched in the field, on the north side of the Castle Peak stock, near point marked "E" in Figure 35. Granodiorite on the right. Figures show elevations in feet and dips of contact surface.

INTRUSION OF SYENITE PORPHYRY.

At Monument 80 the Boundary slash crosses a small mass of hornblende-biotite syenite porphyry. It is intrusive into the Pasayten sandstones. The area of the body as exposed on the present erosion-surface is about one-half of a square mile. The contacts are very poorly displayed and it was found impracticable to determine the structural relation of the porphyry. In ground-plan the body is elliptical, with its longer axis directed N.W.-S.E. This ele-ment of form suggests that the underground relations are those of a true stock, but the steady persistence of a strongly porphyritic structure at all points in the body tends to show that it is an injected, rather than a subjacent mass. If this second view is correct the body must be classed among the chonoliths rather than among the laccoliths, for it cuts across the edges of the sandstone beds along the whole length of the intrusive body.

Petrography.—In the field the porphyry has great uniformity of colour, texture and grain. It is a light-gray, fine-grained, strong rock, breaking with a sonorous ring. Nowhere could evidences of crushing be discerned. The phenocrysts are orthoclase, andesine, brown biotite, and green hornblende. The ground-mass is a microcrystalline, granular aggregate of orthoclase, plagioclase, and quartz, with accessory magnetite and apatite. The feldspars are generally more or less altered. The hornblende is usually represented only by pseudo-morphs of carbonate and chlorite. The biotite is much less thoroughly altered. The specific gravities of the two freshest specimens collected are 2·623 and 2·617.

Chemically and mineralogically, though not structurally, this porphyry is in many essential respects like the younger phase of the Similkameen batholith. As the younger and older phases of the batholith are of common magmatic origin and nearly of contemporaneous age, and since we have grounds for correlating the older phase with the Castle Peak stock, it seems simplest to regard the syenite porphyry as a satellite of the Castle Peak stock and both of those bodies as satellites of the Similkameen batholith.

Correlation.—A tentative correlation of the Similkameen batholith, Castle Peak stock, Lightning creek diorite bodies, and syenite porphyry chonolith may be expressed in the following form:—

Okanagan range—	Hozomeen range—
Similkameen batholith.	Castle Peak stock.
Older phase.	Principal phase.
Basic contact-shell.	Basic contact-shell.
Younger phase.	Syenite porphyry chonolith.

HOZOMEEN SERIES.

General Description.—The ridge culminating in the remarkable double summit of Mt. Hozomeen is wholly composed of pre-Cretaceous rocks to which the name 'Hozomeen series' may, for convenience, be given. These rocks extend from the major fault at Lightning creek, to the alluvium of the Skagit river. Another area of what appears to be the same series is mapped in the Skagit range.

In both areas the rocks are enormously crushed, so much so that all efforts to define the original succession or structures have so far failed. The difficulty of discovering the relations of the series is enhanced by the fact that the eastern or Hozomeen mass is cut off on both sides by faults and the western area is cut off on the west by an intrusive granite batholith, on the south by a master-fault and on the east probably by another great fault. The eastern area was studied at the close of the season of 1905; the western area during the season of 1906. On neither occasion did the plan of the Boundary survey permit of the study of areas more than three or four miles distant from the Boundary slash.

Either to north or to south of the Boundary belt the field-conditions may favour the discovery of the essential geological features of the series; within the belt they are distinctly unfavourable and the writer's results are largely negative.

The Hozomeen ridge is composed of a group of massive greenstones, cherty quartzites, and rare interculations of white to pale gray limestone. Of these the greenstone is dominant wherever outcrops occur, and appears to make up the nearly or quite inaccessible horn of Mt. Hozomeen as well as the higher though accessible summit just to the north. The greenstone has everywhere been crushed and altered, in both respects so profoundly that the writer was unable to secure a single specimen which in character even approached the original material. Minute jointing is extraordinarily developed, making it almost impossible to trim a specimen to standard size or shape; usually the very freshest rock crumbled to small polygonal pieces under the hammer.

The rock has the normal dark gray-green colour and almost aphanitic, massive character of greenstone. Occasionally it shows a brecciated structure which simulates that of a pyroclastic, yet no distinct beds of agglomerate or tuff could be discerned. Everywhere irregular, discontinuous and innumerable planes of fracture, generally heavily slickened, cut the rock in all directions. At one or two points suggestions of an original vesicular structure were encountered but they were too obscure to make the effusive origin of the greenstone perfectly clear. Nevertheless, the writer believes that this rock does represent the altered equivalent of basaltic or basic-andesite flows of great aggregate thickness.

Four typical specimens were examined under the microscope. They were all found to be essentially made up of secondary material,—the usual mat of uralite or actinolitic hornblende, epidote, chlorite, saussurite, omphacite, calcite, zoisite, and quartz, with here and there a granulated, altered plagioclase feldspar. Original crystal forms have been obliterated in all four thin sections.

If all of the greenstone exposed in the Hozomeen ridge is of extrusive origin, its total thickness must be great; 2,000 feet is a safe minimum.

The cherty quartzite is the next most important member of the series. If its whole strength were known it might prove to have a greater thickness than the greenstone. Some colour is lent to this idea by the discovery of other thick sections of the quartzite as it is followed along the Skagit valley trail towards Hope. In that stretch a large quantity of phyllite was found to be interbedded with the quartzite. In the Boundary sections phyllitic phases were comparatively rare; the silicious sediments there are pretty generally gray to greenish gray, compact, cherty rocks. They are thin to thick-bedded, breaking often with subconchoidal fracture. Like the greenstones they are heavily jointed, crushed and veinleted with white quartz. Under the microscope the rock is seen to have the common cryptocrystalline to microcrystalline structure of chert in which minute grains of apparently clastic quartz are embedded.

The limestone beds intercalated in the greenstone were observed only on the long spur running north-northeast from Mt. Hozomeen. Wherever seen, the beds are never continuous for more than a few hundred feet but occur as pods or

2 GEORGE V., A. 1912

lenses from thirty to forty feet thick in the middle and tapering off to nothing at each end. This form of limestone body is that often assumed when the rock-series in which it occurs has been subjected to powerful squeezing and rolling cut. The carbonate acted as if it were plastic, thinning here, thickening there, according as the lines of force were directed. The material pinched out at one point became accumulated in pods elsewhere. The relations are thus parallel to those found in the Pend D'Oreille series of the Selkirk range. It is little wonder that the traces of bedding have here disappeared. The rock is now a fine-grained, white to light bluish-gray marble, charged with concretions of chert. These concretions are most irregularly distributed, giving no indication of original bedding-planes. The pods are always vertical or nearly so and strike rather faithfully in the direction N. 20° E.

Other masses of limestone may occur on the slope down to the Skagit river but the thick brush of the slope prevented their discovery, though the outcrops sufficed to show the predominance of the greenstone and quartzite all the way to the river-flat.

It is not possible to state the relative ages of the different members with confidence. From the analogy with the less disturbed and probably contemporaneous Chilliwack series on the west slope of the Skagit range (see next chapter), it seems best to believe, as a working hypothesis, that the Hozomeen greenstone and limestone are younger than the principal quartzite (phyllite) group and overlie the latter conformably.

Correlation.—Since the series is so far quite unfossiliferous, the search for its equivalents among the determined formations of the Cordillera is aided only by the analogies of stratigraphic relations and of lithological resemblances. On these grounds the provisional correlation of the series, at least in part, with the Carboniferous Cache Creek series has been made. Rocks which are clearly much like those at the Skagit river have been found by Smith and Calkins in their reconnaissance of the Boundary belt and have similarly been tentatively referred by them to the Cache Creek division. Their description may be quoted at length:—

'The supposed Cache Creek series, as represented in this district (upper Okanagan valley), comprises both sedimentary and volcanic rocks. Its lower portion consists chiefly of clay slates and graywacke slates, usually of gray or greenish colour, together with some moderately coarse metamorphic sandstones and fine conglomerates, but comprises no coarse conglomerates. Occasionally the arenaceous portions of the series take on the character of fairly pure quartzite. Material of this sort becomes especially abundant near Mount Chopaka. In the upper portion of the series, as developed at Loomis, there are at least two beds of light-gray limestone, whose areal distribution is indicated roughly in the geologic map. Farther south, to the west and northwest of Riverside, this rock plays a more important role. The western wall of the coulee north of that place is a cliff perhaps 200 or 300 feet high and composed mainly of limestone.

'The upper part of the series comprises large volumes of volcanic material, which it was not found practicable to separate, on the preliminary map, from the slaty rocks. These old volcanics are for the most part extensively developed on the southern end of Palmer Mountain, in the basin southeast of that point, to the west of Blue Lake, and on the hill southwest of Palmer Lake. Lithologically, they were roughly classified in the field as greenstones. In broad, distant views the dark brownish hues of the weathered surfaces and their rugged erosion forms give them a resemblance to basaltic rocks. In hand specimens their original character is found to be obscured by decomposition, but the porphyritic texture is occasionally noted, as well as amygdaloidal structure and brecciated structure suggestive of pyroclastic origin. Microscopic study of these rocks is productive of no very satisfactory results, owing to the advanced decomposition which they have universally suffered, the original materials being almost always completely replaced. The character of the resulting secondary minerals, however, as well as the textural features, confirms the field diagnosis of the rocks. They are basic extrusives, probably for the most part basaltic, though perhaps including some basic andesite. Pyroclastics appear to be fully as abundant as the massive lavas.

'The rocks tentatively referred to the Cache Creek series have suffered various degrees of metamorphism. The sedimentary portions have in general an indurated slaty character in the localities removed from granitic intrusions. In the vicinity of the several intrusive granite contacts, however, much more advanced alteration has taken place, the slates being more or less completely converted to mica-schist. Interesting changes have been produced also in the basic eruptives by the granitic intrusions, the description of which will be deferred to the chapter on petrography.

'The upper part of Jacks Mountain or Mount Nokomokeen, is carved from a series which has the aspect of being much older than the Cretaceous rocks farther east and may be equivalent to the supposed Cache Creek of Okanagan Valley. It comprises both sedimentary and volcanic rocks.

'Most prominent of the sedimentaries are quartzites and bedded cherts. The latter are generally of a light-gray or drab tint and are cut by innumerable veinlets of quartz. Their bedding is their most noteworthy feature. They are built up of distinct laminae, about an inch in average thickness, readily separable from one another. The similarity of their structure with that of the red cherts of the Franciscan series in California is striking. As in the supposed Carboniferous of Okanagan Valley, there are beds of limestone (which are, however, rather thin and lenticular), and the highest portions of the mountain reached was built up largely of altered volcanic rocks, among which amygdaloids were observed. Although obscured greatly by alteration, the constitution and texture of these rocks as observed under the microscope indicate that these old lavas are basaltic...........

'Old schists, slates, cherts, and quartzites are also the principal country rocks in the valley of the Skagit above Ruby Creek, as far north as Jackass

2 GEORGE V., A. 1912

Point, but some 'greenstone' (basalt or basic andesite) also occurs. North of Jackass Point the country rock is mainly granitic, though interrupted by a belt of slate. The impression of the observer was that these sedimentary and volcanic rocks were plainly older than the Cretaceous and might in part be correlative with the Cache Creek series.'*

The results of Messrs. Smith and Calkins are seen to be essentially similar to those of the present writer in his study of the Anarchist and Hozomeen series. The discovery of Upper Carboniferous fossils in the very thick sedimentary series cut by the Chilliwack river canyon, a series which corresponds well lithologically with the Anarchist series and with some of the rocks in Mt. Hozomeen, is further significant. (See chapter XVIII.) These sediments on the Chilliwack river are only about twenty-five miles west of Mt. Hozomeen. It seems probable, therefore, that the Hozomeen series is to be correlated with the Anarchist series, and both of them with Dawson's Cache Creek series as well as with the likewise fossiliferous Chilliwack River series. There is nothing, however, to prove that some part of the Hozomeen series, if not a part of the rocks grouped under the name Anarchist series, is not of Triassic or even early Jurassic age. Yet it should be noted that the fossiliferous Triassic rocks of the lower Chilliwack valley are lithologically unlike any rocks observed in the Hozomeen ridge or in the area of cherty rocks across the Skagit. Finally, this matter of correlation can not be fully understood without reference to Dawson's several descriptions of the original Cache Creek series; to his papers the reader is referred for fuller information.†

STRUCTURAL RELATIONS IN THE RANGE.

In the Hozomeen range, for the first time since leaving the summit of the Selkirks 130 miles to the eastward, we enter a comparatively broad belt where stratified rocks afford horizons which permit of the discovery of the usual mountain structures, folds and faults. The structures are relatively simple and are illustrated in the map and section.

The fundamental feature in the stratigraphy of the Hozomeen range is the erosion unconformity at the base of the Pasayten series, where it rests on the Remmel batholith. Above that horizon all the members of the series seem to be quite conformable. The only pre-Cretaceous sediments are those in the Hozomeen series which are also clearly in unconformable relation to the Lower Cretaceous beds.

As already noted the Cretaceous series forms a great monocline complicated at its top by secondary crumples. The arch-and-trough structure is seen locally on the heights east of Lightning creek. (See general profile-section on map sheet). Elsewhere and thus generally throughout the range, faults are much more important structural features than folds. Profound normal faulting took place

* G. O. Smith and F. C. Calkins, Bull. 235, U.S. Geol. Survey, 1904, p. 22-70.
† See specially G. M. Dawson, Bull. Geol. Soc. America, Vol. 12, 1901, p. 70, where further references.

in the line of Lightning creek valley and, perhaps simultaneously, along the trough excavated by the Skagit river. In each case the faulting was probably normal, with throws as shown in the profile-section. Less important faults are postulated and mapped at Chuchuwanten creek and in the axis of the anti cline traversed by the Castle Peak stock.

CORRELATION.

The fossiliferous character of the Pasayten series renders possible its definite correlation with the Shasta-Chico series of California. On account of the fact that the youngest members of the series are upturned to verticality and otherwise show evidences of deformation much more intense than that usually seen as a result of Tertiary orogenic movements in the Cordillera, it seems in high degree probable that no Tertiary strata are represented in the series. Impressions of ammonites have, in fact, been found well above the base of *member L.* The upturning of the series is believed to have been largely completed during the post-Laramie orogenic revolution.

There is no question that the Hozomeen series is in unconformable relation to the overlying Pasayten series. We have concluded that the former series probably represents the Carboniferous Cache Creek series of western British Columbia.

The pyroclastic beds of the Pasayten volcanic formation bear no fossils but the structural relation of this member to *member B* of the sedimentary series suggests the advisability of dating the volcanic outburst in the Lower Cretaceous. It is hardly likely that *members A* and *B* would show such apparent strict conformity if the volcanics were of Triassic or Jurassic age and a Palæozoic age is almost certainly excluded. If the Remmel granodiorite is truly of late Jurassic age, the Pasayten volcanics cannot be other than post-Jurassic; among other obvious reasons for this conclusion is the fact that the Pasayten agglomerate also occurs in pipe-like form within the Remmel batholith in such relations as to suggest strongly a true volcanic neck.

The intrusion of the Castle Peak stock has been assigned to the Miocene. The argument for that reference is much the same as the one outlined for the dating of the Similkameen batholith (page 469). The stock is   rtainly post-Cretaceous. It shows no sign of such straining as would be ex   ed if it had undergone the squeezing incidental to the late Miocene mountain-building which has so generally affected this part of the Cordillera. The lithological similarity of this stock with the proved Miocene granodiorite of Snoqualmie Pass is some further indication that the stock should be referred to a geological date so relatively recent.

If the writer is correct in considering the syenite-porphyry chonolith as a satellite from the Castle Peak stock, the chonolith should be dated the same, namely, in the Miocene.

The Lightning Creek diorite and the apparently satellitic sills and dike of porphyrite have been subjected to the correlation already briefly discussed.

2 GEORGE V., A. 1912

They may have been intruded during the Eocene, Oligocene, or early Miocene, preferably during the Miocene; in any case they seem to antedate the Castle Peak stock.

The following table indicates the probable correlations for the Hozomeen range:—

Pleistocene............	Glacial and Recent deposits.
Miocene..	{ Syenite-porphyry chonolith. { Castle Peak granodiorite stock.
Miocene?......	{ Lightning Creek diorite stocks, { Porphyrite sills and dikes cutting Pasayten series.
Cretaceous (Shasta-Chico)	... Pasayten series, members *B* to *L*.
Cretaceous, near or at base of Shasta group......	.... Pasayten volcanic formation, agglomerate beds and volcanic 'neck (?)
	Unconformity.
Jurassic..	Remmel granodiorite batholith.
Carboniferous (Cache Creek)	Hozomeen series, quartzite, chert, limestone and dominant greenstone.

SUMMARY OF GEOLOGICAL HISTORY.

The Hozomeen formation represents a part of the Paleozoic formation which was intensely mashed and metamorphosed in Mesozoic, doubtless Jurassic, time.   That crustal revolution was immediately followed by the invasion of the Remmel batholith from below. Rapid erosion followed, during which the cover of the batholith was partly removed. The region subsided just after the erosion-surface had been deeply covered by the mantle of Pasayten pyroclastics. The subsidence continued during the formation of a typical geosynclinal depression.   Keeping pace with the sinking, an enormous thickness of (partly marine) Cretaceous strata was piled on the geosynclinal surface.   This great body of strata was deformed in post-Cretaceous time and, on account of the intensity of the action, it seems best to attribute this upturning to the well-established post-Laramie, early Eocene orogenic revolution.   The penetration of the Cretaceous beds by porphyrite sills and dikes, by diorite stock-like masses and by the Castle Peak stocks with its satellites, probably all occurred in later Tertiary time, with the Miocene assumed as the host date for the largest stock. The great faults about Mt. Hozomeen may date from the early Eocene or from a later, pre-Pliocene time. The possibility is thus recognized that they may be somewhat younger than the folds in the upper strata of the Pasayten series.

CHAPTER XVIII.

FORMATIONS OF THE SKAGIT MOUNTAIN RANGE.

GENERAL STATEMENT.

From the Skagit river to the great gravel plain traversed by the lower Fraser river, the Boundary belt crosses a large number of distinct geological formations which range in age from the Miocene to the Carboniferous, if not to the pre-Cambrian. The oldest fossiliferous sediments so far discovered date from the Upper Carboniferous; these belong to a thick group of rocks (named the Chilliwack series) most of which are believed to be Carboniferous. The as yet unfossiliferous Hozomeen series crops out in the area east of Chilliwack lake; these rocks are probably contemporaneous with certain phases of the Chilliwack series. A very thick andesitic group forms the upper part of the Chilliwack series as exposed near Tamihy creek and will bear the special name, Chilliwack Volcanic formation. A peculiar intrusive, dike-like mass of highly altered gabbroid rock, forming the western part of Vedder Mountain ridge, may be called the Vedder greenstone. Triassic argillites showing great thickness in the region east of Cultus lake have been grouped under the name, Cultus formation. Southwest of Tamihy creek canyon a group of conglomerates and green, massive sandstones, to which the name Tamihy series is given, seems to represent the equivalent of the Pasayten series farther east. On Sumas mountain, north of Huntingdon railway station, fossiliferous sandstones and conglomerates, named the Huntingdon formation, seem to represent the Eocene Puget group. It overlies unconformably a body of intrusive diorite cut by a biotite granite, which will bear the respective names, Sumas diorite and Sumas granite. These intrusives may be contemporaneous with a batholithic mass of greatly sheared granite occurring on and near Custer ridge at the main divide of the range; this body will be referred to as the Custer granite-gneiss. It seems to cut the Hozomeen series and is provisionally assigned to a Jurassic date of intrusion, but this truly old-looking rock may really represent a pre-Cambrian terrane. The eastern slope of the range at the Boundary line forms an area where, possibly in early Tertiary time, vigorous volcanic action built a thick local accumulation of andesitic breccias, associated with flows and with more acid lava; to the whole group the name, Skagit Volcanic formation, may be given. The remaining, specially named bodies in the range are the Slesse diorite and the Chilliwack granodiorite, both of which are in batholithic, intrusive relation to the Chilliwack series. The former occurs on Slesse creek; the latter forms the bed of Chilliwack lake and spreads far out on all sides. Both bodies are believed to be of mid-Tertiary age. (See Maps No. 15 and 16.)

2 GEORGE V., A. 1912

The invention of these many new formation names is intended to facilitate correlation along the Boundary; it is hoped that they may be of service to geologists who, in the future, need to correlate with any of the rock-groups cropping out along the Forty-ninth Parallel.

STRATIFIED FORMATIONS.

HOZOMEEN SERIES.

· A group of rocks believed to belong to the Hozomeen series covers three or more square miles of the Boundary belt north of Glacier Peak and just east of the main divide of the Skagit range. The area presented no geological features of special novelty and its description may be given in few words.

The cherty quartzite is here the prevailing rock, occurring generally in thin, flaggy beds from one inch or less to three inches in thickness. Phyllitic interbeds are commoner here than at Mt. Hozomeen. Near the Custer batholith, the quartzites are micaceous and the once-argillaceous beds are now mica schists. Occasional bands of probably conformable and extrusive greenstones are intercalated, but greenstone nowhere in this area assumes the importance it has east of the Skagit river. No limestone was observed in the main area; a patch of intensely metamorphosed schist and quartzite with included limestone pods occurs on the ridge-summit north of Depot creek, where the older Custer granite makes contact with the Chilliwack granodiorite. This stratified mass formed part of the roof of the older batholith and then a second time underwent metamorphism as it was invaded by the Chilliwack batholith. The limestone will be described in the section dealing with the contact-aureole about the latter intrusive.

At all the outcrops in the western areas the quartzite-phyllite series has steep dips, ranging from 70° to 90°. In the larger area the beds are intensely crumpled but the strike averages about N. 35° W.; the dip is generally about vertical. It is probable that several thousand feet of the sedimentary beds alone are represented in this area but it has proved so far impossible to secure either top or bottom for the series.

CHILLIWACK SERIES.

General Character and Distribution.—From the western limit of the Chilliwack granodiorite batholith to a point about two miles below the confluence of Tamihy creek.—a distance of sixteen miles in an air-line—, the Chilliwack river flows over a great thickness of sedimentary rocks to which the name, Chilliwack series, has been given. These rocks cover the whole width of the Boundary belt (as mapped) throughout most of the distance and extend far to north and south of the belt. They were examined by Bauerman in his reconnaissance of 1859, when he estimated the total thickness of the sediments exposed along

SESSIONAL PAPER No. 25a

the river as about 24,000 feet.* While he did not allow for duplication by fault and fold, his belief that the series is very thick was certainly justified. The Paleozoic section along the Chilliwack river is, indeed, one of the most complete of all those so far recorded on the western slope of the Skagit range, and besides the definitely Paleozoic strata of this section, there is another important group of Mesozoic beds occurring along the Chilliwack river. To the former group only, and particularly to the Carboniferous portion of it, the name Chilliwack series is intended to apply. For the first half-dozen miles westward of the Chilliwack batholith there are heavy masses of old-looking sediments which are so far unfossiliferous and may in part belong to the pre-Carboniferous terranes. From the mouth of Slesse creek to a point about ten miles due westward, and from the river southward to the Boundary line, the Chilliwack series is typically represented and is fossiliferous at so many points that little doubt remains as to the Carboniferous age of practically all the sediments occurring in these sixty square miles.

The eastern limit of the large area of Chilliwack sediments is, within the Boundary belt mapped, fixed by the intrusive contacts of the Slesse diorite and the Chilliwack granodiorite. The western limit is exceedingly difficult to place but is provisionally placed at the outcrop of an assumed master-fault mapped as crossing the belt a few miles west of Tamihy creek. The northern and southern limits of the sedimentary mass have not been determined.

From the fault just mentioned to another assumed fault running along the axis of Cultus lake valley, the Mesozoic (probably Triassic) formation separates the main body of the Chilliwack rocks from a smaller one which forms much of the long ridge known as Vedder mountain. No fossils have been found in this ridge but it seems most probable that its rocks form the lower part of the Chilliwack series and may be, therefore, all of Carboniferous age. On this view the intervening block of Mesozoic strata have been faulted down into lateral contact with the Carboniferous Chilliwack series.

Fossiliferous limestones associated with some shale and with a heavy body of contemporaneous andesite make contact with the Mesozoic formation along a line running nearly parallel to, and just south of the Boundary line. The former group represent a part of the Carboniferous series which has, apparently, been here thrust up over the Triassic rocks. The thrust-plane dips south at an unknown angle.

Finally, the Chilliwack series may be represented in some small areas of poorly exposed quartzites and slaty rocks unconformably underlying the Eocene (?) beds on Sumas mountain.

Notwithstanding a very considerable amount of arduous climbing distributed through part of each season in 1901 and 1906, not sufficient data are in hand to afford a complete idea of the succession of rocks included in the Chilliwack series. The density of the vegetation in these mountains, unparalleled as it is on the whole Boundary section elsewhere, will always stand in the way

---

* H. Bauerman, Report of Progress, Geol. Surv. of Canada for 1882-3-4, Part B, p. 32.

2 GEORGE V., A. 1912

of the full discovery of the facts needed for the stratigraphy of the series and the structural geology of its rocks in these areas. It is to be understood that the following statements should be subject to careful revision through future field work.

. *Detailed Sections and the Fossiliferous Horizons.*—Neither base nor top has been found for the series. Partial sections have been roughly measured and these will be described in brief form. On the basis of these as well as a multitude of details, isolated facts entered in the field note-books, a provisional columnar section embracing the rocks actually observed east of Cultus lake, has been constructed.

SECTION I.

About one mile west-southwest of Monument 48, beds which are believed to be the youngest exposed members of the Chilliwack series are unconformably overlain by grits and conglomerates belonging to the Tamihy Cretaceous (?) formation. From that point to the ridge of Church mountain two miles north of the Boundary line the exposures are unusually good for this region and a partial section of the series has there been made with some degree of confidence. The order is as follows:—

Top, unconformable contact with Tamihy formation.

a.	50 (or more) feet.—	Quartzitic sandstone.	
b.	20	"	Dark gray argillite.
c.	50	"	Light gray limestone, bearing fossils with numbers 1506, 1509, 10.
d.	60 (estimated)	"	Gray calcareous quartzite and dark gray, calcareous argillite.
e.	2,000+	"	Andesitic flows, tuffs and agglomerates.
f.	200	"	Gray and brownish shale and sandstone; thin conglomerate bands; crumbling, thin-bedded; highly fossiliferous. Collection Nos. 1512, 1514.
g.	600 (estimated)	"	Light gray, generally crystalline limestone, with fossils, No. 1513.

2,980 feet.

Base concealed.

For the determination of these as well as of the other collections made in the Chilliwack rocks, the writer is indebted to the great kindness of Dr. George H. Girty, and Dr. R. S. Bassler. The latter determined the bryozoa; the other genera were determined by Dr. Girty. The results may be quoted from Dr. Girty's letter, in terms of the collection numbers:—

No. 1506. About 900 yards south of the Boundary slash and 1,500 yards southwest of Monument 48.
Fossils: crinoidal fragments.

No. 1509. 100+ yards southwest of Monument 48.
Fossils:
Zaphrentis sp.
Campophyllum sp.
Euomphalus sp.

SESSIONAL PAPER No. 25a

Nos. 1510-11. Same locality as 1509.
  Fossils:
    Fucoidal markings.

No. 1512. On top of ridge 1,500 yards northwest of Monument 4.
  Fossils:
    Plant fragments.
    Clisiophyllum sp.
    Crinoidal fragments.
    Fenestella sp.
    Rhombopora sp.
    Cystodictya sp.
    Productus semireticulatus Martin.
    Productus aff. *jakovleri* Tschern.
    Spirifer aff. *cameratus* Morton.
    Reticularia lineata Martin (?)
    Spiriferina aff. *campestris* White.
    Martinia (?) sp.
    Seminula (?) sp.
    Terebratuloid (?)
    Myalina aff. *M. squamosa* Sowerby.
    Aviculipecten sp.
    Pleurophorus (?) sp.
    Orthoceras (?) sp.

No. 1513. On same ridge as 1512, 1,000 yards farther north.
  Fossils:
    Lonsdaleia sp.
    Campophyllum (?) sp.
    Crinoidal fragments.
    Fistulipora sp.

No. 1514. About 1,200 yards west of summit of Church mountain; top of
    ridge.
  Fossils:
    Fenestella sp.
    Pinnatopora sp.
    Rhipidomella aff. *nevadensis* Meek.
    Chonetes sp.
    Productus semireticulatus Martin.
    Productus aff. *wallacei* Derby.
    Productus aff. *jakovleri* Tschern.
    Spirifer aff. *cameratus* Morton.
    Spirifer aff. *dura* Kut.
    Spirifer sp.
    Reticularia lineata Martin (?)
    Martinia (?) sp.

Fossils—*Continued.*
    Spiriferina aff. *billingsi* Shumard.
    Cliothyridina pectinifera Sow (?)
    Hustedia aff. *compressa* Meek.
    Hustedia aff. *meekana* Shumard.
    Pugnax aff. *utah* Marcou.
    Dielasma (?) sp.
    Camarophoria sp.
    Aviculipecten aff. *coxanus* M. and W.
    Parallelodon aff. *tenuistriatus* Meek and Worthen.
    Parallelodon sp.
    Sanguinolites sp.
    Naticopsis sp.
    Orthoceras sp.

Section II.

On the Commission trail running along the Boundary line eastward from the Cultus lake valley and about 1,200 yards southwest of Monument 45, a massive limestone with a fifty-foot interbed of dark gray shale was found to carry fossils (No. 1500). The species were identified by Dr. Girty as:—

    Fusulina elongata Shumard.
    Rhombopora sp.
    Productus (?) sp.

This limestone appears to correspond to member *g* of Section I. It dips under the great volcanic member in apparent conformity. The exposures at this point are too poor to make the section of very great value.

Section III.

One of the most useful sections in the series is one traversed, in 1901, along the west slope of McGuire mountain where it steeply plunges to the bed of Tamihy creek, 6,000 feet below its summit. This mountain is crowned by a very ragged and broken syncline of massive limestone equivalent to that on Church mountain across Tamihy creek and to member *g* of Section I, (Plate 42, A). An infold of the shale overlying the limestone seems to correspond to the shale of member *j* of that section, but the volcanic member seems to have been here entirely destroyed by erosion. Below the massive limestone is a great thickness of sediments which are fairly well exposed in the gulches leading down to Tamihy creek. Measurements are very difficult to make on account of frequent faults and crumples in the bed. An approximate idea of the succession can be obtained from the following table:—

Top, erosion surface.
a. 300 feet (rough estimate).- Shale and sandstone.
b. 600 " " " Massive light gray to whitish, crystalline limestone with numerous crinoidal fragments in places.
c. 90 " Shale, sandstone and grit.
d. 110 ± " Massive light gray limestone with large crinoid stems and same fossils as member 2.
e. 300 + " Dark gray shale with fossils, No. 104.
f. 100 " Massive hard sandstone.
g. 1,400 " (rough estimate).- Hard sandstone, red and black shale, grit and thin bands of fine conglomerate.
h. 800 " " " Hard massive gray sandstone with gritty layers.
Base hidden under talus of Tamihy creek canyon

Dr. Girty found the fossils of No. 104 to belong to the following species:—

Pentremites sp.

Platycrinus sp.

Fenestella aff. *perannulata* Ulrich.

Fenestella sp.

Pinnatopora sp.

Polypora cf. *submarginata* M.

Chonetes sp.

Productus semireticulatus Martin.

Spiriferina sp.

Hustedia aff. *compressa* Meek.

SECTION IV.

At Thurston's ranch nearly opposite the mouth of Slesse creek, to .., northward up the mountain-side, and also westward along the Chilliwack, a very rough section has been run through the dense brush. No great ., is felt in the result, for there is a possibility that strike faul... unsuspected structural complications have repeated members of the .. on the other hand, have faulted some of them out of sight. Attenti... was called to this particular part of the river section by the discovery of .... crinoid stems in a heavy limestone cropping out just north of the ranch. ... limestone seems to be at least 400 feet thick, though its base is concealed, ., probably corresponds to *member e* in Section I. Northward from this outcrop the succession was crudely determined to be:—

Top of section, not well exposed.
a. 125 + feet.—Dark gray and black shale.
b. 150 " Coarse agglomerate composed of dark andesitic or basaltic fragments of large size together with other large fragments of limestone.
c. 90 + " Typical pillow-lava, basaltic; pillows round, up to three feet in diameter, with the spaces between them filled with cherty matter
d. 75 " Brown and gray shale.
e. 300 " Light gray, massive limestone.
f. 50 " Brownish shale.
g. 150 " Coarse feldspathic sandstone with conglomerate lenses.
h. 400 + " Light gray, crystalline, massive limestone with large crinoidal stems quite abundant.

1,340 + feet.

Base concealed.

2 GEORGE V., A. 1912

West of Thurston's ranch the crinoidal limestone seems to be repeated by a strike-fault, the beds retaining their general northeasterly dips of from 20° to tat° or more. This attitude is fairly well preserved in the outcrops along the river trail all the way to the mouth of Tamihy creek. The welter of forest, brush, and moss, as well as a heavy mass of Glacial drift on the river-valley floor, prevent any accurate conception of the nature of the beds crossed in this seven-mile traverse down the river. It is probable that the rocks corresponding to *members c, d, g* and *h* of Section III, are represented in this section or have been faulted out of it and that the very thick, phyllitic argillite seen along the north bank of the river at and just above the confluence of Tamihy creek with the river, is an older member of the series than any of those so far mentioned. Nothing better than a guess as to the thickness of this member is possible but 1,000 feet is apparently a very safe minimum.

The pillow lava and agglomerate of Section IV, seem to represent the lower part of the great volcanic *member e*, of Section I. The adjacent rocks match the respective members of Section I, in a rough way; considering that continuous exposures were not to be found at either locality, an exact correspondence should not be expected.

General Columnar Section. Combining the facts determined in these four sections with the many scattered observations made elsewhere, the following table may be made to express the writer's tentative conclusion as to the anatomy of the Chilliwack series:—

Top, erosion surface at plane of unconformity with the Tamihy (Cretaceous?) formation.

1.	50 + feet.	"	Quartzitic sandstone.
2.	20	"	Dark gray argillite.
3.	50	"	Light gray limestone; fossils, Nos. 1506, 1509, 1510.
4.	60 +	"	Gray calcareous quartzite and argillite.
5.	2,000 +	"	Andesitic flows, tuffs, and agglomerates (pillow-lava probably in this member where locally developed). This member may for convenience be referred to as the Chilliwack Volcanic formation.
6.	200	"	Gray and brownish shale and sandstone, with thin conglomerate bands; shales crumbling and thin-bedded; highly fossiliferous. Fossils Nos. 1512 and 1511.
7.	600 +	"	Light gray, massive, generally crystalline limestone, often crinoidal; with fossils No. 1513 (crinoidal fragments also represented in Nos. 69, 70, 71, 72, 98, 129).
8.	90	"	Shale, sandstone and grit.
9.	110 +	"	Massive light gray limestone, with large crinoid stems and fossils as No. 104 (not collected here).
10.	300 +	"	Dark gray and brown shales, with fossils, No. 104.
11.	100	"	Massive, hard sandstone.
12.	1,400 +	"	Hard sandstone, and black and red shales with bands of grit and thin beds of conglomerate; thickness very roughly estimated.
13.	800 +	"	Hard, massive sandstone with gritty layers.
14.	1,000 +	"	Dark gray to black, often phyllitic argillite with quartzitic bands.

6,780+ feet.

Base concealed.

Geological Age of the Series. As already indicated, the lower members of the Chilliwack series may belong to one or more systems older than that

PLATE 44.

A. Carboniferous limestone, summit of McGuire Mountain. Looking north.
B. Rugged topography at the Boundary, east of Chilliwack Lake and north of Glacier
  Peak. Mountains composed of Skagit volcanic formation. Looking southwest.
C. Horn topography between Tamihy and Slesse creeks. Peaks composed of metamor-
  phosed members of the Chilliwack series. Looking southeast from McGuire
  Mountain.
D. Horn topography on ridge between Slesse and Mobile creeks. Massive crags of
  Chilliwack granodiorite. Looking east.

one represented in the fossiliferous horizons. On that question there is at present absolutely no light. It remains to note in Dr. Girty's general summary of the status of the fossiliferous beds themselves. He writes:

'In the way of explanation I may state that, owing to the imperfect knowledge of most of our western Carboniferous faunas and to the poor state of preservation in which the fossils occurred, it was not possible to make positive identifications in most cases.

'Faunally, I would be disposed to arrange these collections into several groups. Lots 1512 and 1514 are closely related and represent, perhaps, the only strongly marked fauna in the collection. Lot 1500 is also rather diagnostic. Lot 1504 is moderately extensive, but is not strongly characteristic. It seems to differ considerably from either of the two faunas just mentioned. Of the remaining collections, which are faunally very limited, two groups can possibly be made. One of these comprises such lots as consist only of very abundant and very large crinoid stems (lots 69, 70, 71, 72, 129, 1506 and possibly 1503, or crinoid stems and cup corals (lots 1498 and 1513), or cup corals alone (lot 1500). The other group shows only fucoidal markings (lots 1510 and 1511).

'The most natural geologic section with which to compare these faunas is that of northern California. The sequence of the Carboniferous formations there consists, in ascending order, of the Baird shales, the McCloud limestone, and the Nosoni formation (formerly the McCloud shales). The Baird shales have usually been regarded as of Lower Carboniferous age and the McCloud and Nosoni as Upper Carboniferous. All three have extensive and characteristic faunas. There is nothing among your collections which suggests the Baird or McCloud. The most strongly characterized of your faunas (lots 1512, 1514, and 1500), however, have much that is similar to the Nosoni. At present I am disposed to correlate the two horizons. Lot 1504 is less certain, but possibly belongs to the same group. The lots furnishing only corals and crinoids differ widely from 1512 and 1514, but they might readily come from a specialized bed in the same formation. Nothing positive can, however, be stated about them. As to the three remaining lots, the data do not warrant suggesting anything whatever. On the whole, from the little that I understand of the stratigraphic relations and from the relationship manifested by the most marked of your faunas with that of the Nosoni formation, I am disposed to correlate all your beds in a general way with the latter. They may contain measures younger or older than the Nosoni, but from the absence of the well-marked Baird and McCloud facies it seems probable that none of the horizons from which your collections came is as old as the McCloud.'

In conclusion, it may be stated that the great volcanic member, the Chilliwack Volcanic formation, which will be specially described, is of distinctly Upper Carboniferous age. Just above and just below this member are conformable limestone beds containing samples of the fauna discussed by Dr.

25a—vol. ii—33½

2 GEORGE V., A. 1912

Girty. The estimated thickness given above for the sediments of the series—6,780 feet—is the minimum thickness of the Upper Carboniferous sediments in this region.

CULTUS FORMATION.

Stratigraphy and Structure.—In 1859 Bauerman recognized the strong lithological contrasts between the rocks on the two sides of Cultus lake and remarked that the distinctly more metamorphosed sedimentaries on the west side looked geologically older than the shales and sandstones of the eastern shore. The writer is inclined to share Bauerman's view and, as noted above, tentatively maps the rocks of Cultus ridge, as well as the large area of (fossiliferous) argillite to the southeast of that ridge as Triassic, while the beds occurring in Vedder mountain are mapped as Carboniferous. The name, Cultus formation, may be advantageously given to the younger group of strata. It may be defined as the local series of sediments which belong to the same geological system as the thick argillite bearing the Mesozoic fossils of lot No. 1502, hereafter described.

The dominant rock of the formation is a dark gray to blackish argillite, often bituminous in moderate degree. With it there are generally associated thin or thick bands of gray or greenish-gray sandstone and grit, and, more rarely, interbeds of fine conglomerate. The gritty beds are characteristically charged, very often, with small angular fragments of black argillite. All the coarser-grained types tend to be decidedly feldspathic, sometimes suggesting an arkose. These rocks are invariably deformed, with dips running up to 70° or 80°, though those of 30° or 35° are the commonest. The strike is highly variable in many places but the average direction is that parallel to the Cultus Lake valley; the average dip of, say 30°, is to the southeast all across the area where the formation is mapped. The argillites are very often heavily slickensided by local faults but the formation as a whole cannot be described as much metamorphosed. Phyllitic phases, for example, were not discovered. This relative lack of metamorphism was one of the criteria by which the formation was separated from the argillaceous phases of the Chilliwack series. As Dawson found in Vancouver island, the difficulty of distinguishing the Paleozoic and Mesozoic beds is greatly enhanced by the fact that in both, argillaceous types of great similarity in their original composition are the dominant types. Needless to say, the future worker in the geology of the lower Chilliwack valley will not take the accompanying map too seriously but will regard it as simply the first rough approximation in mapping. Incidentally, the present writer anticipates with great sympathy the struggle of such future worker with the jungle beneath which the truth is here hidden.

Two great normal faults and a no less important over-thrust are entered on the map as explaining the lateral relations of the block of Cultus sediments with the surrounding Chilliwack formations. These suggestions will need

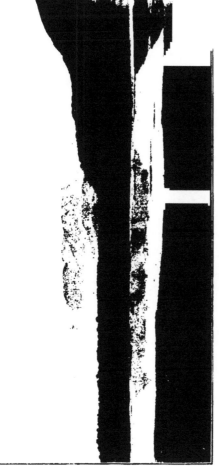

Western edge of Skagit Range, viewed from alluvial plain of the Fraser Valley at Chilliwack. Cheam Mountains in the distance.

The question of the thickness of Triassic beds actually exposed in the Boundary belt cannot be fully answered. A safe minimum is 1,000 feet but there is reason to believe that it is much greater. The great monocline of Cultus ridge alone seems to carry between 6,000 and 7,000 feet of beds, chiefly argillite with subordinate sandy layers. The possibility of duplication in this section makes it advisable to place the minimum thickness at no more than half the apparent thickness, say, 3,000 feet. The exposures both in this monoclinal section and elsewhere are too poor to permit of even an approximate columnar section for the Triassic formation.

Fossils.—The only fossils found in the Cultus formation were discovered in 1906 at a point about 500 yards south of the Boundary and too yards west-south-west of Monument 47. Here the staple black to dark-gray argillite is very homogeneous and carries few lenses of sandstone. Near the bottom of the 800-foot section, where the Boundary Commission trail crosses the creek, the fossils were discovered. Throughout the section the strike averages N. 65° E., and the dip, 45° S.S.E. There is considerable evidence of local slippings, with some brecciation and slickening of the argillite. The fossils are usually much distorted; all of them were found in a thin band close to a plane of slipping.

The writer owes the determination of the fossils, so far as that was possible, to the kindness of Dr. T. W. Stanton, of Washington. He writes:—

'Lot No. 1502 contains:
Arniolites vancouverensis Whiteaves? Numerous, more or less distorted specimens apparently belonging to this species.
Aulacoceras ? sp. A single fragment of a belemnoid which resembles *A. carlottense* Whiteaves.

'The lot numbered 1502, consists almost entirely of ammonites which seem to be identical with some described from the Triassic of Vancouver and Queen Charlotte Islands. Like the original types with which they are compared, they are not well enough preserved to show the septa and other features that are needed for their accurate classification.'

Combining this paleontological evidence with a comparison of the lithology of the Cultus formation and the Triassic rocks of Vancouver island, the writer has come to the view that little doubt need be entertained as to the Triassic age of the Cultus beds. Neither limestone nor contemporaneous volcanic matter have been found in association with the Cultus argillites, but this failure, by which we recognize an important difference from the Triassic sections of Dawson, can be readily explained on the view that these formations, if present, are faulted out of sight in the Cultus lake region. It is, of course, possible, though not probable, that these important members here apparently missing, were never laid down in the Chilliwack region.

2 GEORGE V., A. 1912

TAMIHY SERIES.

For about two miles from the Boundary line down Tamihy creek, the southwestern slope of the deep valley is underlain by an important series of rocks which have not been discovered with certainty at any other point in the Boundary belt this side of the Skagit river. This group of rocks was first seen in 1901 and given the provisional name, 'Green Quartzite Series.' It was unfossiliferous but it was thought to be older than the Chilliwack series.* During the season of 1906 much better exposures were found on the heights west of the creek and especially on the south side of the Boundary line. The relations are such as to enforce the belief that this new series lies unconformably upon the Carboniferous limestone, quartzite, and greenstone, and is very probably younger than the Triassic Cultus formation as well. Instead of the name chosen in 1901 the writer prefers to use the localizing name, Tamihy series, for this younger group of sediments. It is not the intention to name it as if it had been thoroughly analyzed and become stratigraphically understood; the name is chosen for convenience in the present report only, though possibly it may be of service in the hands of the geologist whose duty it will be to investigate this important mass of strata.

The relation of the Tamihy series to the Carboniferous rocks was determined with a fair degree of finality on the summit southwest of Monument 48 of the International line. At that point the quartzite noted as *member 1* of the general columnar section of the Chilliwack series, is overlain by a well exposed body of strata, of which some 400 feet are clearly visible near tree-line. It is a heterogeneous mass of gray conglomerate, black quartzitic sandstone, dark-gray paper shales, gray grit, and green sandstones in rapid alternation. The pebbles of the conglomerates include quartzite, vein (?) quartz, chert, and argillite, with almost certainly a few of greenstone like that of the Chilliwack Volcanic formation in the immediate vicinity. The chert pebbles are apparently of the same material as that so commonly found in the chert nodules of the underlying limestone. A few obscure plant-remains were found in the sandstone. The attitude of these beds is variable but the local dip averages about 30° to the southwest.

What appears to be the same series of rocks was followed for a mile south of the Boundary line as far as the top of a long ridge which runs eastward to Tamihy creek. That ridge is composed of a thick, very massive group of green and gray sandstones, grits and conglomerates like that just described, with little or no argillite. The dominant sandstone is extremely thick-bedded, so much so that it is rarely possible to obtain indications of strike and dip. Where these could be read, as south of Monument 48, the dips were 25°-30° to the southeast.

On that traverse, as along the Tamihy creek section, the minimum thickness is estimated to be 2,500 feet, but there is an indefinite addition to be made to this estimate when the area south of the Boundary line is investigated. No

*See Summary Report for 1901, page 51.

hint of a top to the series was anywhere visible and the total thickness may be several times 2,500 feet.

In the field the writer was much struck with the extreme similarity of the dominant (highly feldspathic), characteristically green sandstones with the sandstones which make up so much of the great Pasayten series farther east. Messrs. Smith and Calkins also note a similarity between 'Mesozoic' rocks occurring at Austin Pass some twenty miles southeast of the locality now being described, and the sandstones of the Pasayten. In 1898 Dr. Stanton reported on some fossils collected by Mr. W. H. Fuller on Cowap creek which lies immediately south of the Tamihy creek area. Dr. Stanton wrote:

'The fossils are evidently all from one horizon, which I believe to be upper Jurassic, this opinion being based chiefly on the distinctly serial form of *Aucella*, identified with *A. erringtoni* (Gabb) of the California upper Jurassic Mariposa beds. This species was collected at both localities. The collection from Canyon (Cowap) creek includes also a fragmentary *Pleuromya* and the impression of a small belemnite.

'The collection from the divide between Canyon creek and the waters of the Fraser river contains the *Aucella erringtoni*, a fragment of an ammonite apparently belonging to the genus *Stephanoceras*, a small similar belemnite like that from the last-mentioned locality, and the phragmacone of a large robust belemnite.' †

It is possible that these fossils were collected from beds which belong to the Tamihy series as here defined, or they may have been taken from beds more directly associated with the Cultus Triassic formation. Though the text and map of the report of Messrs. Smith and Calkins imply that those authors regard the green sandstones as of probably Jurassic age, their statement of the lithological resemblance of the sandstone with that of the Pasayten series suggest also that the Austin Pass rocks are really Cretaceous. The present writer is inclined to take the view that the Tamihy series, as represented on the Forty-ninth Parallel, should be referred to the Shasta division of the Cretaceous. This tentative reference has naturally little value; it invites criticism as a result of much additional field work in the region.

No other occurrences of the Tamihy series have been proved in the Boundary belt, but in the floor of Cultus lake valley above the lake, and, again, on the top of Pyramid ridge near the contact with the Chilliwack granodiorite batholith, certain conglomeratic and sandy beds have strong similarity with certain phases of the Tamihy series.

HUNTINGDON FORMATION.

The southern end of Sumas mountain is underlain by a cover of stratified rocks which pretty clearly belong to a period much later than any other group of consolidated sediments so far seen in the Boundary belt west of Osoyoos lake.

---

* G. O. Smith and F. C. Calkins, Bull. 235, U.S. Geol. Surv., 1904, p. 27.
† From Bull. 235, U.S. Geol. Survey, 1904, p. 27.

2 GEORGE V., A. 1912

In 1901 a brief examination of these beds was made. The exposures are generally small and poor, so that a complete treatment of the series cannot be given. It consists of heavy masses of medium grained, gray-tinted conglomerate alternating with sandstone and shale, the conglomerate being apparently most abundant at the base of the group. Unlike the clastic rocks of the Tamihy series, these are friable to a notable extent. Thin and seemingly unworkable beds of tolerable coal have been found in the upper part of the formation, and it is reported that borings have declared the presence of a valuable bed of fire-clay which was found beneath the ledges cropping out at the edge of the Fraser river alluvium southeast of the main sedimentary area.

The conglomerates contain pebbles manifestly derived from the (Chilliwack series ?) quartzite and from the Sumas granite, with both of which the sediments make unconformable contact. The sand-tones are feldspathic and arkose-like. The shales are sometimes carbonaceous, and at a point about 800 feet above the prairie and near the northern edge of the formation as mapped, Messrs. D. G. Gray and M. McArdle, who were in charge of the boring operations, discovered some fossil leaves in the shales.

The fossils were found in the vicinity of a thin coal-bed which is considerably broken, and seems not to afford a body large enough for economical working. The plant specimens were submitted to Mr. F. H. Knowlton, who reported that the collection was of little diagnostic value. He writes that the material has somewhat the appearance of species regularly found in the Laramie group, but states that much weight should not be given to this impression won from the study of the very poor material. Mr. Knowlton ventures on no specific names for the fossil forms submitted to him.

The general relations of the deposit, its degree of induration, and the evidence of the fossil plants, slender as it is, suggest that the formation should be equated with the Puget group, and thus belongs in the Eocene. The dips are not often to be read, but they seem to be always rather low, with 30° the observed maximum and 5° to 12° common readings. The strike is highly variable. We seem, therefore, to have here a relatively little disturbed cap of strata, laid down at a date distinctly later than that of the post-Laramie orogenic revolution, which so signally deformed the Cretaceous rocks of this general region of the Cordillera. The thickness of the visible beds totals probably about 1,000 feet. To this group of sediments the name, Huntingdon formation (from the name of the neighbouring village), may be given. Rocks of apparently the same nature and age have been long known as coal-bearing in the Hamilton and lower Nooksak Valley districts south of the Boundary line. *

Near the southern end of Wade's Trail over Sumas mountain a bench of the Huntingdon conglomerate and sandstone is cut by thin sheets of a greatly weathered porphyry, apparently a syenite porphyry. The relations are obscure; the porphyry may occur as one or more sills, or as dikes. It is the only eruptive rock known to cut the Eocene formation. The porphyry has not been examined under the microscope.

_____

* See G. O. Smith and F. C. Calkins, op. cit., page 34.

IGNEOUS ROCK FORMATIONS.

CHILLIWACK VOLCANIC FORMATION.

On Tamihy creek about three miles below the Boundary line crossing, a large body of altered basic lava was discovered during the season of 1901. It was followed southwestwardly to the top of a very rugged ridge where the lava was found in close association with strong beds of obscurely fossiliferous limestone dipping under the lava. It was, however, not until the Commission trail west of the creek was opened up that the writer was (in 1906) able to see definite evidence as to the age of the lava and as to its relation to the sedimentaries. A few hundred yards south of Monument 18, the southern limit of the lava was found. It there makes direct contact with a fifty foot bed of fossiliferous limestone similar in habit to that at the lower contact of the lava. From the fossils collected in the upper limestone Dr. Girty has concluded that this limestone is certainly Palæozoic and in all probability upper Carboniferous in age. The limestone at the base of the lava formation is likewise apparently upper Carboniferous. Since the lava is conformably intercalated between the two limestones, it must also be referred to the upper Carboniferous. From its position in the limestone one may fairly conclude that the eruptions were wholly or in part submarine.

Westward from Tamihy creek the band of old lava was followed for a distance of some ten miles. Throughout most of that stretch the northern contact of the lava lies only a few hundred yards south of the Boundary slash. The best exposures are on the ridge southwest of Tamihy creek. The formation may be here called the 'Chilliwack volcanics.' The total thickness can be only roughly estimated but it must be at least 3,000 feet.

The formation consists mainly of thick, massive flows, which are so welded into one another and so altered as to make the individual flows very hard to distinguish in the field. Among the flows a notable, though subordinate amount of ash-bed material is intercalated. At no point, however, were the conditions favourable for working out a detailed columnar section of the formation.

Although every effort was made to secure the freshest material for study it was found that all of the twenty type-specimens collected were greatly altered. The exact petrographic nature of the different flows is therefore obscure. The net result of the microscopic study of nearly all of the twenty specimens went to show that two rock-species are represented—augite andesite and hornblende andesite. The former is probably the more abundant.

The lavas are often amygdaloidal, with calcite generally filling the pores. Very often the andesites have been altered into typical greenstones, or, where the shearing has been particularly intense, into green schists. In a few specimens, the augite has the relations and abundance observed in olivine-free basalts. Olivine was not found in any thin section, but its absence may be due to the profound alteration of the lavas. Some of the specimens carry quartz in the groundmass but it may all be of secondary origin.

MICROCOPY RESOLUTION TEST CHART

(ANSI and ISO TEST CHART No. 2)

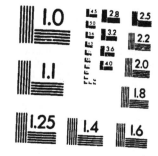

APPLIED IMAGE Inc

1653 East Main Street
Rochester, New York 14609 USA
(716) 482 - 0300 - Phone
(716) 288 - 5989 - Fax

2 GEORGE V., A. 1912

At Tamihy creek about five miles from the Chilliwack river, a 100-foot bluff of rhyolite was discovered. The density of the forest-cover in the vicinity rendered it impossible to determine the relation of this rock to the Paleozoic sediments or to the adjacent Chilliwack andesites. The writer conjectures that the rhyolite is a lava flow occurring at or near the base of the great andesitic series and the rhyolite is tentatively included in the Chilliwack Volcanic formation. The rhyolite seems to be about 100 feet thick or even more but it could not be followed far in any direction. No similar acid rock was found in the sections of the formation farther west.

The rhyolite is peculiar in being coloured almost black by an abundant material rather uniformly distributed through the ground-mass. This substance is quite opaque, dead-black and amorphous and has a dust-like appearance in the thin section. It is certainly not an iron-oxide. The rock decolourizes before the blow-pipe and it seems almost assured that the black dust is carbon. The occurrence of this element in a rhyolitic lava is unusual and the writer can find no record of its having been found in rhyolite or porphyry elsewhere, in anything like the abundance observed in this case at Tamihy creek.

VEDDER GREENSTONE.

The northwestern slope of Vedder mountain ridge is underlain by an altered, basic igneous rock which seems to be intrusive into the Paleozoic argillites and sandstones of the ridge. As exposed the body forms a remarkably long and straight band, running from the head of the Chilliwack river alluvial fan to the International line south of Sumas lake. The body was not followed farther to the southward. As shown on the map the known length of the mass is more than ten miles. On the northwest, for most of its length it is covered by alluvium, so that the exact shape and relations of the body cannot be determined.

At the point nearest to Sumas lake the igneous rock is bounded on the northwest by a narrow belt of dark-gray argillite, cropping out at intervals for about 700 yards along the wagon-road. Here the argillite seems to dip southeastward and thus under the intrusive rock, at an average angle of 65°, while at the southeastern contact on the summit of the ridge, the dip of the argillite is about 40° to the east-southeast. At this point, therefore, the intrusive appears to have the relation of a great sill, injected into a bedding-plane of the sediments. The width of the outcropping igneous mass is about 1,000 yards.

Elsewhere in the ridge the dips and strikes of the strata, always highly variable, show no such simple relation to the intrusive. The singular straightness of the southeastern contact suggests that the body is a gigantic dike, and this view is tentatively adopted. An intrusive character is inferred more from the petrographic nature of the mass and from its position in the sedimentary terrane than from the usual criteria of apophyses, inclusions, and contact-metamorphism. Owing to the dense brush and heavy mat of moss and humus, not a single, actual contact was discovered.

PLATE 46.

Summit of the Skagit Range. Looking southwest from a point about two miles north of the Boundary Line.

SESSIONAL PAPER No. 25a

From end to end of the body the igneous rock is profoundly altered; hence it is almost impossible to ascertain its precise original composition. It is now chiefly a mass of secondary minerals, including serpentine, talc, epidote, zoisite, kaolin, chlorite, and quartz. A pale green actinolitic amphibole never fails among the essential constituents; it also is probably secondary. A darker-tinted green hornblende is commonly present, and may represent a product of original crystallization from the magma. With this hornblende an original plagioclase, probably labradorite, is usually associated. The feldspar is always altered in high degree. A little magnetite, pyrite, and apatite, and much titanite are accessories.

The original rock was probably a basic diorite or a gabbro. It has been greatly sheared and mashed and has degenerated into several secondary types. The commonest of these is a massive greenstone bearing a fair amount of the skeletal plagioclase and dark green hornblende which are regarded as primary in origin. This rock is often intimately sheared and slickened, but is scarcely a true schist. Toward the International Boundary the mass becomes distinctly gneissic, with the field-habit of a medium- to coarse-grained hornblende-diorite gneiss; under the microscope, however, this type was seen to be a hornblende-zoisite-quartz schist, the amphibole being of actinolitic appearance. In certain zones of specially intense shearing the rock has been converted into a garnetiferous talc-quartz schist.

The amount of shearing and alteration undergone by this gabbroid intrusive is of the same order as that seen in the Chilliwack volcanics, which have been referred to the upper Carboniferous. It is possible that this great Vedder mountain ' dike' represents the intrusive phase of the same eruptions which gave rise to the surface flows of the Chilliwack formation. In any case the greenstone is certainly pre-Eocene and probably pre-Jurassic in age.*

CUSTER GRANITE-GNEISS.

On Custer ridge, which locally forms the main divide of the Skagit range, the Boundary belt crosses a considerable mass of crushed and now banded, intrusive granite. Its exposed area is known to be at least twenty square miles, but it may be found to be much greater as the body is followed northward and southward from the Boundary belt. The western limit of the banded granite so far as mapped is fixed by the intrusive contact of the younger Chilliwack granodiorite. The eastern limit is fixed in part by a band of the Hozomeen sedimentary series, into which the banded granite is intrusive; in other part, by the very thick blanket of Skagit volcanics, which are clearly younger than the gneissic granite. From its occurrence on Custer ridge this batholithic body may be called the Custer granite-gneiss.

_____

*During the preliminary examination of this district, in 1901, the relations of this schistose intrusive were not understood and the body was regarded as part of a basal crystalline series. In the Summary Report for 1901 (page 51) this series was given the provisional name ' Vedder Mountain gneisses.' The writer wishes to withdraw this name which should obviously not be perpetuated in the literature.

2 GEORGE V., A. 1912

If the Hozomeen series is of Carboniferous age, the Custer batholith must
have been intruded in late Carboniferous or post-Carboniferous time. The
general relations and metamorphosed condition of the batholith point to a pre-
Tertiary date of intrusion. The similarity in these respects to the Remmel and
Osoyoos batholiths has led the writer to place the date tentatively in the Jurassic,
thus making all three batholiths essentially contemporaneous. It is obvious,
however, that such correlations among the older batholiths must be held with a
very open mind, for they are founded largely on simple conjectures as to the
ages of the metamorphic rocks cut by these batholiths. Until fossils are actually
found in the Hozomeen series there is nothing to compel the view that the
Custer batholith is of late Palæozoic age; it may, indeed, be an uplifted fragment of an old pre-Cambrian terrane.

In the field the batholith has all the appearance of a typical pre-Cambrian
gneiss. It is seldom quite massive, and at no known point has it escaped more
or less powerful crushing and shearing. As a rule, the original rock has been
converted into a well-banded gneiss, very similar to that produced in the metamorphosed Cascade and Remmel batholiths farther east.

Original Rock Type.—The original rock seems to have been a granodiorite. Because of the intense metamorphism of the whole body, it is not
possible to distinguish the primary phases into which the batholithic magma
crystallized. Indeed, there are few places where the crushing and chemical
rearrangement of the mass were slight enough to leave remnants of the original
granite. One such locality was found at the head of Depot creek and about one
mile north of the Boundary line. The rock there is crushed and somewhat
gneissic, but it is not banded. It is of a darkish gray colour and of medium
grain. The hand-specimen shows the presence of much hornblende, less biotite,
and little quartz. From the persistently white to gray tint of the dominant
feldspar one would suspect, from the macroscopic appearance, that one were
dealing with a plagioclase-rock.

That conclusion is corroborated by the study of thin sections. The essential
minerals, named in the order of decreasing abundance, are: plagioclase, varying
from basic andesine, near $Ab_1 An_1$, to basic oligoclase near $Ab_2 An_1$; dark green
hornblende; orthoclase and microcline; quartz, and biotite. The usual accessories, magnetite, apatite, and titanite, are present. The essential minerals all
show straining. The plagioclase lamellæ are often bent or broken, and some of
the orthoclase has been converted, by pressure, into microcline. Nevertheless,
there is no doubt that the specimen just described represents a common phase,
and probably the dominant phase, of the original batholith. With this granodiorite type the materials making up the bands which form the staple rock of
the batholith at present, are in striking contrast.

Banded Structure.—As in the case of the Remmel, Osoyoos, and Cascade
batholiths, the bands are here often of stratiform regularity. They may be
divided into two classes: one acid-aplitic, the other basic in composition.

The acid bands are light-gray to whitish or very pale pink in colour. The
grain varies from rather fine to quite coarsely pegmatitic. In the latter case it

is sometimes not easy to distinguish such bands from the younger pegmatitic dikes (off-shoots from the Chilliwack batholith?) cutting the banded granite In most cases, however, the pegmatitic habit of the light bands is apparently due to some recrystallization of the original rock of the Custer batholith itself.

In thin section these light bands were seen to consist of dominant quartz, microcline, and orthoclase, with subordinate oligoclase (generally untwinned or poorly twinned), and biotite. A few, small, pink garnets, rare crystals of zircon and apatite, and small anhedra of titanite are accessory. Both hornblende and free iron oxide seem to be entirely absent. These bands have, thus, the composition of many acid, aplitic granites poor in biotite. The component minerals are generally strained and the cataclastic structure is usual. The specific gravities of two fresh specimens of the light bands were, respectively, 2.655 and 2.641.

The dark bands are of three kinds, according to the character and proportions of the constituents. The commonest kind is a dark greenish-gray, foliated, medium-grained, highly biotitic rock, composed of dominant plagioclase (basic andesine), biotite and quartz, with rare orthoclase. A few grains of garnet, some magnetite, and apatite are accessory. One specimen showed a specific gravity of 2.732, but many bands, yet richer in biotite, would be heavier. Only one thin section of this type—a biotite-diorite gneiss—was studied. It showed neither granulation nor pronounced straining of the component minerals, and it seems necessary to believe that the material of these dark bands was crystallized in its present form during the metamorphism of the batholith and has not since been subjected to extraordinary orogenic stress.

Dark bands of the second kind differ from those of the first in carrying essential hornblende as well as biotite in large amount. No special study has been made of these, but they doubtless have the same principal features as the biotite-diorite gneiss, excepting for the entrance of essential hornblende. Bands of this class have the composition of basic hornblende-biotite diorite gneiss.

Basic bands without essential biotite are uncommon but were noted at several points. In these green hornblende is the only important femic mineral. Basic labradorite is the only other essential constituent. Much apatite, very abundant, well crystallized titanite, and some pyrite are the observed accessories. The specific gravity of a somewhat altered specimen is 2.888. Bands of this third class seem to range in composition from amphibolite to hornblende-diorite gneiss. A few of them may possibly be sheared basic dikes cutting the batholith, but the majority, like the other dark bands, must be regarded as forming metamorphic phases of the sheared batholithic rock.

The Custer batholith thus includes the following species of rocks:—

Original type: granodiorite.
Secondary, metamorphic types:
  Biotite-aplite gneiss;
  Basic biotite-diorite gneiss;
  Basic hornblende-biotite-diorite gneiss;
  Basic hornblende-diorite gneiss;
  Amphibolite.

2 GEORGE V., A. 1912

The explanation of the banding is here the same as that offered for the banding of the Osoyoos, Remmel, and other batholiths farther east. The light bands represent the intensely granulated diorite from which the hornblende, biotite, basic plagioclase, and accessories have been slowly leached during the shearing of the batholith. The dark bands represent the shear-zones in which the same basic materials were recrystallized. In many cases there has also been some recrystallization of the light bands with the development of new quartz, biotite, feldspar, and some garnets.

SUMAS GRANITE AND DIORITE.

Rather more than one-half of Sumas mountain is composed of plutonic igneous rock. Nine square miles of the central and northeastern parts of the mountain are underlain by a biotite granite which may be called, for convenience, the Sumas granite. . An area of about three square miles is underlain by a plutonic breccia. This breccia consists of a vast multitude of blocks of a dioritic rock cemented by the Sumas granite; the whole forms a peripheral intrusion-breccia on a large scale. The diorite is evidently the older of these two rocks and may be called the Sumas diorite.

On the north and east the plutonic masses disappear beneath the Fraser valley alluvium. On the northwest the granite makes contact with a hard, massive quartzite, into which it is intrusive. On the southwest the granite is unconformably overlain by the nearly flat Eocene (?) beds.

Granite.—The granite is pre-Eocene in age. The date of intrusion cannot yet be more closely fixed with certainty. The rock is nowhere eroded in any notable way. It seems therefore doubtful that it was intruded before the great orogenic revolution of the Jurassic and the date may be tentatively fixed as later Jurassic, or (less probably) Cretaceous. The diorite of the intrusion-breccia is also massive and un-sheared and may belong to the same general period of igneous action, though of course, being older than the granite.

The granite is a light pinkish-gray, fine, to medium   rock, poor in dark constituents. The composition and structure are   aracteristic of mica granites. Quartz, orthoclase (sometimes slightly   itic), basic andesine, averaging $Ab_5 An_3$, and biotite are the essential   green hornblende, magnetite, apatite, titanite, and rare zircons ar   y. In all of the four specimens collected, the rock is seen to be considerably altered. The alteration is so marked even on well glaciated ledges that one may possibly refer it in largest part to the secular weathering which preceded the deposition of the Eocene beds. The feldspar is often much kaolinized and the biotite is generally chloritized to some extent.

The freshest specimen (No. 201) has been analyzed by Professor Dittrich with the following result:—

Analysis of Sumas granite.

		Mol.
SiO_2	71·24	1·187
TiO_2	·42	·005
Al_2O_3	14·11	·138
Fe_2O_3	1·75	·011
FeO	1·24	·017
MnO	tr.	....
MgO	1·07	·027
CaO	2·87	·051
BaO	·09	·001
Na_2O	2·37	·038
K_2O	3·97	·042
H_2O at 110° C.	·11	....
H_2O above 110° C.	·59	....
P_2O_5	·17	·001
CO_2	·28	
	100·27	
Sp. gr.	2·651	

The calculated norm is:—

Quartz	34·86
Orthoclase	23·35
Albite	19·91
Anorthite	13·62
Corundum	·92
Hypersthene	2·83
Magnetite	2·55
Ilmenite	·76
Apatite	·31
Water and CO_2	·98
	100·09

In the Norm classification the rock enters the sodipotassic subrang, riesenose, of the alkalicalcic rang, riesenase, in the persalane order, columbare, but it is very close to amiatose, the corresponding subrang of the order britannare.

In the older classification it is obviously a common type of biotite granite. The specific gravities of two of the freshest specimens are 2·651 and 2·653. On account of the alteration of the rock a useful determination of the actual mineralogical composition by the Rosiwal method is practically impossible. The mode is in this case not very different from the norm.

Diorite.—The dioritic rock of the intrusion-breccia is also considerably altered, as if by weathering. Its essential minerals are green hornblende and plagioclase, averaging basic andesine. Orthoclase is an abundant accessory and some interstitial quartz is always present. Much epidote with some chlorite and kaolin are the secondary minerals.

The smaller diorite xenoliths in the granite have been more or less completely recrystallized and modified in composition by the granitic magma. The hornblende there characteristically occurs in long idiomorphic blades shot through the feldspar and other constituents. In one thin section of a xenolith, potash feldspar and quartz are so abundant as to cause the rock to simulate a granodiorite; in that case it seems probable that the original diorite has been

2 GEORGE V., A. 1912

affected in its composition by the introduction of material from the granite. The metamorphic effects are analogous to those observed about the xenoliths in the Moyie sills of the Purcell range.

In other xenoliths which show in their rounded outlines the corrosive effects of the acid magma, a large number of peculiar round bodies have been developed. These are of the size and shape of a small pea and, because of their special hardness, they project above the general weathered surface of the xenolith.

Under the microscope each of these small bodies is seen to be composed of pure quartz, generally as a single crystal, but sometimes in the form of a coarse-grained aggregate. The quartz nodules are perfectly clear and bear no inclusions of the dioritic material. Between the diorite matrix and the quartz there is usually a narrow aureole of idiomorphic orthoclase and plagioclase crystals. These project into the quartz much as similar crystals project into vugs and miaroles of other rocks.

The origin of these silicious nodules is not clear. They can hardly be regarded as filled amygdules of the ordinary type, but seem rather to represent phenocrystic growths in the xenolith after the latter had been softened by the granite magma and been impregnated with silicious material from that source.

SKAGIT VOLCANIC FORMATION.

From the first summit west of the Skagit river to the summit of Custer ridge (the main divide of the Skagit range), the Boundary line crosses a very thick group of volcanic rocks which may be called the Skagit volcanic formation. These rocks extend over at least twenty square miles in the belt and continue unknown distances in the mountains to north and to south.

The volcanic rocks characteristically weather into jagged peaks, knife-edge ridges, and forbidding precipices, forming the highest and most rugged mountains in this part of the Skagit range. (Plate 44, B.) Glacier Peak and its neighbours are, indeed; among the most inaccessible summits in the whole Boundary belt west of the Flathead river. Small but numerous glaciers and a succession of impassable breaks in the ridges render the study of the volcanic formation difficult even where outcrops are plentiful. Below tree-line it has so far proved quite impossible to find a sufficient number of actual contacts or to work out the succession of the many members of the group. The results of the exploration are, therefore, far from being satisfactory. It is known that the formation is exceedingly thick—apparently at least 5,000 feet thick at the Boundary line— but the writer has been baffled in the attempt to construct a detailed and final columnar section. The great thickness of the volcanic accumulation and the abundance and coarseness of the agglomerates suggest that the major eruptions actually took place in the area of the Boundary belt. It is quite possible that a vast cone of Mount Baker or Mount Rainier proportions was situated over the present site of Glacier Peak.

The lower and greater part of the formation, probably 4,000 feet or more in thickness, is composed of massive breccias and ash-beds, with one layer of coarse conglomerate and with many interbedded flows of compact and vesicular lava.

SESSIONAL PAPER No. 25a

All of the purely volcanic constituents are andesitic. Conformably overlying these rocks and underlain by other andesitic flows and breccias, comes a widely extended layer of white to pale-gray trachytic or rhyolitic tuff, aggregating perhaps 200 feet in thickness. The top of the whole group is not exposed in the Boundary belt and the series remains incomplete.

The following table gives an extremely crude idea of the general relations and thicknesses as estimated in the field:—

Top, erosion-surface.

1,000± feet.—	Andesite flows and breccias.
200± "	Liparitic (?) tuff.
900± "	Andesite flows, ash and breccias.
100 "	Conglomerate.
3,000± "	Andesite breccias, flows and ash-beds.

5,200± feet.

Base, unconformable contact with Custer batholith and Hozomeen sediments.

The andesitic members are always very massive. It is seldom possible to distinguish the contacts between different flows, and even the contacts between flow and breccia are generally obscure. The individual flows seem to be usually very thick; cliffy slopes as much as 300 or 400 feet high do not disclose undoubted breaks in the massive lava.

The more basic material of the breccias, ash-beds, and flows has great uniformity of composition. Nine typical specimens were collected in different parts of the area and at various horizons from near the base upward. Thin sections of all the specimens were studied. Though not crushed they are all more or less altered. Without exception, the flows and lava-fragments of the agglomerates seem to belong to the one common type, augite andesite. The phenocrysts are regularly labradorite, averaging Ab_2An_3, and augite. The latter is generally uralitized pretty thoroughly. Neither primary hornblende nor olivine was detected, though in some cases the former may have accompanied the augite as a subordinate phenocryst. The ground-mass is more altered than the phenocrysts and is a mass of chlorite, uralite, plagioclase microlites, and indeterminable material, perhaps derived from glass, which was apparently a very abundant staple constituent of the ground-mass.

The beds of agglomerate are usually thick, individual ones measuring more than 200 feet in thickness. The blocks are of all sizes up to those a foot more in diameter. At many points angular fragments of cherty quartzite a slate, identical in look with the dominant rocks of the Hozomeen series were found in the breccia. At one section in the deep valley northwest of Monument 68, and about 1,200 yards from the monument, there occurs a layer of r wholly or almost wholly made up of fragments of cherty quartzite and s tine; this bed is at least 75 feet thick. The fragments are angular, ran in size from sand-grains to blocks six inches or more in diameter. There be little doubt that these fragments were derived from the Hozomeen seri. The bed shows no sign of water-action; from its position in the midst manifest volcanic agglomerates, it may best be regarded as a special product

2 GEORGE V., A. 1912

of gas-explosion which operated in this vicinity and blew out a large quantity of the foundation rock. The matrix of this bed was not examined microscopically; it may be a fine andesitic ash.

Sometimes, though rarely, granitic and gneissic blocks appear in the staple breccias; most of those observed seem to have been derived from the underlying Custer batholith.

About one mile northwest of Monument 68 a bed of coarse conglomerate, 100 feet or more in thickness, interrupts the succession of breccias and flows. The pebbles are very well rounded and were unquestionably long rolled by waves or currents. They vary in size but few are over a foot in diameter. They consist of altered andesite (dominant kind), quartzite, chert, slate, and, rarely weathered granite. The matrix is sandy. The bed dips about 16° to the eastward and seems to be quite conformable to the yet more massive volcanic members above and below.

Above the conglomerate the tuffaceous rocks carry several thin conformable lenses of gray argillite, which also appear to have been laid down under water.

The acid tuff was seen at two localities. It crops out on the summit of the rugged ridge 1·5 miles south of Monument 68 and on a much greater scale upon the long ridge running eastward from Monument 69. This tuff covers the latter ridge for one mile of its length and from its white colour is very conspicuous in the landscape. The tuff is extremely jointed, so that it is difficult to secure a hand-specimen of standard size. Some of the rock is vesicular and it is possible that thin flows are represented in the middle part of the 200-foot band. The whole composite mass overlies the andesites, dipping at angles of from 10° to 15° to the north. On the higher ridge on the west the acid tuff seems to be overlain by younger andesites, roughly estimated to be 1,000 feet thick.

The acid tuff is nearly pure white to pale-gray when fresh, weathering white to pale-yellow. Macroscopically it is quite aphanitic for the most part, with only the rarest suggestion of a small feldspar phenocryst. The rock reminds one of porcelain viewed on a broken edge. Under the microscope the phenocrysts of the angular fragments are seen to be few in number and to have the properties of sanidine or orthoclase. The ground-mass is a cryptocrystalline aggregate of quartz and feldspar, with the character of a devitrified obsidian. The matrix of the tuff is optically like the ground-mass of the fragments. The mass has clearly the composition of an acid obsidian and is perhaps nearer trachyte than rhyolite.

The age of the formation has not been determined by direct fossil evidence. The lava-flows, ash-beds, breccias, and interbedded conglomerates are not crushed. The dips are generally low, running from 5° to 30° as the observed maximum. The breccias and conglomeratic beds contain many fragments and pebbles of quartzite, slate, and granite which were without much doubt derived from the eroded Hozomeen series and the Custer gneissic batholith. It seems reasonably certain, therefore, that the vulcanism dates from a period much later than the intrusion of the batholith and, à fortiori, than the folding of the Hozomeen series of sediments. If the latter are of Carboniferous age and the granite is Jurassic, the Skagit volcanics rest upon a late Jurassic or post-Jurassic erosion surface. The relation is somewhat similar to that between the volcanic breccia at the base of the Pasayten formation and the underlying, probably Jurassic Remmel batholith.

There is something, therefore, to be said for the hypothesis that the Skagit volcanics are of Lower Cretaceous age and contemporaneous with the Pasayten volcanics. If, however, the Custer batholith was intruded in the late Jurassic and sheared and metamorphosed during the orogenic revolution at the close of the Laramie period, it would seem certain that the Skagit volcanics are volcanic in the Tertiary. This follows from the fact that the volcanic rock are comparatively little disturbed and are nowhere sheared as anything like the measure shown in the Custer gneissic batholith. It would not be improbable that the basement could be so profoundly affected while the rocks overlying it escaped the deformation. That the Skagit volcanic formation is not younger than the Miocene is probably indicated by the fact that it is cut by a stock of quartz-bearing monzonite, which shows evidence of being essentially contemporaneous with the Castle Peak stock (late Miocene). As present the data on the age of the volcanics cannot be made any closer with definiteness. In the circumstances it is probable the writer will postulate an Oligocene date for them, thus equating the Skagit andesite with the proved Oligocene andesite in the Midway district. The Skagit andesite may, on the other hand, be Eocene or, possibly, Cretaceous.

SKAGIT HARZBURGITE.

On the ridge 2,500 yards north northwest of Monument 67, at the 6,000 foot contour, the Custer gneiss is cut by a large pod-like intrusion of coarse peridotite. This mass is 150 feet or more in width and can be followed along its longer north-south axis about 900 feet. It appears to taper off toward each end. It is probably an irregular dike injected into a schistosity plane of the gneiss. From wall to wall the peridotite is very coarse, showing olivines often reaching 2 cm. in diameter and an abundant pyroxene of similar dimensions. At the ledge the rock is seen to be somewhat altered, but it shows no sign of crushing.

The general colour of the rock is a deep, almost blackish green. Feldspar is entirely lacking. In thin section the composition and structure are seen to be that of a typical, partly altered harzburgite. The only primary minerals are olivine and enstatite, both colourless in thin section. About fifty per cent of the rock is made up of secondary minerals, including serpentine, tremolite, iddingsite, talc, chlorite, much sulphide (probably pyrite), and considerable limonite. Minute inclusions of picotite or chromite could be discerned in the olivine. The iddingsite noted has most of the features described by Lawson for the type material at Carmelo bay, but the optical angle is very small, 2V being well under 5°. The specific gravity of the rock described is 3.083.

The date of this intrusion is apparent only in relation to the period when the Custer batholith was sheared; the shearing seems to have been completed

2 GEORGE V., A. 1912

before the peridotite was injected. The proximity of the Skagit volcanics overlying the gneiss leads one to suspect that these basic rocks belong to the same eruptive period, and that the harzburgite and andesites are genetically connected. The relation is conceivably the same as that connecting the peridotites of the Columbia range and Midway mountains with the basalts and andesites of those regions.

SLESSE DIORITE.

The walls of Middle creek canyon in its lower part are composed of diorite, which extends over the divide past Slesse mountain to Slesse creek. The diorite forms a stock-like mass covering about nine square miles on the Canadian side of the line; it was not mapped to the southward, but it is known to extend several miles into Washington. The diorite body once undoubtedly stretched farther eastward, but it has there been replaced by the younger Chilliwack batholith of granodiorite.

The diorite is very clearly intrusive into the slates on the west and north. These argillites are highly altered, but, as they enclose lenses of crinoidal carboniferous limestone, it seems most probable that the date of intrusion is post-Carboniferous.

The diorite is not crushed or greatly strained except in the immediate vicinity of the great Chilliwack batholith, where such effects might naturally be expected. Elsewhere there are no evidences that the diorite has undergone the severe pressures involved in the post-Cretaceous mountain-building of the Cascade range; it is therefore probable, though not proved, that the diorite was intruded in post-Laramie time.

The contacts of the body are so imperfectly exposed in this densely forested area that its structural relations have not been fully worked out. The diorite certainly cross-cuts the sedimentaries and has metamorphosed them in the thorough way characteristic of most stocks. The intensity of the metamorphism is of a higher order than that usually observed about a laccolith or chonolith, and it seems safer to regard the mass as a true stock or batholith, that is, a subjacent, downwardly enlarging body.

The diorite is in places richly charged with large, slab-like inclusions of crumpled black slate; these often attain lengths of 50 to 100 feet or more. A large number of them, forming a veritable breccia on a great scale, may be seen on both slopes of Middle creek canyon, especially at points about four miles from the confluence of the creek with the Chilliwack river.

Petrography.—The diorite is a dark brownish to greenish gray, fresh rock of normal habit. It appears to have a rather uniform chemical composition. The chief variations are those of grain. At its own intrusive contacts the stock is fine-grained as if by chilling; elsewhere the grain is generally of medium size. Where the diorite contacts with the younger granodiorite the grain is still medium, but the more basic rock has been metamorphosed along a narrow zone. Basic segregations were not observed in the diorite.

The list of essential minerals in the diorite includes acid labradorite, near Ab_1An_1, hornblende, and biotite, named in the order of decreasing abundance.

The hornblende sometimes, though rarely, encloses small cores of nearly colourless augite; the latter mineral also occurs in small independent anhedra, but is clearly only an accessory constituent. The other accessories are quartz, magnetite, pyrite, apatite, and titanite. The structure is the usual hypidiomorphic-granular. The order of crystallization is not very clear, but much of the plagioclase seems to antedate the biotite and hornblende.

A type specimen (No. 54), with the mineralogical composition just described, was collected on Middle creek in the heart of the main mass. It has been analyzed by Professor Dittrich, with the following result:

Analysis of Slesse diorite.

		Mol.
SiO_2	56·90	·918
TiO_2	·84	·010
Al_2O_3	18·17	·178
Fe_2O_3	1·23	·008
FeO	5·88	·082
MnO	·21	·003
MgO	4·36	·109
CaO	6·51	·116
SrO	·18	·002
Na_2O	3·23	·052
K_2O	1·57	·017
H_2O at 110°C	·12	....
H_2O above 110°C	·77	....
P_2O_5	·10	·001
CO_2	·08	....
	100·15	
Sp. gr.	2·793	

The calculated norm is:—

Quartz	8·04
Orthoclase	9·45
Albite	27·25
Anorthite	30·30
Hypersthene	19·08
Diopside	1·36
Magnetite	1·85
Ilmenite	1·52
Apatite	·31
Water and CO_2	·98
	100·14

The mode (Rosiwal method) is approximately:—

Quartz	9·5
Labradorite	58·0
Hornblende	12·8
Biotite	12·5
Augite	4·3
Pyrite	1·5
Magnetite	·8
Apatite	·5
Titanite and zircon	·1
	100·0

2 GEORGE V., A. 1912

In the Norm classification the rock enters the dosodic subrang, andose, of the alkalicalcic rang, andase, in the dosalane order, germanare; but is near the corresponding subrang of the docalcic rang, hessase.

In the older classification the rock is a hornblende-biotite diorite. Mineralogically and chemically it is almost identical with a California diorite described by Turner.* The specific gravities of six fresh specimens vary from 2.786 to 2.863, averaging 2.806.

The apophyses of the body are chemically similar but have the structure of hornblende-biotite diorite porphyrite.

Contact Metamorphism.—The Paleozoic sedimentaries cut by the diorite have been decidedly metamorphosed. The effects were noticeable at all of the observed contacts, but were specially studied on Pierce mountain which forms part of the rugged divide running southward between Slesse and Middle creeks, and again along the contacts on Middle creek. The belt of altered rock seems to average at least 600 feet and may be 1,000 feet or more. (Plate 44, C).

Mineralogically the metamorphism shows nothing very unusual. The sandstones have been converted into tough, vitreous quartzites. Some of the argillites have been changed into dark greenish-gray hornfelses or schists, richly charged with metamorphic biotite and sericite. Other argillaceous beds have been recrystallized, with the generation of abundant cordierite, that mineral forming, as normally, large, interlocking individuals which are filled with inclusions of quartz, biotite and magnetite. A few thin lenses of pale green, felted tremolite and of more granular tremolite and epidote probably represent completely altered beds of limestone; other limestone bands have been changed to white marble. With the limestones much chalcedonic silica is often associated.

The contact-belt is often traversed by small quartz-veins, some of which form fairly high grade, free-milling gold ore. The Pierce claim on Pierce mountain is located on one of these veins, close to the main contact of the diorite. Like all the others seen in the vicinity this vein is quite variable in width, pinching out irregularly from its maximum width of a few feet. At the time of the writer's visit to the claim, in 1901, not enough development work had been done to show the amount or average value of the gold-bearing quartz. A similar, though narrower vein, nine to twenty-one inches wide, cuts the diorite at a point about 700 feet above Middle creek and 3,000 feet or more below the main claim on Pierce mountain; the two veins may be connected, and both were being opened up by Mr. Pierce in 1901. From the writer's experience the veins occurring along this contact must be very high grade if they are to pay for their development; they are much too small and irregular to give hope of profitable low-grade ore.

CHILLIWACK GRANODIORITE BATHOLITH.

Some of the wildest and most rugged mountains in the Skagit range are composed of a massive granodiorite which forms the largest intrusive area in

---

*See Bulletin 228, U. S. Geol. Survey, 1904, p. 234.

PLATE 4.

Typical view of granitic mountains (Chilliwack batholith), near summit of Skagit Range; looking southeast from summit east of Chilliwack Lake.

the Boundary belt west of the Remmel batholith. The basin of Chilliwack lake has been excavated in this rock, for which the name, Chilliwack granodiorite, has been selected. The body has the size and field-relations of a typical batholith. (Plates 44, D, 47, and 62, A).

On the Canadian side of the Boundary line the granodiorite underlies at least 100 square miles of mountains. The formation stretches an unknown distance to the northward of the Boundary belt and also continues a few miles on the United States side.

A couple of miles north of the Boundary line, and a like distance west of the Skagit river, a small granitic stock, of composition probably similar to the more salic phase of the Chilliwack batholith, cuts the Hozomeen series. Owing to bad weather and to other causes, the writer was unable to examine this western slope of the Skagit valley. At his request, Mr. Charles Camsell, of the Dominion Geological Survey, mapped the formations on this slope, and special thanks are due him for this service. He discovered the small stock and has referred its date of intrusion to the Tertiary. As yet the rock has not been studied with the microscope.

The date of the intrusion of the main batholith can be fixed within certain limits. The granodiorite clearly cuts the greatly deformed sediments on the long ridge northwest of the lake. In that region the strata are unfossiliferous but appear to belong to the same group as the definitely Carboniferous beds west of Middle creek canyon. The granodiorite cuts the Slesse diorite, forming a wide belt of intrusion-breccia with the latter where the main contact crosses Middle creek. The diorite just as clearly cuts fossiliferous Carboniferous slates and limestones. It follows that the granodiorite is of post-Carboniferous date. At no point does it show evidence of crushing or of pronounced straining; as in the case of the older diorite, there can be little doubt as to the relatively late date of the intrusion. In field-habit, as in many essential microscopic details, this granodiorite is like that of the post-Cretaceous Castle Peak stock. There are, thus, some grounds for the belief that the Chilliwack granodiorite was, like the granodiorite at Snoqualmie Pass to the southward,[*] intruded at a date as recent as the Miocene.

In the field the batholith preserves great uniformity of colour, grain, and massiveness. It was only after microscopic examination that its actual variation in composition became apparent. Three main phases were recognized from the thin sections.

Petrography.—The most basic phase of the three is a quartz diorite rather than a true granodiorite. It occurs along contacts and also at points two or more miles from any visible contact; so that it is apparently not the product of simple contact-basification. A type-specimen was collected at the Boundary line in the lower of the two cirques occurring in the mountains southwest of the upper end of Chilliwack lake. This point is at least two miles from any lateral contact and probably at least a mile from the original roof-contact. The rock is exposed on a great scale on the steep, 4,000-foot slope to the westward of the

---

[*] See page 469.

2 GEORGE V., A. 1912

lake and upper river. The walls of the tandem-cirques seem to be composed throughout of the quartz-diorite phase.

The typical specimen is a light-gray, medium to moderately coarse-grained, granitic rock poor in quartz and speckled with abundant, brilliant prisms of hornblende and black foils of biotite. The microscope shows that the dominant constituent is an unzoned plagioclase, averaging labradorite, $Ab_1 An_1$. Orthoclase appears to be entirely lacking. The amphibole is a common hornblende with an extinction of about $17°$ on (010) and colour scheme as follows:—

     a = pale yellowish green with olive tinge.
     b = strong olive-green.
     c = olive-green.

       b > c > a.

The mica is a common brown biotite with the usual strong absorption. Quartz is interstitial and in relatively small amount. Magnetite, apatite, and rare zircon crystals are the accessories.

The order of crystallization is: the accessories; then plagioclase, followed in order by biotite, hornblende, and quartz. The structure is the eugranitic.

Professor Dittrich has analyzed this phase (specimen No. 7), with the following result:—

Analysis of quartz diorite, Chilliwack batholith (phase 1).

		Mol.
SiO_2	60·36	1·006
TiO_2	·70	·009
Al_2O_3	17·23	·168
Fe_2O_3	1·93	·012
FeO	3·74	·052
MnO	·14	·002
MgO	3·66	·093
CaO	6·07	·109
Na_2O	3·58	·058
K_2O	1·74	·018
H_2O at 110°C	·06	....
H_2O above 110°C	·55	....
F_2O_3	·11	·001
CO_2	·08	....
	99·95	
Sp. gr.	2·757	

The calculated norm is:—

Quartz	13·56
Orthoclase	10·01
Albite	30·39
Anorthite	25·85
Hypersthene	12·16
Diopside	2·90
Magnetite	2·78
Ilmenite	1·37
Apatite	·31
Water and CO_2	·69
	100·02

SESSIONAL PAPER No. 25a

The mode (Rosiwal method) is approximately:—

Quartz	19·1
Orthoclase	2·0
Labradorite	55·7
Hornblende	11·1
Biotite	10·3
Magnetite	1·3
Apatite and zircon	·5
	100·0

In the Norm classification the rock enters the dosodic subrang, tonalose, of the alkalicalcic rang, tonalase, of the dosalane order, austrare. In the older classification it is a typical quartz-biotite-hornblende diorite.

A rock which appears to be a second phase of the batholith, was collected at the western wall of the canyon of Sriver creek where it debouches on the valley-flat one mile north-northwest of the lower end of Chilliwack lake. This type is a fresh, light pinkish-gray granite with abundant quartz and biotite, but with no hornblende. Orthoclase is an essential plainly visible as such to the naked eye. The dominant feldspar is again plagioclase. It is often zoned, the outer rims reaching the acidity of the mixture, $Ab_? An_?$. The average mixture seems to be near $Ab_? An_?$. The orthoclase is often somewhat microperthitic. Magnetite, apatite, zircon, and a little titanite are the accessories.

Professor Dittrich's analysis of this phase (specimen No. 30) resulted as follows:—

Analysis of soda granite, Chilliwack batholith (phase 2).

		Mol.
SiO_2	71·41	1·190
TiO_2	·34	·004
Al_2O_3	14·38	·141
Fe_2O_3	1·33	·008
FeO	1·17	·016
MnO	·04	....
MgO	1·13	·028
CaO	2·51	·045
BaO	·03	....
Na_2O	4·12	·066
K_2O	2·97	·032
H_2O at 110°C.	·09	....
H_2O above 110°C.	·30	....
P_2O_5	·13	·001
CO_2	·12	....
	100·07	
Sp. gr.	2·653	

2 GEORGE V., A. 1912

The calculated norm is:—

Quartz	29.16
Orthoclase	17.79
Albite	34.58
Anorthite	11.68
Corundum	.10
Hypersthene	3.33
Magnetite	1.86
Ilmenite	.61
A atite	.31
Water and CO_2	.51
	99.93

The mode (Rosiwal method) is approximately:—

Quartz	34.1
Orthoclase and microperthite	25.7
Oligoclase	30.2
Biotite	8.1
Magnetite	1.1
Titanite	.4
Apatite	.3
Zircon	.1
	100.0

In the Norm classification the rock enters the dosodic subrang, lassenose, of the domalkalic rang, toscanase, in the persalane order, britannare. In the older classification it is a biotite granite with dominant oligoclase—a soda granite.

So far as observed, this rock occurs only on the north side of Chilliwack valley and north of the lake. It may conceivably represent a distinct intrusive body, bearing the same relation to the hornblende-labradorite phase of the main batholith as the Cathedral granite of the Okanagan range bears to the Similkameen granodiorite. Yet no sharp contact between the granite and diorite phases was found, and the writer has concluded that both probably belong to the one batholith. It is of some interest to note that the arithmetical mean of the two analyses is almost the exact equivalent of the analysis of the average granodiorite in the Cordillera. The latter average appears in column 4 of the following Table XXXIII, and represents nine analyses from California types, one Oregon type, and two Washington types, all of these being taken from Bulletin No. 228 of the United States Geological Survey.

Table XXXIII.—*Analyses of granodiorites.*

	1	2	3	4
	Phase 1 of Chilli-wack batholith.	Phase 2 of Chilli-wack batholith	Average of Two Phases.	Average of Twelve Types.
SiO_2	60·36	71·11	65·88	65·10
TiO_2	·70	·34	·52	·54
Al_2O_3	17·23	14·38	15·80	15·82
Fe_2O_3	1·03	1·33	1·63	1·64
FeO	3·74	1·17	2·46	2·66
MnO	·14	·04	·09	·05
MgO	3·66	1·13	2·40	2·47
CaO	6·07	2·51	4·29	4·03
Na_2O	3·58	4·12	3·85	3·82
K_2O	1·74	2·97	2·35	2·29
H_2O, at 110°C	·06	·09	·08	·16
H_2O, above 110°C	·55	·30	·43	·9?
P_2O_5	·11	·13	·12	·16
CO_2	·0?	·12	·10	····
	99·95	100·04	100·00	100·00

The third major phase of the batholith is probably the most abundant of the three. Macroscopically it is almost indistinguishable from the phase first described. The microscope shows, however, that orthoclase is here an essential constituent. In the order of decreasing abundance the essentials are plagioclase, near Ab_4An_3; quartz; orthoclase; hornblende; biotite. The two femic minerals are in about equal amount. In optical properties all these minerals are identical with those of the first phase. The accessories are also the same as there but a few grains of allanite are associated.

This phase occurs at many points in the batholith. The type specimen was collected at the mouth of Depot creek on the east side of Chilliwack lake, and thus in the heart of the batholith. The specific gravity of this specimen is 2·678. It is a normal granodiorite. It has not been analyzed, but its analysis would probably be close to the mean of the two analyses of the other phases (column 3).

We may conclude that the average rock of this batholith is a true granodiorite tending towards the composition of a quartz diorite.

The specific gravities of six fresh, type specimens from the batholith vary from 2·626 to 2·757, averaging 2·693.

Nodular basic inclusions occur at various points in the mass. They are seldom very numerous and, so far as observed, never of large size; diameters exceeding 10 cm. are very uncommon. All of these dark-coloured nodules are probably indigenous bodies. They are of two kinds, both of which occur in the staple granodiorite phase.

The one kind has some similarity to the Slesse diorite. It is a rather dark greenish-gray, fine-grained rock, composed of labradorite, green hornblende, and less important biotite as the chief essentials, with magnetite, apatite, titanite, and zircon, a little quartz, and a very little orthoclase as accessories. The struc-

2 GEORGE V., A. 1912

ture is peculiar in being remarkably poikilitic. The larger individuals of each essential mineral contain smaller individuals of each of the other essentials as well as crystals of the accessories except quartz and orthoclase. Those two minerals are as usual the youngest of all. A few of the hornblende crystals contain small cores of colourless augite. The specific gravity of a typical nodule of this kind is 2.791, which is near the average for the Slesse diorite (2.806). The nodule evidently has the composition of a basic hornblende-biotite diorite. Its special structure could be explained on the hypothesis that it is simply an inclusion of the Slesse diorite which has been heated and largely recrystallized in the younger granodiorite magma. Since, however, these nodules occur in parts of the batholith far removed from the diorite contact, and since they show perfect interlocking with their host, it seems at least equally probable that they are true basic segregations. If this second hypothesis could be proved we should have one more illustration of the obvious consanguinity of the two batholiths.

The other kind of inclusion is of a much darker green-gray colour and is also fine-grained. The essential components are a nearly colourless diopsidic augite (very abundant), pale green hornblende, and labradorite, Ab_1An_1. The accessories are magnetite, apatite, titanite, a very little biotite and, quartz, and, possibly, a little orthoclase. The structure is the hypidiomorphic-granular, but at various places in the thin section suggests the diabasic structure. The specific gravity of a typical nodule of this class is 2.908. It has the composition of a gabbro or of a basic augite-hornblende diorite.

Contact Metamorphism.—The thermal metamorphism of the Carboniferous sediments on the divide between Slesse and Middle creeks is intense, and is essentially like that noted as due to the intrusion of the Slesse diorite. A new type of metamorphic product was found on the ridge north of Chilliwack river, about four miles from the lake. This is a hornfels richly charged with phenocryst-like prisms of andalusite, which are shot through a mat of green mica and quartz—a rock clearly derived from a siliceous argillite.

At the main contact of the granodiorite, on the ridge north of Depot creek, a small patch of intensely metamorphosed limestone is cut by basic diorite dikes, by the Custer gneiss-granite, as well as by the Chilliwack granodiorite. It is probable that all three kinds of intrusive rock, especially the more acid ones, have produced the observed recrystallization of the limestone. That rock has the appearance of a typical pre-Cambrian crystalline limestone of Ontario or Quebec. It is a white coarse-grained mass of calcite, bearing numerous scales of graphite, epidote, and zoisite in rounded grains, cubes of pyrite and anhedra of grossularite.

INTRUSIVES CUTTING THE VOLCANICS.

Besides the occasional andesitic and basaltic dikes which have evidently originated in the same magma as the surface lavas, the Skagit formation is cut by a small stock and by several wide dikes of quite different materials.* The

---

* One highly vesicular, basaltic dike cutting the intercalated conglomerate may be of distinctly later date than the Skagit volcanic formation.

stock and most of these dikes have the composition and structure of a quartz-bearing monzonite verging on granodiorite. One great dike has the properties of a typical hornblende-diorite porphyry.

Monzonite Stock.—The stock is intrusive into the volcanics at their fault-contact with the Hozomeen quartzites a short distance north of Monument 69. This stock as exposed has an elliptical ground-plan, measuring about 1,200 yards in its greatest diameter and 800 yards along the minor axis. Like the agglomerate it is devoid of any crush-schistosity and the intrusion appears to have occurred later than the post-Laramie epoch of intense crushing. The intrusion may have been genetically connected with the faulting by which the volcanics were dropped down into their present lateral contact with the old quartzites.

The material of the stock seems to be rather uniform, with the habit of a fresh, light gray, medium-grained syenite. The essential constituents are, in the order of decreasing importance: plagioclase, orthoclase, hornblende, quartz, biotite, augite. The plagioclase is often zoned, with Ab_1An_1 in the cores and oligoclase in the outer shells; the average mixture is an andesine near Ab_1An_1. The hornblende is green in about the same tones as those of the amphibole in the Chilliwack granodiorite. The characters of the other essential minerals and of the accessories are also the same as in that batholith. The structure is the eugranitic.

The rock clearly belongs among the quartz-bearing monzonites and chemically would show the composition also allied to that of a basic granodiorite. The Chilliwack batholith is only five miles distant and it is highly probable that this monzonite stock is its satellite.

Dikes.—On the rugged, glacier-covered ridge south of Monument 68, the Skagit volcanics are cut by two or more great, north-and-south dikes of monzonite, similar to the staple material of the stock but relatively richer in plagioclase and quartz and poorer in biotite. These dikes, which range from 100 to 300 feet or more in width, are doubtless giant apophyses from the magma-chamber of which the Chilliwack batholith was a part.

A half-mile west of these dikes and running nearly parallel to them is a third dike over 100 feet in width. It is composed of a dark gray to greenish-gray, medium to fine-grained, somewhat porphyritic rock of different habit from the monzonite. The phenocrysts are green hornblende which is often in parallel intergrowth with augite; and basic labradorite. The ground-mass is a hypidiomorphic-granular aggregate of labradorite and interstitial quartz. Magnetite, apatite, and titanite are the accessories. Orthoclase seems to be absent. The rock is somewhat altered and is charged with a considerable amount of uralite evidently derived from augite. A small amount of chlorite may represent original biotite, but none of this mineral was discovered in the thin section. The rock is to be classed as a hornblende-diorite porphyrite.

The dike is uncrushed. It has the habit and nearly the composition of the finer-grained phases of the Slesse diorite. The similarity is so great that one

2 GEORGE V., A. 1912

may believe that the porphyrite is an off-shoot of the same magma as the diorite. That relation would be parallel to the one just postulated for the neighbouring monzonite dikes and the Chilliwack granodiorite. In fact, it seems simplest to suppose, first, that all four rock-types belong to one eruptive period, the more basic intrusions antedating the acid intrusions by only a short interval of time; and, secondly, that all four rocks were differentiated from one great magma-chamber.

DIKES CUTTING THE CHILLIWACK BATHOLITH.

Two different kinds of acid dikes cut the Chilliwack granodiorite. One of these kinds seems to be merely a later expression of the same magma from which most of the batholith itself was crystallized. Such dikes are not common and were never found far from the main batholithic contacts. This fact suggests that the batholithic magma first solidified along the contacts and that this early formed shell was injected by dikes from the still molten interior of the mass. Four of these dikes were observed on the ridge north of Depot creek. They are all composed of light gray granodiorite porphyry, somewhat more acid than the staple quartz-diorite of the contact-shell into which they have been intruded.

Acid dikes of the second kind also specially affect the borders of the batholith but occur in the interior as well. They are not numerous and rarely attain widths greater than four feet. They are light pinkish-gray to whitish, fine-grained granites of aplitic habit. The essential constituents are quartz, micro-perthite, orthoclase, andesine (Ab, An,), and biotite; titanite, magnetite and apatite are accessory. The structure is the hypidiomorphic-granular. The rock is an alkaline biotite granite, verging on biotite aplite. Its relation to the granodiorite recalls the similar succession of granites—acid, alkaline and micro-perthite-bearing biotite granite succeeding granodiorites—in the Okanagan and Selkirk ranges, as so often in other granitic provinces of the Cordillera.

Two classes of basic dikes cut the granodiorite. So far as known, the one class is represented only in one 10-foot, nearly vertical dike at about the 5,000-foot contour on the southern slope of Pyramid mountain (the high conical peak northwest of the outlet of Chilliwack lake). This rock is fine-grained, dark greenish-gray, and of lamprophyric habit. Under the microscope it is seen to have the composition and structure of an acid camptonite.

The other kind of basic dikes has been recognized at several points, but always in bodies of small size; no one of them is known to be wider than two feet. Four of these dikes were found at a point on the same southern slope of Pyramid mountain at about the 4,100-foot contour. A fifth was encountered in the gulch running eastward from the southern end of Chilliwack lake and at a point about 2,200 feet above the lake. All of the dikes are greatly altered and their diagnosis is not easy. One of the thin sections showed, however, some residual augite intersertally placed in a web of basic plagioclase, the only other primary essential. The quantities and relations of the minerals as well as their alteration phenomena show pretty clearly that these dikes are normal diabase.

The basic and acid dikes were nowhere found in contact. Judging from analogy the diabase dikes would be regarded as younger than the camptonite or than either of the acid kinds of dikes.

ACID DIKES CUTTING THE CHILLIWACK SERIES.

Apart from the somewhat numerous dikes which are plainly apophysal to the Chilliwack granodiorite (granodiorite porphyry), there are relatively few acid dikes cutting the Palæozoic sediments of the region. The Glacial drift of the valley carries considerable number of boulders of a porphyritic rock which, judging from the distribution of the erratics, should be in place at several points in the Chilliwack river basin between the lake and Tamihy creek. This rock was actually found in place as a dike or sheet at the 3,400 foot contour on the slope north of the confluence of Middle creek and the river. The exposure is poor and neither the exact relation nor the thickness of the rock could be determined.

This dike-rock is of a light-gray colour, weathering a pale brown, with conspicuous white phenocrysts of oligoclase standing out of the rock as prisms. The phenocrysts measure from 0.5 cm. to 1.5 cm. in length. There are smaller likewise idiomorphic crystals of quartz and orthoclase visible to the naked eye. No ferromagnesian mineral could be found either in the hand-specimen or in the thin section. The groundmass is a finely granular aggregate of quartz, orthoclase and oligoclase. The rock is a granite porphyry of aplitic composition; it is, however, a rock of very different habit from the aplitic dikes cutting the Chilliwack granodiorite and there is no evidence that the granite porphyry has any direct genetic connection with that rock or with any of the visible batholiths of the region.

BASIC DIKES AND OTHER INTRUSIVES IN THE CHILLIWACK SERIES.

At a few points the great argillite-andesite series of the Chilliwack river valley and the adjacent region is charged with small bodies of basic and ultra-basic igneous material, all of which is probably intrusive. One of the bodies has the form and relations of a much faulted dike about twenty feet in width; it cuts the sediments close beside the diorite contact on Pierce mountain. The dike has been squeezed and rolled out into a number of more or less perfectly disconnected lenses. Its compact, dark greenish material proved, on microscopic examination, to be a mass of serpentine, original olivine, and magnetite. The rock was doubtless originally a dunite. At this locality considerable masses of tremolite occur in the sediments and may in part at least have been derived from the serpentine through the metamorphic action of the intrusive Slesse diorite.

Close beside this dike of altered dunite, the crumpled argillites are inextricably mixed with similarly mashed, dike-like bodies of amphibolite, which is transitional in a few places into a fairly coarse-grained gabbro. In the gabbro the bisilicate has all gone over to an amphibole of actinolitic habit.

2 GEORGE V., A. 1912

A mass of composition similar to that of the just mentioned gabbro occurs as a sill or great dike, cutting the Paleozoic strata at the high cliff facing the mouth of Middle creek on the north side of the Chilliwack river and about 2,000 feet above the river.

In general relations, chemical composition, and degree of metasomatic alteration all of these smaller bodies are much like the Vedder greenstone and they may be tentatively correlated with it both in age and origin.

STRUCTURAL RELATIONS.

In structure and composition the Skagit range is, in many essential respects, analogous to the Columbia mountain system. Here, however, the Paleozoic rocks are less intensely crumpled and metamorphosed.

The Skagit range is structurally divisible into two part   From the Skagit river to Middle creek it is chiefly composed of intrusive granites or allied rocks, which occur in such large numbers and so differing in age that we may fitly call the whole plutonic group the Skagit composite batholith. The oldest member of the batholith is unconformably overlain by the Skagit volcanic group. A remnant of the Hozomeen formation appears as a second rock-body which is not part of the composite batholith but is a part of its country-rock terrane.

West of the Slesse diorite the mountains are made of dominant sedimentary rocks. The Paleozoics (Chilliwack series) are very thick, the suggested minimum of about 6,800 feet of strata being, perhaps, much below the real thickness of the rocks actually exposed. An unknown additional thickness of conformable strata underlies those beds; thus no base is known to the Paleozoic (largely Upper Carboniferous) sediments of the west slope. The heavy mass of basic (andesitic) lavas and pyroclastics, named the Chilliwack Volcanic formation, is plainly contemporaneous with the fossiliferous uppermost beds of the series. The Triassic argillites and sandstones of the Cultus formation are not well exposed, but they seem also to be of imposing thickness. It is not known whether they are conformable with the Paleozoics, but an unconformity is suspected. Very little of the terrane called the Tamihy series and tentatively equated with the Pasayten series, occurs within the Boundary belt; and it has not been specially studied. It is unconformable upon the Upper Carboniferous and probably upon the Cultus Triassic beds as well. The Eocene (?) Huntingdon formation forms only a small patch on Sumas mountain; it is unconformable upon the Paleozoic quartzite and also upon the Sumas granite, provisionally assigned to the Upper Jurassic.

Throughout the whole width of the range simple folds are extremely rare. A much broken syncline, pitching gently eastward from the summit of McGuire mountain, is one of the very few decipherable structures in the mountains of the Boundary belt. The Chilliwack river, between Slesse and Tamihy creeks, seems to be flowing on the axis of a broken anticline, the east-west axis of which pitches eastward at a low angle. The southern limb of this arch is also the northern limb of the McGuire mountain syncline. The east-west direction of these axes may possibly be connected genetically with the east-west course of the wide Fraser valley to the north.

Elsewhere the only observed structures in the stratified rocks are local crumples, faults, and small thrusts. Of these, normal faults seem to be most important in explaining the actual distribution of the rocks now exposed. As noted long ago by Bauerman, the section up the Chilliwack river seems to be that of a gigantic monocline, showing an almost incredible thickness of Paleozoic rocks. This is probably a deceitful appearance. East of the nose of the supposed anticlinal near Slesse creek a heavy, crinoidal limestone with moderate northeasterly dip appears four times in the river section, besides appearing in the northern limb of the anticline. The writer is inclined to regard this limestone as representing the same horizon throughout; if so, it is best to suppose that the repetition of the limestone, with the associated shales and sandstones, is due to normal faulting. The faults are probably strike faults, running in a general northwesterly direction; the downthrow being on the southwest in each of the four displacements postulated. It should be added that the exposures are so poor that this partial explanation of the great thickness of beds outcropping along the Chilliwack river is in high degree still hypothetical.

Somewhat more certain is the necessity of mapping the faults bounding the Triassic Cultus formation on east and west. The one on the west seems proved rather clearly; the other is not proved as to its actual location, but has been entered on the map to explain in this case the relation of fossiliferous Mesozoic and Paleozoic strata in lateral contact.

The faults limiting the Skagit volcanics on north and west as well as at the Skagit river, have already been mentioned; little doubt is felt as to the existence of all three of these master displacements.

Nothing need be added to the descriptions of the structural relations of the granitic bodies, as already given in the respective sections of the present chapter. The cardinal fact of magmatic replacement of the huge Paleozoic geosynclinal prism as well as the pre-Cambrian basement terrane by the Chilliwack batholith, and also by the Custer batholith if it is of Jurassic date, seems to the writer quite obvious in the field. The relations are precisely the same as those stated for the vast Coast range batholith, described by Dawson, Lawson, and the geologists working in Alaska, except that the Chilliwack batholith is probably younger than its great neighbour. Nearly all of these observers agree as to the fact of the replacement for the Coast range batholith. The significance of their agreement is great, for they have studied the world's greatest post-Cambrian batholith invading one of the world's greatest geosynclinals.

CORRELATION.

The geological dating of the various formations observed in the Boundary belt where it crosses the Skagit range, has already been discussed in connection with the description of the more important rock bodies. Many doubts remain to us to the exact order in which they should be arranged in the geological time-

25a - vol. ii—35

2 GEORGE V., A. 1912

scale. However, as implied so often in the preceding chapters, the writer believes that a tentative correlation made by the geologist who has actually observed the rocks in the field is better than no correlation at all and in most cases will give safer results than the correlations which would be made by systematists who have no personal knowledge of the ground. For the Skagit range we have the advantage of knowing that there are certain definitely fossiliferous bands in the different stratified series; the chances for serious error are not nearly so great as in some of the eastern ranges. Among the more important unsolved problems are those relating to the age of the Custer granite-gneiss, of the older members of the Chilliwack series, of the Skagit volcanics and of the Tamihy series. If the Hozomeen series is really Carboniferous the Custer batholith is almost certainly of Mesozoic, and presumably late Jurassic, date. But it is conceivable that the country-rocks of this batholith are all of much older date and that this gneissic body may be a small fragment of the pre-Cambrian terrane whence the materials of the Rocky Mountain geosynclinal prism were derived. With the exception of this one body it seems likely that we have no other exposure of those ancient rocks west of the Priest River terrane in the eastern Selkirks. Tempting as it seems to regard the granite-gneiss as a part of the missing pre-Cambrian, the writer is inclined to dismiss that hypothesis and to adhere to the correlation given in the foregoing text, with which the table of preferred correlations should be read:—

Correlation in the Skagit Range.

Pleistocene . . . . Recent and Glacial (including the gravel plateau of the lower Fraser river).

Miocene or Post-Miocene ? . Diabase dikes cutting the Chilliwack batholith.

Miocene ? . . . .
- Camptonite dikes cutting the Chilliwack batholith.
- Syenite-porphyry dikes cutting the Chilliwack series.
- Syenite-porphyry (?) dikes (?) cutting the Huntingdon formation.
- Monzonite stock cutting the Skagit volcanics.
- Chilliwack granodiorite batholith.
- Sloan diorite stock (?)

Oligocene (?) . . . .
- Skagit volcanic formation.
- Skagit harzburgite intrusion.
- Dunite dikes and gabbro dikes (in part) cutting Chilliwack series.

Eocene . . . . Huntingdon formation ; unconformable on Chilliwack series, quartzite and granite.

Unconformity.

Cretaceous (?) . . . . Tamihy series.

Unconformity.

Jurassic ? . . . .
- Sumas granite.
- Sumas diorite.
- Custer granite-gneiss (possibly pre-Cambrian).

Triassic . . . . Cultus formation.

Unconformity.

Upper Carboniferous . . . .
- Vedder greenstone (altered gabbroid rock).
- Chilliwack Volcanic formation.

Upper Carboniferous (and older) . . . .
- Chilliwack series.
- Hozomeen series.

. 1912

writer
tually
most
de by
Skagit
ossili-
r are
npor-
neiss,
d of
uster
date,
auch
pre-
lina!
that
iver
lite-
niss
ext,

aser

ite

MICROCOPY RESOLUTION TEST CHART

(ANSI and ISO TEST CHART No. 2)

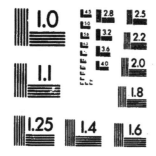

APPLIED IMAGE Inc

1653 East Main Street
Rochester, New York 14609 USA
(716) 482 - 0300 - Phone
(716) 288 - 5989 - Fax